David C. Smith
860 214 9322

AIRBORNE LASER

Bullets of Light

AIRBORNE LASER
Bullets of Light

ROBERT W. DUFFNER

PLENUM TRADE • NEW YORK AND LONDON

Library of Congress Cataloging in Publication Data

Duffner, Robert W.
   Airborne laser: bullets of light / Robert W. Duffner.
   p.   cm.
   Includes bibliographical references (p.) and index.
   ISBN 0-306-45622-2
   1. Lasers—Military applications—United States—History. 2. High power lasers. I. Title.
UG486.D84   1997
623.4′46—dc21
                                            97-21335
                                                CIP

ISBN 0-306-45622-2

© 1997 Plenum Press, New York
A Division of Plenum Publishing Corporation
233 Spring Street, New York, N.Y. 10013-1578
http://www.plenum.com

10 9 8 7 6 5 4 3 2 1

Printed in the United States of America

To Susanne

Your interest, encouragement, and support made a difference.

Foreword

Albert Einstein spent 1916 pondering the quantum theory of radiation. He published the result of his thinking in 1917, including a new effect he called the induced or "stimulated" emission of light that makes the laser possible. *Airborne Laser: Bullets of Light* tells a fascinating story of one of the most remarkable consequences of Einstein's important discovery, which has had a lasting influence on our national security and our ability to keep the peace in a dangerous world.

Today the Air Force is engaged in the development of a major new weapon system known as the Airborne Laser (ABL). The ABL is a specially designed Boeing 747 aircraft equipped with a high-power laser to destroy theater ballistic missiles in their boost stage at ranges of several hundred kilometers. But before the ABL appeared, an incredible number of difficult engineering and system integration problems had to be solved. That feat was accomplished with the Airborne Laser Laboratory (ALL) during the 1970s and early 1980s.

This book is the first to trace the complexity of events that led up to the ALL and the development of cutting-edge technology that ultimately demonstrated that a high-power laser system in an aircraft could destroy supersonic missiles in flight. It is a unique and refreshing treatment of a topic that has too long been ignored. Throughout, the reader is led through a maze of technology in a well-documented and clearly narrated tale. Above all, it

captures an extremely important historical event of far-reaching consequences to the national defense.

But this account is as much about people as it is about technology. It is about the ALL leaders and those who labored in the scientific and engineering trenches to make the first airborne laser possible. You get to meet many of them and to share their disappointments and hardships as the story unfolds.

By 1967, the possibility of generating very high-power gas dynamic laser beams had been demonstrated. I first learned of those developments when I became a member of the Air Force Scientific Advisory Board. The Board had a subcommittee, headed by Abe Hertzberg, to look at the problem of developing laser weapons and I was fortunate to serve with that distinguished group. There was one unforgettable meeting where we considered the possible military applications of placing a high-power laser on airplanes of various types. Edward Teller was present and he was particularly intrigued by the idea of what he called the "aerial battleship." The concept he presented was to put one or more high-energy lasers on a large airplane. A futuristic aircraft of this type could be used as an escort to defend bombers from enemy attack—the powerful airborne lasers would shoot down hostile interceptor airplanes as well as air-to-air and surface-to-air missiles at the speed of light.

Our subcommittee began to consider Teller's idea seriously and we looked at some concepts for a prototype. What eventually emerged from these discussions was a proposal to put a large carbon dioxide gas dynamic laser on a four-engine jet aircraft. A lively debate developed over this proposal among the subcommittee members. One faction felt that the carbon dioxide laser would be the wrong one to use because of the known problems with atmospheric absorption, and that it would be prudent to wait until a high-power laser producing a beam with a shorter wavelength than carbon dioxide became available. The other faction, to which I belonged, believed that it did not matter much what kind of laser was used because there were many system-level engineering and fire-control problems that could be solved with the carbon dioxide laser. Furthermore, it was not clear just when a large laser producing a short-wavelength beam would be available. There were some promising candidates among chemical lasers, but none that could operate at power levels as high as a carbon dioxide laser.

Our subcommittee was closely divided on the issue. I remember a meeting at Kirtland late in 1970 or early in 1971 that Chairman Abe Hertzberg called to prepare a recommendation for the general meeting of the Scientific Advisory Board scheduled for April 1971. At the end of the first day, I was worried that we would lose the vote the next morning and that the subcommittee would recommend we do nothing. In my view, this was an unacceptable outcome, so something had to be done. After the meeting, we found a copy of Billy Mitchell's biography that contained some choice arguments we could use the next day to persuade the wavering members of our subcommittee. We finally won the vote, and our subcommittee recommended to the Board that we go ahead with what would eventually become the Airborne Laser Laboratory.

During my years of service in the Pentagon, I had several opportunities to move the ALL project along. Hopefully, these were useful to the people involved. More important perhaps was a change in my own thinking about the real purpose of the weapon system that would eventually emerge from our work with the ALL. I had come to believe by that time that airplanes carrying high-power lasers would be important in building an effective defense against submarine-based ballistic missiles.

In March 1983 when President Reagan made the speech proposing the development of a defensive system against ballistic missiles, an important consequence was the rapid expansion of research on high-power lasers. One of the reasons why such research was intensified was because of the success of the ALL in shooting down five air-to-air missiles and one U.S. Navy drone target in 1983. It is true that these experiments were somewhat contrived, but so was the bombing of the *Ostfriesland* by Billy Mitchell. This experiment was precisely what we had in mind when the ALL project was started fourteen years earlier.

I would like to offer a speculation that I cannot prove, but which I think is plausible. In his book, *Perestroika* (Harper & Row, 1987), Mikhail Gorbachev wrote about his October 1986 summit meeting with President Reagan in Reykjavik. He was puzzled by Reagan's attitude toward defenses against ballistic missiles. He could not believe that President Reagan was honestly committed to eventually building such a system and making it work. He thought that the United States was merely trying to intimidate the Soviet Union with the threat of deployment, yet he recognized the superb American technological capability and clearly feared that President Reagan meant what be said. There is no doubt in the discussions at Reykjavik that President Reagan was negotiating from strength.

It can be argued that President Reagan's refusal to compromise at Reykjavik was the beginning of the end for the Soviet Union. It is clear from his book that Gorbachev realized that he was up against a determined opponent. Almost exactly three years after the Reykjavik summit, the Berlin wall was torn down, and two years later, Gorbachev and the Soviet Union were history. There is no question that American technical competence was one of the telling factors that led to final victory in the Cold War in 1991.

The world is still a complicated and dangerous place in spite of the end of the Cold War. To ensure that peace is secured, the Air Force is moving forward with the ABL. This system is not an experiment, but a deployable weapon. It is intended to patrol a combat zone and to shoot down hostile intermediate-range missiles, very similar to the scenarios that we used to justify the ALL two decades ago. This program is not controversial because the use of Scud missiles by the Iraqis in the Gulf War makes this threat palpable. Thus, much of what was learned during the ALL program will be employed in the operational world.

This book tells a classic 20th century tale and tells it well. All of the triumphs and disappointments that those of us who were involved experienced are vividly recounted. It is a riveting account that starts with a fundamental scientific insight in the early years of the century, transitions to experimental verification in midcentury, and demonstrates its engineering applications in the century's final years. Above all, leadership, persistence, and dedication of people made the difference for the ALL to be successful. Without the vision and steady hand of Don Lamberson who shepherded the ALL through its early years, the cool resolve and technical expertise of Demos Kyrazis during the tumultuous engineering integration phases of the program, and the take-charge and action-oriented determination of John Otten, who truly made history when he gave the command to fire the laser that shot down the first missile, the ALL would never had happened. It has been a pleasure and an honor to have worked with all of them and to have the opportunity to write this essay about what they and so many others in the ALL program accomplished.

Hans Mark
Secretary of the Air Force, 1979–1981

Austin, Texas, 1997

Preface

When Demos Kyrazis first came to talk to me about the Airborne Laser Laboratory in the summer of 1985, I didn't quite know what to expect. I had spoken to him on the phone earlier to propose writing a book on the airborne laser and had arranged an interview with him to discuss the project in more detail. A history of the airborne laser seemed timely and appropriate as this was a topic gaining more and more attention from the military and the nation at large.

I had known Kyrazis by reputation only when he had worked on the airborne laser program in the 1970s and early 1980s. When he arrived for our first meeting, my immediate impression was that somehow he didn't quite fit the part of one of the most important and influential leaders of the country's pioneering laser program. He was not an imposing figure, but his quiet self-confidence and pleasant way of speaking that never displayed a hint of arrogance made people want to listen to what he had to say. Clearly, he had a firm grasp on all the intricacies of laser technology. When questions were posed, he paused deliberately to take the time to collect his thoughts to make sure he delivered the most accurate response possible. And as I was to find out later, Kyrazis's leadership style was not to project the loudest voice in the room, intimidating people into moving forward. Instead, he relied on his technical expertise and applied the power of logic to calmly persuade his highly

opinionated army of scientists and engineers to proceed in the face of enormous technical problems.

It was during our first encounter that Kyrazis erased any doubts in my mind about the value of moving ahead to write a history of the airborne laser program. From the very start, he was the one who offered constant encouragement and a clear vision as to the lasting contribution of the first airborne laser. He was convinced the story should be told because the airborne laser's unique accomplishment represented a watershed in the advancement of science and technology. In the world of directed-energy systems, the Airborne Laser Laboratory was the Wright flyer destined to take on even more significance and meaning in later years. Here too was an account worth telling because the airborne laser was a prime example of a meaningful research and development effort that resulted in a real-time product created by the Air Force laboratory system that would strengthen the nation's military defense.

One dominant theme that emerged again and again during all my conversations with Kyrazis, and with over 100 other individuals I interviewed for this project, was the tremendous sense of pride and accomplishment by the team of military, government civilians, and contractor personnel who worked relentlessly on the airborne laser program. It was this ingenious and resourceful band that always stayed focused on solving exceedingly complex issues to meet the next technical milestone. And at every level, the emphasis was more on doing than planning. Planning requirements answered the mail to higher headquarters, but the success of the program rested to a very high degree on people conducting experiments and continually tinkering with the hardware at all hours of the day and night to get the system up and running. Theoretical science and applied technology were critical ingredients, but what also prevailed were the immeasurable human qualities of intense desire, vision, and a healthy dose of risk-taking that led to the development of a high-energy laser capable of disabling air-to-air missiles.

Kyrazis was the first to point me in the right direction to meet with and interview those risk-takers and other participants who played key roles in the program. It was the information gathered in these lengthy interviews, combined with an extensive collection of airborne laser documents preserved in the Phillips Laboratory archives, that provided the bulk of the source material for this book. All photos and charts that appear in the book came from the Phillips Laboratory archives except for those photos other organizations provided and were given credit for in the caption.

I am grateful to the 37 individuals who read portions of or the entire manuscript and offered invaluable suggestions to improve the quality of the narrative. A special thanks goes to all those who read the manuscript from beginning to end: Demos Kyrazis, John Otten, Denny Boesen, Keith Gilbert, Darrell Spreen, Steve Coulombe, Bill Dettmer, and Don Lamberson. They were not only extremely generous with their time, but were very patient in explaining all the technical complexities of the airborne laser. Their "tutorials" and "translations" of a diversity of technical reports were indispensable to me in fine-tuning the technical precision essential for this type of book. Equally important were the numerous other readers who were the experts on various portions of the project and who gave freely of their time along with thoughtful advice on how to make the writing more accurate. In particular, Ida Houseknecht deserves a special thanks for her editorial assistance during the numerous revisions of the chapters. Also, I thank Steve Watson and Laurel Burnett for their patience and help in checking and rechecking the final formatting of the footnotes.

Although scientists and engineers furnished the factual basis of this book, three Air Force historians and colleagues proved to be extremely helpful in raising those probing questions to force the issues of balance and interpretation into the narrative. Early on, Dr. Dan Harrington of the Air Force Weapons Laboratory History Office frequently inquired about and presented his thoughtful insights as to the value and contribution of the airborne laser program. His editorial handiwork on the first several chapters improved the tone, style, and content of the writing. Dan never was reluctant to ask, "What does this sentence mean?" to improve clarity. I thank him for that!

Throughout this entire project, I came to rely on Dr. Barron Oder for his invaluable assistance and counsel. He was there to always pose the difficult questions and to emphasize objectivity in the writing. He singlehandedly organized thousands of documents by subject into an accessible and useful airborne laser archives at the Air Force Weapons Laboratory (now Phillips Lab). This collection consisting of a vast array of technical reports, contracts, correspondence to higher headquarters, special studies, briefings, strategic plans, executive summaries, photos, charts, and much more proved to be a gold mine of information. These documents served as the factual foundation of the book and were especially helpful in confirming or refuting information extracted from the interviews. Barron's thorough read of the manuscript with a keen eye for detail detected minor and major errors of content that I had simply missed or overlooked. I am deeply indebted to him for his cheerful encouragement during the low points of the research and writing stages and for his support urging me to "get the manuscript out" so others could read and assess the contribution of the airborne laser.

Dr. Don Baucom, a historian and authority on strategic defense technology in the Air Force assigned to the Ballistic Missile Defense Organization, was very helpful in his read of the manuscript. He too was a strong advocate of getting the history of technology "out there" for others to read and judge.

Finally, completion of an undertaking of this magnitude does not happen without strong and sustained family support. My wife Carol and two daughters, Kelly and Susanne, contributed more to this project than they realize. I thank them.

Robert W. Duffner

Albuquerque, New Mexico

Contents

Contents

Introduction

The development of the Airborne Laser Laboratory (ALL) was one of the most extraordinary achievements in the annals of science and technology. A highly dedicated and diversified team of talented military and civilian scientists and engineers at the Air Force Weapons Laboratory (AFWL), Kirtland Air Force Base (AFB), New Mexico, succeeded in testing and operating a high-power CO_2 gas dynamic laser and precision pointing and tracking system aboard a specially modified NKC-135 aircraft known as the ALL.

In May 1983, the highlight of the ALL program occurred over the Naval Weapons Center Range at China Lake, California, where the laser combined with a sophisticated pointer and tracker to shoot down five AIM-9 "Sidewinder" missiles. Hailed in the scientific community as an event of major proportions with far-reaching consequences, this demonstration proved for the first time that an airborne laser could intercept and destroy an air-to-air missile. Four months later, the beam fired by the ALL intercepted three Navy BQM-34A drones over the Pacific near Point Mugu, California. The success of these two demonstrations marked an unparalleled technical milestone, as the ALL clearly showed the potential of high-energy lasers as airborne weapons.[1]

Although the Air Force had become the military leader in exploring the possibilities of using lasers on aircraft in the 1970s and 1980s, the roots of the ALL program stretched back

to research conducted by private industry in the early 1960s. On 7 July 1960 Dr. Theodore H. Maiman, senior staff scientist at Hughes Aircraft Company's Research Laboratory in Malibu, announced to the press at the Delmonico Hotel in New York that he had succeeded in generating the world's first laser beam on 15 May 1960. Using a corkscrew-shaped xenon flash lamp (GE FT-506), Maiman irradiated a synthetic ruby crystal rod (aluminum oxide with chromium impurities) to produce a dark-red beam of coherent light. The lamp and ruby rod (only 4 inches long) fit snugly inside a hollow aluminum cylinder about the size of a coffee mug. Light from the flash lamp surrounding the ruby rod excited or "pumped" the electrons of the chromium atoms to a higher energy state, a condition analogous to a vibrating tuning fork. As these stimulated atoms relaxed to return to their normal energy level, they emitted photons (energy in the form of particles of light) that collided with other atoms, which in turn produced more photons. This chain reaction generated billions of photons bouncing back and forth between the silver-coated ends of the ruby rod. During this process, the radiation in the form of light was amplified (increasing power of the light signal without changing the wavelength or frequency of the light) to form a slender and harmless low-power laser beam.[*] Although this device generated only a few watts of power, the discovery of the ruby laser signified a revolution in the science of light that brought with it both confusion and promise for future applications.

Dr. Theodore H. Maiman shows a cube of synthetic ruby crystal that forms the "heart" of a laser, which generates the light. (Photo courtesy of Hughes Research Labs.)

A light source (corkscrew-shaped xenon flash lamp) surrounds and excites tightly packed atoms in the ruby which produces laser light in the form of an intense parallel beam. (Photo courtesy of Hughes Research Labs.)

Maiman was one of the first to realize this and was cautiously optimistic about the significance of his groundbreaking research when he commented, "The laser is a solution looking for a problem!"[2]

But the identification of problems and the application of solutions proceeded at an extremely rapid pace. As early as 1963, a medical team at Stanford University performed the first operation using a low-power argon laser (focused by the lens of the eye) to repair a detached retina in the human eye. The beam was focused to a spot about 40 millionths of an inch in diameter, making it ideally suited for delicate surgery. Performed routinely today, this operation is painless, requires no anesthesia, and can be done quickly in a doctor's office;

---

[*]Maiman's laser was highly inefficient, as less than 1 percent of the input energy from the flash lamp was converted into the output laser beam. The other 99 percent of the input energy was dissipated in the form of heat. Maiman's laser was a single pulse of light that lasted for only 1/1000th of a second.

the procedure has prevented thousands of people from going blind. Small hand-held neodymium yttrium aluminum garnet laser devices resembling a dentist's drill make clean, sterile cuts to safely remove cataracts and vaporize tumors by literally exploding cancerous cells. Other medical applications of the laser as a surgical scalpel include sealing leaky blood vessels and erasing unsightly skin disfigurements such as moles and blemishes, as well as removing malignant tumors embedded in previously inaccessible regions of the brain. A skilled surgeon with the help of a powerful electron microscope or endoscope can direct a beam 40 millionths of an inch wide with extraordinary precision to hit and destroy a single cell without damaging surrounding tissue. Dentists are relying more and more on laser drills that quickly irradiate and painlessly remove tooth decay; the beam also hardens the tooth to resist decay.[3]

Commercially, small helium–neon lasers have progressed to the point where they can accurately read in a fraction of a second the bar codes on grocery items at the supermarket checkout counter, the price of the item being displayed instantly on the cash register. Publishing houses no longer struggle with the time-consuming process of typeset printing as lasers are faster and more economical for printing books and magazines. Tied in with the computer, laser printers (first introduced by IBM in 1975) permit office workers to add a professional touch to all types of correspondence and reports. Laser separators produce high-quality color photographs and weather maps to give us our morning newspapers in Technicolor![4]

In the communications industry, the introduction in the late 1960s of fiber optics—durable glass and plastic rods that are smaller than a human hair—has led to the replacement of bulky cables packed with hundreds of individual wires for transmitting all types of information at blinding speeds. Short-wavelength lasers (as opposed to long-wave radio and microwaves used to transmit data) can compact an enormous amount of information into the very small area of an optical fiber. The extremely high-speed pulsing action—turning on and off—of the laser up to several billion times per second allows each pulse stream to carry many separate messages. As a result, thousands of telephone conversations can take place simultaneously by riding the same laser beam as it travels along the fiber-optic line, resulting in improved efficiency, reliability, and service to customers. A bundle of optical fibers about as thick as a toothpick can transmit all of the words in several hundred books in a second.[5]

Especially eager to exploit the marketability of laser products was the entertainment business. Compact laser disks give music lovers a front-row seat at the premier concert halls, allowing them to listen to the world's greatest symphonies—or goriest hard rock—from the comfort of their favorite easy chair. Unlike phonograph records, laser disks offer the advantage of lifelong durability; the thin beam produces exactly the same high-quality sound on every play, whereas the fidelity of an old-style vinyl record deteriorates over time with the continual wear and tear on the needle and record. In the mid-1970s, North American Philips Corporation and MCA, Inc. developed the home videodisk, which uses a laser to transmit both sound and picture for display on a television screen. Inside some hand-held TV remote controls (others use light-emitting diodes) is a tiny laser hard at work to switch channels for us on command.[6]

The art world did not escape the influence of laser technology as the high-intensity light contributed to a new art form, known as holographics, which present images in three-dimensional form. Emmett Leith and Juris Upatnieks, from the University of Michigan, made the first laser holograms in the early 1960s. By the late 1970s, the process had been refined to making large, lifelike holograms. These new holograms project a razor-sharp image and give greater depth to the picture. Viewed at the proper angle, objects appear to stand out several feet—suspended in space—from the flat surface of the picture. More recently, laser engraving uses a laser to carve intricate designs into solid black walnut and a variety of other materials.[7]

The versatility of laser light extended into a variety of other applications. Because the energy from a laser can be focused on such a small area, beams can drill clean-cut holes in the world's hardest materials, such as diamonds and steel. And because of their speed and power, lasers can also drill precision holes in soft material, such as the rubber nipples of baby bottles. Still other uses of lasers include welding metal and cutting fabric to exact shapes and sizes within tolerances of 5/1000th of an inch (the width of a thread) for clothing manufacturers to offer the perfect fit. Making precise measurements of very long and very minute distances, such as the tiny and gradual movement of plates of rock along geographical faults to help predict earthquakes, is another important contribution of laser technology. Surveyors rely on tripod-mounted helium–neon lasers to send a thin beam to a distant point to serve as a center reference line for building tunnels and roads. In sum, over the years the laser has emerged as a unique tool suited for hundreds of practical uses, which has turned into a multibillion-dollar business today.[8]

Since the 1960s, industry has played a vitally important role in leading the way in developing low-power laser devices to improve the quality of life for its customers. Making a profit has been the primary motive driving the private sector to invest in laser research and development. But Maiman's invention also whetted the appetite of the defense community, who envisioned high-power lasers, not as a moneymaking enterprise, but as a new class of weapons for the future that would tip the strategic balance of power in favor of the United States and revolutionize the science and art of warfare in the 21st century.[9]

Military's Early Involvement with Lasers

The idea of using an omnipotent "death ray" on the battlefield is not a new concept. Ancient literature credits the Greek mathematician Archimedes as the first to conceive the idea of using light as a defensive weapon. Hippocrates, commander of the Greek forces, applied Archimedes' concept by focusing the energy of sunlight through a series of mirrors to produce a beam that set fire to the sails of the Roman fleet under Consul Marcus Claudius Marcellus during the siege of Syracuse in 212 B.C.[1]

At the end of the 19th century, H. G. Wells in *The War of the Worlds* painted an eerie vision of invading Martians relying on the "invisible sword of heat" to conquer the unprepared earthlings with their primitive defenses. His science fiction masterpiece portrayed the enemy Martian lowering his heat-ray tube and firing at a helpless ironclad. The destructive power of the beam was immediate and deadly as it drove "through the iron of the ship's side like a white-hot iron rod through paper." And in more recent times, cartoonist Dick Calkins lured a dedicated following of readers to check their daily newspaper and track the continuing saga of Buck Rogers in pursuit of the evil Dr. Pounce. Everyone knew that Buck relied on his deadly "heat-ray gun" to knock his wicked opponent's rocket machine out of the sky![2]

But the Department of Defense (DOD) was not interested in Buck Rogers or Flash Gordon fantasies. Its goal was to build on Maiman's seminal work and develop a high-power

laser—defined as having a power output greater than 20 kilowatts—that could be eventually used as the unbeatable operational weapon. The military envisioned using lasers for ballistic missile defense, antisatellite, and antiaircraft missions. Other potential military applications of lasers included fire control and weapons guidance, laser-surveillance radar, and optical countermeasures. Pentagon officials reasoned the unique characteristics of lasers, including energy transfer to remote targets at the speed of light, made them superior to conventional weapons that relied on the kinetic energy of a projectile colliding with a target as the kill mechanism.[3]

To find out more about the progress and potential military applications of lasers, DOD sponsored the first conference on laser technology from 12 to 14 November 1963 in San Diego. The conference addressed six key areas: semiconductor and glass lasers, techniques (measurements, designing operating parameters), pumping, propagation, systems, and effects. Of 43 papers delivered, only eight were presented by DOD employees, attesting that private industry had indeed pioneered laser research and development programs. DOD was clearly aware of this and over the next two decades would take steps to emerge as a prominent laser player.[4]

LASER FUNDAMENTALS

The word *laser* is an acronym for *l*ight *a*mplification by *s*timulated *e*mission of *r*adiation. A laser is the most concentrated and powerful form of light known. In simplest terms, three conditions are necessary for "lasing" to occur. First, some type of substance (gas, liquid, or solid) is needed to produce a beam. Second, an intense energy source (a pulsed-discharge lamp, a chemical reaction, electricity, or a flow of electrons from an electron gun) is required to excite and alter the condition of the selected material so it is capable of lasing. The primary energy source is generally referred to as a *pump* by laser technologists. Third, a device commonly referred to as a *resonator* with mirrors at each end is required to extract the precise optical energy in the form of a beam.[5]

As external energy is pumped into the lasing media, it causes the media to change its atomic makeup, resulting in a condition known as *population inversion*. Introduction of this additional energy causes the electrons of the gas, liquid, or solid to move out of their stable inner orbits to higher unstable energy state orbits. When more electrons are in this excited (stimulated) condition than at their normal thermal equilibrium (at rest), population inversion occurs. The natural tendency of the excited electrons (each of which has acquired additional energy from the pumped input) is to return to their normal state of equilibrium in the span of only a few milliseconds. Each stimulated atom does this by releasing its excess (absorbed) energy by emitting a photon or "light bundle."* The amount of energy emitted by the photon is exactly the same amount of energy the electron absorbed to jump to a higher orbit. This

---

*After an electron drops from its higher energy level, releasing a photon in the process, the electron returns to its normal energy orbit of a particular atom. That atom is then available again to be stimulated to elevate an electron to a higher energy level to produce additional laser power.

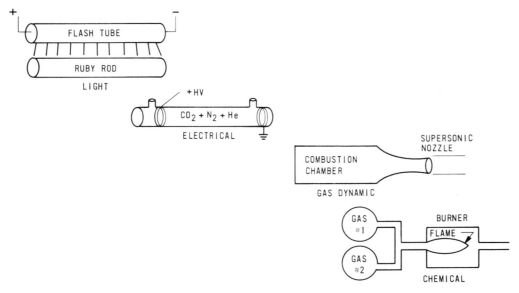

Laser "pumping" methods.

"stimulated emission of radiation" or release of photons represents the conversion of atomic molecular radiation to electromagnetic radiation.[6]

When a single photon is released, it will collide with another excited atom (of the same wavelength) and stimulate that atom to give up a photon. Both photons radiate light in exactly the same direction and exhibit exactly the same wavelength and frequency. These two photons then stimulate other atoms, setting off a chain reaction of released photons that make up the light of the laser beam. Intensity of the beam grows exponentially as more and more photons are produced. For example, two photons will each react with another atom to produce two additional photons, for a total of four photons all traveling side by side in exactly the same direction. Hence, in this case, the beam of four photons is amplified[*] twice that of the beam made up of only two photons. As this process is repeated over and over (4 to 8 to 16 and so on), billions of photons are released to form a beam of greater and greater intensity.[7]

Control and buildup of the beam takes place in a resonator, a chamber with mirrors positioned at each end facing one another. As the stimulated atoms continue to release photons of light at one specific wavelength, the beam reflects off one mirror and travels the length of the resonator cavity (stimulating still more atoms) and bounces off the mirror at the opposite end. (Not all photons travel the length of the resonator; some photons are absorbed by the interior walls of the resonator.) The effect of all this is that as the beam bounces back and forth between the two mirrors, additional photons are released, and the beam acquires more and more energy. Both mirrors are highly reflective, but one has a small

---

[*]A wine glass constantly oscillates at its own natural frequency/wavelength, although this movement cannot be detected by the human eye. However, if an outside source (singer's voice or tuning fork) vibrates at exactly the same frequency of the glass, then "amplification" of the vibration of the glass occurs and the glass shatters.

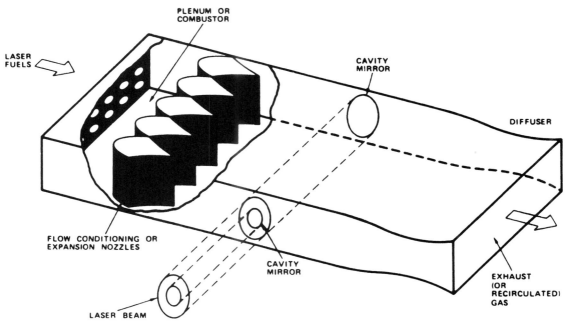

PLENUM OR COMBUSTOR

LASER FUELS

CAVITY MIRROR

DIFFUSER

FLOW CONDITIONING OR EXPANSION NOZZLES

CAVITY MIRROR

EXHAUST (OR RECIRCULATED) GAS

LASER BEAM

Schematic diagram of a gas dynamic laser.

hole through which a portion of the beam passes and can be extracted from the resonator. A second extraction technique is to make one of the mirror's surface partially reflective; the reflective portion of the surface will redirect part of the beam back into the resonator while the unreflective portion will allow the remainder of the beam to escape from the resonator. Ordinary sunglasses work the same way. Part of the light bounces off the reflective surface while a portion of light passes through the sunglasses to reach the eye. A third way to extract the beam is by aligning one of the resonator mirrors at exactly the right angle so the beam can be reflected below the resonator sidewalls to the outside. The bottom of the resonator is open and leads to a diffuser, which vents the gas dynamic laser gas flow to the outside.[8]

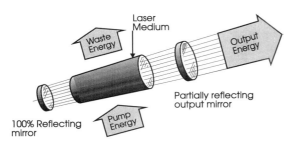

Laser Medium

Waste Energy

Output Energy

Partially reflecting output mirror

100% Reflecting mirror

Pump Energy

Essential elements of a laser.

How quickly a substance lases depends on a number of variables, including atomic bonding and the speed and amount of energy pumped into the resonator to react with the material to start stimulated emission of photons. With sufficient and sustained pumping power, almost any substance can be made to lase, including bourbon whiskey and even Jell-O. One of the problems with lasers, regardless of the material selected, is that they are notoriously inefficient; that is, it takes an enormous amount of input energy to produce a small amount of laser light. Not all of the input energy is converted to laser light. A large portion of the energy remains as heat, which creates major cooling problems for the laser to operate efficiently.[9]

Lasers offer the distinct advantage of delivering large amounts of energy to a very small area by focusing the beam with conventional optical devices (mirrors). A small spot means greater energy density on target, thereby increasing thermal penetration power of the beam.

Ordinary light (the visible span of the electromagnetic spectrum), such as given off by a light bulb, is considered "undirected" because it consists of a conglomeration or random mix of many different wavelengths. Wavelength is the distance from one crest to the next in a light wave. All light of the same wavelength exhibits identical color (monochromatic) and frequency (the number of waves that pass the same point in 1 second).

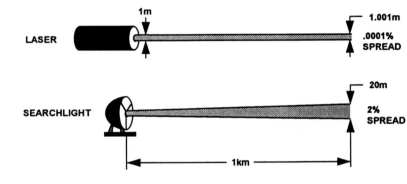

Laser light versus ordinary light.

Undirected or incoherent light randomly spreads out in all directions like a flashlight, which prevents it from being focused to a tight spot. An analogy can be made by observing a handful of pebbles thrown into a calm pond. Each pebble, depending on its size, creates a distinctive ripple or wave. There is no synchronized pattern; each wave bumps helter-skelter into waves formed by the other pebbles. In other words, energy is dispersed. Light emitted by a laser, however, is highly directional because it is made up of exactly the same wavelength, color, or frequency; this allows for the energy of the beam to be highly concentrated in one spot. (Wavelength, color, and frequency are related. A certain wavelength will always have the same color and frequency.) The energy contained in the light bundles, called photons, making up the beam is defined by the wavelength. The shorter the wavelength, the greater is the frequency. Therefore, using the same size optics to focus the beam, short-wavelength lasers (1–2 microns) shine brighter (number of photons per unit area) and deliver more energy over a smaller spot on a target.[*] Beam intensity increases as the inverse square (second power) of laser wavelength. Consequently, a beam produced by using the same size mirror for focusing a 3.9-micron deuterium fluoride (DF) chemical laser would be about eight times more intense (watts per square centimeter on target) than a carbon dioxide laser beam of 10.6 microns.[10]

---

[*]In comparison to the relatively short wavelength of lasers, radio wavelengths vary from about a half mile to up to 6 miles long. Microwaves are a few inches long. Visible light wavelengths are 4 to 8 one hundred thousandths of a centimeter. X rays are only 1 billionth of a centimeter long.

Lasers "lase" at different wavelengths measured in microns. A micron is 1 millionth of a meter, i.e., 1 million microns equals 1 meter. The wavelength of visible light is 0.4 to 0.7 micron. Chemical lasers emit light at 1.3 to 3.8 microns (infrared), excimers at 0.3 micron (ultraviolet), free electron tunable from 0.1 to 20 microns, and CO_2 gas dynamic at 10.6 microns. Wavelength is an indicator of energy. Short-wavelength lasers pack more energy than long-wavelength lasers. Long-wavelength lasers require larger optics than short-wavelength lasers to focus the beam to a small spot.

More importantly, the coherent alignment of photons all traveling "in step" and in exactly the same direction (their crests and troughs coinciding and reinforcing one another) accounts for the compactness of the highly intense energy levels or brightness of the pencil-width beam. Coherency of a laser can be compared to a squad of 10 infantry soldiers lined up in "dress-right-dress" formation (representing the width of the laser beam) and firing their rifles in the same direction and at precisely the same instant. All 10 bullets (like photons) exiting the barrels would also move off in perfect formation—none in front or behind its neighbor—on their way to the target. Compacting the energy of all 10 bullets—or millions of photons—into a small area and impacting at exactly the same time greatly improves the chances for target destruction.[11]

LIGHT BULB

RED FLASHLIGHT

LASER

Coherent = all light waves "in step."

Two other factors convinced DOD officials that lasers were ideally suited as the ultimate weapon. Perhaps the most attractive feature of a laser is it travels over long distances at the speed of light or about 186,230 miles per second. It takes only 6 millionths of a second (6 microseconds) for a laser beam to travel 1 mile. This means large amounts of energy can be delivered to a target at long and short ranges almost instantaneously. For short distances (a few hundred kilometers), an operator aiming and firing a laser does not have to compensate for the movement of the target. However, for longer ranges (several thousand kilometers) he must lead the target slightly. But in either case, time of flight of laser energy to the target is extremely quick. This gives the operator more time to identify targets before committing to firing the laser and virtually no time for the target to evade the beam. The implications of this is that an intercontinental ballistic missile (ICBM) traveling at a relative snail's pace of 5300 miles per hour (about seven times the speed of sound) is no match for the lightning speed of the laser zipping through space. From the time the laser is fired until it strikes the missile at a mile away, the missile will move less than an inch. In the atmosphere, a supersonic jet traveling at twice the speed of sound will only move 1/8th of an inch in the time it takes the laser beam to traverse 1 mile. And once it hits the target, the laser inflicts damage by rapidly heating and burning a hole in the skin of its target, melting structural and electronic components, blinding sensors and detectors, and, in some cases, igniting on-board flammable materials. Lasers fired in short bursts (pulses) can also inflict damage from the explosive shock waves created by rapid heating of the target material by the beam. Shock waves cause structural failure of the target material. Focusing the beam to a small target spot and holding it steady on target without the beam energy "dancing around" (much like holding a magnifying glass steady to focus the sun's rays on a sheet of paper to start a fire) is one of the most formidable technical challenges. This requires building a highly accurate and reliable beam-control subsystem. The laser's near-zero flight time, large field of fire, high number of shots (each shot has no recoil and would only consume a small amount of fuel), and ability—with the tilt of a mirror—to slew quickly from one target to another give it more frequent firing opportunities than conventional weapons.[12]

Besides the technical advantages, a second attractive feature of lasers stems from political considerations. Lasers are not weapons of mass destruction. Unlike nuclear weapons capable of killing millions of people over a large area with blast and fallout, lasers are highly selective because all of their energy and destructive power are concentrated on one small area. Thus, lasers are considered "clean weapons" that do not cause collateral damage to structures near the target as nuclear or conventional bombs do. This greatly reduces the potential for large numbers of civilian casualties. The notion that a laser could be used as a "surgical scalpel" appealed to the top decision makers at Air Staff and DOD and was one of the major reasons for increased funding for laser research in the 1970s.[13]

Technical and political considerations provided two good reasons for supporting laser research. However, there also were clear disadvantages to lasers. It would take an enormous power source to generate a beam of sufficient energy levels (scaling up to higher power was a basic problem) to satisfy requirements for a laser weapon. Even if a powerful enough beam could be produced, the density of laser radiation on target decreases in relation to the square of the distance to the target. For example, a laser aimed at a target 10 miles away would deposit only 1/100th the amount of heat per square inch as the amount of energy deposited by the same laser to a target 1 mile away. Atmospheric absorption of the laser beam posed a second limitation of lasers. As the beam propagates through the atmosphere, carbon dioxide molecules and water vapor in the air absorb the energy of the beam, thereby decreasing the beam's power.[14]

A third basic drawback of lasers was that the beam heats the air and reduces its density. As a result, the burned-through air acts as a "negative lens" which, in turn, defocuses and spreads out the beam, a condition known as *thermal blooming*. Beam defocusing occurs because the heated air in the beam's path changes its index of refraction, which can rob the beam of 90 percent of its power. This distorts and bends the beam, further dissipating its energy and deflecting the beam from its intended target.[15]

Once the beam reaches the target, it must remain tightly focused to cause damage. If the beam is "smeared" or moves around the center of the target spot, then the beam must dwell longer on the target to deposit the required energy to disrupt or destroy the target. Keeping the beam steady on target must occur at the same time the laser platform and target are maneuvering and moving at high speeds. Finally, the enemy can use countermeasures to negate the effectiveness of a laser. By spinning a missile, the beam will not be able to dwell on the same target spot, thereby preventing rapid burn-through of the target. Coating the target with a highly reflective material may cause the beam to bounce off instead of penetrating the target. Also, it would be relatively inexpensive for an opponent to overwhelm a laser weapon defense system by building and launching more missiles and decoys or by deploying submarine missiles with depressed trajectories. Although there were some potential drawbacks to the development of lasers, the Air Force was confident that careful system design could reduce or eliminate these limitations.[16]

Recognizing the advantages and disadvantages of lasers, DOD and the Air Force knew that to proceed with the development of any high-energy laser system required perfecting the technology for a variety of subsystems. At the heart of an airborne system, for example, would be the laser device that would create the correct physical conditions to produce a beam of intense light. But a beam was useless unless it could hit a target. To do this required that a search and acquisition subsystem (fire control) locate and start tracking the target. Once

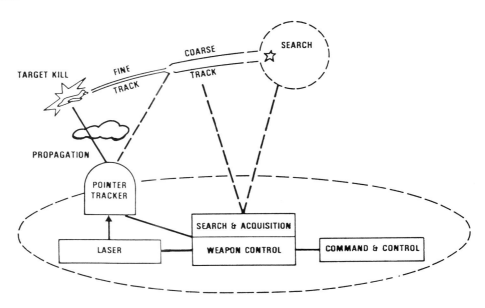

High-energy laser weapons system components.

that was accomplished, the tracking function would be transferred to a precision pointer and tracker. A beam control system consisting of an intricate series of mirrors would transmit the beam from the laser device to the pointer and tracker. The job of the pointer and tracker was to focus the beam with a large primary mirror and aim and direct the beam to hit the target. Although perhaps simple in concept, integrating these distinctly different components into one harmonious system would be an extremely difficult challenge.[17]

GETTING UNDER WAY—INVESTING IN LASERS

Over the years, the advantages outweighed the disadvantages in terms of investing in laser research. Dr. William J. Perry, Undersecretary of Defense for Research and Development, told a Senate hearing in the summer of 1979 the money spent on laser research was "the single largest science and technology program we [DOD] are pursuing." By 1980, the laser market had reached $1 billion. Programs sponsored by the federal government accounted for about 58 percent of the market. The remaining 42 percent went to commercial enterprises. DOD's budget for lasers in 1980 totaled $210.8 million: the largest share, $101.2 million, went to the Air Force,* while the Navy received $40.8 million, the Army $20.5

---

*In 1969 the Air Force budget for high-energy lasers was $4 million. The next year it doubled to $8 million and soared to $26 million in fiscal year 1971. In fiscal year 1972, the Air Force laser budget continued to climb and reached $30 million. This trend continued through the 1970s as the Air Force received $54 million in 1975, $63 million in 1976, and $88 million in 1977. In 1979 the Air Force almost reached the century mark, receiving $98 million for laser research and development. In 1980, the Air Force exceeded $100 million for the first time, when it received $101 million for laser research. By FY83, the Air Force laser budget totaled $103 million.

million, and the Defense Advanced Research Projects Agency (DARPA) $48.3 million. The Department of Energy and the National Aeronautics and Space Administration spent another $200 million on laser research. In the 20 years following the invention of the laser in 1960, the federal government had invested $2.46 billion in laser research and development. The trend in the late 1970s and early 1980s was a steady escalation of money dedicated to laser research. For fiscal year 1983, the Reagan administration spent $500 million for lasers, more than double the $194 million set aside by President Carter for fiscal year 1979.[18]

What had accelerated the pace of government-sponsored programs on laser research in the 1960s and development in the 1970s and 1980s was the concept embraced by the military that lasers would be the high-payoff weapons of the future. Enthusiasm ran high and some scientists were prematurely predicting a laser weapon would be ready to operate in 1966. However, others took a more realistic and cautious approach. Dr. Charles Townes of Columbia University's Physics Department, who had pioneered the seminal work on the maser (*m*icrowave *a*mplification by *s*timulated *e*mission of *r*adiation)—the forerunner of the laser—while working at Bell Laboratories in the 1950s, recognized the likelihood of eventually developing a laser weapon system. A strong supporter of lasers from the start, he piqued the interest of an Air Force/Aerospace Corporation audience in the summer of 1961 by announcing, "Fundamentally, there is no limit to the power which can be obtained by the optical maser [laser]." With this kind of optimistic endorsement, 75 laser research and development programs were under way at government and contractor facilities at a cost of $10 million in 1962 and increased to $15 million in 1963. By the following year, the government's investment in lasers had climbed to $19 million in pursuit of developing the ultimate weapon.[19]

Contracts for laser research funded by all three services (Army, Navy, and Air Force) found their way to a diverse group of companies with expertise in a variety of scientific disciplines. Leaders in the private sector participating in early laser development work included Hughes Aircraft Company, AVCO-Everett Research Laboratory, Optics Technology, Pratt & Whitney, Perkin Elmer, Maser Optics, Texas Instruments, Applied Lasers, General Precision, Honeywell Incorporated, Corning Glass, RCA, Chrysler Corporation, American Optical, Columbia University, Westinghouse, Atlantic Research, General Electric, General Dynamics, Philco, Technical Research Group, Raytheon, Coherent Radiation, Spectra-Physics, and Quantatron. Most of the initial in-house Air Force work (about 36 percent of military expenditures for lasers) ended up at Aeronautical Systems Division (Wright-Patterson AFB, Ohio), Rome Air Development Center (Rome, New York), and the Air Force Special Weapons Center (AFSWC) and Air Force Weapons Laboratory (AFWL) at Kirtland AFB, New Mexico.[20]

Recognizing the potential military payoffs for lasers, DOD's Advanced Research Projects Agency (ARPA) was the first government organization to sponsor laser research and development work. The decision to create ARPA on 7 February 1958 was a direct response to the Soviet launch of Sputnik a year earlier (4 October 1957). To avoid a similar technological surprise in the future, the United States formed this small centralized DOD agency to tap the best scientific and engineering minds in the country to conduct high-risk research that would lead to technological breakthroughs of a "revolutionary nature," as

opposed to the more traditional "evolutionary" approach to weapon development. The Secretary of Defense selected promising research and development projects for each military service and funded government laboratories and private contractors to carry out the work. By dividing new projects among the services, ARPA hoped to avoid duplication of effort. Early projects included defense against ICBMs and satellites, solid-propellant chemistry, undersea surveillance, improved armored vehicles, and lasers. ARPA's investment strategy stressed the development of high-payoff and long-term technology to stay ahead of the Soviets so the United States would be prepared, 10 to 20 years in the future, to meet any new mission requirements demanded by the changing Soviet threat.[21]

ARPA was the logical agency to sponsor laser research, but because this field was so new, there was some degree of confusion as to what organization should perform the work. At the local level at Kirtland Air Force Base, New Mexico, Colonel David R. Jones, head of AFSWC's Physics Division[*] in 1962, was a man at the right place at the right time to take advantage of this opportunity. He had learned from ARPA officials in Washington, D.C. that they were extremely interested in finding out the effects of lasers on nose cones of reentry vehicles. Jones was a streetwise officer always alert to the prospects of attracting new business to his division. Although he readily admitted he barely understood the rudiments of lasers, he was nevertheless confident that his scientists, who were experienced in conducting tests on the effects of X rays on various materials, would have little trouble calculating the effects of lasers on ICBMs. Dr. Arthur H. Guenther, Chief of the Material Dynamics Laboratory in the Physics Division, recognized the value of getting in on the ground floor of an untapped scientific discipline and strongly urged Jones to lobby for the ARPA laser work. Guenther went on to become AFWL's chief scientist in 1974 and served in that position through 1987.[22]

Jones wasted no time in attempting to secure the ARPA laser work. He and AFSWC's Captain Marvin C. Atkins traveled to Washington during the second week of February 1962 to brief ARPA officials on a conceptual game plan for inaugurating laser effects work at AFSWC. Impressed by Jones's proposal, ARPA was encouraged to hear from an organization giving assurance that it had capable workers who were eager and willing to explore a new technical field by performing a variety of laser experiments.[23]

Colonel Jones's efforts paid almost immediate dividends. With the issuance of ARPA Order 313-62 on 26 February 1962, the Air Force took its first tentative step into the world of laser exploration. This document gave AFSWC, the precursor of AFWL (established in May 1963), $800,000 to start a program on "impulsive loading research." ARPA selected AFSWC because its Physics Division scientists were experts on the interaction of pulsed energy with reentry vehicles, specifically X rays generated by a nuclear weapon. Investigating the effects of solid-state (ruby and neodymium) pulsed laser beams on a variety of materials was the broad goal of 313-62. Interaction of an X-ray pulse with a reentry vehicle was thought to be similar to a pulse of energy emitted from a ruby laser. The specific objective

---

[*]Created in 1952, AFSWC conducted nuclear testing and effects research. In May 1963, AFSWC's Physics Division moved to form the fundamental building block of the newly created AFWL.

of the study was to determine how powerful a laser would have to be to deliver enough energy to destroy an ICBM. If these findings turned out positive, then lasers would be a leading candidate in DOD's long-range antiballistic missile (ABM) defense program called PROJECT DEFENDER.[24]

A few years earlier, in 1958, President Eisenhower had directed ARPA to launch PROJECT DEFENDER as a research and development program. The goal was to look at bolder and more speculative types of technology that would lead to more effective ABM systems as potential replacements for Nike-Zeus, which had been rejected by the Congress. The AFSWC work supported a Pentagon study (managed by the Office of Naval Research) started in 1962 called PROJECT SEASIDE—a subset of DEFENDER—that specifically explored the feasibility of ruby lasers as anti-ABM weapons. (Laser research was grouped under GLIPAR, which stood for Guidelines Identification Program for Antimissile Research.) These studies received considerably more attention in October 1962, when U-2 overflights revealed Soviet missile emplacements in Cuba. This crisis, with the potential for escalating

Colonel David R. Jones was an early proponent of lasers in the early 1960s and later served as AFWL commander from 1 July 1967 to 31 January 1970.

into a nuclear war, resulted in President Kennedy upgrading the DEFENDER program to the "highest national priority category." Although PROJECT DEFENDER would not solve the Cuban missile crisis, Kennedy's decision placed increased emphasis on laser research and made funds more readily available, in hopes of developing lasers as ABM weapons to be better prepared for future confrontations.[25]

Work directed by ARPA Order 313-62 gave AFSWC its first opportunity to get its laser feet wet and to build a small cadre of officers who would prove invaluable to AFWL's laser program in the 1970s. Three of the most influential people who had a major role in the initial laser work were Captain Donald L. Lamberson, First Lieutenant John C. Rich, and First Lieutenant Petras V. Avizonis, who had just arrived at AFSWC in March 1962. Avizonis was a welcome addition who appeared "out of the blue," according to Colonel Jones. The lieutenant, who emigrated from Lithuania to the United States in his youth, had received a Ph.D. in physical chemistry from the University of Delaware. Although Avizonis was by no means an expert in lasers, his interest in the subject had been stirred after listening to a graduate seminar lecture delivered by Charles Townes. Consequently, Avizonis readily looked forward to participating in the laser effects experiments when he reported to Colonel Jones's Physics Division. Avizonis's on-the-job training in lasers during these early years laid the foundation for his technical expertise that propelled him to become the highest-ranked civilian in the Weapons Lab's laser program by the mid-1970s.[26]

Dr. Petras V. Avizonis, one of the early laser pioneers who became the leading technical expert on lasers at the Air Force Weapons Lab.

With the initial ARPA work finding its way to the Physics Division, Jones sensed the potential for attracting even more laser programs in the future. To do this, he knew he and his people had to get smart on lasers—AFSWC, the colonel pointed out, "cannot afford to be second to anyone." So he put together a task force headed by Dr. Guenther to learn more about laser research conducted by other organizations. As Jones put it, he wanted "Art Guenther, Marv Atkins, and Pete Moore [to] get on the road to get us up to speed." This team traveled around the country, visiting government agencies and private companies, identifying the key players, and determining the state of the art for lasers. Trips to Bell Laboratories and Fort Monmouth (home of the Army's laser research) in New Jersey, AVCO-Everett Research Laboratory and Moleculon Consultants, Inc. in Massachusetts, Pratt & Whitney in Connecticut, the Optical Society of America and National Bureau of Standards in Washington, D.C., Westinghouse in Pennsylvania, and Hughes in California, collected essential information providing AFSWC scientists a better national perspective on laser research and development.[27]

Although Physics Division personnel were gradually becoming better informed on lasers, the results of AFSWC's initial laser work were disappointing. On 20 July 1963, the Center's two young officers, Captain Lamberson and Lieutenant Rich, who would later lead the AFWL laser program in the 1970s, reported on the progress of laser work authorized by ARPA. Their findings, presented as part of the PROJECT SEASIDE summer conference held at Woods Hole, Massachusetts, concluded solid-state ruby lasers were generally inefficient and not the best choice for ABM weapons. For every 1000 joules (a watt equals 1 joule per second) put into the system, less than 1 percent could be extracted as laser output power. Increasing power output to operational levels would require a monstrous input device capable of generating power comparable to that produced by several Hoover Dams linked together. (Lamberson and Rich relied mainly on basic modeling and structural calculations from Stanford Research Institute to reach their conclusions.) Heat buildup with glass lasers also degraded the quality of the beam and presented difficulties in removing waste energy from the system. Lamberson recalled, "We showed conclusively that you really just couldn't get there. Energy levels were about 4 orders of magnitude too short; the precision of pointing was at least 2 orders of magnitude deficient; and even to get those inadequate capabilities, you had to have all the glass in the world!" Not only was it difficult to reach desired power and beam-quality levels for solid-state lasers to be practical (about 7×10^5 joules per square centimeter), but the enemy, through relatively inexpensive design changes, could sufficiently harden its missiles and reentry vehicles to negate the effects of the laser. A third concern was that even if solid-state lasers led to a magical system that would work perfectly, they would not be cost-effective. The extra missiles the Soviets would have to build to overwhelm any laser ABM system would cost them less than the money required by the United States to build laser weapons. Even though the SEASIDE findings were disappointing, Lamberson and Rich had, in the process, established a reputation for themselves and AFSWC as one of the prominent centers for laser research in the Air Force.[28]

Although the promise of using solid-state lasers as weapons appeared to be at a dead end, important advances in laser technology from 1964 to 1967 revitalized the prospects of developing a different type of laser as a weapon. A Bell Telephone scientist, Dr. C. Kumar N. Patel, discovered in April 1964 that molecular gas carbon dioxide (CO_2), through electrical pumping, could be used as a lasing medium. (Nitrogen served as a source of energy to excite or pump the CO_2.) This was the first continuous-wave CO_2 laser. In June 1965, he announced he had obtained 16 watts of power from a CO_2 laser with an efficiency of 4 percent. His breakthrough touched off widespread interest in CO_2 lasers mainly because of the promise to attain high power and high efficiency. Building on Patel's work, Professor Abraham Hertzberg at the Cornell Aeronautical Labora-

Dr. Arthur H. Guenther headed a team that traveled around the country visiting private companies to assess the potential of lasers for military applications.

tory in 1965 advanced the theory that CO_2 could lase by heating a CO_2–N_2 gas mixture. To prove this principle, Dr. Edward Gerry of AVCO-Everett Research Laboratory (AERL) in 1966 conducted separate experiments—first in a shock tube (April) and then in a combustion chamber (June)—demonstrating new pumping techniques to stimulate population inversion (lasing) by rapid expansion of a hot equilibrium gas mixture passing through a bank of supersonic nozzles. This was the first gas dynamic laser. In 1967, Gerry and his colleague, Arthur Kantrowitz, accidentally introduced water vapor into the mixture of nitrogen and CO_2 gas, which for the first time produced good beam quality at a power of 10 kilowatts. This CO_2 gas dynamic laser (GDL)[*] produced thermal excitation of photons by combustion of gases and also offered the advantage of removing waste heat more efficiently than solid-state lasers. Whereas CO_2 gas laser power depended on building longer and longer devices to increase power, the GDL research indicated smaller devices could achieve higher power levels. (Up until 1964, because the power levels of pulsed lasers were so low, "Gillettes" were used as the standard unit of measurement, that is, power was determined by the number of razor blades a laser could penetrate in about 1/5000th of a second.) All of these breakthroughs were extremely important, because they indicated to DOD officials that continuous-wave GDLs, with their potential for greater efficiencies and higher power

[*]The gas dynamic laser derived its name from the high-speed flow and manipulation of gases resulting in population inversion of photons that produced laser light.

levels, offered the most promise for near-term military applications. AVCO continued to perfect its GDL, generating 138 kilowatts in March 1968.[29]

AFWL TAKES LEAD

With expectations running high for GDLs, the Weapons Laboratory issued on 15 September 1967 its first advanced development plan (ADP). Written by Dr. Avizonis and Captain Harry I. Axelrod, the plan outlined a program to demonstrate laser weapon feasibility for ground-based and airborne applications. AFWL proposed building a GDL in the 200- to 400-kilowatt range and integrating it with a fire control system to direct the beam. Avizonis and others clearly recognized this was a "high-risk" program. Tough technical issues had to be addressed and solved regarding beam propagation in the atmosphere, beam interaction with targets, and engineering problems associated with the design, fabrication, and integration of the laser device with an optical system to steer the beam.[30]

A large share of what went into the 1967 ADP originated from a conversation Avizonis had with Lieutenant Colonel Howard W. Leaf, who worked at Headquarters Air Force, Deputy Chief of Staff for Research and Development. Leaf, a crusty fighter pilot who was always looking for new ways to enhance firepower on airplanes, was impressed with the early laser work that started at AFSWC and carried over to AFWL. He believed a laser would be a good candidate for a weapon system to be installed on the defenseless Boeing 707 Airborne Warning and Control System (AWACS) aircraft. His advice to Avizonis was that the "bits and pieces" of laser technology developed at AFWL should be integrated and demonstrated on an airborne platform. Avizonis recalled it was Leaf who coined the phrase *airborne testbed* that Avizonis and Axelrod wrote into the early ADPs.[31]

The 1967 ADP marked a major turning point for the Weapons Laboratory for two reasons. First, AFWL's plan laid the groundwork to gain the support of higher headquarters to make the Laboratory the center of expertise for all Air Force high-energy laser development work. One of the most effective ways for AFWL to showcase its expertise was to design, build, and fire a ground-based laser capable of engaging static and moving targets. AFWL vigorously pursued this goal over the next few years by taking charge of what was known as the tri-service laser (TSL) program. Second, the ADP was the first document to recommend placing a laser on an aircraft to be tested as a tactical or possible strategic weapon system. Hence, AFWL had planted the seed that would grow into the development of the Airborne Laser Laboratory to perfect the physics and engineering requirements for firing a laser from an airborne platform. Once the TSL had proven itself on the ground, the plan called for AFWL to proceed with development of an airborne laser.[32]

One unusual feature of the AFWL ADP was the way it addressed the mission. That is, the proposed laser work would not support an existing Air Force mission. Instead, the Lab's goals were to first advance a new technology (lasers) and demonstrate its feasibility. Once that had been achieved, AFWL would try to sell lasers to the operational commands—Strategic Air Command (SAC) and Tactical Air Command (TAC)—to gain their support to satisfy a *new* mission requirement. In other words, the technology would drive the mission rather than vice versa. This approach ran counter to the traditional way technology was introduced into the Air Force. The problem in the long run was that it was difficult to get the operating commands to support radically new technology. SAC and TAC were reluctant to give up their missiles and guns on aircraft and replace them with a new and unproven high-risk technology like lasers. Their reaction was, according to Avizonis, "show the technology could work and then we'll talk about putting a laser on an aircraft."[33]

Responding favorably to AFWL's ADP, Headquarters Air Force on 3 October 1968 released its Development Directive for High-Energy Lasers (DD 104-1) authorizing the Laboratory to proceed with building and testing a GDL as described in a revised AFWL ADP dated February 1968. The decision to allow AFWL to go ahead with the GDL work reinforced the Air Force's and Air Force Systems Command's (AFSC's) confidence in the Laboratory's ability to get the job done. This was not simply an "act of faith" but was based on earlier AFWL performance on solid-state laser research and more recently on preliminary investigations into the new field of GDLs. For example, in the summer of 1966 AFWL scientists had successfully operated a small electrical CO_2 continuous-wavelength laser generating 500 to 700 watts of power for approximately 2 minutes. A series of vulnerability experiments showed that this laser, acquired from Raytheon at a cost of $78,000, was capable of burning holes in a variety of materials. The following year, this small laser was consistently reaching an output power of 1000 watts with good efficiency.[34]

As AFWL's reputation grew, more and more laser work filtered into the Albuquerque laboratory. However, during those early years AFWL found itself competing for business with one of its sister laboratories, the Air Force Avionics Laboratory (AFAL) at Wright-Patterson AFB, Dayton, Ohio. AFAL had contracted AVCO in 1966 to build and test a small GDL. By June 1966, AVCO had built a burner system and achieved lasing from a cyanogen combustion-powered device. In May 1967 the contractor achieved a power level of 8 kilowatts for its GDL called RASTA (Radiation Augmented Special Test Apparatus) and reached 138

AVCO's Haverhill research facility where testing of the MK-5 laser took place. Beam propagated from the test facility through an opening in the trees leading to a target on the edge of the Merrimack River in background.

kilowatts with its follow-on version laser (the MK-5) tested at AVCO's Haverhill research facility (40 miles north of Boston) in March 1968. But management squabbles between

AFAL and ARPA (who provided most of the money for the Avionics Lab research, which included a contract with AVCO to continue work on the MK-5) opened the door for the AFWL commander, Colonel Jones, who was prepared to offer a solution. He sent Dr. Avizonis and Colonel C. K. Stambaugh, chief of AFWL's Research Division, to meet with Avionics Lab personnel to work out which organization should conduct future GDL research. After consulting with Avizonis and Stambaugh, Jones recommended to Systems Command's Director of Laboratories, Brigadier General Raymond A. Gilbert, the AVCO contract be transferred to the Weapons Laboratory. Gilbert, who had served as AFWL's first commander from 1963 to 1967, was receptive to Jones's idea because it made sense to him to consolidate most of the laser development programs under one lead laboratory, in this case, AFWL.[35]

Gilbert believed AFWL had more laser in-house "hands-on" experience with its "blue-suiters" and civilian scientists than AFAL, which contracted a large share of its work. It was AFWL's in-house track record that, in Gilbert's mind, tipped the scales in favor of moving laser research from AFAL to AFWL. Gilbert's loyalty to his former organization and confidence in the blue-suit scientists who worked there, no doubt, influenced his decision to make Albuquerque the center of excellence for Air Force laser research. Plus, Lieutenant Colonel Ron Cunningham, who worked for Gilbert, was able to persuade Colonel Glenn Sherwood, an ARPA program manager who controlled

Brigadier General Raymond A. Gilbert served as AFWL's first commander, 1 May 1963 to 31 August 1966, and was responsible for centralizing laser research at the Air Force Weapons Lab.

funding, to support shifting the laser work from the Avionics Lab to Kirtland. In spite of some initial protests from the Avionics Lab, especially its director, Colonel James L. Dick, and Dr. William C. Eppers, Jr., who headed the laser program, Gilbert moved the GDL AVCO work from AFAL to AFWL on 1 July 1968. Gilbert recalled, once he made that decision, Colonel Dick and Dr. Eppers offered their full support and recognized the potential long-term benefits of transferring the laser work to AFWL.[36]

Moving the laser work to Kirtland was clearly one of the critical decisions early on that allowed AFWL to build on its technical expertise and to attract significant funding to reinforce its plan to design, build, and test an airborne laser. Looking back on the laser program years later, Lamberson credited Gilbert for making a courageous decision based to a large degree "on nothing more than gut faith" that AFWL could get the job done. As events unfolded over the next decade, Gilbert's instinct turned out to be a heady decision.[37]

At the time, the Air Force considered AVCO the industry leader in GDL technology development. But two other companies, Boeing and United Aircraft's Pratt & Whitney, also competed for the Air Force's attention. Boeing, mainly because of insufficient technical expertise and funding problems, never entered the mainstream of Air Force laser projects. Basic research in lasers was not Boeing's strong suit. They were more interested in long-term studies leading to applying laser technology to aircraft systems. The story was different with Pratt & Whitney officials. Working out of their newly constructed Florida Test Facility at West Palm Beach, they invested their own money and built the XLD-1 laser device that produced an output beam of 77 kilowatts with an efficiency of 1 percent in April 1968.[*] General Otto Glasser, Air Force Assistant to the Deputy Chief of Staff for Research and Development in the Pentagon, immediately recognized the importance and long-term potential of the XLD-1's progress. At Glasser's urging, ARPA's Dr. David Mann contacted Colonel Jones who, in turn, issued an AFWL contract (F29601-68-C-0109) on 10 May to Pratt & Whitney to continue work on upgrading the XLD-1 at their Florida Test Facility.[38]

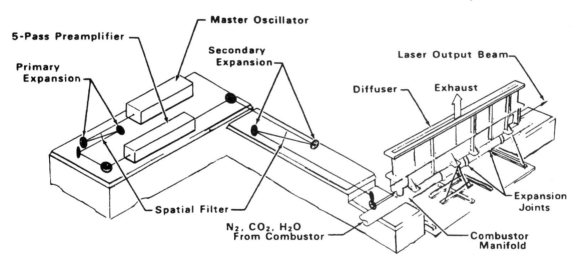

XLD-1 laser configuration.

In the meantime, while AVCO and Pratt & Whitney advanced the state of the art developing large devices to generate laser beams, AFWL was hard at work conducting extensive vulnerability and propagation studies with low-power CO_2 lasers as well as advancing device development. By the summer of 1968, there was no doubt that AFWL was the Air Force's lead laboratory for the majority of laser research and development efforts. A key point was that no *one* laser program dominated at AFWL. Rather, several parallel efforts coexisted at the same time. This was done intentionally by AFWL's leadership, because lasers

---

[*]XLD stood for experimental laser device. Most of the original GDL work by Pratt & Whitney was ARPA funded. However, realizing the potential of lasers over time, the Air Force invested its own money to fund the ALL as a way to control its own program without interference from ARPA.

were only in their infancy. Changes occurred in the technology almost monthly, and no one was sure exactly where the research would eventually lead. So AFWL kept its options open—resisting the temptation to place all of its so-called "laser eggs" in one basket—by allowing AVCO and Pratt & Whitney to work independently of one another, with the hopes that progress would be achieved by sharing each other's successes and failures. At the same time, the Laboratory was in the process of embarking on its ADP approved by Headquarters Air Force. AFWL's next step was to implement a program, which took shape in the creation of the tri-service laser, the first phase of the laser work identified in the Avizonis/Axelrod advanced development plan.[39]

2

First Laser Ground Demonstrations

AFWL's revised ADP, approved in February 1968, proposed building a "laboratory GDL" with a power output of 500 kilowatts. However, it did not specify the detailed design of the GDL, nor did it identify the facilities that would test and operate the laser device. Consequently, the Laboratory awarded separate $10,000 contracts to AVCO-Everett Research Laboratory and Pratt & Whitney to iron out these uncertainties and recommend the size, design, and operating parameters of the GDL, as well as to develop a plan to test the laser, first in a laboratory, then at a test range. Both contractors confidently predicted current GDLs (AVCO's MK-5 and Pratt & Whitney's XLD-1) could be "scaled up" to higher power levels of several hundred kilowatts. To complete testing of the components required to validate the design of the laser, the two contractors recommended the construction of a Laser Optics Laboratory at AFWL and the development of Sandia Optical Range (SOR) as a laser test site located on Kirtland AFB 7 miles southeast of Albuquerque International Airport.[1]

TRI-SERVICE LASER

In the early summer of 1968, John S. Foster, Director of Defense Research and Engineering (DDR&E) at the Department of Defense in the Pentagon, met with the Defense

Science Board (DSB), an independent body of civilian scientific experts outside the government who advised DOD on a variety of technical matters, including lasers. Foster, who had participated in the Manhattan Project, recognized the revolutionary potential of lasers (based on the rapid progress of GDL work) and was willing to gamble on this high-risk technology. As a result, he and the DSB agreed to establish EIGHTH CARD, the code name of the highly classified GDL program to be managed by the Advanced Research Projects Agency (ARPA) and AFWL. (ARPA reported to DDR&E.) The name *EIGHTH CARD* originated from seven-card stud poker, implying the advantage went to the player who held the eighth card. Translated to the political arena, this suggested that if the United States held the eighth laser card, it would hold a distinct advantage over the Soviets in the Cold War or in any future military confrontation.[2]

During the first week of July, Foster discussed with AFWL, AVCO, and Pratt & Whitney officials the goals and direction GDL research should head. AVCO and Pratt & Whitney officials reassured the Director on the feasibility of developing a GDL weapon. They also stressed the emphasis should be on short-range (less than 5 kilometers) tactical applications where power loss resulting from laser beam divergence would not be excessive. Arthur Kantrowitz, who headed AVCO and knew Foster, claimed his company had the technical expertise to build three lasers (one for each service—original estimate for each laser was $800,000) that would "revolutionize the Defense Department." Convinced that lasers were a high priority, Foster's next step was to provide some organizational structure to EIGHTH CARD. He and the DSB endorsed the formation of a tri-service laser (TSL) committee to be responsible for defining objectives, reporting on technical progress, and assessing the overall value of the GDL program. Lieutenant Colonel John Scholtz, who had joined AFWL in April 1968 and replaced Captain Axelrod as head of the laser development group, chaired the TSL committee. Membership included two AFWL officers representing the Air Force (one was Captain Axelrod), two Army civilians from Redstone Arsenal, and two senior civilians from the Naval Research Laboratory.[3]

Foster and the DSB insisted all three services participate in the GDL program to provide an "integrated approach" to GDL research, which would facilitate a sharing of technical knowledge. Also, each service could develop and tailor its laser for its own special mission by operating the laser in defense scenarios against the type of weapons typically used against each service. One of the first jobs of Scholtz's committee was to define the requirements and specifications of a proposed GDL. By August the committee had decided to fund building three identical CO_2 GDLs lasing at a wavelength of 10.6 microns (infrared) with a power output of at least 100 kilowatts. (Because all CO_2 molecules exhibit a wavelength of 10.6 microns, the CO_2 beam also lases at 10.6 microns.) One laser would be given to each service to conduct research and experiments that would best advance laser missions tailored specifically for the Army, Navy, and Air Force. Each service would fund its own TSL. AVCO argued, in the long run, it would be more economical to design and distribute three identical lasers. In addition to receiving its own laser device—named the TSL—each service independently would investigate other related laser technologies, such as optics, effects (how the beam interacted with target material), propagation, pointing and tracking, output window material, and more. ARPA contributed no funding to the TSL. The Air Force wanted to remain in complete control of the first ground-based laser program to avoid any technical interference from and financial dependence on ARPA.[4]

The TSL committee continued to meet in the fall of 1968 and finalized the joint specifications and statement of work for the TSL contract. In the meantime, Pratt & Whitney had rebuilt its XLD-1 device, increasing power from 77 kilowatts (June 1968) to 210 kilowatts of raw power on 2 October. Design of the XLD-1 was based on Pratt & Whitney's extensive experience with high-performance rocket and air-breathing engine technology. In spite of its success with large devices, Pratt & Whitney elected not to bid on the TSL contract, mainly because it did not feel comfortable with its ability to build and scale-up smaller laser devices that would produce good beam quality and serve as the building blocks for the TSL. This left the door wide open for AVCO, the only other qualified contractor with the laser experience and technical expertise, as the leading candidate to bid on the TSL.[5]

AVCO entered contract negotiations with the government in December 1968. The following month, AVCO submitted its initial proposal to build three TSLs at a cost of $4.9 million. Projected delivery date of each laser—one to AFWL, one to the Army's Redstone Arsenal at Huntsville, Alabama, and the third to the Naval Research Laboratory, Chesapeake Bay Annex, Maryland—was set for the summer of 1970. On 10 March 1969 the Air Force let the TSL contract to AVCO for $5.8 million. Major Axelrod managed the entire contract, including the TSLs for the Army and Navy, because the Air Force had the most laser experience and the most dealings of any of the services with AVCO and Pratt & Whitney.[6]

Even before finalization of the TSL contract, AFWL was busy lining up other laser work. The TSL device (later, the Air Force's TSL was referred to as Air Force Laser I)—the hardware that generated the laser beam—was only one component of the overall laser system. An optical system, essentially a series of mirrors, was needed to accept the output beam from the device and then align and steer the beam to a field test telescope (FTT). A large primary mirror in the FTT then focused and pointed the beam to its target 500 to 5000 meters away. Because of the precise tolerances required for the protective coatings applied to thin mirror surfaces to attain almost 100 percent reflectivity, and the precision alignment of the mirrors to turn and steer the beam, the optics was considered the most difficult technical challenge in the development of any laser weapon system. AFWL took the first step to meet this challenge by issuing a contract on 3 February 1969 for $785,000 to Hughes Aircraft to build an FTT that could be mated to the Air Force TSL. Although the Air Force FTT ended up costing almost $4 million, Colonel Scholtz believed it was well worth the money, because it represented the most advanced piece of hardware for pointing and tracking ever built up to that time. And the selection of Hughes was not an overnight decision. Hughes was the AVCO subcontractor developing the internal optics for the TSL device. In addition, Scholtz recalled that "we had seen some Navy optical equipment that was very good research and development equipment that was built by Hughes, and it looked like about the same kind of technology we would need."[7]

Although Colonel Scholtz generally sensed the future importance of lasers, Axelrod and Avizonis were the real men of vision. While preparing for the TSL program, they were the early proponents who consistently had pushed for the airborne laser concept that they made certain appeared in the first advanced development plan in 1967. Air-to-air missiles and

aircraft were relatively soft targets for high-energy lasers, convincing them that airborne lasers made the most sense for operational systems. To start gathering support for this idea, they made sure that the contract to Hughes for the FTT contained a reference to using the

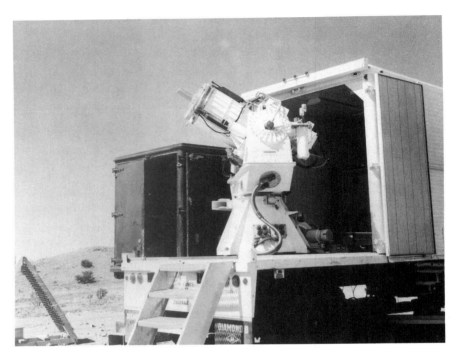

Side view of field test telescope (FTT).

FTT on board an aircraft. Two critical airborne objectives in the FTT contract, which laid the foundation for the follow-on airborne laser program, were:

1. To develop the basic technology required for lightweight airborne GDL optical systems
2. To conduct a system and design study of airborne GDL optical systems

To meet these two contract specifications, Hughes conducted its own airborne study beginning in April 1969 at the same time they were building the ground-based FTT. Colonel Scholtz's plan, at the urging of Axelrod and Avizonis, was for the FTT to prove itself on the ground first and then place the FTT on an aircraft. The original concept was that a lightweight FTT would look out the side of the aircraft to acquire and track targets. As it turned out, AFWL discarded this side-view configuration because of technical problems. In the end, AFWL directed Hughes to design and build a lightweight airborne pointing and tracking (APT) system to go on top of the plane. However, Scholtz, Axelrod, and Avizonis had made an extremely important contribution by focusing more attention on the airborne concept during the early stages of the ground-based program. It was this approach, conducting two key programs in parallel, that Lamberson (who took over for Scholtz in June 1970) adopted to form the bedrock of his management plan as the ALL gained more momentum in later years.[8]

It was during this formative period of the laser program at AFWL in the late 1960s that Avizonis emerged as the most knowledgeable technical expert who set the focus and direction for future laser work. Scholtz and Axelrod were without a doubt prominent players, but they would eventually move on to new assignments. Avizonis, on the other hand, provided strong technical continuity and leadership over several decades as he remained with the laser program into the late 1980s. Not only did Avizonis play an enormously important role in laying the groundwork for laser research, particularly in the early years through his comprehensive and conceptual insight on lasers embodied in the advanced de-

Colonel Dave Jones (left) with Lieutenant Colonel John Scholtz.

velopment plans, but later on he became Lamberson's indispensable associate and right-hand man. From 1969 to 1978, Avizonis served as Technical Director for AFWL's laser program and, in that capacity, proved invaluable to Lamberson. Lamberson and Avizonis forged a unique management team rarely seen in government for that length of time. Clearly, Lamberson, as Program Manager, was the one in charge. However, the relationship was definitely not a superior–subordinate one. It was a team connection where fundamentally all decisions were consensus decisions between the two men. Each had tremendous respect and confidence in the other's expertise and exploited it greatly to the benefit of the entire laser program. It was this bedrock of management, established early on and lasting for a long time, that was one of the main reasons for the success of AFWL's laser program. The first sign of success was the TSL.[9]

Success of the TSL program depended on meeting three critical milestones involving propagation and effects testing. The first was to demonstrate that the FTT could track a moving target and then accurately point a low-power CO_2 beam to hit that target. Next, the TSL device had to produce a high-power beam with good beam quality. Good beam quality meant a tightly focused beam that would deliver maximum energy on a target spot. (The term used to describe this condition was a *nearly diffraction-limited* beam. This meant the beam spread in angle at the minimum rate theoretically possible.) The beam control system had to be of sufficient quality to prevent the beam from spreading out on the way to the target and moving around the target aimpoint. Finally, the most important part of the program was

the integration of the laser device and FTT to show that a high-power (150–200 kilowatt) laser, with good beam quality, could be accurately directed to engage a moving target.[10]

From the beginning, AVCO ran into major design and engineering problems trying to build and operate the TSL. Originally, AVCO was confident that its MK-5B laser, with an output of over 100 kilowatts, had sufficiently advanced the theory and physics of GDLs. Based on this, the contractor believed developing the TSL would be a simple engineering exercise of "scaling up" the MK-5B device to higher output levels. As it turned out, TSL reverted to a "technology development" program that had to first understand and solve complicated physics problems before the contractor could move on to fabricate the laser device. In the end, this led to frequent hardware modifications, schedule slips, cost overruns, and in the words of one Air Force official, "the premature aging of everyone involved."[11]

The most frustrating problems for engineers who worked on building the TSL at AVCO's Wilmington plant outside Boston, and later at AFWL's Sandia Optical Range (where construction had begun in February 1970), were the master oscillator power amplifier (MOPA), fabrication of cooled and uncooled mirrors, reliable start-up of the diffuser, and design of supersonic nozzles. Design and testing of the MOPA, the key component that generated the output beam of the desired quality, power, and wavelength, was considered "duck soup" at first, but it eventually had to be replaced by an unstable oscillator configuration. Hughes, AVCO's subcontractor for the TSL internal optics, ran into problems in the design, material selection, fabrication techniques, and coating of large mirror surfaces. Design modifications also had to be made to the diffuser to properly vent the waste gases from the device, and to the large array of nozzles the combusted gases passed through at supersonic speed to create the right lasing conditions.[12]

Tri-service laser (TSL) at AVCO.

All of these problems and more contributed to about a 6-month delay in the delivery of the TSL to AFWL's SOR. The first components—diffuser, combustor, and nozzle assembly—began arriving in April 1970. By July, the entire TSL had been assembled, minus the optics. There was only one problem. When turned on, the device did not generate a beam. For the next 16 months AVCO personnel worked feverishly to correct the problems and met with some success. By the end of February 1971, cooling panels in the combustion chamber had been fixed and pipes carrying gases to fuel the device had been rewelded to eliminate leaks. From 19 to 26 February, AFWL conducted five hot flow tests. Hot flow tests verified that the gaseous and coolant feed systems had been debugged and that the automatic sequencing worked properly. In April 1971 the first power extraction of 39 kilowatts (from the MOPA) occurred. By the end of April, the TSL diffuser had started in 6 of 10 attempts. Two months later AVCO had solved the start-up reliability problems of the diffuser, which vented the excess heat in the system. And by November 1971, after switching from the MOPA to an unstable oscillator, the device had consistently "approached" the contract specification of 100 kilowatts for beam output power. What was not achieved

was a good-quality beam, that is, a nearly diffraction-limited beam that could be focused and projected over several hundred meters. So after nearly 2 years of work, what the Air Force had was a TSL that resembled a harmless searchlight rather than a tightly focused beam that could instantaneously zap its target.[13]

BLUE-SUITERS TAKE OVER

Colonel Robert W. Rowden, who assumed command of AFWL in 1970, was not satisfied with progress on the TSL. Highly respected by those who worked for him, he was a scholarly type who looked as though he would be more at home in the classroom than out leading one of the most important laser programs in the country. Although quiet and unassuming, he could be tough and demanding when required. Rowden wanted to see results from the costly investment in the TSL, and did not hesitate to criticize AVCO's performance. His main concern was the inferior quality of the beam coming out of the TSL. In August 1970 he visited AVCO's headquarters in Wilmington, Massachusetts, and complained about "poor engineering practices" and cost overruns by the contractor. Numerous letters followed over the next year, and one

Colonel Robert W. Rowden closely monitored the TSL work while AFWL commander, 1 February 1970 to 29 June 1973.

from the Kirtland Procurement Office even threatened to place AVCO on the government's "blacklist" of contractors. However, the contractor pleaded not guilty to all charges, claiming setbacks were to be expected because of the physics and uncertainties in the state of the art

of new and complex laser technology. This resulted in tighter manufacturing tolerances for TSL components and, in turn, led to increased costs.[14]

Although highly disappointed by AVCO's performance, for the most part, AFWL accepted the contractor's explanation as to why the TSL was not making more rapid progress. For 16 months, AFWL remained tolerant. But on 8 November 1971, Colonel Rowden again wrote to AVCO stating that he was "deeply concerned about the successful demonstration of the tri-service laser" and, especially, about the failure to demonstrate desired beam quality at high power. Rowden pointed out that AFWL had been "extremely patient and understanding," but "in recent months we have gotten the feeling that the program [TSL] is losing momentum." Lieutenant Colonel Lamberson, who had taken over from Colonel Scholtz in

June 1970 as the head of AFWL's laser program (Scholtz departed for a 1-year assignment at the Air War College), also suspected that the brilliant minds at AVCO had gradually lost interest in GDLs. He was convinced that AVCO had taken the position that electric discharge lasers were the wave of the future in terms of advancing the technology and attracting a more profitable customer market. AVCO simply was not providing all of the experimental data requested by AFWL. Plus, when data were reported, they were often incomplete. By that time, Lamberson had expected AVCO's results to be further along.[15]

In December 1971, Colonel Lamberson made a bold recommendation to Colonel Rowden to break the AVCO logjam so as to revive the momentum on the TSL. Frustrated by endless delays, Lamberson suggested AFWL accept the Air Force TSL as a "nonconforming item" and take over technical direction of the program from AVCO. Based on his confidence in the technical capabilities of his "blue-suiters" at AFWL, Lamberson gambled their "tinkering" could fix the stubborn TSL.[16]

Anxious to get the program moving, Rowden accepted Lamberson's innovative proposal with formal approval taking place at the TSL Technical Direction meeting held on 8 December 1971. Along with the members of the TSL committee present, the

Lieutenant Colonel Don Lamberson prior to taking over Laser Division.

list of attendees, as one AFWL project officer recalled, read like a "laser Who's Who." Representatives came from the highest ranks of government: DOD's Deputy Director for Research and Engineering (DDR&E) and Advanced Research Projects Agency (ARPA), Office of the Secretary of the Air Force, Headquarters Air Force, Systems Command, the Naval Research

Laboratory, and the Air Force's Scientific Advisory Board (SAB). Those in attendance, in an effort to revive the TSL program, voted in favor of AFWL taking over the Air Force TSL. Similarly, AVCO would turn over the other two TSLs to the Army and Navy. It was further decided that the Tri-Service Laser Committee would disband on 26 January 1972, the same day the Air Force would "officially" accept the TSL.[17]

AVCO was in a weak position to keep the TSL program and offered only a perfunctory protest to the Air Force takeover. After all, the contractor was already 18 months behind schedule, the equipment still did not operate according to specifications, and costs had skyrocketed. Final cost of the TSL program jumped from the original projection of $5.8 million to $12.5 million by the time AVCO received its final payment in April 1972. Of this total, the Air Force contributed $4.7 million, the Army $4.7 million, and the Navy $3.1 million. (The Navy's price was substantially lower because once it received its TSL, the Navy did not contract with AVCO to assemble and test the device.) AVCO did not go off licking its wounds. It received its money even though it had not met all contract specifications.[18]

At the same time the 70,000-pound TSL was experiencing problems, Hughes made steady progress on the FTT. Hughes delivered the FTT to AFWL on 16 October 1970. Before it moved out to SOR in July of 1971,* the 2000-pound and 6-foot-tall FTT was stored in a special trailer parked on the west side of Kirtland AFB adjacent to Albuquerque Airport's north–south runway. There AFWL scientists and Hughes technicians had an opportunity to conduct preliminary checks to work out the bugs on the

Head-on view of field test telescope.

---

*Construction of SOR began in February 1970. It became fully operational in March 1971 and was the first high-power long-range laser propagation and test range in the country. Located in an isolated area 5 miles southeast of Kirtland AFB, adjacent to the foothills of the Manzano Mountains, SOR was bordered by national forest to the east and by the Isleta Indian Reservation to the south. Areas north and west of SOR belonged to Kirtland. The facility consisted of a transmitter site housing the laser and three receiver or downrange target sites. Total area of SOR was approximately 3 square miles.

sophisticated device. The location of the FTT gave the AFWL/Hughes team a chance to test the FTT by tracking commercial airliners as they made their final landing approach to Albuquerque International Airport.[19]

Shortly after the arrival of the FTT, AFWL encountered an extraordinary stroke of good luck. General George S. Brown, the new AFSC commander who would rise to the position of Air Force Chief of Staff in August 1973, happened to be visiting Kirtland. Lamberson saw this as a golden opportunity to solidify Brown's support for the AFWL laser program by giving him a firsthand demonstration of the FTT. On the afternoon of 19 November 1970, Brown, Lamberson, and Rowden watched intently as the FTT slewed to track a New Mexico Air Guard T-33. The pilot, Major Hal Shelton, who later served as Lamberson's ALL integration specialist, made a pass over the north–south runway. The FTT, operating out of its trailer van located 100 yards east of the runway, performed almost flawlessly. Shelton recalled the FTT locked on to the T-33's fuel filler cap and didn't break lock until the plane pulled up and accelerated. Brown walked away impressed by what he saw. Lamberson had added another notch to his ALL gun, knowing the successful FTT demonstration was one more step in winning over one of the most influential players at the highest level to support the AFWL laser program.[20]

T-33 aircraft tracked by field test telescope.

TECHNICAL MILESTONES

In October 1971, a year after its arrival at Kirtland, the FTT completed the first of three TSL milestones established by the Air Staff. Mounted in a garagelike structure (officially named the transmitter site) atop a hill at SOR, the FTT proved its worth by tracking a North American T-39 aircraft as it flew up and down the remote valley below the transmitter site. Sensors in the passive imaging tracker swept back and forth across the field of view, detected the infrared radiation (3–5 microns) of the target, and projected the target image on an 8 × 10-inch television screen. On the screen, the tracking gate followed the T-39 by centroid tracking (lining up the center of the target in the tracking gate) or by tracking on the edge of the target. An instrumented pod slung under the center fuselage of the T-39 carried five quartz iodine lamps that served as the infrared radiation source for the 3- to 5-micron tracker to follow.[21]

The T-39 flew two trajectories. For the transverse flybys, the plane flew by the FTT right to left at a speed of 200 knots at a minimum range of 1500 meters. In this scenario, the FTT tracked the T-39 with an error less than 25 microradians.* For the head-on pass, where the plane flew directly at the FTT and then pulled up, the tracking error was higher. The smaller size of the target—head-on view of the nose of the plane—was the main reason for the larger tracking error. At the 1760-meter downrange site from the FTT was a rotoplane target, a 30-foot device resembling a single propeller blade that carried an infrared (IR) source. The FTT was able to track the IR target attached to the rotoplane arm, rotating at speeds of 10, 20, and 33 revolutions per minute.[22]

Although demonstrating tracking accuracy of 25 microradians was the objective of the first FTT milestone, the FTT also managed to focus a low-power (1000-watt) CO_2 laser (not the high-power TSL) on the target pod attached to the T-39. This was a tremendously important event, because it was the first time a laser beam had hit an aircraft in flight. Using a low-power laser offered several advantages. It was economical, plus the laser could be left on for hours to allow precise alignment of the mirrors without the worry of safety hazards. Seven silver Pyrex glass mirrors in the FTT received the laser beam and directed it through the telescope. Because the power of the beam was so low, uncooled mirrors could be used to reflect and steer the beam. As the 10-centimeter beam struck the next-to-last (secondary) mirror, the beam expanded to 30 centimeters onto the primary mirror, which focused the beam down to a small spot at the target. Not only did the beam hit the T-39 and a stationary rotoplane target board, but test

T-39 served as diagnostic aircraft for ground-based laser testing at Sandia Optical Range.

---

*The distance the beam moved off the target aimpoint is a small angular measurement expressed in microradians. A radian is an angle of just less than 60 degrees (57.295 degrees, to be more precise). A microradian is a millionth of a radian or 0.000057 degree. Twenty-five microradians is an angle of 0.001425 degree, an angular resolution not detectable by the naked eye. Another way of looking at this is that a microradian is the equivalent of trying to focus on a dime on the ground from 33,000 feet.

officials were able to measure the exact location where the beam intersected the target. With these data, autoboresight corrections could be made to move the beam to hit the exact center of the desired target spot. Larry Sher and Major Ben Johnson, two of the most capable people on the FTT program, explained that some of the beam reflected off the target back into the IR tracker imager. They further stated:

> A two-element 'chevron' array of mercury cadmium telluride (HgCdTe) detects the position of the beam spot with respect to the cross hairs. If the beam spot does not lie directly under the cross hairs, the azimuth and elevation scan mirrors are offset to realign the boresight of the tracker with respect to the FTT tube.

As part of a three-gimbaled telescope stabilized by two gyroscopes, the inner gimbals (hydraulically actuated) allowed for fine tuning of azimuth and elevation. The outer gimbal provided coarse azimuth and elevation pointing adjustments. Mechanical misalignments and atmospheric effects accounted for most of the boresighting errors. Overall, the gimbaled structure of the FTT allowed it to "slew" smoothly to maintain "lock-on" as the plane changed direction. At this stage of the game, the results of the FTT experiment went counter to the traditional notion that the optics, as opposed to the laser device, was the most difficult technical nut to crack. In short, the performance of the FTT was a pleasant surprise.[23]

As successful as the FTT demonstration was in tracking a moving target at 25-microradian accuracies, it could not help AFWL scientists overcome the numerous technical obstacles still plaguing the high-power TSL device. But the ever-optimistic Colonel Lamberson had a plan. He called on his "ace in his pocket," Major John Rich, to take charge of the TSL and get it working as soon as possible. Armed with a Ph.D. from Harvard in astrophysics, Rich had earned the reputation as one of the leading laser minds in the Air Force. While an anxious group of onlookers peered over his shoulder, the young major immediately set to work with a small AFWL cadre of physicists, engineers, and optics personnel, who worked virtually around the clock to tackle the uncooperative TSL device.[24]

The two most pressing problems were the aerodynamic flow and optical configuration in the TSL. Frank Wells, a civilian with an abundance of common sense and analytical skills, solved the first problem. A long array of hoses fed the CO_2 gas into the combustion chamber, but because of the way the hoses were set up, the gases did not reach all of the injector ports in the combustor simultaneously. This resulted in an uneven combustion wave causing the gas flow to reach the nozzle bank at different times. (In essence, optimum lasing conditions did not exist throughout the entire cavity, because there were pressure variations caused by the uneven gas flow.) It also made it difficult to efficiently vent all of the waste gases through the diffuser. After many trial-and-error adjustments, Wells tailored the length of each hose so all of the gases arrived at the combustor at exactly the same time, which resolved the aerodynamic flow problem.[25]

The second major problem centered on devising a reliable technique for extracting the laser beam from the resonator, the area of the device where lasing took place. Originally, AVCO had designed a "dogleg" extraction system to remove the beam from inside the resonator, but after a lengthy test and evaluation program it proved to be unacceptable. The higher air pressure outside the device tended to be sucked into the cavity and interfered with the beam as it exited the cavity. Captain Dale A. Holmes, a brilliant young Ph.D. officer with postdoctoral experience in optics at the University of Rochester, replaced the dogleg configuration with an aerodynamic window (a curtain of nitrogen) which produced the desired beam results. In April 1972, less than 4 months after Rich had assembled his ad hoc team of troubleshooters, the 10-centimeter (diameter) aerodynamic window extracted a high-power beam of good quality from the resonator. This event marked completion of the second TSL milestone and represented a critical turning point. It was the first time a high-power laser device had been operated by a military team. From that point on, the device proved to be highly reliable and was used during the next few years as a research tool for irradiating various targets to better define laser effects and propagation characteristics of the beam. Once it became operational, Lamberson praised the TSL device, commenting that it was the "only operational gas dynamic laser . . . that can produce power in the 100-kilowatt region on an on-call basis."[26]

Lamberson promptly reported to higher headquarters the success of extracting a high-quality beam from the TSL at SOR. Shortly afterwards, Dr. Herb Bennington, Deputy Director for Research and Engineering, asked to see burn patterns the laser produced. Over the weekend of 8–9 April 1972, AFWL personnel fired the laser at a Plexiglas target and collected the information Bennington had requested. On 11 April, Lamberson flew to Washington to deliver the burn patterns to Bennington. Impressed by this firsthand evidence of AFWL's progress with lasers, Bennington briefed Secretary of Defense Melvin Laird. The Secretary was pleased with this latest laser success story and instructed Bennington to draft a congratulatory

Tri-service laser at Sandia Optical Range.

memo to AFWL. Several weeks later, the memo worked its way down the chain of command and reached the workers at SOR who were the ones most deserving of the credit.[27]

Lamberson's gamble to go with his blue-suiters to fix the TSL had paid handsome dividends and attested to the technical capabilities of the highly talented work force at the Weapons Laboratory. Although the first two milestones had been achieved—tracking and pointing a low-power beam on target and generating a high-power beam of good quality—the

TSL could not be declared a complete success until the FTT and TSL were combined to work as one integrated system. This was perhaps the most important part in the evolution of the airborne laser, because the third TSL milestone would demonstrate that a *high-power* beam (150 kilowatts) could be placed on a moving target.[28]

The introduction of a high-power beam required Hughes to modify the optics of the FTT. Mirrors originally installed in the FTT to direct a low-power beam were not capable of handling the high-power beam, because the intense heat deposited on the mirror surfaces would damage them. Consequently, the same type of water-cooled mirrors that Hughes pioneered during the development of the TSL were fabricated for the FTT mirrors to accept the high-power beam. Water flowed through a series of tiny channels etched on the backside of the mirrors that served as a heat transfer system to carry away the laser heat absorbed by the mirrors. The reflective surface of these mirrors had to remain within 30 millionths of an inch so that their shape (flatness or curvature) would not be distorted by the heat of the high-energy beam. All of the FTT mirrors were water cooled except for the primary mirror. The surface of the primary was large enough (increased from 30 to 50 centimeters for the high-power laser experiments) to disperse the heat deposited by the laser, thereby eliminating the requirement for cooling. By the summer of 1972, Hughes had installed and tested the new mirrors, concluding that each one could handle up to 250 kilowatts of beam power for 15 seconds.[29]

While Hughes was busy installing the new mirrors, work proceeded on upgrading the TSL device. Workers at SOR continued to fine-tune it, making improvements to obtain just the right aerodynamic flow, reworking the supersonic nozzle array, and fixing leaks in the diffuser. By the fall of 1972, the TSL was consistently producing a good-quality beam and was operating better than ever. All components were now ready to be tested as one system.[30]

In testing the system, AFWL wanted to assess two performance functions of the FTT tracker imager that later could be applied to the advanced pointer and tracker proposed for use on an aircraft. The first question had to do with accuracy. Could the imaging sensors in the tracker provide a clear enough picture to identify a vulnerable aimpoint for the laser beam to strike the target? Once the aimpoint was selected, could the autoboresight system accurately direct the beam to that target aimpoint? The second issue had to do with precision. Ideally, the focused beam would hit the target so the aimpoint was squarely in the center of the beam footprint, which at 1 kilometer was a circle measuring 5.2 centimeters in diameter. Captain Ed Duff, who performed extensive analysis of the beam during the TSL and follow-on DELTA and Cycle I experiments, explained:

> You want that tracking precision to be about a third of the footprint of the beam on target.
> If you make it more than a third, then you're wasting a lot of energy with the beam moving
> to another part of the target.

In other words, the goal was to have the beam not move more than a third of its diameter so it remained on the aimpoint long enough (2–3 seconds) to cause damage to the target. [When the relative motion between the beam centroid (center of beam) and the target aimpoint was held to a fraction of the beam spot size, this was considered "accurate pointing."] For the TSL experiments, the beam could not wander more than 25 microradians. For the follow-on

Cycle II testing on the ALL, there was a more ambitious goal to have the beam not move more than 10 microradians.[31]

Mating of the high-power laser and FTT took place in October 1972. As a safety precaution, to prevent catastrophic damage to the system, SOR workers gradually increased the power of the beam as it moved from the device to the FTT. After these initial shakedown tests, the FTT directed a 100-kilowatt, 10.6-micron CO_2 laser beam at static and moving targets for 3 seconds without any damage to the FTT or the TSL device. The first firings of the laser sent the

Sandia Optical Range—site of TSL testing. Laser device and FTT in lower-level building fired to targets downrange. Upper building was FTT lab.

beam to a stationary target (Masonite board) at the 350-meter and then the 1760-meter downrange site at SOR. Results demonstrated the beam could be focused and accurately directed to the desired target aimpoint by the autoboresight system of the FTT. Most importantly, the beam could be held on target. The next series of tests was to determine if the beam could hit a moving target board mounted on a 30-foot rotoplane (acquired from a carnival ride manufacturer to save money) at the 1760-meter site. Resembling a windmill contraption, the arm of the rotoplane rotated the target 360 degrees along its vertical plane 25 revolutions per minute. Again, the FTT pointed the laser beam to hit the moving target (roughly the size of a wallet-size photo) over a mile away and held the beam on target for several seconds. The best single result from all of the testing confirmed pointing and tracking accuracies of 16 microradians.[32]

Accomplishment of the third milestone in December 1972 was the fitting climax of over $3\frac{1}{2}$ years of plain hard work punctuated with seemingly unrecoverable setbacks. Lamberson's risky decision to have AFWL take control of the TSL from AVCO proved to be the turning point for the struggling new laser technology program. Lamberson later stated he never regretted that decision. The dedicated group of scientists at AFWL [which had grown to 180 people assigned to the Advanced Radiation Technology Office (ARTO) formed in 1972] not only fixed the TSL but integrated it with the FTT so the beam could hit a moving target. Measured by anyone's technology yardstick, this was an accomplishment of enormous proportions, considering the word *laser* did not even exist a mere 12 years earlier. Plus, TSL

Rotoplane at Sandia Optical Range.

attested to the technical competency of Air Force scientists, which made it easier for Lamberson to decide to run the airborne laser program as a "blue-suit" operation.[33]

Lasers certainly were gaining more respect and visibility in the upper echelons of the Air Force after AFWL completed the three TSL milestones. A memo from Headquarters Air Force praised the Laboratory for its dramatic progress with lasers over the past 2 years. Success of TSL also reaffirmed the wisdom of Air Force's earlier decision to initiate the ALL program and reinforced the position that the "potential of laser weapons to Air Force operations still appears great." Consequently, Systems Command continued to support a strong laser research program to develop the technologies required to eventually field the laser as a short-range, air-to-air missile defense weapon. At the local level, Colonel Rowden naturally endorsed Systems Command's optimistic outlook. He summed up his enthusiasm for the future of lasers by patting his own people on the back for meeting the third TSL milestone. This turning point, Rowden stated, "represents confirmation of the technology incorporated in the Airborne Laser Laboratory and envisioned for airborne laser weapon prototypes."[34]

Lamberson was not as quick to claim that laser weapon prototypes were just over the horizon. He had no doubts that the testing at SOR clearly demonstrated the ability of the ground-based FTT to track a moving target and point a high-power laser beam at 25 microradians. Based on the successful performance of the integrated TSL system, Systems Command gave approval to AFWL to develop hardware for an airborne demonstration system (the ALL). Although pleased with these results, Lamberson looked to the future with some degree of apprehension. Anchored to the ground, the FTT operated effectively because it had a fairly stable base to eliminate high vibrational levels. But the high-risk plan to put an FTT device in an aircraft, which vibrated and flexed in flight, posed a new set of precision pointing and tracking problems. The value of the TSL program was that it developed a technology database which served as a vital stepping stone to come up with eventual solutions to the more complex problems of pointing and tracking that would be encountered in the airborne environment. These were formidable physics and engineering challenges that lay ahead.[35]

PROJECT DELTA: TSL AFTERTHOUGHT

Fresh from the success of TSL, Lieutenant Colonel Dick Feaster, who served as AFWL's TSL project officer, strongly urged Lamberson to capitalize on that work by taking one more bold step. Now that AFWL had shown an integrated laser system could work on the ground, what was to be done with the system? Feaster proposed using it to shoot down a drone flying over SOR. Lamberson knew that such an experiment would not bring much new in the advancement of technology, mainly because the TSL had already demonstrated the technology would work. (Avizonis believed the DELTA program was not driven to achieve some lofty technical goal, but the reason for moving ahead with DELTA was based on an "internal gut emotional issue." Avizonis recalled, "Hey, we've got all this technology playing together, all of these different subsystems, and we know they all work, but we have never shot down anything in the history of optical technology—shouldn't we see if we can shoot something down with it?") Therefore, shootdown of a drone would be a "low-risk" technology test. But Lamberson and others also were thinking more of long-term political benefits that could be gained from a dramatic display of laser power against an aerial target. If the beam could knock a drone out of the sky in true Buck Rogers fashion, then it might win over the "doubters" in Washington to lend their support to ongoing and future laser programs.[36]

Work officially started on Project DELTA (Drone Experimental Laser Test and Assessment) on 13 August 1973. The objective of the experiment was to demonstrate that the integrated TSL and FTT ground system could acquire, track, and destroy an aerial target in a realistic, dynamic environment. Weighing 248 pounds and measuring over 12 feet long with a wing span of 11 feet, a Northrop MQM-33B Radio Controlled Aerial Target (RCAT), or drone, flew at approximately 200 miles per hour over a 4-mile racetrack course at SOR. The drone actually followed a course between the SOR site and the Manzano Mountains, which served as a natural backstop for the laser radiation in case the beam missed the drone. Army Air Defense Artillery personnel from Fort Bliss's McGregor Range Camp, New Mexico, flew the highly maneuverable drone by radio-remote control, typically in a counterclockwise direction.[37]

An IR precision tracker, mounted on the FTT, tracked the target by "locking on" to the IR heat source given off by the four-cylinder internal combustion gasoline engine located in the nose of the drone. The IR signature (image) of the drone was displayed on a screen in front of the test operator to monitor. For the initial tests, the laser aimpoint was located on the fuselage below the wing root, a calculated offset from the tracking point. The tracker encountered several problems attributed to signal fades, caused by the drone's turns which blocked the line of sight to the engine track point, and by background noises and sun glints, which on numerous occasions caused the tracking gate on the target to break lock. Eventually, several test runs of the tracker against an F-4 aircraft and the fabrication of an optical acquisition device, consisting of a six-power telescope installed on a hand-directed tracking

mount capable of "handing off" the position of the drone to the tracker, made the system more reliable.[38]

There were a total of 14 drone test runs during DELTA. Six gave the operators practice in acquiring and tracking the target. For six other tests, the drone was fitted with a Plexiglas or metal plate to measure beam stability. Data collected in these tests helped to make corrections to maintain a separate tracking point (the drone's engine) and laser aimpoint. When the laser hit the aimpoint (first the wing root and then the fuel tank in later tests), the energy created by the beam spilled over into the tracking gate. This caused the 3- to 5-micron IR tracker imager to shift its lock-on from the engine to the more intense IR signal reflected from the laser aimpoint. Because the laser aimpoint was a fixed distance from the tracker spot on target, as the tracking point moved, so did the laser aimpoint move off the most vulnerable spot on the target. Adjustments were made to solve this "gate stealing" phenomenon for the last two drone flights.[39]

The last two, and most dramatic, DELTA tests took place in November 1973. To make the tests as realistic as possible, the fuel tank selected for the drone resembled that of an F-4 aircraft in combat. The tank was made of 0.06-inch steel, pressurized to 15 psi, and filled with half a gallon (2 liters) of JP-4 jet fuel. Another auxiliary tank in the drone carried about 4 gallons (15.1 liters) of aviation fuel for the drone engine. During the rehearsal to make one final assessment of the pointer and tracker accuracies, the beam missed the drone and hit a weather tower downrange from SOR. The laser produced a dramatic flash of light as it struck the tower. No damage was done, although the beam left scorch marks on the tower's metal supports. A group of excited technicians hurried to make corrections, increasing the distance between the laser aimpoint and track point for the next drone run. Shortly after noon on 13 November, the beam hit the aluminum fuselage aft of the fuel tank. The beam remained on the drone long enough to burn through the skin, causing the drone to lose control and make one last diving left turn before crashing into the desert floor. Inspection of the debris revealed the beam had burned and shorted out the internal electrical control cabling, forcing the drone into a rolling pitch-down maneuver. Although the experiment disabled the drone, it suffered only minor damage. A recovery parachute was deployed as soon as the drone went out of control to reduce the impact of the drone hitting the ground.[40]

The next day the men at AFWL directed the beam squarely on the center of the JP-4 fuel tank located inside the drone fuselage just aft of the engine. Heat created by the beam remained on target for only 1.2 seconds, but that was enough time for the laser to sufficiently raise the temperature of the steel fuel tank to ignite the fuel vapors inside, producing sufficient overpressure to rupture the tank. In a spectacular blaze of fire and smoke, the disabled drone tumbled 200 feet before hitting the ground. Prior to impact, the engine separated from the drone. A parachute deployed to float the engine to earth as a safety precaution; the engine landed approximately 300 meters from the rest of the drone. First Lieutenant Tom Dyble, who worked with the ground recovery crew, remembered this was an astounding event. The bolts holding the tank together severed, he recalled, and the "main weld seam along the length of the tank blew," turning the tank into an almost flat sheet of metal.[41]

Those who witnessed the historic events of 13 and 14 November 1973 realized Project DELTA was a major milestone, because for the first time a high-energy laser beam shot down

a flying target. It had clearly demonstrated that all of the functions of an integrated system—acquisition, tracking, and pointing the beam to the aimpoint on target—worked against an airborne target. In essence, AFWL believed they had solved the basic physics problems of generating a highly focused beam, processing it with optical surfaces to direct the beam to its target. Based on this scientific achievement, AFWL scientists were confident they could scale up the power level and propagate the beam in the atmosphere to distances of several kilometers. Success also verified the concept that operational aircraft (simulated by the drone) were extremely vulnerable to lasers. And all of this took on even more meaning, considering lasers had only been on the scene for 13 years.[42]

Having scored the first shootdown of an aerial target by a laser, Project DELTA clearly established that the Air Force was ahead of the Army and Navy in terms of laser demonstration programs. It wasn't until the summer of 1976 that the Army used an AVCO electric discharge laser to disable slow-moving (300 miles per hour) 15-foot-long Beech Aircraft winged (MQM-16A) and helicopter drones at Redstone Arsenal, Alabama. (The laser was housed in a mobile test unit mounted on a modified Marine Corps LVTP-7 tracked amphibious assault vehicle.) The Navy did not succeed in a similar laser demonstration until March 1978, when it fired a 400-kilowatt deuterium fluoride high-energy chemical laser built by TRW and a precision pointer tracker built by Hughes to shoot down an Army TOW (tube-launched, optically tracked, wire-guided) antitank missile (6-inch diameter) launched near San Juan Capistrano, California. This is not to suggest the Army and Navy shootdowns were not important. Indeed, they were major accomplishments, but it was the Air Force that had struck first. The small AFWL group at the remote SOR site in 1973 had made

Radio Controlled Aerial Target (RCAT) drone engaged by TSL on 13 November 1973.

a valuable technical contribution by reaffirming the Air Force was the leader in the overall high-energy laser program. The spectacular results achieved by DELTA helped AFWL to crack the door open to attract more attention, support, and monies for future laser research and development programs.[43]

But Lamberson was not content to crack the door open. He wanted to knock it down to ensure that there would be no turning back on programs investigating the feasibility of lasers as weapons. In his opinion DELTA was an unqualified success, but small potatoes for what lay ahead for the future of lasers. Lamberson's vision, since the early days of TSL, had always been an unwavering commitment to place a laser on an airplane to further demonstrate the potential of lasers. He did not need the success of DELTA to convince him to move ahead with an airborne laser. The real value of DELTA was the effect it had on convincing others. Armed with the stunning DELTA film that had captured the successful shootdown in Technicolor, he briefed at the Pentagon, Air Force Headquarters, DARPA, SAB, Systems Command, SAC, TAC, Aerospace Defense Command, and even Congress to give them firsthand visual evidence of the damage lasers could do. After watching the film, few could dispute that AFWL had taken the laser to the field and had applied it in a real-world scenario. In short, DELTA was an important technical accomplishment, but, more importantly, it served as a political rallying point to sustain laser work in general and, more particularly, to persuade the purse-string holders and the doubters, unfamiliar with the new technology of lasers, of the critical importance of moving forward with the ALL. This was the logical next step.[44]

Launching the Airborne
Laser Laboratory

As work progressed on TSL and Project DELTA, the groundwork for the Airborne Laser Laboratory (ALL) was already under way. During the late 1960s, the airborne laser concept did not command the high-level attention of TSL, but the idea of placing a laser on an aircraft had been pursued by the small laser group at AFWL as early as 1965. These initial discussions eventually led to the writing of the AFWL 1967 Advanced Development Plan and Systems Command's Five-Year Plan for the development of lasers, which included the goal to build and test an airborne testbed. For both plans, Lamberson and others had intentionally structured laser research to proceed along a number of parallel paths to take advantage of different technical breakthroughs as they occurred. With a number of irons in the fire, laser technologies that showed little promise could be eliminated, while those that offered practical applications would be pursued more vigorously. This approach explained why AFWL supported a diversified laser program that included work on solid-state, gas dynamic, and electric discharge lasers, as well as a variety of projects involving optics, beam diagnostic techniques, mission analysis, and conceptual design studies. The synergy of all of these investigations eventually led to the ALL.[1]

The success of TSL and DELTA certainly reinforced the commitment to an airborne laser program by Lamberson and his energetic team of laser pioneers. Demonstrating that a

ground-based laser could shoot down an aerial target confirmed the technology had reached the stage where it was ready to be placed on an aircraft to shoot down air-to-air missiles. The important point was that the momentum of TSL and the later success of DELTA carried over to the ALL.[2]

But several years before the proof of concepts in TSL, FTT, and DELTA were completed, a number of Air Force and AFWL long-range laser planning and feasibility studies were already under way that supported the idea of eventually putting a laser on a plane. What these early studies did not do, however, was lay out a detailed blueprint of the kind of laser that would be the best for operating in an airborne environment. What AFWL did do, beginning in 1967 through the writing of advanced development plans, was set up a dual technology program aimed at determining which technology was best suited for the airborne laser concept. As a result, one team at AFWL dedicated its efforts to examine the feasibility of solid-state glass lasers. A competing group, led by Avizonis and Axelrod, zeroed in on the assets and liabilities of the GDL.[3]

In the mid-1960s, AFWL began studying the effects of solid-state lasers. The purpose of the solid-state work was to conduct basic research to gain a better understanding of the physics of a laser beam interacting with different materials. Solid-state lasers were notoriously inefficient (as Rich and Lamberson had noted in their 1963 SEASIDE report), and many scientists did not think they could be turned into weapons. (Since the 1960s, solid-state lasers have made rather spectacular progress. Today, for example, solid-state diode lasers exhibit efficiencies of over 50 percent, which is extremely good.) However, the early solid-state lasers were an extremely useful research tool to determine the effectiveness of a laser beam for burning holes in metal, glass, bricks, aircraft canopies, and so forth. With this type of result in hand, AFWL could then judge how vulnerable different weapon systems were to laser energy. AFWL used solid-state lasers in its experiments, because they were the *only* lasers available in the early and mid-1960s. The Laboratory acquired most of these low-power solid-state lasers from commercial vendors, such as the Westinghouse Corporation.[4]

In December 1965, AFWL first suggested that a program be started to investigate the possibility of installing a solid-state glass laser on an aircraft. Initial work in this area was fragmented, with a number of small projects taking place simultaneously that explored the basic physics of solid-state lasers. This work received a boost in January 1967 when Systems Command's West Coast Study Facility completed a comprehensive evaluation of "laser weapon technology."[*] The findings, contained in a report titled "SCIENCE CUBE," concluded that the feasibility of using a solid-state 500-kilowatt laser as a tactical weapon appeared "highly encouraging." A special SCIENCE CUBE study group also established in 1967 by Major General Glenn A. Kent, Deputy Chief of Staff for System Analysis at AFSC, investigated the tactical application of pulsed and continuous-wavelength CO_2 lasers, which offered great potential for scaling to higher-power beams that could be propagated through the atmosphere.[5]

---

[*]The West Coast Study Facility was an independent study house with offices near the Los Angeles airport. It was located there for easy access to people with scientific expertise who worked for aerospace contractors, RAND, and universities.

Based on the optimistic predictions of the SCIENCE CUBE report, Headquarters Air Force issued a Requirements Action Directive [RAD 7-120-(1)] in April 1967 directing AFSC to prepare a long-range ADP for solid-state work. This was a landmark decision, because it formalized the Air Force's commitment to develop a laser for use in the air. (Systems Command suggested laser weapons could be used to suppress ground and antiaircraft artillery fire and interdict supply routes.) Systems Command tasked AFWL, as it had the most experience with solid-state lasers, to write an ADP outlining the broad goals and schedules for solid-state research. AFWL completed the ADP in September, which established Program 644A-Task I giving the Laboratory the lead management responsibility for solid-state research. Each year from 1967 to 1971, AFWL revised the ADP to redirect the program based on recent technology breakthroughs and guidance from higher headquarters.[6]

Over the next couple of years, AFWL tested a variety of glass lasers in the laboratory, on the ground, and aboard a C-130 aircraft. In general, results proved disheartening. Under the rigid conditions imposed by laboratory experiments, solid-state lasers made little headway regarding the suggestion that they would make good weapons. There were many problems, but the main one was that glass rods used in the solid-state laser to produce the beam generated so much heat that the glass shattered. Consequently, a high-power beam when extracted could not be sustained very long. In spite of these setbacks, funding for solid-state work continued into the early 1970s in hopes that the heatbuildup problem could be solved. Those expectations were never realized. Yet the exhaustive studies and investment in solid-state lasers were not wasted, because they told the Air Force that solid-state lasers did not appear to be feasible as tactical weapons, thereby eliminating one scientific option.[7]

As it became clear rather early on that solid-state lasers faced too many insurmountable obstacles to make them realistic candidates for an operational weapon system, AFWL scientists and engineers turned more and more of their attention to GDLs.* At the same time the solid-state laser ADP was issued, AFWL produced its 1967 GDL ADP, which served as a technology road map for the

Major Harry I. Axelrod, one of the authors of AFWL's first laser advanced development plan for high-power gas lasers and manager of the TSL from its inception through early testing at SOR.

---

*AFWL operated its first gas laser in 1966 that burned through a high-grade firebrick in 5 seconds. A gas mixture of carbon dioxide, nitrogen, and helium contained in a 44-foot-long glass tube was electrically zapped with 5000 watts to produce a continuous beam of 500 watts of power.

development of a GDL weapon under Program 644A-Task II. The GDL work proved much more successful than the solid-state project. A large share of the GDL resources ended up supporting the TSL and DELTA programs, which produced the first "workable" integrated laser system.[8]

The early GDL ADPs were vague in their description of how a laser, once it passed the proof-of-concept test phase, would be developed and deployed as a weapon. Most of the early documents referred to "integrated laser weapon systems" and "weapon prototypes" to stimulate interest in airborne weapons throughout the Air Force and DOD. However, these types of systems would only be built after experiments had been conducted in an "airborne testbed" to verify proof of concept that a laser could operate aboard a flying laboratory. This meant obtaining clear answers to fundamental physics and engineering questions about laser devices, optics, pointing and tracking, beam propagation and quality, and so forth. The term *airborne testbed* first appeared in the AFWL GDL ADP dated March 1970. At first glance, it might have seemed odd to proceed with the laser "application" or mission study before finishing the "testbed" work. But on closer examination, it reflected a logical progression in the GDL program. In selling such a new concept as an airborne laser, most of the proponents at the laboratory level agreed an essential first step was to show there was a future Air Force mission for lasers. Otherwise it would be virtually impossible to convince people to invest in a costly testbed research and development program. Consequently, AFWL's first serious effort to look at how lasers could best be deployed to meet real-world mission requirements occurred in 1969, when the Lab let contracts to Hughes and Boeing to conduct an airborne and ground-based laser applications study. Boeing evaluated the ground-based applications of lasers, and Hughes assessed the airborne applications.[9]

AIRBORNE AND GROUND-BASED APPLICATION STUDIES: HUGHES AND BOEING

Hughes began its airborne applications study on 3 April 1969 and issued its findings in October of the same year. The basic goal of the study was to evaluate the technology to determine the feasibility of installing a futuristic GDL weapon aboard an aircraft. Three aircraft were identified as the leading laser platforms to cover a variety of mission scenarios and target engagements. Hughes studied the C-5A, B-1A, and F-111D in terms of their ability to defend against high-performance aircraft and missiles. The company also assessed the suitability of a laser-equipped F-111D for use against ground targets. These aircraft represented the spectrum of large, medium, and small planes, each with a different mission scenario: Airborne Warning and Control for the C-5A, deep interdiction for the B-1A, and tactical strike for the F-111D.[10]

The Hughes study restricted its focus to the GDL system and did not attempt to make a comparison with other lasers, such as chemical or electric discharge. Dividing the work into two basic parts, the contractor first devised a feasibility model and then tailored those findings to each of the three aircraft. The model served as a starting point to identify general airborne roles that best suited GDL technology. To define the most appropriate missions, the model addressed five critical factors contributing to the performance level of an airborne

GDL system: the laser, pointing and tracking, response time, atmospheric effects, and target vulnerability. Under each of these general categories, the Hughes team then refined the model by assessing narrower phenomenology issues influencing the design of a proposed airborne GDL system to include laser efficiency, beam quality, mirror flux loading limits, thermal distortion, aerodynamic effects on an external turret, and missile vulnerability.[11]

As expected, the weight and size of a GDL system presented the greatest penalty regarding placement on an aircraft. Although no airborne GDL system had ever been built, the Hughes team of analysts claimed its calculations confirmed it was feasible to install a laser on board several types of aircraft without serious degradation to either the aircraft or laser performance. As the C-5A was the largest aircraft, it was best suited for carrying a heavy laser estimated to take up 50 percent of the C-5A's payload weight. Findings for the B-1A and F-111D were not as optimistic, mainly because they were smaller aircraft. The B-1A was within the "marginal" feasibility limits to fly a GDL system. However, the contractor warned the bomber aircraft probably would have to undergo some redesign modifications to improve flight control stability. With no margin for technology projection errors, the F-111D was the least likely candidate to accommodate a GDL system. The airborne applications study also concluded it was feasible to build an airborne pointing and tracking (APT) system capable of keeping the beam on target long enough to kill airborne targets (missiles and aircraft). Mirror size presented a potential technical obstacle, but the results of the study confidently predicted that mirrors could be built measuring more than 1 meter in diameter to make an airborne laser system work.[12]

Hughes went one step beyond its position that it was feasible to go airborne with lasers. Well aware that the TSL was already under way, it recommended work should proceed on development and testing of a ground-based laser in 1970 and 1971. With the technical knowledge gained from eventually getting a ground-based laser up and running, a stronger case could be made for developing a first-generation airborne prototype some time in 1972.[13]

The seed of a GDL "airborne concept," first planted in AFWL's 1967 ADP, had begun to germinate with the completion of Hughes's airborne applications study. Clearly, the conclusions from this assessment represented an important step forward in the evolutionary process to build an airborne laser. Equally important was that first Scholtz and then Lamberson were carefully laying out a program that would not fail. To sell an airborne laser, they knew they would have to establish technical credibility. As work proceeded on the TSL in the early 1970s, Lamberson counted on the success of that program to give him the ammunition to show the technology was sufficiently advanced to make ground-based lasers work. At the same time, he never lost sight of the long-range goal to go airborne. Overlapping the ground-based TSL laser research was Hughes's thorough investigation lending initial support to the airborne laser concept. Lamberson realized the combination of payoffs from these two distinctive programs—ground-based technology hardware advances derived from the TSL and the paper "applications" analytical studies—would put him in a strong position to silence critics who might try to label the airborne laser program as just another harebrained scheme.[14]

Boeing published its final report on ground-based laser applications in October 1969, the same time that the Hughes study appeared. After studying the report, Scholtz and

Lamberson rejected the Boeing position pushing for ground-based lasers, mainly because they believed the mission and first priority of the Air Force was to fly, fight, and gain air superiority. Also, basing a laser on the ground presented some severe technical problems in terms of beam propagation limitations. It was more difficult for a ground-based laser to project its beam, because the atmosphere tended to spread out the beam in a process called *thermal blooming* and also to absorb it. At higher altitudes the atmosphere was less dense, reducing the degrading effects of absorption. Plus, aircraft and wind motions translated to less thermal blooming of the beam at higher altitudes. Finally, vulnerability was a critical issue. A fixed ground-based laser facility offered an easy target for the enemy as opposed to a highly maneuverable aircraft.[15]

SALESMANSHIP: RECRUITING SUPPORT

Once Scholtz and Lamberson decided to give their full support to the findings of the airborne applications study, they set out to obtain the blessings of higher headquarters and other commands. In December 1969, representatives from Strategic Air Command, Aerospace Defense Command, Electronic Systems Division, Aeronautical Systems Division, Space and Missile Systems Organization, Rome Air Development Center, and Tactical Air Command attended the Advanced Technology Laser Applications Study (ATLAS) meeting at Kirtland. The main purpose of this gathering was to give leaders throughout the Air Force an opportunity to evaluate the two recently published Boeing and Hughes laser applications studies. Lamberson and his team briefed the merits and drawbacks of the ground-based and airborne concepts, led the discussions, and fielded questions from the audience on the current state of the art and laser progress made at AFWL. By the end of the meeting, the majority of the attendees took the position that airborne lasers appeared to offer substantial promise for future weapon systems.[16]

Although the ATLAS meeting pushed for the demonstration of an airborne testbed as the next logical progression of laser research, there was no commitment on the type of laser to be used in the airborne environment. The leading contenders were the gas dynamic, chemical, and electric discharge lasers. Of the three choices, most experts leaned toward the GDL, because its technology was the most advanced and offered the least time and risk to develop it into an airborne system.[17]

Hughes's airborne applications study and the subsequent ATLAS meeting represented two important mile markers on the road to building an airborne laser laboratory. However, these two developments were not isolated events that by themselves aroused interest and backing for airborne lasers. They simply contributed to a larger evolutionary process that Lamberson and his versatile work force had been nurturing over the past several years. At the same time the Hughes study was under way, Lamberson expended a great deal of time and energy trying to convince the Systems Command commander and the Chief of Staff of the Air Force of the urgency of devoting more resources to the high-energy laser program at AFWL.[18]

Lamberson's first opportunity to sell the potential of lasers occurred when General James Ferguson, AFSC commander, chaired a commanders' conference at Kirtland in April

1969. Lamberson, at that time a lieutenant colonel who had just come on board at AFWL a month earlier as Scholtz's deputy, recognized that he had to make an extremely strong presentation to convince the general of the importance of laser research for the future defense of the nation. Acquiring Ferguson's support was crucial, because he decided what technology programs would be approved in Systems Command's annual budget before submitting it to the Air Staff. Without Ferguson's support, AFWL could not expect its high-energy laser program to prosper over the next few years.[19]

In preparation for his session with Ferguson, Lamberson recalled he "worked very, very hard. Haven't worked so hard on a briefing since!" But his work paid off handsomely as he explained to Ferguson the tremendous technical progress lasers had made in the last year. Not only had AFWL moved forward by developing a data base to better understand the physics and effects of solid-state and gas lasers, but the Laboratory's TSL program was under

way to demonstrate the accurate pointing of a ground-based GDL to an airborne target. But the most persuasive piece of evidence Lamberson had to offer was the performance of Pratt & Whitney's XLD-1 laser operating at West Palm Beach. Initially developed exclusively with company funds, the XLD-1 over the years had progressed by leaps and bounds in terms of power output. Although beam quality had always been a problem with the XLD-1 device, raw power was not. The Air Force defined a high-energy laser as one with a power output of 20 kilowatts or higher, which the XLD-1 easily exceeded in April 1968, when after only a month in operation it delivered 77 kilowatts. Barely had the excitement of this breakthrough subsided when the device cracked the 200-kilowatt barrier in October. By May 1969, the XLD-1 reached 455 kilowatts, a power level Systems Command believed high enough for near-term tactical applications. (The performance of the XLD-1 represented a power output increase of a factor of one billion over Maiman's ruby laser of 1960. Near term for AFSC meant deploying a laser in the mid-1980s.) Lamberson capitalized on the Pratt & Whitney work, driving home the point to Ferguson that at this power level airborne lasers could easily destroy almost any aerial target at the speed of light.[20]

General James Ferguson, Commander, Air Force Systems Command, September 1966–August 1970, was an influential supporter of lasers at the higher echelons of the Air Force.

Lamberson also made a strong pitch stressing that advances in the state of the art were rapidly moving lasers from the realm of the theoretical to real-world applications. Effects testing helped to specify the vulnerability of

Pratt & Whitney's XLD-1 laser.

a variety of targets to include metals, ablators, and ceramics. AFWL referred to these as "physics of interaction" tests designed to reveal how lasers of different wavelength, beam intensity, and spot size affected target materials. Other issues studied were the effects of continuous and pulsed lasers on the same target and how well experimental data matched computer-generated models that predicted laser effects. Vulnerability experiments at AFWL's Laser Optics Laboratory, AVCO's MK-5 research facility, and Pratt & Whitney's West Palm Beach facility proved that a GDL possessed the required killing power to make it a prime candidate for a revolutionary weapon. Laser beams with a spot size of 1 square centimeter—about half the area of a dime—consistently vaporized holes in a variety of materials in less than a quarter of a second. Targets included flat plates of aluminum, titanium, stainless steel, Plexiglas, carbon phenolic, graphite, and fused silica. AFWL selected these typical materials because they represented the outer skins and heat shields used on aircraft, missiles, satellites, and reentry vehicles. Most of these experiments took place indoors with the laser (20 kilowatts) about 75 feet from the target. Burn-through times and mechanisms were studied in detail. AFWL used these data to size laser systems and to maintain interest by filming and showing "smoke and fire" footage of these important experiments. Years later, Lamberson commented on the significance of these early laser vulnerability experiments:

Pratt & Whitney's West Palm Beach Research & Development Center.

It was those earliest experiments on fuel cells, on airplane wings [MIG-21], on fuselage structures, as well as pieces of tanks, I mean Army tanks, and pieces of jeeps and trucks and things like that which convinced us . . . and we didn't need much convincing, that airborne targets are significantly more vulnerable than the things that are around on the ground. Therefore, and still today, the main Air Force missions are air-to-air or space missions, as opposed to air-to-ground.[21]

Essentially, this was the same message a confident and enthusiastic Lamberson conveyed in his briefing to General Ferguson. Indeed, Ferguson was impressed with Lamberson's technical expertise, his unswerving determination and sense of urgency for pushing ahead with a full-fledged research program, and the vision he conveyed in laying out a long-range development plan to show a laser could work on an airplane. Although the Lamberson briefing did not trigger a sudden laser revelation in Ferguson's mind, it did sufficiently whet the general's

Laser damage inflicted on radome of AIM-7 Sparrow missile.

appetite to explore lasers more fully. So by the time he left Kirtland to return to Washington, Ferguson carried with him a renewed outlook on lasers that would eventually make him one of the most influential supporters of the high-energy laser work at AFWL. It was this initial encounter with Ferguson that was one of the critical early steps leading to the decision to make AFWL the launching site for the ALL. Ferguson was absolutely the early driver of the ALL program. "If it hadn't been for Ferguson," Lamberson pointed out, "there's no question in my mind that the program [ALL] would have not been supported anything like the way it was."[22]

Once back in Washington, Ferguson's interest and commitment to lasers continued to grow, especially because of more dramatic progress made on the XLD-1. Throughout the summer of 1969 the Pratt & Whitney device consistently generated 100–200 kilowatts of power *with good beam quality*. With this kind of evidence in hand, Ferguson believed the time was ripe to present a strong case for airborne lasers to the Chief of Staff of the Air Force, General John D. Ryan. As a result, Ferguson called on Lamberson and Colonel Glenn

Sherwood, the laser point of contact in Systems Command's plans shop, to be on hand at the briefing to answer any of the mission applications or funding issues the Chief might raise. After meeting early one autumn Saturday morning with Lamberson and Sherwood to put the finishing touches on the briefing to Ryan, Ferguson led his small contingent of laser experts across the Potomac River to argue his case before the Chief in the Pentagon. The thrust of Lamberson's remarks was establishing credibility that it was technically possible to scale lasers to higher powers and, with the right optics, to focus the beam on target. Ryan

and his senior aides listened attentively and then fired off a number of technical questions that Lamberson fielded more smoothly than an all-star third baseman snagging vicious shots down the line.[23]

Once again, the outcome of this high-level briefing produced no dramatic or immediate results in terms of infusing more money into the high-energy laser program or elevating lasers to a higher-priority research program. But it did establish technical credibility and managed to get the right people to begin paying more attention to the fact that lasers should be considered as strong candidates for future weapon systems on aircraft. The briefing to Ryan had not made him a laser convert yet, but it did have a powerful and lasting effect on Ferguson. Watching the top Air Force scientists in Ryan's office grill Lamberson, who in turn responded flawlessly with rational answers to difficult technical questions, reinforced Ferguson's confidence and commitment to work toward a program designed to demonstrate an airborne laser.[24]

Lamberson's first encounter with Ferguson at Kirtland, the meeting with General Ryan and his key advisors, AFWL's decision to back the findings of Hughes's airborne applications study over the conclusions offered in Boeing's ground-based la-

General John D. Ryan, USAF Chief of Staff, August 1969 to July 1973, gave the go-ahead to increase funding to accelerate the ALL program.

ser report, and the endorsement of that decision by a wider Air Force community at the ATLAS meeting in December, all contributed to overcoming inertia and to building up steam to get the ALL off the ground. The unwieldy bureaucratic machine, greased by these key and, at first glance, seemingly unrelated events of 1969, began to cough and sputter as part of the slow start-up process. But the important point was that AFWL had succeeded in injecting momentum into the slow and cumbersome process that would lead to the development and testing of a laser on an airborne platform. Sustaining and accelerating that momentum was the greater challenge that lay ahead for the 50 people assigned to AFWL's laser group in 1970.[25]

Before Systems Command would commit any money to fabricate hardware for the ALL, Lamberson understood that he would first have to show that the technical milestones of the TSL ground-based program had been achieved. Planning ahead was key. While he waited for the TSL to be built and to perform according to contract specifications, he worked hard to maintain the ALL momentum by hiring contractors to conduct competing airborne mission analysis "paper" studies. He recognized time would be wasted if he waited to investigate possible airborne configurations until *after* the TSL performed. In the interval, while work proceeded on TSL, he intended to complete and select the technically superior paper study of a recommended airborne system. This would allow him to immediately begin building the hardware for the ALL on completion of each of the TSL milestones. In this way, he knew that time would not be wasted waiting for the TSL results.[26]

AIRBORNE TESTBED CONCEPTS:
PRATT & WHITNEY AND AVCO

In April 1970, AFWL stood fast to its ALL course by issuing contracts to Pratt & Whitney and AVCO to conduct the first conceptual design study for an airborne "testbed." Pratt & Whitney and AVCO each were to put together the first preliminary design of a GDL on an aircraft for general testing of airborne high-power laser technology. AFWL directed each contractor to devise a GDL system that would produce a 10.6-micron, near-diffraction-limited beam of approximately 500 kilowatts of power and to come up with a recommendation for packaging this system on either a C-135 or C-141 aircraft flying up to 40,000 feet. Beam quality, power, weight, and volume of the total system were the most important considerations in designing the major components of the system, which included the fuel supply storage system, the injector and combustor, the distribution manifold, the laser cavity, the pointer and tracker, the diffuser, and the laser output window. The results of these studies, AFWL reasoned, would form the basis of the design–parameter selection for the GDL Technology Confirmation Program.[27]

Work by the contractors began in April 1970 and finished in the fall, with final reports published the following spring. The two proposed designs for an airborne GDL system were remarkably similar. Both contractors recognized that system integration of all of the components inside an aircraft was a much bigger problem than they had anticipated, as was the design of the APT. As a first cut, Pratt & Whitney and AVCO design configurations placed the computer-driven command and control consoles that issued commands to the GDL system and monitored its performance in the rear of the airplane. The midsection of the plane would house the fuel storage tanks, the laser device, and the pointer and tracker (patterned after the field test telescope under development by Hughes for the TSL program), which was positioned to fire out an open window on the left side of the aircraft. Spent gases were vented through a diffuser out the top of the fuselage. Pratt & Whitney estimated the weight of the entire system at 20,000 pounds.[28]

The two contractors differed in their recommendations for the type of aircraft best suited to carry the GDL system. Pratt & Whitney favored the C-135 over the C-141 because the

former had a low-wing profile allowing for a larger field of fire, its cargo door provided easy access for working on the GDL/APT hardware, it would be less costly to modify the C-135 to accept the GDL, and more of these aircraft were available in the Air Force inventory. Perhaps the most important drawback of the C-141 was its wings were supported by the wing box on top of the fuselage. To cut through that area and install a turret (which eventually won out over pointing the laser out the side of the aircraft) would be a major structural design problem. AVCO leaned toward the C-141 as its first choice mainly because of its larger flight deck, a wider fuselage that allowed more room for expansion of the next-generation GDL system, and its ability to carry a heavier payload above 40,000 feet. However, both contractors were quick to point out that *either* the C-135 or C-141 would be sufficient as a testbed for high-energy laser experiments.[*][29]

Lamberson and his staff closely monitored the progress of the Pratt & Whitney and AVCO studies and at the same time had formed their own "in-house" conceptual design study group to investigate the best options for an airborne laser testbed. One of the key members of this in-house study team was Major Hal Shelton, a former Thunderbirds maintenance chief with extensive experience in modifying F-4s, who served as Lamberson's aircraft integration advisor. Two other technical experts were Major Harry Axelrod and Lieutenant Colonel Edgar O'Hair, who evaluated the design of the lasers proposed by AVCO and Pratt & Whitney contracted studies. Every Monday morning they met with other laser technical managers, who made up Lamberson's advisory council, to apprise him of their progress and to offer the pros and cons of various airborne laser configurations. By the fall of 1970, using the input provided by Pratt & Whitney, AVCO, and his in-house group, Lamberson had to decide which preliminary design of an airborne GDL system was most feasible so work could get under way on the initial design and subsequent fabrication of hardware to fit aboard some type of aircraft.[30]

Although Lamberson carefully assessed the recommendations of the two contractor studies, he ended up relying heavily on the advice of his local study group. After weighing all of the facts and opinions discussed in numerous Monday morning skull sessions, he came to the conclusion that the C-135 was the best aircraft to meet all of the needs of an airborne laser platform. Technical and practical considerations swayed his judgment. First, preliminary calculations provided by his in-house study team strongly suggested that there were major problems in getting a beam to propagate through the turbulent air boundary layer next to the open port on the left side of the fuselage. (As it turned out, data collected later showed that propagation of a 10.6-micron beam through the air boundary layer was a "nonissue.") As a result, they recommended removing the pointer and tracker from inside the aircraft and placing it in a turret on top of the fuselage. With the beam exiting from a $6\frac{1}{2}$-foot-high turret mounted on top of the plane, the beam avoided passing through the violent aerodynamic flow region along the side of the aircraft fuselage. Later in the program, a fairing placed

---

[*]Early in 1969, Colonel Jones and his AFWL EIGHTH CARD specialists had proposed using the older B-58 as a "reasonable candidate" for the airborne testbed. Both contractors rejected the B-58 as an inferior aircraft when compared with the C-135 and C-141.

around the turret served to reduce and smooth out the airflow between the top of the turret and top of the aircraft.[31]

There were other advantages to using an external turret for the pointer and tracker. A side door configuration limited the laser's field of view for acquiring, tracking, and engaging aerial targets, especially air-to-air missiles approaching from the rear. Mounting a turret on top of the plane greatly increased the field of fire for the pointer and tracker, as it could slew to cover all angles on the left side of the aircraft from the tail to slightly right of front center. However, this system would work better on the C-135. The position of the high wings of the C-141, attached to the top of the fuselage, blocked the path of the laser beam when aimed downward at targets off the left rear of the plane. This was not the case with the C-135, because its wings were connected near the bottom of the fuselage and its turret would be placed farther forward in relation to its wings than the turret on the C-141, allowing sufficient clearance for an unobstructed line of sight for the beam. (The turret on the C-135 still would block the beam when firing at certain angles. However, the overall firing options for a clear line of sight to target were better on the C-135 than on the C-141.) Although Lamberson's in-house study group also considered mounting the turret underneath the aircraft as another option, they rejected this idea mainly because of engineering complications. A belly-mounted turret would create excessive drag that would adversely affect the aerodynamic performance of the aircraft in flight. Also, the turret would have to be retractable so it would not be disturbed during takeoffs and landings. AFWL optics experts objected to this scheme because, as Shelton pointed out, "they were already having enough trouble trying to get stable optics with fixed mounting points, let alone one that is retractable."[32]

One other important element factored into the choice of aircraft. Because of the war in Vietnam, it would be difficult to acquire any type of aircraft to be used strictly for research experiments. The chances of obtaining an aircraft normally devoted to operational missions were next to zero. And although there existed only a small fleet of research aircraft in Systems Command, mostly assigned to the Aeronautical Systems Division at Wright-Patterson AFB, more C-135s were available than C-141s. So in the final analysis, mainly because of technical considerations and availability of aircraft, Lamberson had no other choice than to select the C-135 over the C-141 for the laser testbed.[33]

Besides the selection of an aircraft, two other decisions emerged from AFWL's assessment of an airborne testbed by the fall of 1970. Originally, the AVCO and Pratt & Whitney conceptual studies operated from the premise that the pointer and tracker inside the aircraft would be modeled after the FTT used in the TSL program. But the change to mount a turret on the aircraft required modifying the size and weight of the pointer and tracker so it would fit inside the turret. To get this work started as soon as possible, AFWL in December 1970 let a contract to Hughes Aircraft Company to come up with a preliminary conceptual design of an airborne turret. This was an extremely important turning point, because it marked the first time the Air Force had committed funds to design a major hardware component (the APT) tailored exclusively for the C-135 airborne testbed.[34]

Redesign of major components was not limited to the APT. A thorough AFWL evaluation of the laser devices proposed by AVCO and Pratt & Whitney led to a second major decision before moving ahead with the airborne testbed. (Pratt & Whitney proposed using a

parallel-stage device—two identical lasing devices connected—for packaging versatility to save space on the aircraft.) The Axelrod and O'Hair in-house group expressed some technical "doubts" about the designs of the GDLs recommended by the two contractors. Instead of endorsing either of the laser devices, AFWL proposed moving on to design a "next-generation" GDL device that would be lighter in weight, smaller in volume, and more reliable in terms of power output and beam quality. The consensus of the Lab scientists was that Pratt & Whitney had taken the lead in the development and proof-of-concept testing of large GDLs. AFWL reached this conclusion based on Pratt & Whitney's earlier work advancing the state of the art in achieving higher power levels under two contracts (F29601-68-C-0109 and F29601-69-C-0069) let by the Lab in the late 1960s.[35]

Between April 1968 and March 1970, the contractor showed steady progress by increasing the power output of its XLD-1 device from 77 to 518 kilowatts, which represented the highest value achieved by any laser in the free world. Plus, the XLD-1 exhibited good beam quality. But the XLD-1 was too large to fit inside a C-135. As a result, in May 1971 AFWL awarded a contract to Pratt & Whitney to design and build the split stage demonstrator (SSD), which would be half the size of a GDL designed to go aboard a C-135. Once the SSD demonstrated desired output and beam quality, AFWL planned to issue a follow-on contract to build a full-size airborne GDL.[36]

Because of its ability to demonstrate steady increases in the output power of GDL devices over the years, Pratt & Whitney by 1970 had emerged as the leader in the field for developing high-energy lasers for the military. Pratt's progress and potential confirmed earlier predictions that it would only be a matter of time before lasers could be turned into weapons. The Air Force seemed to agree. Commenting on the status of lasers in January 1970,

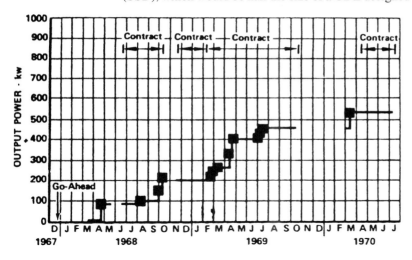

XLD-1 power output history.

Dr. William L. Lehmann, Deputy for Laboratories in the Office of Assistant Secretary of the Air Force for Research and Development,[*] stated, "We are pretty sure this [laser] is a winner. Things are going right. It is pretty much straightforward engineering that we are doing, rapidly building a strong technical capability."[37]

As AFWL worked on refining the conceptual testbed configuration in the summer of 1970—deciding on the type of aircraft, the APT turret, and the next-generation laser—Lamberson was hard at work to sustain interest and support from higher headquarters for the final

---

[*]Dr. Lehmann became the first civilian Director of AFWL on 1 July 1978 and served in that position until 28 February 1981.

design and follow-on fabrication of the airborne platform. Again, it would be General Ferguson who played a key role. In August 1970, a month before Ferguson's retirement, Lamberson recalled that Ferguson "was really starting to get himself worked up about all this [laser] business." P—t of the reason for his excitement was the XLD-1 produced 625 kilowatts only a m——lier. With such massive power output demonstrated, Ferguson became convin——————— was right to accelerate the high-energy laser (HEL) program at AFWL.

But it would—————erguson's desire and commitment to accelerate the HEL program. The ultim——————ided with the Air Force Chief, General Ryan, who ——————two guns an—————the process along, Ferguson, on very short notice, ————————————————with a briefing explaining why the laser program ————————————at if he liked the briefing, "we'll take it over to ————————————efore Ferguson's retirement in August 1970, ————————————(who had replaced Colonel Sherwood at

Colonel Orpha R. Cunningham, one of the early laser proponents who played an important role in acquiring the ALL aircraft and served as AFWL's commander, 30 June 1973 to 30 May 1975.

————————————————n. ————————they an————————'s ————————the Civil————f ————————Eight——————erals

————————ful brie————g rivaled any ————————duction———ng three screens to ——————— vis———ms, with the music of ———— Ninth ———phony playing in the back- ———— the AFWL colonel made an unforgettable ———————— ie led off explaining the physics and ————— —y of the airborne laser program during the ————— half-hour. Cunningham followed with a quick review of the potential applications of lasers as weapons. Ryan listened attentively for an hour and was impressed by what the three Systems Command officers had to offer, especially their optimism that it would only be a matter of time before improved beam quality could be combined with power to conduct a meaningful airborne demonstration. After a few minutes of thoughtful silence, Cunningham remembered, Ryan turned to everyone in the room and said, "Well, we're going to go with it; put the money in." Ryan's go-ahead increased AFWL's HEL funding from approximately $18.7 million in fiscal year 1971 to $28.1 million for fiscal year 1972. Lamberson recalled that "from that point on, we never lost momentum on the ALL. That was the major thing which gave it to us."[40]

In the eyes of AFSC and Headquarters Air Force, a large portion of the accelerated program was directed at the "weaponization" of a laser. Specifically, as a first step to meet

this goal, Systems Command accelerated the airborne testbed (ATB) demonstration to be completed by fiscal year 1974, a year ahead of the original demonstration schedule. Verifying the APT and laser could be mated and operated in the ALL aircraft was the purpose of this effort. General Ryan's decision to move ahead more quickly also gave AFWL funding and authority to look beyond the ATB demonstration. At the same time work on the ATB progressed, AFWL was to start applications studies which would lead to the engineering design and building of a laser weapon system that could be installed in an operational aircraft in 1978 rather than 1980. In effect, AFWL was looking beyond the anticipated success of an ATB demonstration. Systems Command viewed this approach as a "timesaving" action, as the engineering design would occur prior to rather than after the ATB demonstration. However, this prediction to have an operational laser ready by 1978 turned out to be overly optimistic. By the late 1970s, the ALL was encountering numerous technical and engineering problems, underscoring the complexities and difficulties of getting a laser to work on an airborne platform. Without sound empirical data from the ATB demonstration, the chances for the paper application studies to come up with a workable engineering design of an operational laser weapon diminished significantly.[41]

Once back in Albuquerque, Lamberson immediately set to work to put together an AFWL program memo to explain how the additional monies would be spent on an accelerated HEL program. A substantial share was earmarked to support the development and experiments on the ATB, which AFWL personnel expected to last for about 4 years. The balance of the budget would go to support basic and exploratory research in fundamental laser disciplines (e.g., optics, beam quality, devices, electric discharge and chemical lasers) to advance the state of the art. On 7 December 1970, Deputy Secretary of Defense David Packard signed a memo approving the accelerated laser program at AFWL. DOD released Program Budget Decision 338 in December 1970, officially implementing the funding increase approved by Ryan at the August briefing. Development Directive 104, dated 3 February 1971, issued by HQ Air Force followed and notified AFSC and AFWL to begin an accelerated demonstration program for the ALL.[42]

Systems Command also made it clear that successful completion of each of the TSL milestones had to be accomplished before any "hardware" could be built for the ATB. This meant the TSL FTT had to demonstrate its pointing and tracking accuracies before AFWL could authorize building an APT. However, conceptual design studies for the APT could take place *prior* to meeting the FTT milestone. Work was allowed to proceed on these paper studies because they were relatively inexpensive compared with experiments and they would also reduce the technology lead time for developing the APT once the FTT had performed. In other words, as soon as the FTT worked, AFWL with its design of the APT already in hand would be immediately ready to start building the APT, thereby saving valuable time.[43]

Although the various DOD levels of bureaucracy endorsed General Ryan's decision to accelerate the HEL program, many were still skeptical of the performance of any type of airborne laser system. For example, Dr. Eberhardt Rechtin, ARPA Director and Deputy Director for Research and Engineering in 1970, cautioned about the astronomical engineering costs to develop a laser weapon. Because of the engineering problems, he believed, "the enthusiasm of the theorists is a good deal greater than that of the potential users." A year later, Rechtin stated that the major engineering stumbling block was building a "compact" laser power plant to generate high-power beams. So it was not uncommon for Lamberson to hear some experts in DDR&E and other high offices in the Pentagon say, "Well, if you silly guys want to go fly this airplane, OK, we don't think you can do it." This type of attitude only served to redouble AFWL's resolve to prove it was possible to fly an airborne laser, even though Lamberson would later admit that essentially he was proceeding at that time on an "act of faith" because of the many technical unknowns associated with lasers. In the end, AFWL would prove the lingering critics wrong.[44]

Lamberson and others countered the critics by arguing it was important to investigate airborne lasers, even if they failed, because of the Soviet threat. The doubters in Washington who preached, "Just give 'em [AFWL] enough money to let them go hang themselves and show that they really can't get there from here," provided the very rationale for going ahead with an experimental airborne laboratory to prove or disprove the physics of lasers for airborne systems. In the early 1970s, U.S. intelligence agencies had very strong indications that the Soviets had picked up the pace of their laser research but could not pinpoint, with any degree of certainty, how much progress they had made. Lamberson contended that if AFWL could conclusively show the laws of physics ruled out the possibility of airborne lasers, then the United States would have "a lot less to worry about in terms of Soviet defenses in the 1980s and 1990s." On the other hand, if the U.S. investment in lasers showed it was possible to develop the first laser weapons, then we would derive large dividends by preventing another Sputnik-type Soviet surprise.[45]

4

Gaining Momentum

Just prior to the end of World War II, Henry H. "Hap" Arnold, Commanding General of the Army Air Forces, met with Dr. Theodore von Karman, one of the world's leading aerodynamicists. Arnold was confident of the outcome of the war, but he worried about the future of airpower. Where was it going in the next 20 or 30 years and what technologies would be needed to get it there? To help him wrestle with this question and proposed solutions, he turned to his trusted friend, Dr. von Karman, a scholar whose scientific credentials were unquestioned.[1]

Arnold wanted von Karman to recruit the leading civilian scientists in the country to forecast and study those technologies that would have the greatest influence on the future of the Air Force. Topics of immediate concern included jet aircraft, radar, atomic energy, advanced electronics, solid propellants for rocket propulsion, pilotless aircraft, gas turbines, high-temperature materials, and any other promising technology far out on the frontier of science. The soft-spoken scientist, who had been Arnold's technical advisor during the war, was anxious to return to the California Institute of Technology in Pasadena to resume his duties as a full-time academician. However, recognizing the long-term implications of Arnold's goal to get prominent scientists more involved in planning the nation's military strategy, von Karman agreed to stay on working for the government. In December 1944 he and his small staff formed the Scientific Advisory Group (SAG), set up office in the Pentagon, and began their long-range study. As a first step, von Karman and his colleagues traveled to Germany, the Soviet Union, and ten other European countries to interview scientists to obtain information on what they predicted would be the most promising research and development programs for the future. Eight months later (August 1945), the SAG issued its first interim report, *Where We Stand*. Von Karman released a more comprehensive 33-volume report in December 1945 entitled *Toward New Horizons*, which many considered to be the blueprint for future aerospace development.[*] The message in both reports was the United States would have to be willing to take "high risks" and make an unwavering commitment to invest in an ongoing research and development program so the nation would have the most advanced technical resources to fight the next war.[2]

As the initial findings of the SAG attracted more attention, Arnold was able to wrangle permanent status for von Karman's group by abolishing the temporary SAG and replacing it with the U.S. Army Air Forces Scientific Advisory Board (SAB), established in February 1946. This was the forerunner of the USAF SAB formed in May 1948, eight months after the Air Force became a separate service on 18 September 1947. The job of the SAB was straightforward. It was to advise the Air Force on the most promising technologies it should explore and develop that would lead to the building of the most capable and cost-effective weapons of the future. Membership of the SAB would consist of the top civilian scientists and engineers in the country, who had no connection or special interests with the military. They were mostly college professors and scientists in the aerospace industry who would be willing to provide an independent evaluation of proposed Air Force technology programs that offered the most promise for modernizing airborne weapon systems.[3]

Usually, the SAB did not determine what technologies would be evaluated. Rather, from 1948 on, the board responded to requests from the Chief of Staff of the Air Force. The way the SAB operated was to set up a number of technology panels and assign members with the most expertise in a particular scientific discipline to serve on the appropriate subject panel. During the year, each panel would study and assess the potential of a specific technology. It then would report its recommendations to the SAB on the feasibility of applying that technology to a weapons system, which normally occurred at general membership meetings held twice a year. These meetings essentially served as open forums to further discuss the strengths and weaknesses of a particular new technology under investigation. Arriving at a consensus, the SAB would then report its recommendations to the Chief of Staff of the Air Force. Membership in the SAB had grown from a handful of people during von Karman's days to about 60 in 1971.[4]

---

[*]The 33 volumes covered 13 major topics. *Where We Stand* was the second volume in *Toward New Horizons*. The other 12 major topics addressed in the 32 remaining volumes were: *The Key to Air Supremacy*, *Technical Intelligence Supplement*, *Aerodynamics and Aircraft Design*, *Future Airborne Armies*, *Aircraft Power Plants*, *Aircraft Fuels and Propellants*, *Guided Missiles and Pilotless Aircraft*, *Guidance and Homing of Missiles and Pilotless Aircraft*, *Explosives and Terminal Ballistics*, *Radar and Communication*, *Weather*, and *Aviation Medicine and Psychology*.

Paralleling the work of the SAB was the Defense Science Board (DSB) made up of private consultants. Both organizations had similar charters. The main difference was that the DSB operated one level above the SAB and reported to the Deputy Director for Research and Engineering (DDR&E) in the Office of the Secretary of Defense. John S. Foster, who headed DDR&E, appointed Dr. George P. Millburn in July 1968 to chair a special task force to assess "the current and projected status of all relevant technology" pertaining to the GDL. (Dr. Millburn was General Manager, Office of Development Planning, at the Aerospace Corporation located in Los Angeles.) From 23 July to 9 August, Millburn's committee gathered information. They visited AVCO and United Aircraft's Pratt & Whitney facilities to see firsthand the operation of their GDLs and also heard briefings presented by AFWL on its advanced development plan for lasers.[5]

Millburn's report to Foster was cautiously optimistic. Some rather "spectacular advances" had been achieved with GDLs, he noted, but the major stumbling blocks were reaching higher power levels and improved efficiency. Nevertheless, Millburn heartily endorsed the selection of AFWL as the best choice for the lead service laboratory to run the TSL program just getting under way. He felt the Air Force was heading down the right road in its investigation of GDLs. And his initial reaction was that answers found in the Laboratory would eventually lead to applications of lasers in an airborne environment. Another early application that received a great deal of attention was using a laser weapon on board a C-5 aircraft as an antisatellite weapon. Applying lasers to a bomber defense role appeared "quite interesting," according to the DSB study, but to reach that point would require significant reductions in laser weight and improvements in tracker technology.[6]

At the same time the DSB was making its independent study of the overall laser program for the DOD, the SAB had established its own ad hoc panel on lasers in June 1968. It was chaired by George P. Sutton, whose full-time job was Assistant to the President of Rocketdyne Division in Canoga Park, California. Sutton and his colleagues had access to the August 1968 DSB report, but the SAB committee's charter was to look only at plans and programs limited to the Air Force high-energy laser programs. Special emphasis was given to the future of GDL work. The SAB and DSB panels were to reach their own conclusions separately to dispel any notion that the two groups had collaborated on their findings. However, there was some unavoidable overlapping. Because of the relative newness of lasers, there was a scarcity of civilian experts in the field. Consequently, some individuals who sat on the DSB also served as members of the SAB.[7]

Sutton issued his draft report on 2 January 1969 and the final report in May 1969. It was a carbon copy of the earlier DSB commentary. The basic SAB recommendation was that it concurred with the "well-conceived" approach the Air Force was taking on high-energy lasers. The Weapons Laboratory's laser program was extremely "worthwhile," according to Sutton, who advised that this work should not only be "supported" but "augmented." Sutton argued more money should be infused into the Air Force laser program to collect what he referred to as "urgently needed" data on beam propagation, vulnerability, and devices for improving pointing and tracking accuracy. Once these problems were solved, it would be just a matter of time before a laser weapon could be built.[8]

The SAB and DSB conclusions reinforced the DOD's and the Air Force's commitment to a high-energy laser research program: the military was correct in its pursuit to build a laser weapon. These early studies were the rudimentary beginnings of that process. And at that early stage, AFWL deserved much of the praise and credit for taking the lead in investigating ground-based lasers to gain a better understanding of the physics of lasers. Although most of the immediate attention focused on ground-based lasers, the SAB and DSB members' vision included airborne and space lasers. Both groups were enthusiastic about the possibilities of developing GDL weapons on aircraft for bomber and AWACS defense, air-to-air attack, and satellite intercept.[9]

Over the next couple of years, there was a gradual shift in interest from ground-based to airborne lasers. As the scientific breakthroughs occurred, it became obvious that lasers offered the most promise of operating in the air where the beam would be subjected to less distortion than on the ground. Professor Abraham Hertzberg, Director of the Aerospace Laboratory at the University of Washington, was highly respected in the scientific community and one of the first to come out strongly in favor of applying laser technologies to weapon systems. In a keynote speech delivered at the SAB general meeting in 1970, he went on record urging the government to spend more money and brainpower on developing an airborne laser. His remarks appealed to scientific logic, generated a renewed sense of anticipation and excitement, and conveyed to his audience that they were about to enter an era that would move lasers out of the fantasy world and into the real world of operational systems. Hertzberg's speech, the continual support for an aggressive laser research program by the SAB and DSB, and the general advancement of the state of the art in diversified laser technologies, all worked together to set the scene for a unique SAB meeting in the spring of 1971.[10]

A BOOST FROM THE SAB

The 1971 general meeting of the SAB took place at Kirtland AFB, on 15–16 April. What was different about this meeting was its focus. In the past, the general SAB session reported on the status of a number of technologies the working panels had been investigating over the last 12 months. The meetings had been intended to be broad in scope. But for the first time ever, the SAB gathering at Kirtland restricted its discussion exclusively to one theme, "laser technology." This departure from the normal agenda underscored the increased importance and sense of urgency the Air Force placed on lasers.[11]

Over 200 people attended the Kirtland meeting. Besides the 60 SAB members, there were a number of high-ranking government officials present. The Honorable Grant L. Hanson, Assistant Secretary of the Air Force for Research and Development, stayed for the 2-day session. So did General John D. Ryan, the Air Force Chief of Staff, who held a special stake and interest in this meeting. Eight months earlier he had approved an additional $10 million to accelerate the high-energy laser program at AFWL. He wanted assurance from the SAB that this extra money had been a wise investment. Ryan brought with him his Deputy Chief of Staff for Research and Development, Lieutenant General Otto J. Glasser. Representing Systems Command was its commander, General George S. Brown, and Colonel

Orpha R. Cunningham, who would play an important role in the acquisition of the ALL aircraft. Cunningham went on to become the Weapons Laboratory's fifth commander in July 1973. Although retired at the time, General James Ferguson, who had shepherded the laser program through its early rocky days, wanted to be available to lend his support. Ferguson was one of two retired officers present. The other was the highly respected General Bernard A. Schriever, whose vision and willingness to invest in high-risk technology had helped usher in the ICBM age during the 1950s.[12]

Equally impressive were the credentials of the SAB members and other invited civilians. Many looked on Dr. Edward Teller, Associate Director of the Lawrence Radiation Laboratory in Livermore, California, as the father of the thermonuclear age. Teller was joined by Professor Abraham Hertzberg and Dr. Arthur Kantrowitz, President of AVCO-Everett Research Laboratory, who had worked hard to convince the Air Force to move ahead with the TSL program. Rounding out this elite group were Mr. Robert R. Everett of AVCO-Everett, Dr. Ray E. Kidder from Lawrence Radiation Laboratory, and Dr. Paul Kelley from MIT-Lincoln Laboratory in Cambridge, Massachusetts. And, of course, Colonel Lamberson and his staff were on hand to give the AFWL workers in the trenches an opportunity to promote the potential of the GDL and to answer any technical questions on the progress of lasers at the Weapons Laboratory.[13]

The significance of the 1971 SAB meeting was that it provided a tremendous boost to the airborne laser program. With the most brilliant laser minds in the country, both inside and outside the government, confirming the almost unlimited potential for lasers, the work at AFWL took on a new sense of credibility. Gaining support from the august SAB was welcomed, but Lamberson, somewhat out of character, cautioned against what he labeled as too much "overenthusiasm" and moving ahead too quickly in committing large amounts of money to laser research. At the meeting, for example, Dr. William G. McMillan, professor of chemistry at University of California Los Angeles, suggested the possibility of setting up a national laboratory for laser development. It would be modeled after the Manhattan Project that developed the atomic bomb during World War II.[14]

Lamberson rejected McMillan's proposal. The political climate simply wasn't right. He felt taking such a bold step would be very damaging and ill-advised. Summarizing his thoughts in a memo after the April session, Lamberson stated, "To my knowledge, no good has come from the 'Manhattan' talk of the Spring meeting. In fact, many DDR&E men still refer to it ludicrously." Lamberson's point was that the laser program had to be paced by sound technology based on achieving specific step-by-step milestones outlined in his long-range program management plan. A moderate and methodical approach made more sense—gradually add money to expand the program each year based on technical accomplishments. His recommendation, which was followed, was that "we should vigorously oppose Manhattan or multi-hundred megabuck programs until technology is indicated."[15]

Another reason why Lamberson did not endorse a Manhattan-type plan was he wanted to avoid any possibility of removing AFWL laser projects to a national laboratory. He and his staff had worked too long and hard to get AFWL's laser work under way and were not about to allow it to slip away to another agency. He continued to take a firm stance to remain in complete control in keeping with the "single management" philosophy, which had already been approved by Systems Command.[16]

Moderation, reason, and Lamberson's influence prevailed over committing to a crash Manhattan-type program. However, this did not dilute the enthusiasm and urgency for supporting a long-range laser program among the SAB members. Over a dozen formal presentations were made at the 2-day meeting. Teller's opening remarks resembled a sermon on scientific salvation to the attentive SAB congregation, expounding how important it was for the Air Force to "pay early attention to lasers and the changes that the existence of this instrument could bring about." Later he would get up and walk around the room, almost in a frenzy, waving his arms and painting a vision of a "great laser battleship in the sky" that no enemy in the world could defeat. Teller proposed creating a laser weapon development program in the 1970s on the same scale as the ICBM program so ably led by General Bernard A. Schriever in the 1950s. The retired Schriever was sitting quietly in the first row of the meeting room. Suddenly, Teller bellowed, "I have just the right man to lead that program," pointing a menacing finger at the unsuspecting Schriever, "There! There! Bernie Schriever." Everybody in the room knew of Schriever's lasting contribution to the ICBM program, but the general who now was comfortably retired had no desire to take on such a formidable new challenge. Although he knew that he could not lure Schriever out of retirement, Teller had succeeded in setting the tone from the outset as one of confident optimism and unbridled enthusiasm for a new laser program.[17]

The speakers who followed Dr. Teller presented strong cases for the potential development of gas dynamic, electric discharge, and chemical lasers. They did not glibly ignore the seemingly insurmountable problems of optics, pointing and tracking, atmospheric propagation, beam limitations caused by weather (fog, clouds, and rain), excessive weights of laser devices, and the transition of technological advances to operational weapon systems. As men of science, most believed that with enough money and brainpower these problems could eventually be solved. Lamberson's talk reinforced this notion as he covered the steady progress AFWL had made on the TSL. Once over the TSL hurdle, he assured his audience AFWL would be prepared to move on and demonstrate an airborne laser. In his judgment, selection of a prototype weapon system should occur only after a complete assessment of the scientific data collected from tests aboard the ALL.[18]

To dispel any thoughts that it was foolish to pursue lasers as weapons, some of the older and more experienced attendees offered words of encouragement. General Schriever's comments were especially persuasive. He turned back the pages of history to illustrate that innovative ideas in science often were attacked by critics who simply lacked vision. Schriever reminded his audience that in the 1950s there were those irreconcilable naysayers who insisted that nuclear weapons could never be sufficiently reduced in size to be carried on ICBMs to produce the desired yield. Yet this was done by the early 1960s with the introduction of the first ten Minuteman missiles into the operational force at Malmstrom AFB, Montana, on 24 October 1962.[19]

Professor Hertzberg agreed wholeheartedly with Schriever's analogy. Hertzberg stressed one of the most valuable lessons of history was that overachievers made the difference in science. He went on to point out that:

General Bernard A. Schriever, Commander, Air Force Systems Command, April 1959–August 1966, who encouraged laser research to proceed at AFWL.

In the long run, we have always succeeded in doing far more than we had even had the courage to state we thought was possible in the beginning. The accuracies, the weights, the capabilities of the ICBM force structure that we have now, if they had been proposed in the early days, would have been laughed off the stage. Overachievement will come with the laser.[20]

In addition to the ICBM, there were plenty of other examples in the historical record that proved the faultfinders wrong. Thomas J. Watson, chairman of IBM, predicted in 1943, "I think there is a world market for about five computers." Development of the atomic bomb under the Manhattan Project took less than 5 years and even at the conclusion of that program a pessimistic Admiral William D. Leahy told President Harry S. Truman in 1945, "That [atomic] bomb will never go off, and I speak as an expert in explosives." And Dr. Vannevar Bush, who directed the government's World War II science effort, said after the war that he rejected the talk "about a 3,000-mile rocket shot from one continent to the other carrying an atomic bomb . . . we can leave that out of our thinking." Defying all odds, Chuck Yeager on 14 October 1947 broke the sound barrier with his first supersonic flight in Bell's rocket-powered X-1 over the skies of Edwards AFB, California. Many scoffed at John F. Kennedy's announcement on 25 May 1961, pushing the United States to accelerate its space program for a manned lunar landing by the end of the decade. Eight years later, those voices of dissent were silenced when Neil A. Armstrong set foot on the moon at 10:56 P.M. EDT on 20 July 1969.[21]

The lessons of the past for the SAB were that pessimism had to be deleted from the scientists' thinking. An attitude had to be fostered that the impossible was possible. One of the most important and lasting verdicts derived from the April meeting was a consensus that it was possible to develop a revolutionary laser weapon. There was no doubt, based on the initial technological breakthroughs, that lasers offered almost unlimited potential for application on the ground, in the air, and even in space. The United States had no choice but to invest in a sound laser research and development program not only to explore new defensive weapon concepts but also as an insurance policy to prevent any technological surprises from the Soviets. By the end of the second day session, Schriever best summed up the position shared by the entire group of distinguished scientists: "We need to greatly expand and speed up the [laser] program."[22]

Lamberson realized immediately that his AFWL programs stood to gain substantial credibility and visibility as a result of the SAB's strong endorsement of lasers. The first indication of this occurred just before the meeting ended. Professor Hertzberg stood before the group and singled out the Weapons Laboratory for its pioneering work in lasers. He announced, "Of all the services that had an opportunity to respond to the laser and work with ARPA, I think we have to state, it was the Air Force that responded first. It responded the most imaginatively, and I think that the Air Force Weapons Laboratory should have some kind of ribbon for this response." With that type of support, Lamberson must have felt a sense of quiet contentment, knowing that the chances for steady and, hopefully, increased funding for the ALL and other AFWL laser programs were extremely good.[23]

What were the effects of the pivotal SAB meeting in the spring of 1971? First, the leading civilian and military scientists joined ranks and declared that there was an urgent need for the Air Force to establish a first-rate and long-range laser research and development program. They applauded General Ryan's foresight for increasing AFWL's laser budget for fiscal year 1972. Not content with just near-term progress, they committed to pushing for more money and to expand the program over the next 5 years. There was no immediate infusion of additional funds, but the SAB's strong endorsement provided a renewed sense of legitimacy to lasers and helped to pave the way for progressive annual budget increases for laser research throughout the 1970s. With the SAB in his corner, Lamberson was in a much more favorable position for making a strong case to Systems Command and Headquarters Air Force to release money to support the airborne laser program.[24]

A second major effect of the meeting was that the SAB placed the existing laser programs, especially AFWL's TSL and ALL efforts, on more solid footing. AFWL's work stood out as a shining beacon of progress on the relatively barren landscape of laser technology. The SAB sent a clear message to the highest levels in the Air Force that AFWL was indeed heading in the right direction by managing a diversified program designed to come up with basic answers to the complicated problems of laser physics. General Ferguson's willingness to fight for lasers early on, followed by General Ryan's acceleration of the ALL turned out to be heady decisions. General Brown's challenge was to build on this legacy by supporting and increasing funding for laser research in the years ahead.[25]

A year after the SAB meeting, Brown did succeed in attracting more interest and support for lasers at the highest levels of the Air Force. He was influential in having the Air Staff's Directorate of Doctrine Concepts and Objectives sponsor a high-energy laser applications conference at AFWL on 7–9 March 1972. The purpose of that meeting was to get the operating commands more involved in evaluating the potential of laser weapons. Representatives from Strategic Air Command (SAC), Tactical Air Command (TAC), and Aerospace Defense Command (ADCOM) led the panel discussions. Members of the panels consisted of scientists and weapon experts from Electronic Systems Division, Aeronautical Systems Division, Space and Missile Systems Organization, RAND, Mitre, and the Aerospace Corporation.[26]

SAC attendees left the conference expressing strong feelings that lasers offered great promise for future strategic missions on B-52 and B-1 aircraft. TAC was less enthusiastic, taking the position that laser state of the art would have to advance considerably before any laser could be put on a fighter aircraft. However, TAC and ADCOM representatives agreed that lasers had more chance of success applied in an AWACS self-defense and antisatellite

role. Although none of the operational commands signed up to replace guns and missiles with lasers on their aircraft, they endorsed research and development efforts aimed at proof-of-concept demonstrations for lasers. That was the job of Systems Command and AFWL. The clear message to AFSC was to first perfect the laser, then come back to TAC and SAC to sell the credibility of a beam weapon.[27]

One other important result came out of the SAB meeting. That was the creation of a new AFWL laser office named the Laser Engineering and Applications for Prototype Systems (LEAPS). Most of the scientific discussions had centered on two themes. One addressed strictly technological issues. What breakthroughs in devices, optics, pointing and tracking, and window configurations were needed to get a laser built and working properly? A second important concern fell in the area of applications. How would the laser be used once it was up and running? Many at the meeting suggested applications studies deserved much more attention. Specifically, the Air Force needed to conduct more in-depth studies to determine the best way to prototype and eventually deploy a laser. For example, were lasers best suited for bomber/AWACS defense, or did it make better sense to use lasers in an offensive fighter role? These would be the same key issues discussed at the high-energy laser applications conference in March 1972.[28]

Just because a laser worked did not mean it was practical. Weight and volume, for example, had to fit specific parameters before a laser could be deployed on an aircraft for bomber defense or offensive fighter roles. Technology and application studies overlapped and complemented one another. Mission drove the technology and vice versa. On the one hand, the scientist needed to know what mission the laser would be best suited for so that he could tailor his investigations in the laboratory to develop those technologies best satisfying that mission's requirements. Similarly, the person conducting an applications study had to be aware of the most likely technical limitations of a particular type of laser before he could make a sensible decision on how and where a laser weapon should be deployed. Because of the importance of the connection between technology and applications work, Colonel Rowden established the LEAPS Division at AFWL in July 1972. That office's main job was to conduct weapon applications studies, that is, determine the most practical operational uses for lasers in the Air Force. Conceptual studies would explore the feasibility of using lasers in an offensive fighter role or for taking out selected space targets. Lasers might also be used to defend bombers and AWACSs as well as intercepting sea-launched ballistic missiles. LEAPS personnel were to develop prototype weapon systems on paper through computer simulations and advise the Laser Division on targets best suited for low- and high-power lasers to engage in future weapon system feasibility demonstrations. Colonel John Scholtz, who was "invited" to return to AFWL following a tour at the Air War College and who had formerly served as the first director of AFWL's Laser Division, was Lamberson's and Rowden's choice to head up the new LEAPS organization.[29]

MANAGING THE PROGRAM: LAMBERSON'S GAME PLAN

The infusion of an additional $10 million for AFWL's high-energy laser research, the enthusiastic endorsement from the SAB, and the decision to go with the KC-135A as the

ALL aircraft, strengthened Lamberson's determination to keep the momentum moving on the airborne laser. AFWL had turned a major bureaucratic corner with higher headquarters's firm commitment to back an airborne laser. Now more than ever, Lamberson recognized the Lab had to deliver on advancing the technology to make the ALL work. There was no turning back at this point. Practical considerations dictated that he forge ahead and continue to sell and maintain a high level of visibility for the ALL. To do this effectively, he set out to maintain a credible development schedule based on achieving specific technical milestones.[30]

There always had been a long-range schedule embodied in the advanced development plans (ADPs) written by AFWL, which had to be approved by higher headquarters. The process was straightforward. Each year the Laboratory stated the goals, justification, and projected cost for its proposed laser work in the ADP. Eventually, the ADP ended up at Headquarters Air Force and the DOD, who either approved or disapproved the proposal. If approved, Headquarters Air Force sent a development directive to Systems Command authorizing work to proceed. Systems Command then issued its own program management directive to the Laboratory, spelling out how much money was available and where it would be spent on specific projects assigned to AFWL's high-energy laser (HEL) program.[31]

From 1967 through 1971 AFWL had prepared ADPs, but no two were exactly the same, mainly because of the rapid changes taking place in the state of the art for various laser technologies. All plans did agree that the purpose of AFWL's HEL program was to demonstrate the feasibility of using lasers as tactical and/or strategic weapon systems. However, the earliest ADPs, because of the immaturity of the technology, were forced to address broad issues such as establishing parameters for power, beam propagation, vulnerability levels of targets, and type of laser device, as a first step leading to the fabrication of any laser weapon prototype. As some of the questions to laser unknowns were solved, more specific milestones appeared in the later ADPs. By 1971, the AFWL ADP made it clear that specific TSL milestones had to be met before moving ahead with building an airborne laser system.[32]

Lamberson used the TSL benchmarks as a forum to continue to sell the airborne laser program. He was shrewd enough to know it was politically expedient and simply street smart to link the building of an airborne laser with success on the ground at SOR so as to keep funds flowing into the Laboratory. Building up his credibility depended on meeting the three TSL milestones: (1) demonstration of the field test telescope (FTT) to track a target and point the beam, (2) demonstration of a high-power laser with good beam quality, and (3) integration of the FTT with the laser device to show these two major components worked together. Tying these ground-based laser milestones to the airborne laser was the bedrock of Lamberson's program management plan. Although he recognized the importance of TSL, he knew ground experiments could never duplicate the problems of vibrations, beam propagation, and pointing a laser from an aerial platform to a moving target. It was for this reason he always kept his sights on proof-of-concept experiments for lasers in an airborne environment.[33]

Wanting to follow a similar procedure to mirror the TSL milestone checklist, Lamberson laid out three "cycles" for developing the ALL. Cycle I was to show an airborne pointer and tracker could accurately track an aerial target. Aligning a low-power laser beam with the APT and then directing the beam out of the turret on the top of the aircraft to an aerial target

was the goal of Cycle II. The most difficult part of the program was Cycle III. Its goal was to combine a high-power beam with the APT so the beam could shoot down air-to-air missiles. Using this evolutionary building-block approach of system capability would achieve the overall ALL objectives of demonstrating a beam could be propagated from an aircraft with sufficient power to kill an aerial target.[34]

The key to success, in Lamberson's mind, was connecting the TSL and ALL milestones as a management mechanism to transition lasers from the ground to the air. Proving the TSL on the ground as a requirement before moving to an airborne demonstration made the most sense from a proof-of-concept and cost-effectiveness point of view. This meant, for example, before money could be spent to build and install the tracker on the plane, the tracking goals of the first TSL milestone had to be demonstrated on the ground. However, Lamberson's guidance provided ample room for flexibility. As mentioned earlier, money was committed for paper design studies of a pointer and tracker before TSL goals were satisfied as a way to be prepared to immediately move ahead with fabrication of hardware on completion of the tracking milestone on the ground. Systems Command concurred with this plan. But it was Lamberson who determined when each of the TSL milestones had been "officially" accomplished. He was the one who decided when fabrication of the ALL components would begin.[*35]

By insisting that successful demonstration of each of the TSL milestones had to occur before AFWL could "proceed to develop hardware for an airborne demonstration system," Lamberson sent a clear message to various working groups to reduce some of the anxiety associated with such a high-risk program. To the engineers, scientists, and contractors, Lamberson was telling them they had to be productive by advancing the technology, at least on paper, before he would allow fabrication of any hardware for the airborne laser. And probably even more important, his firm stance on meeting TSL milestones conveyed to the doubters and purse-string holders in Washington that AFWL was not heading off in a haphazard manner. The impression Lamberson wanted to communicate to this group was that AFWL had thought long and hard in developing and following a reasonable road map with specific checkpoints along the way before committing a laser to the air. Lamberson's point was technology and money would not be squandered. He believed demonstrating the technology would take time, but this was the inherent nature of the program; taking shortcuts at this phase would jeopardize the airborne laser.[36]

Although establishing very precise technical milestones was the heart of Lamberson's management program, there were other important aspects of his long-range game plan. As

---

[*]Lamberson pointed out that he, as opposed to Systems Command, determined when milestones were met. He stated, "I'm sure I didn't come up here [to Washington] and say I met a milestone—I want to award the next contract. It was strictly a local thing that was done there [at AFWL]. That the power, authority delegated to the program manager in that way, that is the meeting of milestones and proceeding on, closely coupled to the notion that I could move money, I could move money anywhere in that program element and report it after the fact to headquarters. That held true even when we were running $100 million a year in the '75, '76 time period on the ALL. Unheard of today, you couldn't do that."

there was no single industrial contractor who had the expertise in all phases of laser technology, Lamberson took advantage of this opportunity to convince Systems Command to place him in complete control of the laser program. This kind of total dominance of a weapons program by a single individual was unheard of in the Air Force.[37]

But Donald Leslie Lamberson was truly unique. He fit in a special category, yet if he walked into a crowded room few heads would turn. A massive specimen of a man who spoke in a deep throaty voice, he projected an imposing and almost stern demeanor suggesting here was a serious, no-nonsense individual. Plain hard work had earned him the reputation of a man of scholarly interests and unquestioned technical competence. Yet there was another dimension to his personality. Don Lamberson had charisma. He was a modern-day P. T. Barnum who could work the crowds—a hearty laugh, a firm handshake, an uncanny ability to ease the most apprehensive stranger into the conversation, and a habit of listening intently and genuinely considering the other person's point of view won him more friends than enemies. Lamberson was a good ol' boy and intellectual all wrapped up in the same package. He had a wide range of interests, foremost of which was music. At age 5 he began studying the piano and sang in church choirs throughout his youth, a practice he continued on active duty. No one ever accused him of being a phony and people trusted him—his word was as binding as his signature on the most official document. And his great sense of timing and judgment to know when to turn on and off either personality, depending on who his audience was, brought out one of his strongest attributes—a supersalesman. Those closest to him affectionately referred to him as the "Snowman," because when briefing programs, he had a gifted flair for converting the most ardent opponent into one of his strongest supporters. Indeed, Lamberson was different from the average officer. Typically in the Air Force, pilots were the ones to move quickly through the promotion ranks. Lamberson never flew a plane, but his combination of personality and knowledge to get the job done in the R&D arena propelled him to the rank of full colonel before he turned 40. He had all of the wrong credentials for general—he never commanded a unit. Yet his technical abilities did not go unnoticed, and he retired as a major general.[*38]

One officer who had worked with Lamberson over the years summed up his positive management style:

> Lamberson brought you into his confidence; he made you part of the team. He didn't go around kicking asses, lighting fires under people. The guy had more leadership than anyone I ever worked for. On balance, he was the best leader this program [ALL] ever saw and this Laboratory ever saw.[39]

Avizonis, who had worked with Lamberson longer than anyone at AFWL, explained why his boss was so influential:

---

[*]Lamberson received his bachelor's degree in chemical engineering from Purdue University in 1953. He entered the Air Force in April 1954. He earned his master's degree in nuclear engineering in 1961 and a Ph.D. in aerospace engineering in 1969, both from the Air Force Institute of Technology at Dayton, Ohio.

Colonel Lamberson with a helium–neon alignment laser.

Lamberson was able to convey in his briefings his vision from a technical point of view, and his ability to articulate that technically in a way senior officers and DOD officials could understand it. He didn't talk down to them technically. So he had a fantastic combination of abilities, which was the reason the Air Force for some 10 years stood solidly behind the ALL demonstration.[40]

Lamberson's persuasive powers and spirited personality accounted in part for Systems Command designating AFWL as the prime integrating contractor for the airborne laser. (AFWL was responsible for systems integration, testing, and evaluation.) He reasoned that his group of blue-suiters had more experience and expertise than any one contractor in the country for pulling together all parts of the various laser and beam control technologies to fit and operate on an aircraft as one complete system. Under this arrangement, Lamberson realized a high percentage of AFWL funding would go to contractors. As it turned out, approximately 80 percent of the ALL budget supported contracting work on design studies and hardware. Lamberson never discounted the critical role contractors played in performing a large share of the ALL research and development work. But he also stressed AFWL's in-house expertise as an integral part of the contract control process. An Advanced Radiation Technology Office (ARTO) memo dated 28 October 1975 best described what Lamberson had practiced since the early 1970s:

> ARTO does not . . . contract for an area in which it does not have in-house research. This procedure promotes the technical expertise of ARTO, sharpens the technical direction, and prevents contractors from fairly fundamental errors. In other words, a check and balance approach is used. Because contractual efforts are such a large portion of the ARTO program, procurement and contract management procedures are very important in achieving the overall program objectives.[41]

A reorganization of AFWL in June 1972 helped consolidate Lamberson's single-manager approach for controlling the ALL program. Previously, the Laser Division was on an equal par with seven other Laboratory divisions. Because of the urgency to get lasers to work, and nudged ahead by Ryan's watershed decision to accelerate the HEL program, AFWL's reorganization elevated its laser work to be managed by the newly created ARTO.

(Since 1969, the Advanced Development Directives issued by Headquarters Air Force had assigned an "Importance Category of I" to AFWL's laser research and development programs.) Not only did the AFSC commander, General George S. Brown, push for the formation of ARTO in the summer of 1972, but he endorsed the single manager philosophy by delegating "full responsibility and commensurate authority" to Colonel Lamberson, the new director of ARTO. To make this concept work, Brown pledged adequate funding and manning to prevent delays in the notoriously slow government procurement process to issue contracts. The general was totally committed to the importance of moving ahead with laser research. His position was made clear in January 1972 when he stated to the press that "within the Department of Defense today, there is one technology program which truly stands out with the potential to make a profound and lasting impact on military operations. This technology advancement, the high-energy gas laser, is of great national significance."[42]

General Brown inherited a strong preference for an airborne laser from his predecessor, General Ferguson. He believed the most responsive and effective management arrangement was to place total control of the laser program in Lamberson's hands as a convenient way to circumvent inevitable bureaucratic roadblocks. To streamline management procedures within AFSC to ensure the earliest possible operational airborne laser, Brown issued a Program Action Directive (effective 1 July 1972) giving Lamberson additional authority as director of AFWL's newly created ARTO. The most unusual feature of this new scheme was a deviation from traditional military reporting procedures. Because of the high priority assigned to the airborne laser program, Brown gave Lamberson permission to bypass the AFWL commander and report directly to Brown. This "blue-line" reporting gave Lamberson special privileges. In essence, the military chain of command did not apply to the head of the ALL program. But at the time, the AFWL commander, Colonel Robert W. Rowden, had no problem accepting this arrangement, because he, too, was convinced lasers offered more promise than any other new technology on the horizon. Lamberson was obviously the best qualified to lead AFWL to take lasers from the theoretical realm to an operational system.[43]

General George S. Brown, Commander, Air Force Systems Command, September 1970–July 1973, assigned a high priority to AFWL's pioneering laser work.

Rowden realized that, in the case of laser program decisions, he was a middleman between Lamberson and Brown who might only slow down the process for making quick decisions by Brown. Under the new streamlined rules, Lamberson could get to see Brown on a moment's notice, inform him directly of progress and problems (usually contract coordination, potential cost overruns, and deviation from test plans and schedules), and return to Kirtland knowing the general's decision would be carried out without a hitch. Or a simple phone call to Brown's office would usually provide the information needed. In the

long run, Rowden and other AFWL commanders who followed voiced no serious objections to being cut out of the chain of command, because Lamberson, as a courtesy, always made it his practice to tell the AFWL commander whenever he dealt directly with the Systems Command commander. Besides, in only a few cases did Lamberson take advantage of this direct-line reporting privilege. But the fact that this system existed served to elevate the laser program to a special category in the Laboratory, underscoring the importance the Air Force placed on fielding an airborne laser.[44]

Another salient feature of AFWL's HEL program was that Lamberson managed two parallel activities simultaneously. One was the ALL. The other work focused on investigating generic laser technology (e.g., pointing and tracking, optics, device technologies) that could be sufficiently advanced to eventually be used to build a prototype laser weapon. There was a subtle distinction between the two. The ALL was exactly what the name implied, a flying laboratory to prove that theoretical physics and practical engineering integration concepts could be tied together to show a high-energy laser could be fired from an airplane. The goal was not to build a weapon but to build an integrated system to serve as an experimental bench. Indeed, the ALL turned out to be a long series of experiments—a first step—to prove the physics and feasibility of airborne lasers before the Air Force would ever approve moving on to build a weapon prototype. Many incorrectly used the terms *ALL* and *weapon prototype* interchangeably, but the scientists and engineers closest to the program were always careful to refer to the ALL as primarily a laboratory that would provide valuable scientific information to expand and advance the laser technology data base. Part of the confusion was there were different categories of prototypes. AFWL had no problem identifying the ALL as an "experimental" prototype designed to explore technologies to bridge the gap between theory and applications. However, many people usually thought prototype to mean building first-generation hardware to eliminate engineering uncertainties before moving on to fabricating a preproduction prototype. The preproduction prototype resembled the final product to allow any final engineering changes before approving the expensive process of mass production.[45]

To eliminate any misunderstanding about the concept of a laboratory versus a weapon, General Brown approved a change of the name *airborne testbed* to *Airborne Laser Laboratory* in May 1972. To many, the term *testbed* implied an engineering model or the first step in building a weapon prototype. Brown wanted to get away from the "engineering" connotation and emphasize the "science" of the airborne laser program. He believed the first priority was to fly an experimental platform that served as a laboratory for experiments collecting scientific data to gain a better understanding of the physics of lasers. Brown's decision simply reinforced what Lamberson had been advocating all along—to downplay the weapon aspect of the ALL. Lamberson did not want to set expectations too high by claiming the ALL was a weapon. "Since the ALL was not designed as a weapon system but will be used to evaluate laser weapon feasibility," he reported, "the demonstration objectives should be recognized as being ambitious and not without difficulties in accomplishment." Clearly he was saying it was difficult enough to perform the physics experiments on the ALL; claiming the ALL was a weapon would be misleading and result in disappointing people who thought it was. Plus, Lamberson realized he had a better chance to keep money flowing to support a research "laboratory" than fighting for funds to support a "prototype" system. Many thought twice about committing monies to support expensive and long-term prototype projects that, once started, would be difficult or impossible to stop.[46]

As ALL efforts involved designing and fabricating hardware, it fit in the category of advanced development work. This was balanced by exploratory development projects that looked at new ways to advance the broad laser technology base. Although the exploratory investigations might support the ALL work, most of the projects headed off in a different direction. By the summer of 1973, Lamberson had broken down the laser work load into 18 separate tasks. Seven of these related exclusively to ALL work. The remaining tasks spanned broader areas to include new laser concepts such as electric discharge lasers and chemical lasers, as well as studies in optics, pointing and tracking, high-power window material, beam propagation and effects, and innovative mirror coatings. There was some unavoidable overlap in ALL and the broader laser technology efforts; breakthroughs discovered as a result of exploratory research would be applied to improve the ALL technology. By overseeing

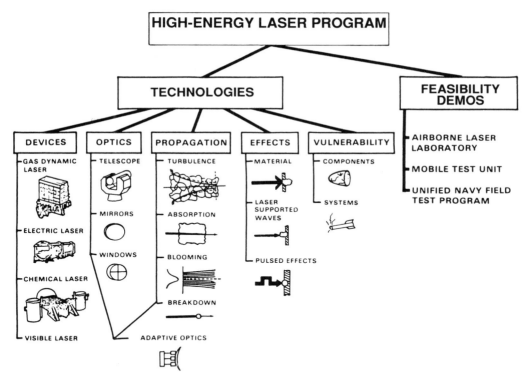

Technologies of high-energy laser program.

this two-pronged parallel approach to laser research, Lamberson was able to hedge his bet. If the ALL failed miserably, he would still be in business. At the same time, he argued money would be well spent in building up the technology base over the years. It would provide valuable insight to allow AFWL to explore the feasibility of other laser options, such as chemical and electric discharge lasers if the GDL turned out to be unproductive. However, Lamberson was supremely confident the GDL in the ALL would work and considered the exploratory and advanced laser programs as complementing one another.[47]

In sum, the success of the ALL depended to a large degree on the management abilities of Lamberson. His strong personality, technical competence, finely honed communicative skills, and power to motivate people enabled him to direct a large and complex research program and to keep it on the right course in spite of numerous adversities. In short, he was an effective leader.[*] Lamberson rightly earned the reputation as the "father" of the ALL because of the vision and guidance he projected. This is not to suggest the ALL was a one-man show. It was not. Lamberson was the first to admit that the execution of the ALL game plan depended almost entirely on the day-to-day activities of his staff of highly qualified scientists and engineers. But there was no doubt in anyone's mind that it was Lamberson who was the conductor of the ALL symphony orchestra. He was the leader out in front, waving the baton. And the AFWL musicians responded in harmonious unanimity.[48]

---

[*]Lamberson had emerged as one of the leading laser experts in the nation. In the fall of 1971, he received the Air Force Association's coveted Citation of Honor for his "superb managerial foresight and ability" and "brilliant leadership" in directing AFWL's laser program. Lamberson and the laser had become synonymous and both received ample recognition and praise. An article in *The New York Times* (16 October 1971) reported Lamberson was, indeed, one of the most influential leaders in the Air Force development of lasers.

5

Moving Ahead with Hardware

Once Lamberson had decided to go with the KC-135 (the equivalent of the commercial Boeing 707) as the airborne laser testbed, the next immediate concern was to find an aircraft that someone would be willing to give up to the Weapons Laboratory. This was no easy task, as the acquisition of any aircraft was a bureaucratic nightmare that could take up to 8 months before anything close to a decision could be made. But the triumvirate of Lamberson at the Laboratory, Colonel Orpha R. Cunningham, Director of the Weapons Division at AFSC, and Colonel Russell K. Parsons, Program Element Manager (63605F) for HEL programs at the Air Staff, combined their efforts to cut through the endless tangle of red tape. The three colonels, who were friends and had worked together before, developed a highly streamlined management system based mainly on telephone calls to one another. As Parsons recalled, "We had an informal system. We all would get on the telephone with a conference call and resolve things right there. We didn't worry about paperwork." It was this informal networking and team effort developed among these three officers that succeeded in reducing the time in pushing the aircraft request through the time-consuming bureaucratic maze.[1]

ACQUISITION OF THE AIRCRAFT

In many ways, Cunningham (who later became the fifth AFWL commander on 1 July 1973) was the key figure as AFSC controlled the type of aircraft desired for the airborne

Lieutenant Colonel Russell K. Parsons was an AFWL ally at the Air Staff who was highly efficient in directing funding through the bureaucracy to ensure it reached the ALL program at AFWL.

testbed. Aeronautical Systems Division (ASD) at Wright-Patterson AFB in Dayton, Ohio, actually owned a fleet of test planes used to conduct scientific experiments in support of various Air Force research and development programs. Some of the jobs performed by these planes included monitoring space launches, conducting tracking research, serving as highly instrumented airborne platforms to detect nuclear testing in the atmosphere by other nations, and practicing nuclear test missions to be prepared, in the event of the cancellation of the Nuclear Test Ban Treaty, to immediately resume nuclear testing in the atmosphere. These aircraft, mostly KC-135s, C-141s, and tactical fighters, came under the operational control of ASD. However, as its next higher headquarters, AFSC had the authority to direct ASD to allocate aircraft to other organizations in support of special research programs. It was Cunningham who first queried Colonel R. R. Greenley at ASD as to the availability of NKC-135 aircraft to support the ALL.[2]

Greenley replied to Cunningham's request by informing him that two KC-135s might be available. One was an A model and the other was a newer B model that ASD's commander, General James T. Stewart, closely guarded to retain for his own special projects. Cunningham relayed ASD's position to Lamberson, cautioning that it might be an extremely time- consuming battle to try to pry the KC-135B loose from the ASD general. Before making a final decision on the choice of the aircraft, Lamberson told Cunningham he wanted to consult his staff to get their input.[3]

Lamberson turned again to his aircraft integration officer, Major Hal Shelton, to help resolve the dilemma. The first question Lamberson asked was whether the Laboratory could complete all of the required experiments to validate the physics and engineering considerations of an airborne laser using the older-version airplane. Shelton recognized the magnitude of the task before him, because the wrong choice of aircraft at this early stage of the program might lead to major technical problems resulting in the premature death of the airborne laser at some later date.[4]

To answer Lamberson's question, Shelton quickly established an AFWL informal study group consisting of Major Ed Laughlin, Major Harry Axelrod, Lieutenant Colonel Edgar A. O'Hair, Captain Keith Gilbert, Dr. Barry Hogge, and others. The group provided data on each of the major components of the proposed airborne laser so Shelton could come up with a weight and volume estimate of the total system. Laughlin was the expert on optics and pointing and tracking. Axelrod and O'Hair had the most experience on the GDL device. Gilbert had initiated preliminary studies on the aerodynamic effects of placing a turret on the plane, and Hogge had been busy developing a computer code on beam propagation that would shape the design of the optical components. The group made their final recommendation based on what they judged would be the most likely design requirements for an airborne laser to shoot down air-to-air missiles.[5]

Shelton took all of the weight and volume data provided by his associates, calculated how these figures would affect drag on the performance of the aircraft, and came up with his best estimate (roughly 125,000 pounds) of how much weight the older KC-135 could carry. As a safety precaution, he doubled his initial weight calculation and concluded the extra weight would not degrade the performance of the KC-135A. He also took into consideration the advantages and disadvantages of the A and B models of the KC-135.[6]

Major Edward N. Laughlin, one of AFWL's bright young officers who headed up the early optics and pointing and tracking work.

The NKC-135A (tail number 55-3123) that ASD was willing to relinquish to AFWL had some definite drawbacks. Built in 1955, it was the sixth of ten hand-built preproduction airplanes, making it one of the oldest aircraft in the fleet. (The first was tail number 55-3118.) The Air Force purchased KC-135s to use as tankers for aerial refueling, but 55-3123 was an NKC-135 model, which meant it had been modified to conduct a variety of scientific experiments. Designated as an NKC-135 (the "N" stood for nonreturnable), the plane would never be returned to the tanker fleet. It would be strictly used as a research aircraft.* Besides being old, the A-model's turbojet engines were not as powerful on takeoff as its sister B-model's turbofan engines. Also it had the reputation as a hangar queen, having been parked for long periods of time at Eglin and Wright-Patterson AFBs in between research programs.[7]

There were liabilities with the NKC-135A, but it had certain advantages over the more modern B model. Even though 55-3123 was one of the oldest aircraft, it had extremely low flying hours because it was used as a research rather than an operational airplane. This meant there was less wear and tear on the airframe. In addition, another big plus was the A model was readily available—it would soon complete a test program at ASD and no other

---

*The Air Force accepted its first KC-135 on 31 August 1956. Fourteen KC-135s were permanently converted to be used as special scientific and test platforms. The ALL had four Pratt & Whitney J57-43WB turbojet engines of 13,750 pounds thrust (each with water injection). Maximum speed was 606 miles per hour (527 knots) with a flight ceiling of 50,000 feet. Cost of the original ALL aircraft was $3,398,000.

organization in the Air Force was seriously competing for it. Low hours, availability, and confident weight and volume estimates all made Shelton's decision to endorse the NKC-135A as a capable airborne laser platform a relatively easy one.[8]

Lamberson accepted Shelton's recommendation that the KC-135A would satisfy all of the requirements for an airborne testbed. He really had no other choice, unless he wanted to engage in a lengthy battle with the ASD general to try to acquire the more advanced B model. Time would be wasted and the odds of beating the ASD general were slim at best. Consequently, Lamberson realized that to expedite the acquisition process, he would take the safer bet of bidding for the older NKC-135A. Early in 1971 he called Cunningham and Parsons to inform them of his decision and for them to help get his request for the aircraft through the system.[9]

Major Hal Shelton.

By the end of February Cunningham and Shelton had arranged for the release of the NKC-135A to AFWL. At that time, the ASD aircraft came under the control of the 4900th Test Group supporting the Air Force Special Weapons Center (AFSWC) at Kirtland AFB. (The 4900th Test Group was part of AFSWC and reported to the AFSWC commander. Ownership of the plane belonged to AFSWC. The 4900th maintained and operated the plane at Kirtland.) AFSWC's commander, Colonel Martin H. Brewer, offered his support by writing a letter on 4 February 1971 to Systems Command, endorsing AFWL's request for a KC-135. He agreed AFSWC would be responsible for aircraft command and maintenance, while AFWL would issue direction for flight requirements related to conducting experiments on the airborne laser. On 1 March 1971 Lieutenant Colonel Carl L. Rucker, Chief of Plans and Requirements at AFSWC, wrote AFWL confirming that AFSWC would provide NKC-135 (55-3123) to support AFWL's airborne testbed requirement. Originally, the plane would be released after it completed its current research program scheduled to end in December 1971. However, because of unforeseen delays, Rucker advised AFWL on 10 November that the transfer of the NKC-135 would not take place until March 1972. That did not upset AFWL's timetable, as the contract for General Dynamics to modify the aircraft in preparation for Cycle I was not let until 15 March.[10]

HUGHES DESIGNS THE AIRBORNE POINTING AND TRACKING (APT) SYSTEM

Consistent with his long-range game plan to demonstrate the feasibility of an airborne laser, Lamberson set out to construct and test the ALL in three distinct phases. This building-block approach began with Cycle I testing in 1973 and ended with the completion of Cycle III in September 1983. Three issues were at stake as the program progressed. The APT system was the main focus of attention for Cycle I. Tracking an aerial target with the APT was the goal of Cycle I. Cycle II was to combine the APT with a low-power electric discharge laser to verify a beam could be accurately pointed to hit and remain locked on the target aimpoint. (Focusing an intense spot of photon energy on a vulnerable target surface, by responding to error signals furnished by the tracker, was the job of the pointing system.) Mating a high-power GDL beam with the APT to determine if the integrated system could shoot down air-to-air missiles was the basic goal of Cycle III.[11]

Flight testing of the APT during Cycle I occurred from May to November 1973. The specific objective of Cycle I was to determine if the IR tracker mounted on top of the APT could track an airborne target. During these experiments there was no laser aboard the aircraft. Although testing of the APT on the aircraft began in May 1973, the origins of Cycle I went back $2\frac{1}{2}$ years to November 1970. Because of the complexity and precision demanded of the APT, AFWL and Hughes Aircraft Company required a long lead time before work could start on the actual building and installation of the device on the NKC-135 aircraft. The key issues that first had to be completed were a number of APT design studies, a series of wind-tunnel tests to verify the turret would not adversely affect the aerodynamics of the airplane, and modifications of the plane to accept the APT. Once a hole was cut in the top of the NKC-135 for the APT, an intensive series of flight certification tests had to be completed before conducting the airborne Cycle I experiments. Lamberson and his staff knew they had to juggle all of these activities at the same time to produce the first hard technical evidence to prove that the tracker technology was feasible. The key here was to ensure "precision" tracking. For example, a fighter tracking an enemy fighter might track on the exhaust plume, which presented a large target. But in the case of the ALL, it had to track with great precision (within microradians) so the calculated aimpoint offset would keep the beam exactly on a small target spot. Control of the gimbals that rotated the APT was critical to ensure the tracking signal did not move off the small track point on target. Plus, all of this precision had to occur while the APT and the ALL were being buffeted by a turbulent airstream caused by the ALL moving through the atmosphere. Without successful Cycle I tracking experiments, momentum on the ALL program would come to an abrupt halt.[12]

In November 1970, AFWL was still trying to get the TSL at Sandia Optical Range (SOR) to work. It would be almost a year before the TSL's field test telescope (FTT) would meet its first milestone of tracking an airborne target in October 1971. Lamberson stood firm on his position of not committing funds for fabrication of the APT for the ALL until after the FTT at SOR met its tracking milestone. But during this waiting period the colonel was not idle. He was busy laying the groundwork to ensure AFWL and Hughes would be ready to build the APT as soon as the FTT performed up to specifications.[13]

To get the preparation process for Cycle I under way, Lamberson's first step was to amend contract F29601-69-C-0058 to allow Hughes to start work on the APT design in December 1970. This contract had been awarded in 1969 and was the centerpiece of AFWL's Optics for High-Power Laser Systems program investigating techniques for developing the

optics for an airborne and ground-based weapon system. As the recognized expert in the optics field, Hughes had the advantage of already having designed and built the first-generation pointer and tracker. This was the FTT used in AFWL's tri-service ground-based laser program. The contractor now was tasked to upgrade this first generation pointer and tracker. Essentially, Hughes's job boiled down to developing on paper a lightweight APT capable of microradian-level pointing accuracies (10 microradians)—keeping a high-power beam on target without moving it around on the aimpoint—that would fit aboard the ALL aircraft. In comparison, the APT tracker operating in the airborne environment had to be two-and-a-half times more precise than the FTT that had been anchored to the ground.[*14]

By the end of February 1971, Hughes had completed its conceptual design study. The final report identified components needed to make the APT work and established performance parameters. Conceptual designs were provided on each subsystem: the pointing assembly, optical train, autoalignment and focusing systems, tracker imager, and control and display consoles. The heart of the system was the APT turret consisting of two subunits: the protective turret assembly and the fine pointing assembly. The turret was the outer dome shell that fitted on top of the forward fuselage. It functioned as an aerodynamic shroud that covered and protected the fine pointing assembly, a 60-centimeter (24 inches) Cassegrain precision telescope consisting of a large ellipsoidal primary and a small spherical secondary mirror. By slightly moving the secondary mirror to adjust for different distances to targets, the laser beam could be focused to deliver maximum energy on target. The turret also supported a TV camera (with a 10 to 1 zoom lens) to help initially acquire targets, a laser range finder to measure distance to targets, and an IR tracker imager to precisely track airborne targets. Pyroelectric detectors, which made up part of the beam diagnostics equipment, measured position, size, and motion of the 10.6-micron beam. Recorders preserved these data for extensive computer analysis.[15]

The APT was suspended in a four-gimbaled configuration combined with a gyro system that served as a stable platform. Commands sent by a precision IR tracking and imaging processor to a hydraulic drive system allowed the turret and telescope to automatically rotate in unison. Both could move left or right off the aircraft nose 110 degrees to scan the sky for a 220-degree horizontal field of view. In addition, the elevation mechanism of the gimbal assembly allowed the turret and telescope to look up 70 degrees and down 30 degrees to give a wider vertical line of sight. Hughes anticipated the turret would weigh 2300 pounds and would be made of steel and aluminum to provide structural stiffness and lightweighting characteristics. One of the most important aspects of the APT was that it had to be strong enough to withstand the constant pounding and buffeting caused by the turbulent airflow as the plane moved through the atmosphere at 350 miles per hour.[16]

After Lamberson reviewed and approved the conceptual design of the APT, he mapped out a two-phase plan for translating the APT from the design to the hardware stage. Phase I

---

[*]The FTT was a workhorse during the TSL program. But its value went an important step beyond the TSL. The FTT served as a precision instrument "for testing advanced and unproven concepts in trackers, optics, and control systems" before AFWL "committed to airborne applications or prototyping."

consisted of a preliminary design followed by a detailed engineering design of the APT. Phase II involved the actual construction of the APT at Hughes's plant in Culver City, California. AFWL issued a $1.5-million letter contract (F29601-71-C-0058) to Hughes on 1 March 1971 to start the preliminary design of the APT. The contract specifications remained unchanged from the conceptual design: develop a pointing and tracking system that could direct first a low-power (a few hundred watts) and then a high-power GDL beam (up to 500 kilowatts) from a C-135 to target aircraft and missiles.[17]

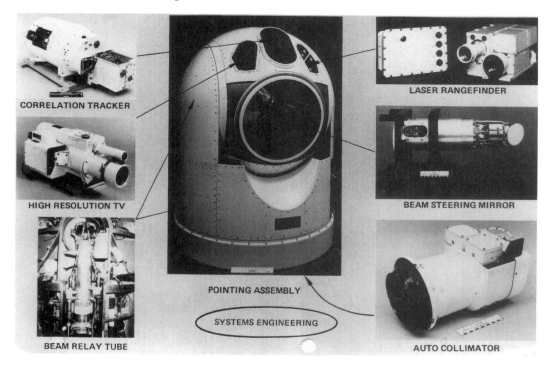

Major components of airborne pointing and tracking system.

Hughes issued monthly reports to apprise AFWL of progress on the preliminary design. During the second week of July 1971, representatives from Hughes and AFWL met at Kirtland for the critical design review of the APT. On the surface this could easily be interpreted as just another routine meeting. But it was much more than that. Lamberson knew the discussions with the contractor would not attract a great deal of attention or document spectacular breakthroughs at that point, as the technology for the APT was basically a continuation of what was done on the FTT. Yet he was fully aware of how vitally important it was to validate the contractor's basic design of the APT. If the contractor and AFWL stumbled over agreeing on the critical design, then the ALL program would suffer a major setback both technically and politically. And as the APT was the most sensitive component of the ALL, any slowdown in the design phase might send the wrong message up the chain of command and jeopardize the future of the program.[18]

Wooden mock-up of airborne pointing and tracking (APT) system prior to building real APT.

There were no surprises at the Kirtland meeting. AFWL did suggest the contractor should devote more attention to the off-axis gimbal design and selection of the computer to run the APT, but overall was more than satisfied with Hughes's technical design approach for the entire pointing and tracking system. The end result was the contractor left the meeting with AFWL's approval to move ahead and complete the final engineering design details of the APT as quickly as possible. Once again, another milestone had been crossed and Lamberson's game plan was on schedule.[19]

From July through October Hughes and AFWL representatives met several times a month to document progress and to ensure contract specifications were accomplished. A large share of this work involved putting the finishing touches on the APT engineering drawings detailing the size, shape, weight, and all of the mechanical and electrical interfaces of the numerous subsystems. This was no easy task. Modifications had to be made to improve the design by trial and error as the APT work progressed. But by the beginning of November, the final engineering design of the APT had been completed and blessed by AFWL. More importantly, the windup of the APT design coincided with another extraordinary event that made Lamberson's game plan appear that it was working like clockwork.[20]

In October 1971 the Hughes-built, ground-based FTT demonstrated it could accurately track a T-39 aircraft as it flew by SOR at the southern edge of Kirtland AFB. This marked the first major TSL milestone. It also signaled to Lamberson that the technology built into the first-generation FTT (pointer and tracker) produced a device that would actually work in the real-world environment. Now over that FTT proof-of-concept hurdle, Lamberson was confident to give Hughes the green light to proceed with the actual building of the APT. As a result, on 7 November 1971 he authorized approval of a $7.3-million contract (F29601-72-C-0029) to Hughes to complete the final design of the APT followed by fabrication of the hardware.[21]

For the next 8 months, Hughes engineers devoted their time refining the design of the APT, preparing blueprints, and acquiring components so construction of the pointer and tracker could begin. Minor interfacing details had to be finalized on the APT's environmental control unit, hydraulic power supply, and wiring harnesses. Also during this time at Hughes's Culver City plant, model builders combined their talents with engineers and completed construction of a full-scale mock-up of the APT pointing assembly. This mahogany masterpiece was the first step for moving the APT off the blueprints to a life-size device. The mock-up gave Hughes personnel an opportunity to determine if all of the components and subsystems of the real APT would fit together as planned. Also, building the wooden mock-up made sense from a cost-effectiveness point of view. By starting off with the mock-up, Hughes and AFWL had a chance to assess design parameters and to make any changes before moving ahead with the more expensive procedure of casting and assembling the final APT device. Minor modifications to the APT suggested by AFWL were agreed to by Hughes and incorporated into the design plans in early 1972.[22]

The partnership between Hughes and AFWL on the APT had worked out extremely well and attested to the close working relationship between the contractor and Air Force bluesuiters, a trend that would continue throughout the ALL program. Lamberson was extremely pleased with Hughes's progress and technical competence. He fully understood the complexity of the APT enterprise and realized the contractor was breaking new ground in designing a piece of hardware that no one had ever attempted to build before. Consequently, it did not take him long to approve the design of the APT and to instruct Hughes to get on with the fabrication of this intricate device that was so essential to the success of the airborne laser. Although the planning and design of the APT took a long time, the actual building of the device only took a few months. Hughes started "bending metal" in November 1971 and by the end of July 1972 had most of the APT completed. From August through October Hughes performed initial subsystem integration and evaluation tests on the APT. Full system-level integration and checkout followed in November. The next step was to make sure the APT fit on the airplane.[23]

GENERAL DYNAMICS DESIGNS AIRBORNE TESTBED

Before the NKC-135A aircraft could accept the APT, it had to undergo major modifications. That planning process began back in March 1971 at the same time Hughes had finished its conceptual design study for the APT. Again, Lamberson intentionally had planned for work on the APT and aircraft modifications to go on simultaneously. In keeping with his long-range game plan, these two parallel programs would progress and, hopefully, reach completion at the same time. In that way, once the APT had been built and checked out at the contractor's plant, modifications to the airplane would have been wrapped up. Culmination of both of these schedules would then allow the mating of the APT with the plane to begin almost immediately. Time would not be lost, and the ALL would be right on track as Lamberson had predicted.[24]

Boeing had designed and built the NKC-135 aircraft but did not receive the contract to modify it. On 23 October 1970 AFWL mailed a Request for Proposal document to 18

companies soliciting input on cost and technical procedures for customizing the NKC-135 to meet the requirements of an airborne laser laboratory. By 4 December, six companies responded by submitting proposals to the Kirtland contracting office. An Air Force technical evaluation group rejected the proposals of North American Rockwell (Los Angeles), LTV Electrosystems, Incorporated (Greenville, Texas), and Lockheed Aircraft Corporation (Burbank, California) as too costly and not technically competitive. Projected costs ranged from a high of $4.45 million by Lockheed to a low of $1.77 million offered by North American Rockwell.[25]

Three companies remained in the running: The Boeing Company (Seattle), McDonnell Douglas Corporation (St. Louis), and General Dynamics Corporation (Fort Worth). Meetings were held during the last week of February 1971 between the Air Force evaluation team and each company to discuss the merits and shortcomings of their technical proposals. Various suggestions by the Air Force were presented to show where costs might be reduced. The three contenders were given another week to add any revisions to upgrade their proposals. Another round of negotiations with each of the companies took place in the first week of March. All of the companies submitted reduced cost estimates to make their proposals more competitive. General Dynamics (GD) settled on $886,503. McDonnell Douglas came in with a bid of $3,047,958. Boeing unknowingly split the difference and came up with a price of $1,736,739.[26]

Not surprisingly, the Air Force selected GD to perform the aircraft modification work. Ronald H. Stephens, Kirtland's contracting officer, explained the evaluation committee ranked GD's proposal the highest technically of all of the competing companies. Plus, the Fort Worth company's cost was "some $850,000 lower than the next lowest proposal." The Air Force called GD back for meetings on 9 and 10 March to work out the details and to agree on the final cost of the contract. One additional requirement emerged from those negotiations. AFWL wanted to expand the scope of the work specified in the original statement of work pertaining to detection, measurements, and effects of vibrations on the NKC-135 aircraft caused by the modifications to the plane. An extra $25,000 covered this effort, increasing the final price of the contract to $912,403. On 15 March the Air Force formally awarded the design and fabrication of the airborne testbed (ATB) contract (F29601-71-C-0064) to GD. GD also maintained a cadre of approximately 29 engineers and technicians at Kirtland AFB. Although these people supported a number of other Air Force programs, some performed on-site liaison between the contractor and AFWL to provide quick reaction on decisions affecting modifications to the ALL.[27]

Extensive experience in modifying Air Force aircraft was another important reason GD received the contract. The company's Convair Aerospace Division's Special Projects team was responsible for the entire ATB modification and reported to the ATB Engineering Project Office. Describing the Special Projects team as "a close knit, quick reaction team of specialists," the contractor stressed few in the industry could match their highly successful performance record. Over the years the team had modified or repaired "over two hundred and sixty peculiar configurations of over seventy-five type models and series of cargo and tactical aircraft." The list of aircraft included the B-36, C-54, C-130, B-57F, B-58, F-111, and F-4. Since 1961 Convair had invested over 3 million man-hours making extensive structural changes to various models of the C-135 aircraft involving installation of avionics and optical equipment, relocating and adding antennas and radars, and fabricating in-flight fuel delivery and dump systems. They also had gained a wealth of practical knowledge on how to anchor "large protuberances on the top of the fuselage," which would be applicable to mounting the APT turret on AFWL's NKC-135. In short, GD knew more about modifying the C-135 than any other contractor.[28]

Not only did Convair have uniquely qualified personnel to make the required bodyshop changes to the NKC-135, but they also had the most up-to-date facilities to get the job done. With 117 buildings and 3.3 million square feet of manufacturing and support area floor space, the contractor had the capability to tailor-make the most complex aircraft parts to accomplish any aircraft modification. Most of the work on the ALL took place in the Special Projects Hangar (Bay 4) involving modification, final assembly, and installation of new components. Nearby, workers in the Special Projects Feeder Shop performed detailed fabrication, subassembly, and testing of parts before installing them in the ALL.[29]

AFWL stipulated that the contract to GD would be divided into four phases. Phase I involved the preliminary design of the ATB. This covered changes to the aircraft that would integrate the APT, a low-power laser (to be used in Cycle II testing), an acquisition radar, an optical acquisition device, fairings to reduce the effects of turbulence and airflow around the turret and fuselage, control consoles, electrical wiring, and support equipment. All of these components had to be tied together to make up the total system.[30]

Phase II built on the results of Phase I. Developing a final "detailed" design integrating all of the major components of the ATB was the primary goal of Phase II. Completion of this phase would allow the Air Force to issue a contract to begin work on gutting out the aircraft to make the necessary modifications in preparation for Cycle I testing. At the same time technicians and engineers proceeded with physically changing the configuration of the aircraft, Phase III and IV design studies were under way to plan additional modifications to the aircraft in preparation for Cycles II and III.[31]

Phase III consisted of three parts. Part 1 required a preliminary design of the testbed so it could accept two new components essential for Cycle II experiments. One was a laser and the other was an automatically aligned mirror train (AMT) to steer the beam from the floor of the aircraft up through the APT so the beam could exit into the atmosphere. Part 2 was the final design. It took into account other issues involving the testbed design, such as the effects of mechanical and acoustical vibrations on the AMT.[32]

Part 3 (preliminary design) of Phase III and all of Phase IV (final design) efforts focused exclusively on modifying the aircraft for Cycle III. The most important design consideration during this time was to make sure the aircraft would be able to fly safely with a large, high-power laser (GDL) installed in its midsection. Developing an efficient exhaust system to vent the spent laser gases out the belly (as opposed to earlier proposals to vent the gases out the side or top) of the aircraft also had to be addressed.[33]

From the very beginning, AFWL had taken extra pains to stress to the contractor the interrelationship of all four phases of the design of the ATB. Because of the scope and expense of the ALL program, AFWL could not afford to have the contractor become bogged down in any one design phase. Although the immediate task at hand always seemed to be

the most critical, Lamberson insisted the contractor exercise vision in progressing from Cycle I to Cycle III modifications. The key to success, in Lamberson's mind, was to make as many design changes to the C-135 as early as possible during Cycle I preparation. In that way, Lamberson believed the amount of retrofit and remodification to install the GDL during Cycle III would be kept to a minimum.[34]

On 16 March, a day after the contract for the design of the ATB was let, AFWL and GD representatives held their first technical exchange meeting at Kirtland. The following day, AFWL personnel traveled to Fort Worth to inspect GD's facilities where modification to the ALL was to take place. A day later, AFWL and GD officials looked over the proposed ALL aircraft for the first time at Wright-Patterson AFB in Dayton, Ohio. (At that time, negotiations were still under way to acquire the ALL, but in anticipation of eventual receipt of the aircraft AFWL and GD made a preliminary inspection.) These initial meetings between AFWL and GD representatives kicked off a two-phase ATB program lasting over the next 12 months in preparation for Cycle I hardware modifications to the aircraft that began in March 1972. The Phase I preliminary design involved reviewing concepts and techniques, evaluating advantages and disadvantages of different technical approaches, and considering possible trade-offs. More specifically, the end product of this work was for GD to come up with a recommended design for the integration of the APT, an acquisition radar, an optical acquisition device, associated controls, and instrumentation on the ALL aircraft. In addition, looking ahead to Cycle II, the Phase I effort took into account changes to the aircraft that would allow it to accept a low-power laser scheduled to be used during Cycle II. This was done as an efficiency measure to save time later on so the aircraft would spend a minimum of time undergoing excessive modifications in preparation for Cycle II. However, after completion of Cycle II, the aircraft would return to Fort Worth for extensive modifications so it would be prepared to accept the large GDL for Cycle III testing.[35]

Progress on the preliminary design of the ATB appeared in monthly reports issued by the contractor. These updates reflected the close cooperation between GD and AFWL in making joint technical decisions affecting the future of the ALL. As an example, one of the early concerns was where to locate the acquisition radar on the plane. Originally, GD planned to place the radar in a fairing on top of the fuselage and then mount the APT on top of the fairing. However, in that stacking configuration the APT protruded too far from the fuselage, which presented possible aerodynamic problems affecting flight safety and performance of the aircraft. After consulting with AFWL, the contractor dropped the idea of mounting the radar on the fuselage and, instead, relocated it in the nose and tail section of the ALL. Critically important also was determining the "exact location" of the APT on the fuselage. After making several trade-off studies assessing the pros and cons of fuselage stations 490, 570, and 750 (roughly front, mid, and rear sections of the top of the fuselage), GD and AFWL agreed on station 490. This offered the best field of view for the APT and provided good stability for the pointer and tracker.[36]

Besides settling on the best location for the acquisition radar, selection of a radar that was capable of handing off target information to the APT had to be made. GD conducted studies on six different radars to determine their suitability. These studies focused primarily on radar performance in terms of target acquisition, antenna scan limits, reliability, and range error. Practical considerations, such as availability of personnel to perform timely maintenance and support, and compatibility with mission requirements, were also taken into account. The net result was that GD and AFWL decided to go with the AN/APQ-109 (the radar used on the F-4 fighter), because it provided the "best APT [target] acquisition over the widest range of mission requirements" and was "found to be superior from a maintenance and support aspect," as Air Force maintenance personnel for this system were already stationed at Kirtland to provide quick turnaround maintenance and repair service.[37]

Lamberson assigned several of his most technically competent officers to closely monitor the contractor's work during the preliminary design stage of the ATB. They were constantly looking over GD's shoulder to make sure they complied with the contract specifications and to advise them on technical considerations to be included in the final modification of the aircraft. Major Shelton was responsible for the engineering integration of the entire ATB. His main concern was whether all of the structural changes made would still allow the plane to fly safely and within its performance limitations. Major Edward N. Laughlin's focus was more narrow, as he was the expert on the APT. He worried about making sure the APT would fit snugly into the 5-foot-diameter hole cut in the top of the fuselage. Still unanswered was whether the support structure holding the APT in place would be strong enough to keep it from bouncing around and in fact reduce the vibrations to a minimum so the tracker and gimbals could perform up to their maximum potential. Finally, Lieutenant Colonel Edgar A. O'Hair, the third member of the AFWL integration team, looked ahead to Cycle III to plan how the modifications would affect placing a large GDL (under development by Pratt & Whitney) into the midsection of the ALL.[38]

All three of these officers consulted with the contractor almost weekly through technical interchange meetings or visits to GD's plant to eyeball hardware modifications planned for the aircraft. Hughes also routinely attended these meetings because they were responsible for building the APT. They especially had to work hand in hand with GD and constantly keep the Fort Worth contractor updated with engineering drawings and any changes to the size and weight of the APT. This information was critical, because the fabrication of the APT at Culver City, California, and modifications to the aircraft at Fort Worth took place halfway across the country from one another. Eventually the day would arrive when the APT would have to be mated to the aircraft. And Lamberson had made it crystal clear to both contractors, far in advance, that when that day came, the APT had better fit on the airplane![39]

Much of the attention during the preliminary design of the ATB rightfully focused on the APT, because it was the centerpiece for the planned Cycle I testing. But the APT could not operate independently. It had to depend on a number of finely tuned support subsystems to allow it to track airborne targets. Designing a hardware configuration containing a variety of control consoles and instrumentation racks to be installed in the rear of the ALL was another one of those essential tasks requiring extremely close coordination between Hughes and GD. The control console consisted of individual test stations for the Test Director; Safety Systems Operator; Acquisition, Pointing and Tracking Operator; Laser Operator; Instrumentation Operator; and Optical Acquisition Device Operator.[40]

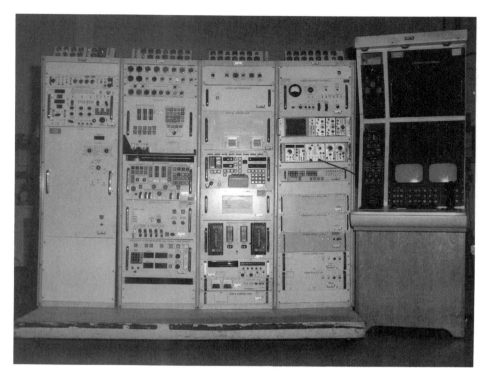

Consoles for controlling APT.

Sitting in front of and operating their designated stations, AFWL personnel would be able to carry out all aspects of the Cycle I experiments at their fingertips. The consoles allowed the test operators to send commands, in the form of electrical signals, to all major components of the ALL. For example, operators in the back of the plane would be able to monitor the performance of the acquisition device to ensure vital tracking data were "handed over" to the APT so it was pointed in the right direction to "lock on" the target. They also would, during Cycles II and III, be responsible for starting up the laser and monitoring it to ensure it operated within strict safety parameters. Still another function was to collect scientific data to be analyzed later to tell how effectively the APT and laser worked in terms of pointing accuracy, beam power and quality, and jitter.[41]

By the summer of 1971, GD had completed work on the initial design of the ATB. But before the contractor could proceed with the final design, GD first had to inform AFWL on the design status of the ATB. The contractor met this obligation by conducting a preliminary design review from 10 to 13 August. During this 4-day session, GD people went over in detail the design of all proposed modifications to the ATB. AFWL personnel had an opportunity to ask questions and to give their technical input. However, they knew ahead of time what the contractor would present, as AFWL had been working hand in hand with GD over the past several months. As a result, the preliminary design review amounted to a formality that fulfilled the Phase I contract specifications. By the end of the last day of meetings, AFWL concurred with GD's preliminary design plan and had authorized them to move ahead with the final design (Phase II) of the ATB.[42]

For the next 4 months after the preliminary design review, GD and AFWL devoted their time to improving on the design of the ATB. On 30 December 1971 the contractor issued a 345-page report covering all aspects of the final design. This document addressed not only how the integration of the APT, acquisition radar, optical acquisition device, and the control and instrumentation panels would be accomplished, but also provided analyses of a variety of subsystems that were absolutely essential for conducting Cycle I experiments.[43]

Installation of these major components into the aircraft, plus test support and airplane equipment, resulted in an increase of 19,117 pounds caused by Cycle I modifications. The APT weighed about 3600 pounds, but it took another 2000 pounds of support equipment just to keep it running. This included such items as the hydraulic power supply to slew the APT, an environmental conditioning unit to keep the turret at the proper operating temperature, and the operator console in the rear of the plane to run the APT. Test support equipment, such as instrumentation racks and consoles, an aircraft intercom system, a high-frequency radio system, and other associated equipment accounted for almost another 2000 pounds of extra weight. But most of the increased weight reflected installation of basic aircraft equipment: an air conditioning system for all three compartments of the plane, the forward and aft radar systems, 960 pounds of interconnecting wiring, an aft escape hatch, a generator bolted onto the number 4 engine to provide additional electrical power, and even a 20-man life raft (160 pounds) in case the ALL had to ditch over water.[44]

Considering all of the Cycle I modifications planned, GD estimated the operating weight of the ALL would be 120,798 pounds. Tacked on this was 5581 pounds of water for injection into the four engines on takeoff and 132,122 pounds of fuel, making the final gross weight of the plane 258,501 pounds. This was well within the maximum allowable gross weight of 275,000 pounds for the ALL to fly safely. Although the plane had the capacity to carry an additional 7160 pounds of fuel in the forward fuel tank, the contractor ruled against this. The reason for this decision was that the placement of the APT on top of the fuselage had moved the center of gravity of the aircraft forward. This could present a safety hazard on takeoff (the nose would pitch down too much). Consequently, the plane flew with less than the maximum amount of fuel it could carry. (Fully fueled, the ALL had a range of 8673 statute miles or 7537 nautical miles.) However, this never became a real problem, because the ALL missions never extended beyond a few hours of flight time. Major Shelton pointed out that the relatively short airborne time of the ALL did not require it to carry a full load of fuel to perform its in-flight experiments.[45]

The most radical design change for Cycle I involved cutting a 5-foot-diameter hole in the fuselage to accept the APT. Once the hole was cut, the contractor had to then build a steel and aluminum bridgework structure strong enough to support the weight of the APT and properly transfer these loads to the rest of the fuselage structure. AFWL carefully evaluated GD's proposal, because these alterations required cutting through four airframes and seven fuselage stringers. Without the APT securely fastened to the top of the fuselage to track

91

Initial ALL Weight Predictions	
Aircraft baseline	125,000 lb
GDL FSS	11,000 lb
GDL	9,000 lb
APT	4,000 lb
GDL fluids	2,000 lb
ADAS	2,000 lb
MISC	22,000 lb
Mission fuel	65,000 lb
Maximum takeoff	240,000 lb

targets, all of the other changes to the plane would be meaningless. However, AFWL by January 1972 concluded GD's design to support the APT was technically sound.[46]

Three other hardware components were high-priority items for Cycle I. Before the IR tracker on the APT could begin tracking, it had to receive target location data (azimuth, elevation, and distance) from either the acquisition radar housed in the nose or in an added tail extension of the aircraft. An alternate or backup to the radar system was a less sophisticated, hand-operated, optical acquisition device (OAD) designed to hand off target data electronically to the APT. The OAD consisted of a zoom telescope, basically a rifle scope, mounted on a precision movable pedestal that was capable of accurate simultaneous elevation and azimuth measurements. Locating an airborne target depended on the keen eye of the operator—such as Captains Mark Rabinowitz and Steve Coulombe during Cycle III—looking through the sights of the OAD pointed out an optical window on the left side of the aircraft. Using left and right pistol grips, the operator could sweep the rifle scope 90 degrees horizontally, up 45 degrees, and down 30 degrees. Once he had the target in view, he continued to move the OAD to follow (track) the target. The OAD measured the target azimuth and elevation and sent signals to an electrical interface unit, which, in turn, instructed the APT to rotate so it was lined up with the target. Originally, AFWL counted on the nose and aft radars to do all of this work. But recurring operating and maintenance problems with these radars caused AFWL to discard them by the end of Cycle II. As a result, the low-tech, but nevertheless effective OAD became the primary system for acquiring airborne targets for the remainder of the ALL program.[47]

A second major modification to the aircraft for Cycle I was the installation of an additional generator. The plane already had three generators (one on the number 1, 2, and 3 engines) to furnish electrical power to run all of the subsystems on board. Although these three generators were capable of providing 108 kilovolt-amperes of electrical power, this was not enough to accommodate all of the power demands of the ATB equipment. GD and AFWL solved this problem by agreeing to mount a fourth generator on the number 4 engine. They selected a 70/90-kVA generator, the same one the Air Force used on its B-52G airplane. Before the NKC-135 number 4 engine could accept the additional generator, rework of the engine nacelle, nose cowl, and cooling ducts had to be accomplished. Details of these hardware changes appeared in GD's engineering drawings completed in December 1971.[48]

One of the jobs of the extra generator was to furnish sufficient electrical power to run the aircraft's environmental control system (ECS). AFWL stipulated that all three sections of the aircraft—flight deck, device compartment, and aft crew compartment—had to be maintained at a temperature of 70 degrees. To meet this specification, GD proposed using a vapor-cycle refrigeration system to dissipate the heat given off by the APT and its hydraulic power unit, plus the laser device, control consoles, instrumentation racks, nose and aft radars, and other equipment on board. During the preliminary design of the ATB, the contractor underestimated by about 13 percent the amount of heat that would be generated in the aircraft. By early 1972, GD had made some headway on the design of the ECS, but was still wrestling with nailing down the final design plan that would be acceptable to AFWL.[49]

6

Aerodynamics and Safety

At the same time General Dynamics was busy settling on the final design and developing blueprints of the aircraft modifications in preparation for Cycle I, AFWL wanted confirmation that the installation of these new components would not present any safety hazards affecting the flight performance of the aircraft. What the Air Force especially worried about were the unanswered questions regarding the aerodynamic effects of placing a bulky turret on the top of the fuselage. The NKC-135 had never been designed to carry a protruding piece of equipment like the APT and some believed the plane might crash. The basic question was whether the aircraft would be airworthy while flying with a turret. Almost everyone predicted the turret would create a harsh aerodynamic environment, especially when flying at high-subsonic Mach numbers. As the airstream enveloped the turret, the airflow would separate, causing extreme buffeting and a violent downstream wake that could inflict severe damage on the fuselage and vertical stabilizer of the airplane.[1]

Lamberson fully realized the potential dangers in this critical early stage of the program. An aircraft accident at this point would only serve as unwelcome testimony that the critics were right that the ALL had been an ill-conceived project from the start. Consequently, to avoid any danger of placing his people in a life-threatening situation, he made safety his top priority throughout the entire ALL program. In this case, he insisted on obtaining hard scientific data from wind-tunnel experiments that would confirm or deny that the plane would fly with the turret installed before moving on to Cycle I testing. To make this happen, he turned to two technically competent Air Force officers for help. One was Lieutenant Colonel Demos Kyrazis, who at the time was working as a Research Associate at the National Aeronautics and Space Administration (NASA)-Ames Research Center at Moffett Field near San Francisco. The other was John Otten, a young captain assigned to AFWL's newly formed High Energy Laser Division.[2]

WIND-TUNNEL TESTING

The selection of Kyrazis and Otten was no accident. Lamberson would settle for no less than the most able officers in the Air Force to usher the ALL through the critical upcoming wind-tunnel testing, even if this meant looking for the most qualified candidates outside the Laboratory. In the end, he turned to his friend and colleague, Dr. Hans Mark, who at the time was director of the NASA-Ames Research Center. Mark was a first-rate scientist with a Ph.D.

in physics from Massachusetts Institute of Technology who later became Secretary of the Air Force in July 1979. Lamberson respected Mark's judgment, because under his direction the Center had gained the reputation as one of the leading institutions in the country for its work in theoretical and experimental aerodynamics. Mark did not disappoint Lamberson. He knew exactly the right man to fill the job at AFWL. That man was Lieutenant Colonel Demos Kyrazis, who had built a solid track record while working for Mark at Ames. Firsthand observation of his performance had convinced Mark that Kyrazis was one of the most able aerodynamicists in the Air Force.[3]

Lamberson never regretted recruiting Kyrazis, who carried impressive academic credentials. Enlisting in the Air Force in 1952, he quickly enrolled in the Aviation Cadet Program, where he graduated and received his commission in 1953. Next he earned a bachelor of science degree in electrical engineering from the prestigious Massachusetts Institute of Technology in 1959 and a master's 4 years later from Ohio State University. In 1977 he capped off his formal education with a Ph.D. in applied science from the University of California, Davis/Livermore. His dissertation on fluid dynamic stability attracted almost as much attention as two of the

The Honorable Hans Mark, Secretary of the Air Force, July 1979 to February 1981, and an avid lobbyist of the Airborne Laser program.

distinguished men who served on his thesis committee, the renowned Drs. Edward Teller and Hans Mark.[4]

Although he had proved himself in the classroom, Kyrazis was much more than simply a one-dimensional scholar. Like any good scientist, he had an unlimited appetite for exploring the unknown. And on his journey to conquer the most vexing scientific problems, this unassuming but highly articulate officer had wrangled some of the most challenging assignments in the Air Force. Before ending up at NASA-Ames in the early 1970s, he had worked in the technical intelligence area analyzing signals transmitted from a variety of emitters. In the mid-1960s, he moved among the nation's premiere scientists at the highly respected Lawrence Livermore Laboratory. There he ventured into the relatively unknown world of designing and testing advanced fusion weapons concepts. These assignments served as a training ground, giving the intently determined Kyrazis an opportunity to apply and test his theoretical knowledge against real-world problems. He was the type of individual who was always striving to sharpen his scientific street smarts, which would prove invaluable to him in even more demanding positions—first as one of the principal investigators on wind-tunnel research, and later as Chief of the Laser Development Division at AFWL responsible for the day-to-day activities of preparing the ALL for Cycle II and III testing.[5]

Colonel John Otten was one of the prominent "hands-on" scientists throughout the ALL program who later became the AFWL commander from 25 August 1988 to 12 December 1990.

The quiet and supremely confident Kyrazis was the perfect complement to temper the unbridled enthusiasm of Captain Otten. As a junior officer in the AFWL Laser Division, Otten naturally wanted to prove himself and move up through the ranks. Little did he know in the early 1970s, when he was struggling to validate calculations of turbulent transonic airflow around the ALL turret, that he would in 1988 be the final commander of the very organization to which he was assigned as a captain. On the road to becoming commander, Otten gradually gained the reputation of a respected scientist. A large share of the recognition he received came out of the wind-tunnel calculations he and Kyrazis developed to prove the ALL aircraft would not become unstable and break apart in flight if a turret was placed on top of its fuselage.[6]

Otten, like Kyrazis, had persuasive credentials. He graduated from the University of New Mexico with a bachelor of science degree in mechanical engineering in 1967; he also was a distinguished graduate of the Air Force Reserve Officer Training Corps on receiving his commission. The following year he earned a master's in aerospace engineering from the Georgia Institute of Technology. Later in his career, he enrolled in graduate study programs at Stanford University and the University of Tennessee Space Institute. His early work on aero-optics and turbulent flow, combined with his research in GDLs, made him ideally suited to serve as one of the ALL test directors during Cycle III.[7]

It was these two men—Kyrazis and Otten—who were destined to play a leading role in literally getting the ALL aircraft off the ground. They faced an enormous challenge in that they had to eventually tell Lamberson whether or not the ALL could fly safely. Their decision would be based primarily on data derived from computer models and a series of experiments conducted in three different wind tunnels. Lamberson never lost sight of the importance of collecting accurate wind-tunnel data, because he knew without this information he would be forced into a position where he might have to cancel the entire ALL program. He could not risk the loss of crew or aircraft by pressing ahead too quickly—the stakes were simply too high at that juncture. Instead, he opted for a more rational and patient course of action, waiting for Kyrazis and Otten to come up with their best estimate as to the flight safety characteristics of the ALL.[8]

Kyrazis was a lieutenant colonel in the spring of 1971, who was steadily gaining a reputation as a superb aerodynamicist. Hans Mark especially was impressed with Kyrazis's dissertation, which presented new data on the numerical simulation of airflow around fixed bodies. Here was an up-and-comer whose talents could not be squandered on some inconsequential desk job in the Air Force. Mark was to see to that. He managed to get Kyrazis assigned to NASA-Ames's Computational Aerodynamics Branch, where he worked on a relatively new supercomputer called Illiac. Using this highly sophisticated machine, Kyrazis had the resources and freedom to test his own theories and to develop codes to predict the effects of aerodynamic flow. In essence, he was building a mathematical wind tunnel that, when plugged into the computer, would tell how an aircraft or any other suspended body would react to turbulent airflow.[9]

Kyrazis was completely content at NASA-Ames. He was deeply involved with his work and believed he was making an important contribution to the Air Force's research effort. He had never heard of the Airborne Laser Laboratory. All that was to change in May 1971 when he received a telephone call from Captain Harry Windsor from the Weapons Lab. The energetic captain wanted to know if Kyrazis could assist in providing some critical data on the ALL's flight characteristics.[10]

Years later Kyrazis vividly recalled that telephone conversation, which ended up as a major turning point in his career.

Harry asked, "Can you calculate the flow around a large bump on top of an airplane?"

And I said, "Like a turret on a B-17?"

"Yeah, like a turret on a B-17," Harry replied.

Then I answered, "Hell, no! No one can do that accurately, although you could roughly figure it out."

Harry seemed disappointed with that answer, which he obviously didn't want to hear.

"Well, we have a problem here with that." Harry persisted, "Do you think you could come down and see what our problem is, and maybe you could still help us anyway?"[11]

Captain Windsor's phone call intrigued Kyrazis. After all, he had never been to Albuquerque and this might be an opportunity to apply some of his theoretical knowledge and find out more about one of the Air Force's mysterious aircraft programs.[12]

A week after the phone call, Kyrazis was in Albuquerque getting his first exposure to the ALL program. Windsor arranged an informative overview briefing, and Kyrazis, for the first time, met with some of the top players—Lamberson, Rich, and Lieutenant Colonel Ed O'Hair, who was the branch chief in charge of development of the GDL to be placed in the belly of the aircraft. These men explained two immediate concerns that worried them about the flight safety of the aircraft. One was whether the plane's stability and control would be adversely affected by installing a blunt-body APT on top the fuselage. The assumption was the turret would shed vortices resembling a storm of small tornadoes that would pound on the plane. These buffeting forces had to be measured to help determine the amount of drag and possible damage that might be inflicted on the panels of the aircraft. More importantly, would the airflow around and behind the APT turret move violently downstream with enough force to tear off the tail of the aircraft?[13]

The second major concern had to do with the open port in the APT. Some initial studies were under way to identify the best type of transparent material to use to cover the opening in the APT where the beam exited into the atmosphere. The material had to exhibit very unique qualities to allow the laser beam to pass through without fracturing the window or distorting the beam. One of the most promising materials was zinc selenide. But in 1971 the biggest piece of zinc selenide that could be manufactured was about the size of a person's thumbnail. This meant until better manufacturing techniques could be perfected, the APT would have to be flown open port on the ALL. However, open port forebode dire conse-quences. Air would have an easy entry to rush into the APT and cause catastrophic damage to the delicate optics inside. The flight condition of the APT was similar to one blowing across the opening of a Coke bottle, producing a distinctive hollow-whistling sound. A plane flying several hundred miles per hour with a hole cut in the APT, many thought, would encounter severe turbulence to produce a loud screeching and cause acoustical damage to the APT. Fairings positioned in front of the turret offered one possible solution for cutting down air circulation inside the turret. Another option was to inject a low-velocity airstream through nozzles inside the turret to counter the airflow rushing over the port from the outside. What Lamberson and the others posed to Kyrazis was, could the ALL fly safely open port?[14]

Lamberson explained he had recently formed a special committee to study the aerody-namic effects of flying the APT aboard the ALL. He wanted impartial opinions on this critical issue and gathered experts from around the country to serve on the committee. Members came from the Air Force Flight Dynamics Laboratory and Aero Propulsion Laboratory at Wright-Patterson Air Force Base, Ohio; General Dynamics; the Aerospace Corporation in El Segundo, California; the USAF Academy; and AFWL. Because of the wind-tunnel expertise at NASA-Ames, Lamberson wanted a representative from that organization to be

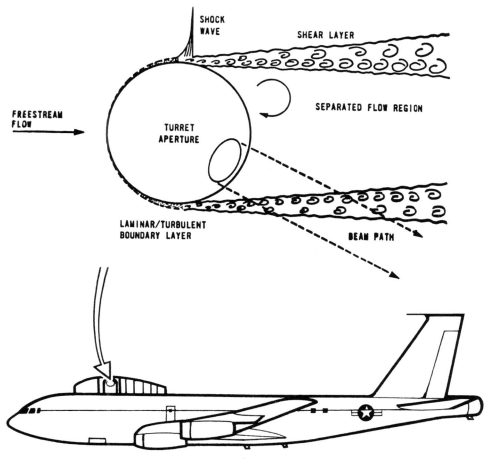

Aerodynamic flow around ALL turret.

on his special committee. Mainly because of Mark's endorsement, Lamberson extended an invitation to Kyrazis to serve as the NASA-Ames consultant to the committee. Kyrazis accepted and on his return to California made a note on his calendar to remind him to attend the first committee meeting scheduled for 21 July at AFWL.[15]

Opinions differed as to the aerodynamic effects of placing a turret with a hole in it on top of an airplane. Lamberson's trusted right-hand civilian technical advisor, Dr. Petras Avizonis, predicted the turret would not make much difference on the performance of the aircraft. After all, B-36s and B-17s had flown with gun turrets protruding from their fuselages without causing any ill effects on the handling of the aircraft. But those propeller-driven behemoths were slow movers compared to the 0.4 to 0.7 Mach speed of a jet-powered NKC-135. Also, several C-135 research aircraft had low-profile turrets installed, which were equipped with cameras and sensors to track reentry vehicles and to detect nuclear radiation in the atmosphere. But it would be an abrupt change to install the ALL turret, which was

higher and wider than any existing turret. This caused concern because AFWL would have to conduct extensive studies to determine the effects of severe buffeting and turbulence produced by the larger and bulkier turret as the ALL plowed through the atmosphere at greater than half the speed of sound.[16]

Kyrazis's gut feeling was different. He believed specially designed fairings would have to be installed around the turret to smooth out the airflow as it passed over and around the turret. The fairings, shaped like upside-down canoes and placed in front or behind the turret, would significantly reduce the air turbulence as it moved from the front to the rear of the aircraft. Kyrazis also had known about a special access ARPA program known as HAVE CHARITY that involved wind-tunnel testing at the Arnold Engineering Development Center's 16-foot transonic wind tunnel in November 1968 and January 1969. As part of that program, General Dynamics (GD) had cut a 26-inch-wide hole in the top of a C-135B to accept a turret that housed and allowed a telescope and other optical hardware to peer outward. A low-angle ramp (fairing) fastened to the fuselage just forward of the opening diverted the airflow over the opening sufficiently to reduce turbulence and potential damage to the optical equipment. However, test results also revealed a large turret would considerably reduce the directional stability of the aircraft in all flight conditions, but especially during landings. This project essentially failed, because the large telescope never fit properly into the hole cut in the aircraft.[17]

When Kyrazis returned to Kirtland to attend the first aerodynamic committee meeting on 21 July, he brought Don A. Buell along. Kyrazis admired and recruited Buell because of

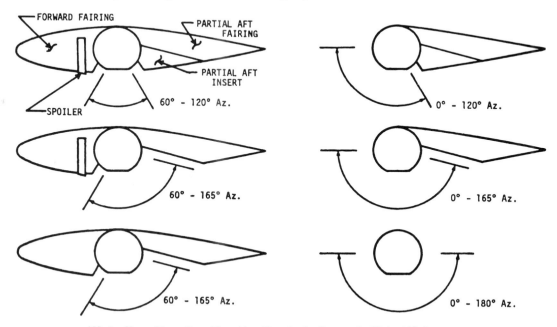

All Configurations Have Elevation View Angle Range of -30 to +45 degrees

Fairing configurations for ALL.

his demonstrated competence as a research scientist at NASA-Ames. He had considerable experience in the fields of blunt-body aerodynamics and wind-tunnel testing. Buell had been a key player in NASA's modification of a specially designed C-141 called the Kuiper Lab. It had a hole in its fuselage for a long-wave infrared 1.5-meter telescope to look at the stars and track reentry vehicles. Buell had labored extensively on this project performing numerous calculations of the airflow effects around the opening. It was Buell, through his precision wind-tunnel experiments and persistence, who succeeded in devising a way to suppress the vibrations and acoustics around the opening in the plane. Kyrazis recognized Buell's contribution and knew his expertise would be invaluable in planning and conducting wind-tunnel tests on the ALL.[18]

The purpose of the 21 July meeting was to settle on a plan to collect test data on the boundary layer flow disturbances interacting with the fuselage and turret. Obtaining measurements of aircraft drag and both static and dynamic forces and torques colliding with the external and internal turret surfaces were the basic goals of the experiments. Buell's presence at the meeting had an immediate impact. He recognized that some initial small-scale wind-tunnel tests already completed and planned in the near future were inappropriate. To gather meaningful scientific data, Buell proposed wind-tunnel tests using larger turret models. What was missing in the small-scale tests were data reflecting the correct Reynolds number, a complicated calculation to show the wind-tunnel model mirrored the same proportional forces as the ALL aircraft turret would face in flight.[19]

For 3 months prior to the July meeting at AFWL, GD had been busy designing and conducting preliminary testing on a variety of fairings in its own small wind-tunnel facility at Fort Worth. This work was part of a last-minute $25,000 add-on to the aircraft modification contract (F29601-C-71-0064) requiring GD to design, build, and test models of the APT and fairings as a first step toward confirming the flight safety of the aircraft before it underwent modification.[20]

On 10 May GD reported its progress to AFWL. The contractor stated it took into account five factors in its design of any proposed fairing: the APT's telescope viewing angle, aircraft/APT drag, buffet forces acting on the aircraft and APT, unsteady forces acting on the telescope internally, and the local turbulence through which the laser beam propagates. What this meant was that a fairing was only effective if (1) it could reduce drag and buffeting caused by external aerodynamic forces, (2) it could reduce turbulence inside the APT, and (3) it could offset the forces operating where the cavity turbulence converged with the external airflow to produce a potentially hazardous environment. In sum, without suitable fairings, the drag created by the turret would limit the plane's range, altitude, and speed. More importantly, severe buffeting would cause structural fatigue and damage to the fuselage and tail of the aircraft.[21]

The contractor considered 14 fairing configurations. These ranged from a simple aft fairing that offered no protection to the front of the turret to a full-forward/full aft-fixed fairing. The latter sandwiched the APT between the two canoelike fairings to minimize drag and buffeting. However, this configuration severely restricted the field of view of the APT. The fairing blocked the line of sight to the front and rear, allowing the APT to look only off to the left side of the aircraft. One of the most bizarre designs was a saucer-shaped

configuration surrounding the entire turret. Another was a snout-nosed barreled shroud covering just the front horizontal half of the turret. In the end, the more exotic fairings fell by the wayside and the contractor concentrated its efforts on the more simple designs of the full/partial front and aft combinations.[22]

After building 1/30th (0.035) and 1/45th (0.022) scale models of the APT and various fairings, GD tested them in its subsonic free-jet facility at Fort Worth. This was a small square chamber measuring only 9.75 by 9.75 inches. Testing began on 11 June and was completed on 25 June. As no scale model of the NKC-135 was used during this series of tests, the tiny models sat atop a curved metal plate instrumented to take measurements at various points on and around the APT and fairings when exposed to 0.5 and 0.7 Mach numbers. By changing the elevation and azimuth of the APT, GD collected a variety of data on turbulence and noise levels and flow velocities. Some tests collected data when no fairings were used. Comparing this information to data gathered from many fairing combinations, with different APT orientations, gave an indication of the effectiveness of the fairings. To keep Lamberson and the others back at Kirtland informed, Captain Otten served as the AFWL project officer who monitored the planning and performance of all of the tests. Most of the exchange of information took place between Otten and Major Shelton, who Lamberson was counting on to carry out the aircraft integration work.[23]

Kyrazis and Otten reported the preliminary GD test results showed the aerodynamic fairings significantly reduced the turbulence and noise level in the APT cavity. A follow-on series of tests supported these findings. This second set of tests took place from 21 to 23 July at Arnold Engineering Development Center's 16-foot transonic propulsion wind tunnel in Tallahoma, Tennessee. During those experiments, a 0.035 scale model of the KC-135 aircraft (built by Boeing) was introduced for the first time. Five different fairings affixed to the top of the model aircraft were tested: full forward, full aft, saucer, partial aft, and coanda wing aft. The turret, when combined with or without the full forward fairings and either the full or partial aft, proved to be the most aerodynamically acceptable configuration from aircraft stability and performance viewpoints. The poor performance of the saucer and coanda wing eliminated them from further consideration. A porous spoiler rake set in front of the APT opening was effective in reducing pressure fluctuations in the cavity.[24]

Although the contractor seemed pleased with the results from both test series (there was also some small-scale testing to develop future concepts of turret designs conducted at the Air Force Academy, the Air Force Flight Dynamics Laboratory, and Ohio State University wind tunnels), Kyrazis, Buell, and Otten expressed doubts about the accuracy of the data. Their main concern was the 0.035 models were too small and provided false data because incorrect Reynolds numbers were used. They argued the real ALL airplane in flight would experience a Reynolds number between 4 and 15 million, depending on its speed. Because of the small size of the 0.035 model, it could not simulate the correct Reynolds number. (Incorrect Reynolds numbers also caused changes in drag and separation points of the flow which was a key issue to be resolved.) In other words, the wind-tunnel data could not be scaled from the small models to match the in-flight Mach and Reynolds number the life-size aircraft would experience. Mainly through the urging of Buell, Kyrazis recommended to Lamberson that the contractor build and test bigger models to acquire more reliable data.[25]

Model of ALL with APT and forward/partial aft fairing in Arnold wind tunnel.

Not everyone felt it was necessary to build bigger models to validate the Reynolds number of the wind-tunnel data so it could be scaled confidently to the actual ALL aircraft Reynolds number. GD claimed their analysis showed the 0.035 data were satisfactory. But Kyrazis was not easily persuaded. He and Major Hal Shelton, the ALL aircraft integration manager back at Kirtland, continued to push for testing larger models. They insisted the safety of the aircraft and flight crew were the crucial issues at stake. Lamberson agreed. He was not about to take a chance at that point in the program to accept at face value the validity of the wind-tunnel data collected so far. If the data proved faulty and the plane crashed after installation of the real APT, then that would be the final curtain call for the ALL. Lamberson could not afford that risk. Consequently, he accepted the advice of the men he trusted and directed GD to build 0.30-size models of the turret and fairings.[26]

As the contractor proceeded with the fabrication of the larger-size turret and fairings, Otten was busy planning the next series of tests. Because they were approximately one-third the size of the real turret (70 inches high and 56 inches in diameter) and fairings, the new models could only be set up in large wind tunnels. Since 1970, Systems Command's policy was all testing would take place in its 16-foot transonic wind tunnel at Arnold. General James Ferguson, the AFSC commander, had put out a letter directing all of his units to maximize use of the Arnold facility. He needed customers to pay for use of the tunnel to keep the facility running from year to year. To use a non-AFSC facility would require a waiver approved by AFSC.[27]

AFWL intended to use the Arnold facility, but changing circumstances prevented this from happening. In spite of Ferguson's guidance, Arnold's attitude was not very receptive

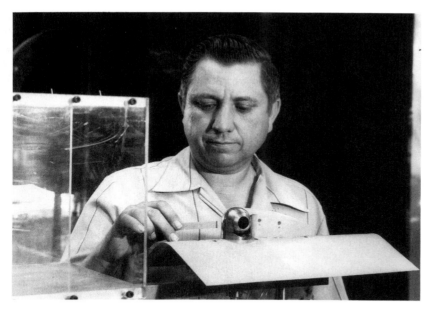

Model of basic turret and partial aft fairing.

to turning its tunnel over to AFWL for long periods of time. Otten and Kyrazis anticipated they would need 2–3 months to set up, install the right instrumentation, and conduct the testing and the rather unusual diagnostics on the 0.30 models. "We'd be using too much time in their [Arnold's] tunnel," Otten recalled, "and they weren't set up to do it—plus it was costing us a fortune to go there." Otten's boss, Lieutenant Colonel Ed O'Hair who ran the Laser Technology Branch, complained that Arnold wanted "millions of dollars" whereas Ames was willing to give us a "bargain." Looking further down the road, AFWL planned to conduct experiments on beam interaction with the turbulent boundary layer in the wind tunnel in preparation for Cycle II work. But Arnold did not have the capability—at least that seemed to be their position as a ploy to discourage the AFWL newcomers—to put a beam in their tunnel to perform optical diagnostics. Ames could accommodate a beam in their facility, plus by running experiments between midnight and 4:00 in the morning, AFWL would be able to draw on more electrical power.[28]

All of the reasons that made Arnold an unsuitable test facility, in AFWL's view, worked in favor of moving the model testing to Ames's 14-foot transonic wind tunnel. Hans Mark also peddled his influence to attract the ALL aerodynamic work to his facility. He laid out a proposition that AFWL could not afford to turn down. Arnold and Ames had different outlooks. Testing at Arnold was done to basically verify the flight characteristics of "preproduction" aircraft destined to enter the Air Force inventory. NASA-Ames, as Mark pointed out, was more "research" oriented. They were more interested in solving the scientific challenges of the ALL: defining the airflow on and around the turret, characterizing the interaction of the boundary layer with a laser beam, and eventually devising the best way to vent the laser exhaust gases without causing the aircraft to pitch and roll violently.[29]

The NASA-Ames director not only conveyed an extremely positive attitude toward the ALL program, but he was unwavering in his commitment to give AFWL as much time as they needed in the wind tunnel. Plus, there was no problem using a laser inside the Ames tunnel to conduct the required beam propagation studies. The other obvious advantage was that two of the experts, Kyrazis and Buell, resided at Ames (Otten also moved there in 1973) and could keep close watch over the testing program.[30]

Originally, the first test of the 0.30 models was scheduled to take place at Arnold in the fall of 1971. However, that testing never got under way. Lamberson felt that the advantages of Ames made it a more attractive test site over the long haul. Yet, moving to Ames was not automatic. Lamberson first had to convince General Ferguson that this move was in the best interest of the Air Force, especially as the general had made it quite clear his organizations would test at Arnold. While Arnold depended on the payment of its customers to support its wind-tunnel operation, NASA-Ames had its own annual budget to operate its tunnel and charged its customers only a nominal fee for actual test time in the tunnel.[31]

When Lamberson went in for a waiver, he knew he had a staunch supporter in Ferguson who a year earlier had been responsible for providing additional funding to accelerate the ALL program. Lamberson turned up his salesmanship burner a notch, explaining to the general all of the unparalleled advantages of testing out at NASA-Ames. Ferguson did not need a great deal of convincing as he looked on the ALL as one of his pet projects. He granted the waiver in September.[32]

Shortly after the waiver went through, Mark's enthusiasm and endorsement of the ALL escalated. He wanted to help make the program a success. One way to do that was to make sure the right people were put in the right jobs to keep the technology moving. Mark had met with Colonel Rowden, the AFWL commander, in October 1971 to discuss Kyrazis's future. Rowden sensed the importance of having a person of Kyrazis's caliber on the ALL team. The Laboratory commander asked Mark if he would be willing to release Kyrazis from his full-time duties as a research associate at Ames and allow him to devote his talents to support the upcoming AFWL tests on the 0.30 model tests. Mark had no problem with that arrangement. A few days later in a letter to Lieutenant Colonel O'Hair, Mark clarified Kyrazis's new role. No longer would he be a part-time consultant. Instead, Mark wrote that Kyrazis's participation in the ALL program would "take precedent over other work that he may be doing at Ames."[33]

Although Kyrazis emerged as one of the dominant players, he was still not officially assigned to AFWL. Lamberson continued to rely on Captain Otten to make sure the wind-tunnel tests came off as planned. And because of Otten's technical competence, Lamberson made it clear to Kyrazis that Otten would be in charge of the first series of 0.3-scale tests at Ames scheduled for January 1972. This was somewhat of an odd relationship because Kyrazis outranked Otten by two grades. Nevertheless, this was a pattern Lamberson chose to follow throughout the ALL program. His prime concern was to make the ALL work as soon as practical and, if that meant colonels working for captains, then that's the way the program would proceed. As it turned out, this was not a serious problem. Rank never interfered in the work relationship between Otten and Kyrazis; they realized, because of the uniqueness and urgency of the program, the mission superseded military protocol. Otten and Kyrazis formed a compatible team to get the job done and each developed a deep respect for the other's technical competency. Over time, they became the best of friends despite the differences in age and rank.[34]

With Kyrazis as the point of contact at Ames and Otten, with help from Captain Ronald H. Puent, working the AFWL end, several meetings during October and November laid the groundwork for the large-model testing scheduled for January 1972. AFWL tasked the Aero Propulsion Laboratory to put together an overall test plan. The plan covered test objectives, instrumentation and data acquisition, test run schedule, test article configuration, and readying the tunnel for testing.[35]

Although the Propulsion Lab had a big job developing and delivering a plan for AFWL's final approval, the work load for carrying out the series of experiments in the Ames tunnel was spread evenly among all of the test participants. GD was to build and maintain the 0.3-scale turret and fairing models. The turret and fairings consisted of high-temperature fiberglass shells attached to an aluminum support structure. Fabricating the inner portions of the cavity in the turret, and instrumenting that area to measure forces and torques on the optical hardware in the open-port configuration, were mainly accomplished through Otten's efforts. (Otten was awarded a patent for his torque measurements for the forces impinging on the optical surfaces.) The Air Force Flight Dynamics Lab designed an air injection system to improve the open-port performance. AFWL's main job was to assemble the optical instrumentation to measure the disturbance of a light beam passing through the airflow boundary layer next to the open port. Finally, Ames was to provide the tunnel and to recommend which fairing configurations would best reduce the pressure fluctuations inside and outside the turret.[36]

Model of ALL APT and fairing mounted on splitter plate in Ames wind tunnel.

In the last week of December, test personnel began installing the model and instrumentation in the 14-foot-wide wind tunnel. The turret and fairings were mounted on a 1-inch metal splitter plate (148 × 54 inches), which resembled a large sheet of plywood. Three-foot pipes extended from the side wall that held the splitter plate perpendicular to the floor of the wind tunnel. Suspended in the tunnel, the splitter represented the aircraft fuselage. The turret centerline was 60 inches above the tunnel floor. As the high-speed wind moved down the tunnel, the airflow passed over the splitter plate and fairings, smacked into the turret and open cavity, and traveled on over the downstream portion of the splitter. Data were collected from 64 high-pressure transducers and 109 steady-state pressure taps installed on the turret, fairings, and splitter plate.[37]

Time was creeping up as the critical element. Lamberson knew the ALL aircraft would be going to Fort Worth in the spring of 1972 to undergo major modifications so it would be ready to perform the Cy-

Closeup view of ALL model fairing and APT used in Ames wind tunnel.

cle I experiments. He was nervous because he didn't want any show-stoppers in the wind-tunnel testing. Without good data to tell him if the turret would be a safety hazard, he could not give the go-ahead to the contractor to start remodeling the ALL aircraft. Recognizing the critical importance of this phase of the program, Lamberson kept a keen eye on the activities out at Ames, mainly by the progress reports fed to him by Otten.[38]

The team at Ames came through in fine fashion to relieve Lamberson's apprehension. Testing started in early January and finished up by the middle of February. It collected data in three areas: turret aerodynamics (torques and pressure distributions), open port testing, and optical propagation of a light beam (a visible helium–neon laser with an output of only a few milliwatts) moving through the turbulent boundary layer in front of the turret. Approximately 60 test runs were made in the tunnel to measure pressures on six fairing configurations. The objective was to determine the trade-off between view angle and degree of flow unsteadiness. At the time, as no suitable window material was yet available, AFWL had to proceed under the assumption that for Cycles II and III the plane would have to fly open port.[39]

Test results were encouraging. The most important conclusions were (1) aircraft performance with the turret and fairing showed no serious effect on aircraft stability and control, (2) pressure fluctuations in the open cavity could be reduced by the porous fence so as not to damage the optical components, and (3) beam spreading as the beam moved through the airstream was within acceptable limits when the turret faced left in the 80- to 90-degree azimuth range (0 degrees was the nose of the plane). Using all three of the most effective

fairings—forward, aft insert, and partial aft—restricted the turret telescope's field of view to looking out the left side of the aircraft in an azimuth range of 60 to 120 degrees.[40]

What had most worried Kyrazis, Buell, and Otten was reproducing a representative Reynolds number when using the larger models. That issue disappeared after the Ames testing. As a relieved Kyrazis pointed out in one of his status reports, "The close matching of both Mach and Reynolds number gives greater confidence that the conclusions drawn from the wind-tunnel tests would be applicable to the flight tests." Follow-on testing would be required to refine the Reynolds number as a double-check against the initial data collected.[41]

Although AFWL was obviously pleased with the Reynolds number findings, the tests also confirmed that it would be impossible to fly the plane without fairings, because there would be a good chance that the wake produced by the bare turret would damage or even remove the vertical stabilizer. That had become abundantly clear to Otten and others after conducting one of the first wind-tunnel tests using no fairing with the APT model. The violent wake shed by the turret ripped apart a 1/2-inch steel rake located downstream. Otten recalled that the ferocity of the wake was so great that it caused an "emergency tunnel shutdown," a highly unusual occurrence for any wind-tunnel operation. Testing also showed that when the aft insert fairing was removed, the skin of the aircraft in that region of the fuselage would be subjected to high fluctuating pressure that could lead to fatigue failure of the fuselage. In addition, reliable torque values inside the turret cavity could not be determined. Additional testing would be required.[42]

In spite of these minor disappointments, the good news was that noise and pressures inside the open port could be sufficiently suppressed by placing a porous fence completely around the lip of the opening. Optical beam propagation measurements telling how much the beam spread out were positive. On the average, the beam spread 10 microradians, which was an appropriate value for the turbulent airstream environment. However, Kyrazis cautioned to use the data judiciously. The experiment was set up so the beam did not fill the exit aperture of the telescope. Because of this, he warned, the data showed only a localized (the area outside the turret) effect of the beam spread.[43]

The completion of the first large-scale series of wind-tunnel tests proved to be a major turning point in the ALL program. This initial set of data was by no means perfect, but it was good enough to confirm that a plane could fly safely with a turret, when combined with various fairing configurations designed to smooth out the airflow. A second comprehensive entry into the Ames tunnel for 7 weeks was scheduled for October 1972 to gather more data, especially in defining the forces inside the open port. But for now, Lamberson could act in good conscience and feel good about telling GD to go ahead modifying the ALL aircraft at their Fort Worth facility.[44]

DEVELOPMENT OF THE LOW-POWER WINDOW

At the same time wind-tunnel testing was under way, AFWL had been busy for several years trying to design, build, and install a window in the APT so it would be ready for Cycle II testing. The origins of the window research program went back to the summer of 1970, when the Lab's optics personnel recognized the need to come up with a highly transparent window material for the laser beam to pass through as it exited the APT. Initially, Lab scientists reasoned a window was required to protect the internal optics of the APT during flight of the ALL.[45]

To identify the most desirable window material, Colonel Lamberson set up a trilaboratory program. Members who served on this evaluation committee came from AFWL, the Air Force Materials Lab at Wright-Patterson AFB, and the Air Force Cambridge Research Laboratories (later renamed the Air Force Geophysics Laboratory) at Cambridge, Massachusetts. The group's charter was to investigate all potential window materials and then recommend the best material suited for the APT window. (AFWL provided the bulk of the funding for the Materials Lab and Cambridge to come up with a suitable material.) As part of this process, the key issues to be addressed were: strength of the material, availability of the material in large sizes, low optical absorption, and high optical uniformity of the entire window surface.[46]

Strength of the window was crucial because it had to be strong enough to withstand the severe pressures and temperature fluctuations encountered when the aircraft flew at different speeds and altitudes, as well as in changing weather. Ensuring the window material could be manufactured in large pieces so it could cover the 60-centimeter APT opening was another critical consideration. But the most important feature of the window was its high degree of IR transparency. This meant the window, ideally, had to allow the laser beam to pass through without absorbing any of the beam's energy or distorting the quality of the beam. In essence, because of the window's high degree of transparency, the beam would be tricked into thinking there was no barrier in its way.[47]

Although selecting the right material was the most difficult part of the research effort, fabrication and mounting techniques of the window frame to the APT were two other important considerations. To find solutions to all of these problems, AFWL in the spring of 1971 issued an amendment to Hughes's initial APT contract (F29601-71-C-0058). AFWL directed Hughes to propose a conceptual design of a low-power window capable of transmitting a laser beam with a maximum of 1 kilowatt of energy. This would be the low-power window used during Cycle II with the 150-watt electric discharge laser. Looking beyond Cycle II, the contractor was also tasked to begin evaluating the engineering technology that could be used in designing a "high-power" window for Cycle III testing with the multi-kilowatt GDL laser.[48]

While Hughes worked on designing the basic configuration of the window, the Materials Lab worked closely with Raytheon conducting studies and experiments on a number of possible window materials. At the top of the list were three promising candidates: potassium chloride (KCl), germanium, and an IR chalcogenide glass (TI-1173). In spite of the initial optimism, none of these materials exhibited all of the desired qualities for a suitable window. KCl proved unable to withstand either the pressure loads or the thermal stress predicted for the window. Poor optical quality of TI-1173 and germanium eliminated these two materials from contention.[49]

Raytheon, who primarily had concentrated on high-power window materials for Cycle III under contract F29601-71-C-0147 with the Materials Lab, had better luck with another

type of material. Initially, Raytheon had focused a large share of its work testing semiconductor materials. One of these was cadmium telluride, but it, too, failed to meet all of the desired window parameters. But one of the spin-offs of this work led to a series of experiments using a process known as chemical vapor deposition (CVD). Raytheon found that the chemical reaction between pure zinc in vapor form and hydrogen selenide gas (H_2Se) produced zinc selenide (ZnSe) in a solid form. Zinc heated in a furnace created zinc vapor that entered a chamber where it combined with the hydrogen selenide gas. The resultant mixture rose upward through a graphite mandrel (a chimney) where it cooled and deposited zinc selenide along the walls of the mandrel. Once the ZnSe reached the desired thickness, it was removed from the walls and cut into panes for the window. This process turned out to be the key to the vexing window problem, as the contractor was able to grow crystals of uniformly high-quality zinc selenide. It took about a year to perfect the CVD technique. During that time Raytheon had made substantial progress by growing increasingly larger pieces (from 70 square inches to 450 square inches) of ZnSe.[50]

Convinced that Raytheon could eventually produce large enough sections of ZnSe to fit Hughes's window design, AFWL in August 1972 decided to accept ZnSe as the material for the low-power window. A few months later, Raytheon received another contract (F29601-72-C-0119) to grow zinc selenide in large pieces exclusively to comply with the Hughes's design specifications of the low-power window. Work started in November 1972 and reached completion in September 1973. Some problems still existed with the CVD process during this time causing Raytheon to conduct numerous trial-and-error runs. If the temperature and flow rate of the Zn and H_2Se were not tightly controlled, then the resulting ZnSe would turn out to be the incorrect thickness and/or the surface would be cloudy because of impurities entering the system. Eventually, the contractor fine-tuned the CVD process to meet acceptance standards. With delivery of the zinc selenide plate (450 square inches) to Hughes at the end of the summer of 1973, that ended Raytheon's involvement in development of the low-power window. But it did not end the contractor's participation in the development of the high-power window. A follow-on contract (F29601-74-C-0069) let in the fall of 1973 funded Raytheon to fabricate even larger pieces (30 × 30-inch plates) of ZnSe to be cut to fit the three-pane window planned for Cycle III.[51]

At the same time Raytheon worked on growing large pieces of ZnSe for the low-power window, it was also providing small samples of that material to the University of Dayton Research Institute (UDRI). As part of a contract (F29601-72-C-0122) awarded in January

Summary of Characteristics of Candidate Window Material

	Size	Absorption	Yield Stress	Optical Suitability	Hygroscopic Behavior
Potassium chloride	6 Sectors	Lowest	Very low	Good	Yes
TI-1173	3 Sectors	Acceptable	Marginal	Poor	No
Germanium	1 Piece	Acceptable to 250 watts when cooled	Highest	Good	No
Zinc Selenide	6 Sectors	Low	High	Good	No

1972, UDRI designed and built a laser window test apparatus to conduct testing on the small ZnSe samples. Over the next couple of years, UDRI conducted a comprehensive program (thermal, pressure, and stress proof-of-concept tests) to develop an extensive data base to confirm that ZnSe was indeed the best window material. Colonel Kyrazis recalled that these experiments ended up "breaking literally hundreds of ZnSe samples" to determine the mechanical and optical properties of this unique window material. UDRI's procedure was to vary the test conditions to see how the ZnSe stood up to a wide range of changes in temperature and pressure. The results of this program provided an invaluable data base that proved especially useful for predicting the feasibility of using even larger segments of ZnSe for the high-power window.[52]

While Raytheon was fabricating zinc selenide in large sizes, Colonel Lamberson on 21 December 1972 signed a sole source justification purchase request to hire Hughes to fabricate and install the low-power window in the APT. Because Hughes had built the APT and completed the conceptual design of the low-power window, he believed it was only logical to select the Culver City contractor. In Lamberson's words, "Hughes . . . is the only source technically acceptable to the Air Force Weapons Laboratory."[53]

Although the decision had been made to hire Hughes, there was a 6-month delay before AFWL awarded the contract for building the low-power window. The reason for this was uncertainty among the subcontractors as to how workers would handle this new material, which was easily scratched and extremely toxic and brittle. Potential health hazards were a top priority, requiring the contractors to take extra safety precautions to protect their workers. This meant establishing strict procedures to ensure good ventilation and reliable air monitoring techniques to eliminate selenium compound dust particles from contacting the skin and eyes or being ingested into the lungs and stomachs of workers. Biological monitoring in the form of periodic urine samples from workers also had to be set up. All of these issues created problems for the contractors in establishing fair pricing estimates. By June 1973 most of the funding problems had been resolved. Hughes subcontracted Perkin-Elmer of Costa Mesa, California, to perform the grinding and polishing of the zinc selenide window. Optical Coating Laboratory, Incorporated of Santa Rosa, California, received the subcontract to apply antireflective coatings on the window surface.[54]

The low-power 24-inch-diameter window designed by Hughes was actually two windows—a Thermopane configuration where one window laid flat on top of a second window with a spacer in between to separate the two. Each window consisted of six pie-shaped segments (0.3 inch thick) held in place by a lightweight aluminum circular window frame. The frame consisted of a hub in the center of each window with six equally spaced radial struts connected to the circular outer frame. This Thermopane design left a gap of only 0.03 inch between the two windows. But that was enough room for the thermal control unit to inject nitrogen gas* to circulate in this area so as to maintain the proper temperature control and pressure—halfway between the temperature and pressure inside the turret and outside

---

*The nitrogen flowed radially inward from the periphery of the window into the area between the two panes.

the plane—for the optimum performance of the two windows. Temperature across the window could not vary by more than 2 degrees centigrade if optimum beam quality were to be maintained.[55]

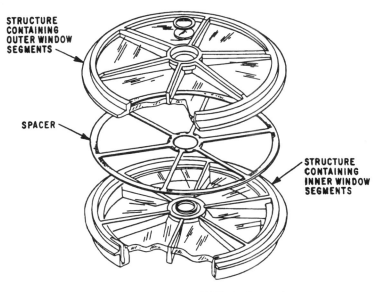

STRUCTURE CONTAINING OUTER WINDOW SEGMENTS

SPACER

STRUCTURE CONTAINING INNER WINDOW SEGMENTS

Exploded view of APT window.

Before the low-power window could be assembled, each pane had to be polished and coated for the zinc selenide to meet the contract specifications. Perkin-Elmer encountered problems in polishing the window material. Zinc selenide, because of its chemical composition, did not respond favorably to standard polishing techniques. There were problems mounting brittle pieces of ZnSe to a polishing block as well as cleaning the surface of ZnSe. Initial polishing produced scratches in the window surface. Also, because of the high diameter-to-thickness ratio of the ZnSe, the material tended to curl at the ends. A series of independent tests conducted from December 1973 to March 1974 at the Air Force Cambridge Research Laboratories warned of these potential shortcomings. Cambridge Lab scientists were especially concerned with the large amounts of impurities in the ZnSe samples which might have a "deleterious effect on the transmitted laser beam quality." Raytheon, in refining the CVD process, eventually solved the impurity problem. After experimenting with a number of new polishing compounds and procedures, Perkin-Elmer gradually was able to achieve the desired surface specifications. Work on depositing the antireflective coating on the window by Optical Coating Laboratory, Incorporated went much more smoothly.[56]

In May 1974 Optical Coating Laboratories, Incorporated began shipping individual panes to Hughes so assembly of the window in its frame could start. Three windows, consisting of six segments each, were assembled: two made up the low-power window and one spare was put together in case problems developed with one of the two other windows. Each segment measured 3.5 inches at its narrowest point and 12 inches at its outer radius. It was somewhat of a tedious job installing the panes in the six segments of the window support structure as workers had to be especially careful not to damage the zinc selenide. The work went along on schedule throughout the summer without any major incidents occurring. By the end of August, Hughes had installed all of the window segments and had completed preliminary acoustical, vibrational, and stress testing of the system as well as checkout of the operation of the thermal control unit. Delivery of the 217-pound low-power window by Hughes to AFWL took place in September. The zinc selenide of the window weighed 60 pounds. The frame was 135 pounds. Associated support equipment (e.g., thermal control unit, ducting) brought the total weight to 217 pounds.[57]

Once the window was at Kirtland, Hughes technicians worked closely with AFWL personnel mounting the window over the opening on the APT to make sure it fit properly. AFWL, in conjunction with the UDRI, subsequently conducted its own series of independent tests on the window to verify it had met all of the material characteristics stipulated in the contract. Many of these thermal and pressure experiments took place in UDRI's Environmental Test Chamber. Lamberson and others were especially satisfied with the results of this testing. The zinc selenide demonstrated low optical absorption and good antireflective characteristics as well as great strength. With an absorption and reflective quality of less than 0.5 percent, this meant that the zinc selenide allowed 99 percent of the laser beam energy to pass through the window. The power of the beam transmitted through the window was between 400 and 500 watts, depending on the type of experiment conducted. The window could also survive pressure loads up to 26 psi.[58]

Follow-on comparison tests for the window were scheduled for Cycle II. Plans called for studying the effects of the window on beam quality during ground-based, in-flight, and laboratory tests. Ground testing involved passing a low-power laser through the window mounted on the APT to a downrange target as well as evaluating the window when removed from the APT. For flight testing, the window would be installed to obtain beam measurements which were then to be compared to data collected when the APT flew open port. Finally, the window would spend a lot of time in AFWL's Laser Window Evaluation Laboratory to periodically check how much the window was degraded by operational use.[59]

In a little over 3 years, AFWL had made tremendous strides in developing the first airborne laser low-power window. What evolved from a fuzzy idea in the summer of 1970 was the discovery of zinc selenide, a new material that exhibited those unique physical characteristics that led to the fabrication of a one-of-a-kind window by the fall of 1974. It was this low-power window that enabled AFWL to collect vital beam quality data during Cycle II. These data provided the foundation for moving ahead with building an improved window that could handle a high-power laser beam during Cycle III. Raytheon was already under contract and hard at work trying to understand the effects of higher energy deposition on zinc selenide and refining the CVD process to grow larger sections of zinc selenide for fabrication of the Cycle III window.

7

Preparing for Cycle I

With the completion of the initial series of wind-tunnel tests at NASA-Ames predicting the ALL could fly safely with a turret, preparations for delivering the aircraft to General Dynamics (GD) began. A year earlier (March 1971), mainly because of the persistent efforts of Colonels Cunningham and Parsons, Lamberson had received confirmation that the 4950th Test Wing at Aeronautical Systems Division (Wright-Patterson AFB, Ohio) would release one of its NKC-135 research aircraft (tail number 053123) to the Air Force Special Weapons Center to serve as the ALL. Flying and maintaining the aircraft would be the responsibility of the Center in supporting AFWL.[1]

GENERAL DYNAMICS MODIFIES THE AIRCRAFT FOR CYCLE I

The Special Weapons Center made good on its promise as Lieutenant Colonel William G. Krause of the 4900th Test Group at Kirtland flew the NKC-135A to GD facilities at Fort Worth on 18 March 1972 to be overhauled and configured into an airborne laser testbed. As this was a Saturday, the contractor did not officially accept the plane until Monday. At that time GD went through its initial acceptance inspection procedures. This involved checking the engines, testing for fuel leaks, draining the fuel tanks, and washing the plane. Three days later, GD work crews had jacked up the aircraft in one of the contractor's hangars so work could start on modifying the plane in preparation for the Cycle I testing.[2]

Originally, the contractor anticipated the customized work on the ALL would be completed by October at a cost of $2.3 million. But because this was such a unique project, GD encountered a number of unexpected engineering problems, such as perfecting the environmental control system, the crew escape chute, and the covering for the radar protruding from the rear of the plane. Several major amendments to the original contract, to fix these problems, delayed the program. Consequently, GD did not deliver the modified ALL to AFWL until January 1973. By the

Airborne Laser Laboratory aircraft prior to modifications at General Dynamics plant in Fort Worth.

time the government officially closed out the contract on 30 April 1974, the final cost of the Cycle I modification had doubled to $4.6 million.[3]

Though the price of the contract climbed, this was not the result of ineptness by GD. The contractor had devised a logical plan, dividing the work on the ALL into two distinct parts. GD first conducted a systematic check of the entire aircraft known as Inspect and Repair as Necessary (IRAN). This was essentially a depot level examination of all parts of the plane to ensure the aircraft had not suffered any degradation that could affect its flightworthiness. Every few years the Air Force required each of its KC-135s to undergo this routine IRAN at its Oklahoma City Air Material Area (OCAMA) shops as a safety precaution. An exception was made to this policy authorizing GD to perform the IRAN on the ALL as it would save time and money by having the work done at Fort Worth rather than OCAMA. Earlier in the planning stages, AFWL had explored the possibility of using OCAMA to modify the aircraft. However, OCAMA lacked the specialized labor pool and precision manufacturing facilities needed to make the one-of-a-kind modifications to the ALL.[4]

The contractor went down a lengthy checklist as part of the IRAN process. Technicians carefully looked over the plane's exterior and interior for such defects as corrosion, panel fatigue, faulty wiring, improper operation of the flight deck's instrumentation panels, access doors, the hydraulics system, and much more. Two of the major deficiencies GD discovered on the ALL were that it had numerous fuel leaks, and it showed signs of corrosion and

excessive wear and tear on parts of the fuselage. Considering the age of the aircraft, this was not unusual. The contractor isolated and repaired the numerous engine and fuel tank leaks and solved the corrosion problem by replacing several of the fuselage panels and giving the plane a new coat of paint ($28,000), which also upgraded the aerodynamics of the aircraft. Repairs of this type improved the performance of the aircraft but also contributed to increasing the total dollar value of the original GD contract.[5]

Making major changes to the aircraft was the second and most time-consuming part of the contractor's tasking. AFWL specified in the contract that GD would alter the physical, electrical, and mechanical characteristics and capabilities of the aircraft to satisfy the Cycle I requirements. This meant building from scratch and installing a variety of "new" systems into the ALL. To coordinate and perform all of the modifications was an enormous undertaking. As an example, GD relied on 767 engineering drawings ranging from cutting the hole in the fuselage to accept the APT to installing three new observation windows on the left side of the aircraft. Teams of engineers and technicians worked side by side on the hangar floor and in GD's fabrication shops perfecting their own specialized piece of the ALL puzzle to be fitted into the overall modification mosaic.[6]

ALL undergoing Cycle I modifications in General Dynamics hangar in Fort Worth.

Making sure that GD accomplished all of the aircraft modifications on schedule and within its budget was a responsibility that fell squarely in Lamberson's lap. Lamberson had faith in the contractor's ability to refurbish the aircraft. But he was not content to rely on faith alone. To monitor GD's progress, he set up a system where his technical and program experts met regularly with contractor personnel at Fort Worth. The two most important GD contacts were T. Peyton Robinson, head of the ALL Engineering Project Office, and E. C. Johnson, who served as the ALL Manufacturing Coordinator. These two men supervised the day-to-day operation of the ALL modification and were the most qualified to answer questions in a moment's notice on any facet of the "hands-on" portion of the work in the hangar or specialty shops.[7]

The two AFWL people who most often traveled to Fort Worth to consult with GD representatives were Major Shelton and Captain Puent. Puent kept a watchful eye on funding and any deviations to the original contract. Shelton's focus was on quality control. He checked on the refurbishing of the plane and was at the contractor's plant two to three times a month. He sat in on many of GD's weekly coordinating meeting to gauge the contractor's progress on the fabrication and installation of equipment and to serve as Lamberson's spokesman providing guidance and direction on technical matters. Shelton was a scrappy watchdog who had an uncanny ability to smell trouble. And he didn't hesitate to tell GD when something was wrong and needed to be fixed. In more cases than not, Shelton and the contractor worked well together to resolve differences. His frequent visits to GD cemented that relationship by providing the continuity essential for establishing a strong communications link between the contractor and the Laboratory. Through Shelton and Puent, Lamberson was able to make sure GD was meeting its contract obligations.[8]

As work on the aircraft proceeded, more and more Laboratory personnel who had a vested interest in one special part or another of the modification eventually made their way to Fort Worth. Kyrazis and Otten showed up in September and October to inspect the recently completed fiberglass fairings and to see how they fit on the plane. Lieutenant Colonel Ben Johnson shuttled between Hughes's Culver City plant—where the construction of the APT was under way—and GD to verify the dimensions of the completed APT base plate would match the drilled holes in the fuselage so the APT could be bolted in place directly over the 5-foot opening cut in the top of the aircraft. Planning ahead for Cycle III, Lieutenant Colonel Ed O'Hair made several trips to confer with the contractor in preparing future modifications so the plane would be ready to carry the large high-power laser for Cycle III testing.[9]

Lamberson recognized his staff was extremely capable of solving any of the day-to-day problems that might arise at GD. He also wanted to impress on the contractor that the head of the ALL program was maintaining a close watch on GD's performance. To drive home this point, Lamberson made three trips to Fort Worth to lend his visibility to the 10-month modification project. His first visit occurred on 20 April, less than a month after the plane arrived at the contractor facility. GD gave him a review of the entire program, informing him that the fiberglass extension fairings to protect the aft radar had been fabricated, the hole in the fuselage for the APT had been cut, and construction of longerons and bulkheads to support the APT were moving ahead on schedule. The one major potential problem was the difficulty GD was having in obtaining drawings for the personnel escape system, but OCAMA was trying to help resolve this issue.[10]

Lamberson came away from the first meeting pleased by what he heard and saw. He returned on 26 June to find out the program was 2 weeks behind schedule and most likely would escalate in price because of the extra man-hours invested to meet the contract specifications. He didn't appear disturbed. Any government program of this magnitude and complexity would encounter inevitable delays and price increases.[11]

On his last visit in August, Lamberson brought along with him Colonel Rowden, the Lab commander, and Brigadier General D. C. Nunn, from the Air Force Contract Management Center. Nunn supported AFWL by administering the GD contract. Reassuring them that their monies were supporting a productive program, Lamberson proudly showed off the airplane to let them see firsthand the progress that had been made so far on the ALL. An entourage of 12 other AFWL people followed Lamberson as he toured the ALL, pointing out along the way the dummy APT, the new access door, the domed-shaped fairings, and

many other improvements. All came away highly impressed with the quality of the contractor's work. No doubt Lamberson smiled as he returned to Kirtland telling Rowden they had made the right decision and were moving closer to the Cycle I testing.[12]

The reason Lamberson and other AFWL visitors felt confident was the steady progress shown by the contractor. GD had made several substantial changes to the aircraft without any major setbacks. Measuring and cutting the 5-foot hole in the fuselage for the APT was a relatively easy task. Once all of the Cycle I modifications were completed, AFWL didn't want to risk putting the real APT on the plane until it had conducted initial flight testing of the ALL. So AFWL instructed GD to build a full-scale mock-up of the APT that simulated the exterior shape and whose weight and center of mass were identical to the real APT. In essence, the mock-up or "dummy" APT was the turret without any of the expensive internal optics. The first flight test of the mock-up would either prove or disprove that the wind-tunnel findings had been correct, that is, the plane could fly safely with a turret.[13]

GD fabricated a metal fuselage pressure "plug" to cover the APT opening when the ALL was flown without the mock-up or the real APT. On 17 July GD also received from Hughes a plate with holes drilled around the edges. GD manufactured a duplicate of the plate and sent the original back to Culver City. The plate or template fit over the APT opening, allowing the contractor to align the holes in the plate in the exact position where mounting holes for the APT would be drilled in the fuselage. Once this was done, either the dummy or real APT could be bolted to the plane. When the plane was not using the APT, it was removed. The contoured fuselage pressure plug was then inserted and bolted down to cover the APT opening during normal flight operations.[14]

After drilling the mounting holes in the fuselage, GD wanted to make sure the real APT would fit. In December 1972, Shelton arranged for an Air Force C-130 to pick up the APT from Hughes and fly it to Fort Worth. Captain Don Pulley, who was AFWL's point of contact monitoring Hughes work on the APT at Culver City, was to coordinate the shipment of the APT. After waiting 2 hours for the Hughes forklift driver to arrive, Pulley became impatient and jumped on the forklift. A surprised Shelton standing nearby asked the creative captain if he had a current forklift license. Pulley smiled and replied, "Yes, a Kirtland forklift license!" Shelton, right on cue, barked, "That's good enough, let's load it!"[15]

This situation could have ended in disaster. But it didn't. Pulley skillfully maneuvered the multimillion-dollar APT into the waiting C-130. However, this incident did not go unnoticed. The forklift drivers' union at Hughes filed a grievance against the enterprising captain, but as time passed nothing came of the incident. No harm was done, except for a few tempers flaring.[16]

The APT arrived safely at Fort Worth on 10 December. Three days later, with the help of Hughes technicians, GD was installing the APT on the plane. The APT aligned perfectly with the mounting holes. For the next few weeks, GD and Hughes conducted a series of electrical compatibility tests. Every cable and connection was thoroughly checked to make sure the APT interfaced with the control console in the rear of the ALL. Verification of the APT fit and its electrical compatibility was an important step that confirmed to AFWL that the plane and APT were ready for Cycle I testing. On 7 January 1973, AFWL returned the real APT to Hughes.[17]

Hole cut in top of ALL fuselage to accept APT.

Lowering "dummy" APT components into fuselage opening.

Lowering cap on dummy APT fitted into fuselage opening.

Completed installation of dummy APT.

To protect the APT and mock-up in flight, GD built the forward and aft aerodynamic fairings modeled after the ones used in the wind-tunnel testing at NASA-Ames. Four fairings were molded. Directly in front of the APT was the forward fairing. Directly behind the APT was the partial aft (sometimes referred to as the center fairing) which consisted of two pieces. A section of the left portion of the partial aft could be removed to give the APT a wider field

119

of view looking to the rear of the plane. With both sections of the partial aft in place, more protection was given to the APT, but its field of view was restricted to only looking off the left side of the aircraft. Behind the partial aft was the full aft fairing. When the partial and full aft fairings butted up against one another, they formed a smooth ramp that reduced the turbulent airflow moving from the front to the rear of the plane. By the second week of October, all of the fiberglass fairings had been fitted on the ALL. They were then removed, painted, and shipped to Edwards AFB where they awaited the arrival of the modified ALL in January 1973 to begin flight testing.[18]

Almost as soon as the fairings arrived at Edwards, they were sent back to Fort Worth. Relatively minor engineering design changes required the fairings be fitted to a new doubler

Molding of fairing for ALL—fairing fit behind the turret.

and the APT. Once that was accomplished, AFWL directed GD to ship the fairings and dummy APT by truck to Kirtland during the last week of January 1973. There AFWL engineers were able to observe firsthand that the fairings and dummy APT fit together exactly according to specifications.[19]

To remotely operate the APT from the back of the airplane, GD built four equipment racks and an APT control console. The racks were shipped to Hughes's Culver City facility—where the APT was being assembled—on 7 April by commercial air; military air delivered the console to Hughes on 19 May 1972.

Hughes inspected the racks and console to ensure proper electrical interfacing with the APT, installed the internal electronics, and shipped the racks and console back to GD by the middle of September. A month later, GD had installed all of the racks and APT console in the ALL, plus the consoles for the test director, device operator, and radar.[20]

Before the APT could lock on to any target, it had to receive and process data from the ALL's acquisition radar. To detect missiles coming at the ALL from the front, rear, and left side of the aircraft, AFWL selected two radars to be installed in the plane—one in the nose and the other in a tail extension that looked like a big snout protruding from the rear of the plane. Working together, the radars in the nose and tail for acquiring targets provided 180 percent coverage off the left side of the plane.[21]

Originally, the ALL came equipped with an AN/APN-59 weather radar housed in its nose section. GD removed this forward radar and replaced it with a fire control radar

(AN/APQ-109 radar used on F-4 fighters) to track targets. To make enough room for the radar's antennas, gyros, electronics, cooling system, hydraulic lines, and wiring harnesses to fit in the nose of the plane required fabricating a new radome. GD subcontracted this job to Brunswick-Balke-Collander who had the radome built and fitted to the plane by the middle of July. GD fabricated and installed the tail extension to house the aft radar. For the aft radar extension to fit, the contractor had to cut 2 inches off the inboard elevators. After initial fit checks, both the forward and aft radomes were shipped to San Diego to undergo electrical interface testing to determine transmission efficiency, reflected power, boresight shift, and boresight shift rate. Once this testing was completed, the radomes were returned to Fort Worth for final installation on the ALL.[22]

As a backup to the two primary radars, AFWL opted to add a simple visual target acquisition system. This was the optical acquisition device (OAD) which was nothing more than a high-power deer rifle scope. Its gimbaled mount allowed the operator to look through and swivel the scope in azimuth and elevation to scan the sky for targets. The OAD operator peered through an 18- by 18-inch viewing window (glass for the window did not arrive from the vendor until 28 December) the contractor installed on the left rear of the aircraft. If the operator spotted a target, he would visually track and "hand over" its exact location electronically to the tracker in the APT. The APT

Forward radar in nose of ALL—APT and aft fairing on top of fuselage.

would almost instantaneously digest this information and slew left or right to look in the direction of the target. Once the target was in view, the IR tracker would lock on and automatically track the target.[23]

To furnish enough electrical power to run the APT, the two radars, the air conditioning, and all of the equipment racks and consoles for system controls in the back of the plane required another substantial power source. To provide the additional power, the contractor reworked the number 4 engine to accept a 70/90-kVA generator and all of the associated wiring that went with it. This more powerful generator, larger than the standard KC-135 generator, was the same type the Air Force used on its B-52 strategic bombers. To make room for the larger generator, a modification of the cowling was required which resulted in a protrusion of the basic cowl shape. By October, the extra generator was installed and operating on the ALL's number 4 engine.[24]

Most of the ALL modification work at Fort Worth went according to plan. However, two problems plagued the contractor and delayed the delivery of the ALL to AFWL beyond the 23 October scheduled date. One had to do with installing an escape chute for the test crew in the rear of the aircraft in case of an in-flight disaster. The other problem involved getting the air conditioning to work.[25]

There were two ways for the aft test crew to exit the aircraft. One way was through a personnel door the contractor fabricated and installed on the left rear of the plane. The other way out was through an escape hatch in the floor. In April, GD reported to AFWL it was having "some difficulty" obtaining the engineering drawings of the escape system from the Air Force. The initial drawings GD received were incomplete and gave two different escape system configurations. Although GD went ahead with cutting the hole in the floor, by mid-June it still lacked five drawings needed for installation of the escape system configuration AFWL selected. Also, another holdup was that GD had to get approval from AFWL to use a substitute material for fabricating the "detail parts" of the escape system.[26]

After evaluating and deciding the contractor's substitute material met all stress specifications, AFWL gave GD authority to proceed with fabrication and assembly of the escape components. However, a critical shortage of its on-hand stock of material for the escape system slowed GD's progress. By October, the contractor had to impose a "two-shift extended work week" to complete the fabrication and installation of the escape system in the ALL by the end of November.[27]

At the same time the contractor was trying to resolve the aft escape system issue, a second major problem arose over the aircraft's proposed air-conditioning system. GD purchased and modified a vapor-cycle air-conditioning system to remove excess heat from all three compartments of the ALL. Three air-conditioning packages consisting of a cooling coil, fan, and an electrically operated refrigerant pressure regulator distributed cool air throughout the plane. Existing ducts had to be torn out and replaced with a new system to handle the increased heat load generated by the equipment racks and consoles in the aft crew compartment. The new duct system ran the length of the right side of the aircraft and fed cool air into the flight deck and the rear section of the plane. A separate air distribution arrangement cooled the center section, which eventually would house the laser device. A fourth air-conditioning unit located near the front of the plane cooled the forward radar.[28]

Removal of the old air-conditioning system, redesign and installation of the new system, and the difficulty of acquiring parts for the new system all added up to a time-consuming operation. Although the installation of Freon lines and air ducts moved along fairly well, the contractor experienced recurring frustrations trying to get the system to operate. Inferior parts and long response times from outside vendors hindered GD's progress. As an example, GD notified AFWL that a shortage of seven pressure switches and a pressure relief valve was "delaying operation of the air-conditioning system." Even when the pressure switches finally arrived, they were of such poor quality that they had to be returned to the vendor for rework. And although GD had received the repaired switches by the beginning of November, other critical components, such as the air-conditioning hinged doors, did not arrive until the second week of December. The contractor eventually resolved the nagging problems of the air-conditioning system and aft escape system. Getting the air-conditioning unit to operate correctly required constant modifications over the next year and a half. It wasn't until the summer of 1974, after numerous flight tests, that AFWL officially accepted the cooling system from the contractor.[29]

GD delivered the ALL to AFWL in the last week of January 1973. Although the delivery date had slipped by several months, GD had done a superb job in this first-of-a-kind modification. It had succeeded in installing the APT mock-up, aerodynamic fairings, numerous equipment racks and consoles, two radar systems, and many other changes in preparation for Cycle I testing. But before Cycle I could begin, the modified ALL first had to pass a series of stringent flight tests to verify the plane was airworthy.[30]

FLIGHT TESTING FOR CYCLE I

When GD finished modifying the ALL, it had turned out an airplane that was radically different from the one that had arrived at Fort Worth in March 1972. The problem was no one was certain that this one-of-a-kind flying machine would be airworthy. To prove the ALL's performance met acceptable operating standards, the plane had to undergo a series of flight tests. Testing took place in two phases. Even though the plane technically belonged to GD while in Fort Worth, on completion of Cycle I modifications the ALL had to be flown by its assigned AFSWC crew to conduct the functional flight check evaluations. Once the ALL passed this initial series of tests to verify its flightworthiness in terms of aerodynamic and structural integrity, AFSWC would then formally accept the plane back from the contractor. Later, the Air Force Flight Test Center at Edwards AFB, 70 miles northeast of Los Angeles, carried out a second and more comprehensive set of flight tests beginning in February 1973.[31]

The Air Force's main concern was to determine if cutting the opening for the APT, adding the tail radar extension, equipment racks, consoles, air-conditioning inlet ducts on the lower sides of the forward fuselage, aft escape hatch, and all of the other changes, adversely affected the flight performance of the aircraft. Originally scheduled for late October, the start of functional flight tests at GD was postponed by bad weather until the first week of November. By 2 November, most of the major modifications had sufficiently progressed on the ALL to allow the first flight test to start. During that test and the subsequent one, the ALL took off and started to climb, but unexpectedly its main landing gear failed to retract, forcing the plane to land. Maintenance crews determined the cause of the problem as malfunctions in the electrical and hydraulic system, which should have been discovered during the IRAN check of the plane back in April. Although the landing gear retraction failure was serious, mechanics fixed the problem within a few days so the plane was ready for its third test flight.[32]

The situation did not improve much for the third test. Although the ALL was able to retract its landing gear and reach its cruising altitude, one of the environmental air-conditioning doors (one on each side of the forward fuselage) snapped in two, damaging the aircraft at the wing root. This was identified as a high-frequency fatigue problem that was later solved by installing doublers to add strength. The Air Force flight crew on board also noticed an undesirable in-flight "rumble," which they attributed to the broken door. At speeds

ALL at Edwards AFB prior to flight testing. Note tail radar extension.

greater than 294 knots, there was a sudden onset of high vibrations. When the ALL reduced its speed, the vibrations diminished significantly. After new environmental control doors were installed and rigidly secured, the rumble noise disappeared during the fourth flight. That eliminated one problem, but a new one surfaced. The flight crew detected a constant low-frequency vibration described as "mild pulsations."[33]

GD did not at first want to acknowledge these vibrations even existed. AFWL believed differently, based on complaints from the pilot. As a result, AFWL "insisted" that GD go back and investigate the vibration problem. After a second look, GD finally agreed that there were abnormal vibrations which AFWL and GD personnel traced to the rear of the aircraft. They pinpointed the tail fairing extension (added during the modification) housing the aft radar as the source of the problem. However, GD and AFWL disagreed over the magnitude of the in-flight vibrations and what should be done to fix the problem.[34]

Kyrazis placed strain gauges on the extension neck (the narrow metal piece connecting the rear of the fuselage and the protruding radar extension) to determine how much it moved on the ground when subjected to shock. When a technician struck the extension neck with his fist, the aft radar structure moved side to side with enough force to create potentially a dangerous situation. However, after studying the problem and making his own calculations, Kyrazis concluded there would be less shaking of the radar extension in flight than on the ground when the radar extension was exposed to the force of a man's fist. Kyrazis sketched his theory on the blackboard before GD's 24-man aerodynamics department, but no one was convinced his numbers were correct.[35]

To test his theory, Kyrazis and the AFWL team installed two accelerometers and four strain gauges to the tail which would record measurements (vibrational frequencies) of the tail radar extension. A chase plane flying behind the ALL provided photos showing the extent of lateral movement of the tail radar extension. The photo chase film and recorded data from the instrumentation confirmed that the tail section moved less than an inch during flight. Measurements of how much the tail radar moved proved Kyrazis's theory was correct. The vibrational frequency was lower in the air than on the ground. However, because of airflow separation, different pressures were exerted on different sections of the radar extension, causing it to move or vibrate sideways. Even though these vibrations were not as great as originally anticipated, there was still the chance (depending on flight conditions) they could upset the normal control and stability of the aircraft. As a safety precaution, Kyrazis advised installing an aerodynamic fence to change the aerodynamic flow as a solution. GD concurred and installed a 6-inch aerodynamic aluminum fence with straps (for strength) around the narrowest section of the tail radar section fairing to smooth out the airflow and reduce the low-frequency vibrations induced on the radar tail extension. The new fence was tested during the seventh and final flight test at GD on 19 January 1973 with good results.[36]

Tail radar extension on ALL. Note aerodynamic fence—notched-like configuration—surrounding center section of tail radar extension.

Between 16 and 20 January, GD completed all of its functional flight tests. By this time, the contractor concluded the ALL was safe to fly and turned the plane over to AFWL on 23 January. AFWL accepted the aircraft knowing that GD's preliminary flight test results had certain limitations. After GD released the ALL to AFWL, the plane was flown from Fort Worth to Kirtland. There AFWL personnel, in preparation for the upcoming flight testing, performed preliminary ground checks to make sure the dummy APT and the various fairing configurations fit together properly on the plane. Once this was completed, the fairings were disassembled and, along with the dummy APT, loaded on the ALL for its trip to Edwards AFB, home of the Air Force Test Pilot School, the Rocket Propulsion Laboratory (renamed the Astronautics Laboratory in March 1987), as well as the Flight Test Center. The Center's main function was to certify that all planes owned by the Air Force were safe to fly within their design parameters. The ALL arrived at Edwards on 14 February and the Center began its preflight inspection of the aircraft and checked the fit of the dummy APT and fairings.[37]

At Fort Worth the plane had been flown with the aluminum covering (pressure plug) contoured to fit over the APT opening in the top of the fuselage for all tests. Neither the dummy APT nor the fairings, two of the most critical design changes to the ALL, were flight tested by GD. That changed at Edwards. The major job of the Flight Test Center was to evaluate the effects of the APT and fairings (four configurations) on the flight performance of the ALL. Major George E. Luck, project manager and test pilot at the Center, stated the ALL had to pass a rigid set of performance standards before it would be certified to fly its Cycle I mission. Specifically, the ALL had to demonstrate it could fly safely with the APT, four different fairing configurations, and the aft radar fairing extending out the back end of the plane. The basic question was, with all of these hardware alterations, could the pilot safely control the plane when flying it at the speed and altitudes of its designed flight envelope?[38]

Nine test flights, beginning on 20 February and ending on 8 March 1973, provided the answer to that question. AFWL personnel were an integral part of the flight crew for these tests. Kyrazis and Otten took real-time measurements of the in-flight aerodynamic forces

operating on the turret, fairings, and tail as a safety precaution. (The turret, fairings, tail, and top of the fuselage were instrumented so Kyrazis and Otten could on the spot collect and analyze in-flight data.) Based on their measurements, they advised the pilot when to increase his speed in 20-knot intervals. At each interval, they rechecked to make sure the aerodynamic forces on the turret and fairings did not have a detrimental effect on the aircraft. Once satisfied, they directed Major Luck to speed up by another 20 knots. A second important reason for Kyrazis and Otten serving as part of the flight team was to confirm that their earlier wind-tunnel data matched with real-world in-flight measurements. Their analysis showed a very close match between the two sets of data. Based on the in-flight performance and input from Kyrazis and Otten, Major Luck and his team of project engineers certified the modified ALL was airworthy. But they also placed some restrictions on how the plane would be flown. These restrictions were determined by comparing the flight test data with the airspeed and Mach number limits described in the current KC-135A Flight Manual.[39]

The Flight Test Center crew flew the ALL at various speeds and altitudes conducting upset maneuvers, Dutch rolls, and stalls to see how the plane would react. This was done in combination with each of five APT/fairing configurations, which fit together like a jigsaw puzzle. Pieces could be removed or added, depending on the fairing profile desired. Configuration 1 flew without the APT and fairing. The APT with full forward and full aft fairing was the second configuration. A full forward and partial aft (a pie-slice piece was removed from the left portion of the aft fairing's centerline closest to the APT) coupled with the APT made up the third combination. Only the full aft fairing was used with the APT for configuration 4. Finally, for number 5, the pie-slice piece or insert was removed again, leaving only the partial aft fairing with the APT.[40]

Each fairing combination caused the ALL to respond differently. Configurations 4 and 5, with the insert removed, experienced higher buffeting levels than 2 and 3. Airflow tended to get trapped in the opening formed by the APT and the fairing (when the insert was removed), resulting in substantial pounding on the APT, fairing, and the airframe. Tufting, or pieces of nylon cord taped to the fairing and APT surfaces, provided a visualization of the airflow, which was recorded by the photo/safety chase plane. Because of the early onset and rapid buildup of high speed buffeting with the APT and fairings on board, the ALL's airspeed was restricted to 325 knots with configurations 4 and 5. For all other configurations, the ALL's normal cruising speed was not to exceed 350 knots.[41]

The test crew also found the cruising range of the ALL aircraft was reduced by 13 to 26 percent, depending on the APT/fairing combination. Configuration 2 sustained the least degradation, and configuration 5 the most. When cruising and during stalls, the pilot had to be aware that the ALL exhibited a tendency to yaw right and roll right. This warning and other idiosyncrasies, such as the stipulation that pilot transition (when a new pilot was assigned to the ALL) would take place only when the APT and fairings were removed, were incorporated in a flight manual written exclusively for the ALL.[42]

Besides determining the in-flight handling characteristics of the ALL, AFWL wanted to measure the pressure fluctuations of the aerodynamic wake caused by the APT and fairings as the airstream moved toward the vertical stabilizer. Small transducers placed at various points along the top of the fuselage, vertical stabilizer, and on the APT and fairings collected

APT with full forward and aft fairing.

APT with full forward/partial aft fairing.

APT with full aft fairing only.

APT with partial aft fairing only.

pressure readings. This information would be used to estimate the amount of fatigue, especially on the tail section, the aircraft would suffer. Otten and Kyrazis, as mentioned earlier, wanted these data to verify the flight instrumented readings were consistent (matched) with the wind-tunnel test data. Aerodynamic data collected during flight testing confirmed Otten's and Kyrazis's earlier wind-tunnel results. They also were looking ahead to Cycles II and III when a laser beam would be projected out the APT. Pressure measurements derived from flight testing would be invaluable in defining the turbulent environment around the APT. Simulating these same conditions in upcoming wind-tunnel tests would help them predict how the beam would propagate through this blustery environment. The main question was, would the high pressure variations distort and degrade the beam as it exited the APT? Otten's work at NASA-Ames over the next few years would provide the answer to this fundamental question.[43]

One other part of the flight testing involved collecting data to determine the sources and magnitudes of interior noise levels in the ALL. A team from the Vehicle Dynamics Division at Wright-Patterson AFB conducted the acoustics tests. Fourteen microphones positioned throughout the aircraft crew, device, and operator compartments of the ALL picked up and recorded noise while the plane flew with various turret fairing configurations at different altitudes and airspeeds. Results revealed that, on the average, noise levels inside the ALL were "significantly higher (13 decibels to 26 decibels) than those normally expected in the KC-135 type aircraft." Major modifications to the aircraft accounted for the higher noise levels. The large environmental control system with three recirculating fans, the tail protrusion enclosing a radar system, and the addition of the turret and fairings were the three main contributors to increased sound pressure levels. Kyrazis disputed the accuracy of the Vehicle Dynamics findings. He claimed their experiments were faulty and borrowed a portable sound level meter to collect his own data to show the noise level was not that much higher than normal. In a few cases, the interior noise exceeded recommended human tolerance standards by as much as 20 decibels. However, even though the noise level was higher than normal (as determined by both the Vehicle Dynamics Division and Kyrazis), the final test report made no mention of any potential medical damage to an individual's hearing. The ALL was obviously noisy inside, but this was not grounds for preventing people from conducting on-board experiments.[44]

By the end of the last test, the Flight Test Center team had amassed 50 hours in the air putting the ALL through numerous flight scenarios. The result of all this was the modified NKC-135A was given a clean bill of health. Although the plane suffered some performance degradation, it was not serious enough to prevent the ALL from moving ahead with Cycle I testing. Because it was such a unique aircraft, the final Test Center report advised AFWL to conduct frequent inspections of the upper fuselage and tail section for wear and tear that might occur because of the buffeting created by the APT and fairings.[45]

Once over the flight test hurdle, the ALL returned to Kirtland on 8 March. Lamberson and his staff recognized the importance of this unpublicized event. In the span of only a few short years, the laser team at AFWL had moved the concept of an airborne laser platform off the drawing board to an operational aircraft parked at the side of the Albuquerque International Airport runway. Travelers departing on commercial airliners must have wondered what the strange-looking, humpbacked aircraft with its conspicuous bubble was doing in the desert of New Mexico. But there was no bewilderment in the minds of AFWL scientists who were feverishly at work preparing the ALL for its first mission.[46]

8

Tracking Aerial Targets
Cycle I: May–November 1973

While the ALL was finishing up its flight certification testing at Edwards, a team of optical experts headed by Major Ed Laughlin and Lieutenant Colonel Ben Johnson departed AFWL to spend time at Hughes's Culver City plant to inspect the APT. Even though Johnson outranked Laughlin, the major was in charge.

Lamberson, consistent with his game plan to place the best-qualified man in the job, had appointed and entrusted Laughlin to run the Optics Technology Branch under the Laser Division. Laughlin was a proven asset as demonstrated by his earlier work on the FTT for the TSL program. Lamberson was not about to disrupt a well-oiled machine just because a new officer needed a position commensurate with his rank. Instead, Johnson would first have to prove himself by demonstrating his technical competence to Lamberson before being considered for a more critical job. As time wore on, Johnson passed Lamberson's test with flying colors and became one of Lamberson's most highly regarded scientists and contributors. A team player all the way, Johnson abided by Lamberson's unorthodox rules and pitched in to assist Laughlin as much as he could. Their main job was to check on the final progress of the Hughes-built APT and recommend to Lamberson whether or not the government should accept this second-generation pointer and tracker, which was the first flightworthy APT.[1]

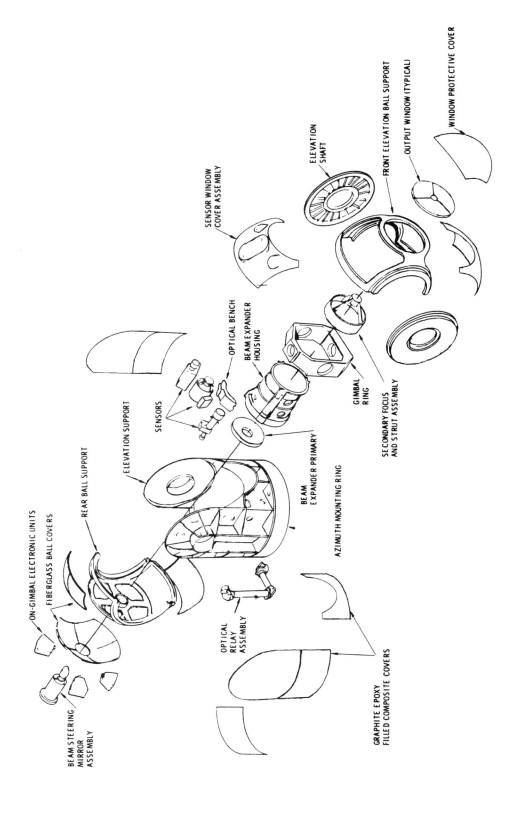

ON-GIMBAL ELECTRONIC UNITS

FIBERGLASS BALL COVERS

REAR BALL SUPPORT

ELEVATION SUPPORT

SENSORS

OPTICAL BENCH

BEAM EXPANDER HOUSING

SENSOR WINDOW COVER ASSEMBLY

ELEVATION SHAFT

FRONT ELEVATION BALL SUPPORT

OUTPUT WINDOW (TYPICAL)

WINDOW PROTECTIVE COVER

BEAM STEERING MIRROR ASSEMBLY

OPTICAL RELAY ASSEMBLY

BEAM EXPANDER PRIMARY

AZIMUTH MOUNTING RING

GIMBAL RING

SECONDARY FOCUS AND STRUT ASSEMBLY

GRAPHITE EPOXY FILLED COMPOSITE COVERS

Exploded view showing major components of APT pointing assembly.

The APT was essentially a next generation version of the FTT. Both devices were similar in principle, but the APT was a more sophisticated optical device that was lightweighted, had more resolution to provide a clearer image of the target, and had more precise pointing even though operating on an airborne platform. During January and February, Hughes completed the system integration of the APT, verifying all of the electrical components plugged into the right connections and all of the mechanical parts operated according to contract specifications. By the time Laughlin and Johnson arrived, Hughes was ready to conduct the preliminary acceptance tests of the APT. Accompanying Laughlin and Johnson were two of AFWL's most technically proficient captains, Don Pulley and Don Pearson. This small band of AFWL blue-suiters watched intently, taking mental notes as the contractor went through a step-by-step procedure demonstrating the basic operation of the new optical device.[2]

Hughes carried out a series of operational tests on the pointing assembly of the APT consisting of several major subsystems—TV, laser range finder, the IR tracker imager, gimbals, and secondary and primary mirrors—to demonstrate the operating proficiency of the APT. For the tracking demonstration, the contractor ushered the AFWL officers into a hangar near the Hughes runway, where Hughes had set up a Barnes collimator. This was essentially a box with an IR heat source simulating a target. The optical system inside the collimator made it appear to the tracker mounted on the APT that the target was a long distance away. Although the APT tracker sensed the target at real-world ranges, in fact, the target was only a few feet away from the APT. The APT also looked outside the hangar to track airplanes taking off and landing, as well as trucks, cars, and anything else that moved in the vicinity. Jitter, tracker noise, and friction created by the movement of the inner gimbals for controlling the elevation and azimuth alignment of the APT telescope, all measured higher than predicted. Based on these results, Laughlin pointed out in his subsequent status report, "the anticipated performance of the APT is a factor of two worse than the design goal."[3]

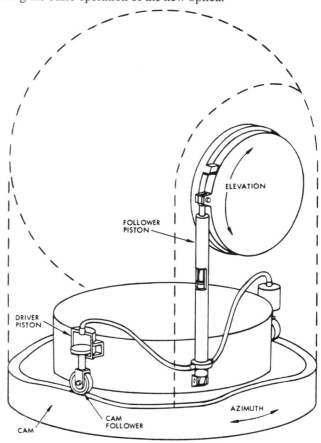

Schematic of airborne pointing and tracking system (APT).

The origins of these problems were complex. Because the APT was a four-gimbal system, the tracker electronically commanded the outer gimbal to rotate the turret to point in the general direction of the target and to make coarse azimuth and elevation adjustments.

Inside the turret were the inner gimbals whose job was to fine-tune azimuth and elevation angles so the tracker line of sight was precisely aligned with the target. One of the technical culprits in the system was the excessive friction caused by the hydraulic actuators of the inner gimbals, which rotated the telescope inside the APT turret. This friction, along with noise and a shaky stabilization subsystem, introduced jitter which prevented the tracker from remaining exactly on the track point on the target. Part of this problem the contractor attributed to the less than ideal laboratory setup for testing. Hughes did concede that the hydraulic actuators would have to be redesigned, plus improvements to the stabilization subsystem were needed. However, the contractor considered these deficiencies could be readily fixed and were not show-stoppers for the ALL program. The tracker did not meet the optimum performance goals of the contract, but data produced from these bench tests were within acceptable limits. With the APT being the first-of-a-kind device, no one expected the APT to work perfectly.[4]

Satisfied with the technical checkout of the APT, the AFWL reps felt confident the APT was now ready to be tested in an airborne environment. AFWL accepted the APT from Hughes during the first week of March 1973. This was no surprise or a sudden decision. During the fabrication of the APT, Laughlin recalled, "Pearson and Pulley virtually lived out at Hughes to keep up on the progress of the pointer and tracker." They weren't the only ones keeping a close eye on the APT. The AFWL laser hierarchy—Lamberson, Avizonis, Rowden, and many others—made numerous visits to Hughes to watch the maturation of the APT and to identify any technical problems that needed fixing. Based on several months of these eye-witness accounts and Laughlin's "final" on-the-scene observations, Lamberson accepted the APT and directed Hughes to ship it to Kirtland. On 6 March 1973, minus the assistance of Captain Pulley, Hughes forklift workers loaded the bulky 4500-pound APT into a Lockheed L-100 (a commercial version of the Air Force's C-130), which flew the APT to Albuquerque for initial ground checkout and evaluation.[5]

Air Force/Hughes team left to right: Stan Novak (first Hughes APT program manager), Bill Spehar, Sgt Mike Applegate, Capt Don Pulley, Capt Don Pearson, Tom Welch (Hughes APT program manager), and John Gudikunst prior to shipment of APT (in background) to Kirtland AFB.

An anxious group of workers greeted the lumbering L-100 as it landed with its precious cargo in a cold rainstorm at Kirtland. Blustery conditions, combined with an impatient Kirtland forklift driver, almost led to disaster. The APT pallet started to sway as the driver off-loaded it from the plane to a waiting truck. But an alert Captain Pearson, who had helped raise the APT since its birth, was not about to allow an accident to happen at this point. He rushed to the rescue, extending his arms just in time to steady the wobbly cargo to prevent a catastrophe. Others came to assist and shortly the unwieldy shipping crate was safely aboard a blue Air Force stake-bed truck.[6]

A small group of AFWL and Hughes personnel escorted the APT to an old World War II wooden storage building, which also doubled as a makeshift optics lab, on the west side of Kirtland, where it was secured for the night. Such an impressive piece of equipment seemed to deserve more elegant quarters. To an uninformed observer, the unceremonious treatment of this one-of-a-kind optical device and the gloomy weather might have signaled an ominous welcome for the precision-crafted APT. But inclement weather did not dampen the spirits of the waiting AFWL scientists eager to get on with the preliminary testing of the device.[7]

It would be another 3 weeks before installation of the APT on the ALL began. In the meantime, Lamberson's scientists were poring over the new device housed in its shabby new home. Hughes optics experts, Ernie Endes and Otis Dodd, had accompanied the APT to Kirtland to serve as technical advisors to assist in the preliminary testing and follow-on integration of the device into the aircraft. There were still plenty of questions to be answered about the APT as AFWL project officers tried to become more familiar with all of the technical intricacies of the new pointer and tracker. Almost immediately, the combined talents of the military, government civilians, and contractor personnel forged a spirited and highly capable work force that began ground checkout and evaluation of the APT. It was this diverse and enthusiastic team of players, cooperating with one another and investing seemingly endless hours with few complaints over the years, that became one of the

APT ($6\frac{1}{2}$ feet high and $4\frac{1}{2}$ feet in diameter) at Kirtland prior to installation on ALL aircraft.

enduring trademarks of the ALL program. And it was this joint effort that was responsible for solving the extremely difficult physics and engineering problems to get the APT to point

and track within desired accuracies, as well as eventually to hold the beam on target long enough to cause catastrophic damage.[8]

After numerous laboratory tests on the APT's hydraulic, electronic, optical, and environmental control systems, AFWL and Hughes technicians were ready to install the device on the plane. Moving the APT from the wooden storage building to the waiting ALL at Hangar 1002 next to the runway a mile away was not a simple operation. Extra precautions were taken to prevent any damage to the Hughes device. A specially built blue cart, affectionately referred to as the "blue goose," cushioned the ride of the APT so as not to upset its sensitive assembled components. The $9-million cargo arrived at its destination in good shape, was carefully unloaded, and preparations began for lifting it on top of the plane. It only took a few hours in one afternoon using an overhead crane in the hangar to hoist the APT, carefully position it, and bolt it to the top of the fuselage. Pearson recalled the APT "fit perfectly." By the first week of April, the APT was ready to undergo more ground testing. The purpose of these tests was to make sure the APT was compatible with the aircraft and could be operated without any problems from the control consoles in the rear of the plane.[9]

Installation of APT on ALL in Hangar 1002.

Testing to get all of the bugs worked out of the APT while the plane was on the ground lasted until the end of April. Colonel Johnson, Captains Pearson and Pulley, and Larry Sher, who Lamberson had managed earlier to snatch from Draper Labs in Cambridge, Massachusetts, played the lead roles in conducting these experiments. Tom Welch, the Hughes program manager, relied on ten of his experts to assist in the installation and checkout of the APT. The AFWL/Hughes team focused on assessing the performance of three key components: the TV that looked through an opening carved out on the top right of the turret, the laser range finder mounted on the top left of the turret, and the IR tracker centered between the TV and range finder. For most tests, the TV was not turned on.

It was used mainly as a backup unit for showing video (television) images of the target. The IR tracker was the workhorse for imaging the target.[10]

Locating targets to test these critical parts of the system before going airborne was not a problem. A stationary target consisting of a quartz iodine heat lamp mounted on a telemetry tower several hundred meters away was one source the tracker could lock on to. Moving targets to track were military and commercial planes taking off and landing in full view of the ALL at all hours of the day and night. The TV camera consistently picked up and projected clear pictures of the heat lamp and various aircraft on a monitor in the rear of the ALL. A small, flightworthy low-power YAG laser (1.06 microns) in the range finder proved reliable for measuring exact distances from the APT to the target. Accurate distance to the target would be used to produce precision focusing of the beam by moving the secondary mirror to direct the beam to the primary mirror. (The configuration of the secondary and primary mirrors was referred to as the beam expander.) The IR tracker detected the heat signature emitted from each target and projected that IR image on the APT console screen next to the TV screen in the rear compartment of the ALL.[11]

Tracking was the most difficult of the three operations (locating, range finding, and tracking), mainly because of vibrations induced when the system was turned on and by the rotation of the APT to follow the target. These were the same problems that had surfaced during the earlier ground testing. Vibrations, noise, and the inability of the hydraulics to rotate the turret precisely and smoothly to keep up with the target were the major shortcomings of the system. As the APT rotated, friction within the system also caused the tracker line of sight to "lag" behind the moving target. As a result, this motion induced on the APT tended to move the tracker line of sight off its desired azimuth and elevation angles. This, in turn, prevented the tracking gates from locking and staying on the exact calculated tracking point on the target. In other words, the vibrations of the entire APT system caused the tracker's line of sight to "jitter" or "dance around" the predetermined track point on target. These tracking errors (measured in microradians) were relatively minor. But they were important. An accurate target track point would be required to calculate the laser aimpoint on target later in the program. Movement of the tracker line of sight would correspondingly shift the laser aimpoint on the target—the laser aimpoint was a fixed distance (computer generated) from the tracking point on target. Too much movement of the tracker line of sight meant that when the beam was fired, it would also move or "smear" over its target aimpoint. This smearing would result in the beam not being able to dwell and deliver its energy on the target aimpoint long enough (measured in seconds) to cause damage. In short, the effective intensity (power delivered to the fixed target point area) of the beam would be decreased.*

---

*The tracker/beam pointing process can be compared to a high-power scope mounted on a rifle. If a shooter keeps the rifle perfectly still while looking through the scope to align his target (tracker line of sight) and squeezes the trigger, then the bullet will strike the target bull's-eye (laser aimpoint). However, if the shooter is shaky and the rifle moves slightly as he looks through the scope—the aimpoint moves back and forth over the bull's-eye (smears)—the bullet will miss the bull's-eye because the shooter's line of sight is misaligned. Therefore, if the tracker is slightly off its target track point, then the laser beam line of sight also will be slightly off its target aimpoint.

Overall, AFWL and Hughes personnel were pleased with the test results when the APT was on the plane while parked on the ground. Most of all, they felt confident the system would perform satisfactorily in the air, but they could not confirm that until they completed the Cycle I airborne experiments.[12]

With ground testing completed, AFWL began Cycle I in May 1973. The goal of Cycle I was to demonstrate the APT could accurately track (within 10 microradians) a maneuvering target when the ALL was airborne. Cycle I was strictly a tracking experiment to make sure the APT was airworthy. There was no laser on the plane. Consequently, the plane flew closed port, meaning a glass window was installed to cover the opening in the turret where the beam were designed to exit. A low-power laser would not be installed on the ALL until Cycle II testing. The high-power GDL would not be fitted on the plane until Cycle III. Also, there were no sophisticated optics on board the ALL during Cycle I except for the uncooled glass primary and secondary mirrors in the APT.[13]

Major John Gromek served as the test director for Cycle I testing. He came to AFWL in 1972 and worked for Major Shelton. Here again was another case where rank was downplayed in keeping with Lamberson's policy of placing the best and most experienced man in the job. Gromek reported to Shelton, even though Gromek outranked his boss. When Shelton departed for an assignment in Korea to fly F-4s in the spring of 1973, Lamberson elevated Gromek to test director. He was responsible for the airplane, made sure maintenance was performed, and coordinated with range command to schedule flight times over White Sands Missile Range where Cycle I testing took place. White Sands was only about 20 minutes flight time from Kirtland. Operated by the Army, the 4000-square-mile range in southern New Mexico was used by the Air Force, Navy, and NASA.[14]

Security was another issue Gromek had to contend with. The ALL, parked in the open for all to see near the Albuquerque runway, bothered a few people. Most AFWL officials suspected Soviet reconnaissance satellites passing overhead would take photos of the plane with the odd-looking bubble on top, but they did not believe the photos would reveal much in terms of intelligence information. Gromek recalled one occasion early in Cycle I when "some guy from the Pentagon told us we had to keep a big cover over the APT so no one could determine the size of the output aperture. He was completely wet, but we did it anyway for awhile just to keep him off our backs!"[15]

Gromek was a jack-of-all-trades, but his primary duty was to be responsive to the scientists and technicians by having the ALL ready to fly when they needed it. He worked closely with Colonel Johnson's optics group in arranging the 60 flights during Cycle I testing. But before any hard-core airborne scientific experiments took place, Gromek's and Lamberson's first priority was to confirm that the plane could fly safely with the APT. Cycle I was the first time that the real APT would be flown; the earlier Edwards testing used only the dummy APT to verify flight certification. As a result, the first few Cycle I tests were flown to confirm flight safety of the ALL and APT. These initial flights, with the full forward and partial aft fairings installed, demonstrated it was safe to fly the ALL with the Hughes APT installed.[16]

For the remaining tests, Gromek arranged for fighter-class targets to show up over White Sands for the ALL to track. Target aircraft included F-4s, T-38s, and T-39s. The aircraft used most often was the T-39, which carried a square target board affixed to the fuselage below the cockpit to provide a steady heat source for the tracker to follow. For the majority of flight scenarios, the nose or engine exhaust of the target aircraft were tracked as the plane moved through the air. Flying beside the ALL, each fighter maintained a "station-keeping pattern" so the tracker would have the same angle of view for each test. However, the range of the ALL from target varied greatly (from less than 1 to several kilometers) to determine the accuracy of the range finder and tracker. Finally, for ten tests at White Sands, an F-4 launched an AIM-9B air-to-air missile (Sidewinder), parallel to the ALL, but for safety reasons it never

came closer than 10,000 feet to the ALL. The goal was to track the missiles through the boost, rocket engine burnout, and glide portions of their flight. Missile firing was conducted in preparation for the future escort defense scenario for the ALL, when the crew expected to track either on the nose or the side of the missile.[17]

Two target boards attached to underside of forward fuselage of T-39.

Most of the 7 months of flight testing (May to November 1973) during Cycle I involved tracking the modified T-39 with the IR target board at White Sands Missile Range. Not only was the T-39 readily available from the 4900th, but it flew a known heat source to simulate a missile or airplane. At each corner of the square target board was a quartz lamp, each about the size of a projector lamp used in a viewgraph machine. The ideal tracking spot was the center of the target, an equal distance from all four lamps. Diagnostic data derived from the ALL instrumentation showed that jitter was the major problem in the airborne environment.[*] Jitter originated from three sources. First, the airplane hardware—e.g., motors, engines, generators, fuel pumps—caused the plane to vibrate. These vibrational waves transferred to the APT and shook its internal components, i.e., the mirrors, tracker, TV, and range finder, making it difficult to acquire and hold the tracker on target. Even a relatively small motion in the plane manifested a very large tracking

---

[*]The ALL on-board computer determined the exact center of the target board. By calculating the tracker error signal—how far the tracker missed the center—adjustments could be made to move the tracker line of sight so it aligned with the computer-calculated center of the target board.

error on target. A small error in the APT measured in microradians could cause the desired track point on target to be off as much as several inches, depending on the range to target.[18]

Air turbulence was a second contributor of jitter that disrupted the tracking accuracies of the APT. Outside airflow causes the "bumpity-bump" ride one experiences on a commercial airliner. The effect of this turbulence is evident to the passenger trying to steady and drink from a glass when the plane bounces around in the air. In addition, the plane naturally flexed in flight, causing even more vibrations. A third factor affecting the APT jitter, and probably the major source of vibrations, was the turret housing the APT on top of the fuselage. Vibrations induced by the airflow around the turret as the plane pushed through the atmosphere were always a concern. Also, noise and vibrations caused by operating the APT were still another source of vibrations. Mechanical vibrations generated by the hydraulics used to rotate the APT and noise produced by the electrical subsystems (sensors in the tracker and the computer processing unit) induced motion on the APT. All of these factors—vibrations, flexing, and noise—upset the stability of the APT by producing jitter in the tracker, which in turn threw off the tracker's line of sight.[19]

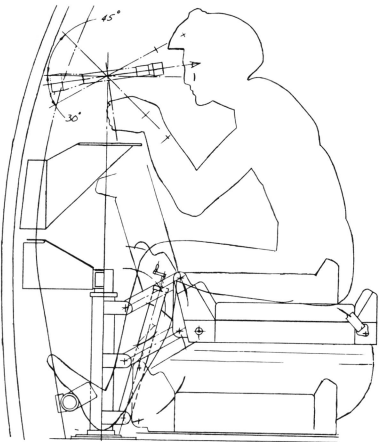

Optical acquisition device (OAD) operator's station seat.

Determining the stability of the tracker to track accurately (showing a clear image of the target) was the fundamental goal of Cycle I testing. To do this, the AFWL/Hughes team had to measure how much the tracker drifted off the target track point. This meant ensuring there was a reliable way to find, acquire, and lock on to the target. The first step in this process was to locate the aerial target with the aft and forward radars installed on the ALL. It immediately became clear during the initial Cycle I flights the radars were not capable of doing the job because they did not work. Most of the malfunctions were attributed to poor maintenance. Time after time the radars were fixed, only to break down again. Consequently, AFWL discarded the radars for Cycle I testing in favor of a simpler and more dependable rifle scope officially named the optical acquisition device (OAD).[20]

An eagle-eyed lieutenant usually served as the OAD operator. He looked out the left rear window of the ALL to scan the sky for a target. Once he aligned the rifle cross hairs on target, the coordinates of the target were computed and electronically handed off to the APT. The APT responded by turning in the direction of the target so the IR tracker could scan the target and display the image on a screen in front of the APT console operator. Using the IR image as a reference point, the IR tracker zoomed in on the target. Precision was the key here as the tracker only had a small field of view (8–9 milliradians or about a half of a degree) to see, acquire, and lock on to the target. The tracker operator could use any of three different sizes of rectangular tracking gates on the screen in front of him to follow a specific point on a target. Once the operator locked on to the target, the APT took over and automatically tracked.[21]

The target emitted invisible, IR light. An array of tiny sensors (60 indium antimonide detectors cooled by liquid nitrogen) in the tracker picked up the IR light and converted it electronically to a heat image displayed on the tracker screen. (The image was scanned across the vertical row of detectors, so that each detector element produced one line of video.) This heat image provided a view of the target (plane or missile), allowing the console operator to place the tracking gates—first manually and then electronically—on the desired spot on target. The tracker sensed heat radiating off the target only in the wavelength range from 3 to 5 microns, referred to as the mid-IR region of the light spectrum. Light in this region came off the "hottest" and best tracking spots on the target (engines, engine exhaust, and the nose of the plane or missile). These "hot spots" were beacons of light easily picked up against the relatively cold target background of the sky. Light outside the 3- to 5-micron range was filtered out so it was not seen or processed by the tracker.[22]

T-39 aircraft tracked by ALL.

One of the lessons learned from Cycle I was that 3 to 5 ("3–5") microns was not the ideal tracking region. Flight tests revealed target signatures were so intense in the 3–5 range that they "were swamping the detectors in that wavelength region," as reported by AFWL's Larry Sher. In effect, the detectors in the tracker were too overloaded

(interference from plumes and hot spots) to process the strong heat signal. The result was a blurred image resembling more of a fuzzy silhouette than a razor-sharp picture on the tracking screen. Trying to take a picture of a plane flying in front of the sun would be a good analogy; the bright sun would obscure the plane and produce an unclear photo. This made it difficult to accurately align the tracking gates on the right target spot. Plus, the hot tracking points (i.e., plume, engine exhaust) did not stay in one spot, but darted about making it more difficult to track. Sher and others recommended that Hughes develop and build an advanced tracker capable of capturing heat signatures in the longer-wavelength 8- to 12-micron range. They predicted that by showing the tracker a cooler signal that was not continually moving about, a sharper image would appear on the tracking screen. Sher explained he preferred the longer wavelength (8–12) "because of the larger dynamic range between hot exhaust gases (or burning target material) and the cool vehicle skin." This meant the tracker could more easily pick up the cooler parts of the airframe, giving the tracker more target spots to select and track.[*][23]

Lamberson favored an improved tracker and authorized Hughes to proceed building one in 1974. Proof-of-concept experiments on the advanced tracker, to determine if the 8 to 12 microns was the optimum tracking range, took place as part of the Oscura Peak Tracker Investigation and Comparison Series (OPTICS) program in 1975. The 3- to 5-micron tracker was most sensitive to picking up the high temperature plume at long ranges. But acquisition never really became a problem. What was needed was a tracker to see the skin of the aircraft (cooler temperatures) to be able to calculate where the laser would hit the target. The laser aimpoint would not be on the hot plume, but would be directed at a cooler and more vulnerable spot on the target. Tracking in the 3–5 region tended to "wash out" the target and prevent the operator from seeing and placing the beam on the desired target spot; and hence, the need to track in the longer wavelength region to pick up the cool target signature. Preference for the 8–12 tracking was explained in an AFWL technical report:

> The less severe dynamic range in the longer wavelength [8–12] region is the reason for this choice and is driven by the requirement to track, and designate for laser kill, the lower temperature skin of the target vehicle in the presence of the high radiation from the engine exhaust and the burning material during the laser kill.

Using the field test telescope from the TSL program, tests were conducted at North Oscura Peak at White Sands Missile Range. Hughes personnel removed the original FTT

---

[*]The advantage of tracking in the 8–12 region was that wavelength provided a steady and consistent heat source emitted by the target. Tracking on the plume—3- to 5-micron region—created problems because the combustion of the plume interacting with the air caused the consistency, temperature, and shape of the plume to constantly change, making it difficult for the tracker to track in that region. The solution was to block out the 3–5 return so the tracker would see only the 8–12 energy reflected off a cooler, but much more stable track point, such as the nose or engine of a plane or missile. The nose or engine track points with a steady return of 8–12 energy—hot, but cooler than the plume—afforded the tracker a better opportunity to track in a stable and precise mode.

tracking sensors and replaced them with a variety of other sensors to determine which were the best ones to eventually put in the advanced tracker. Positioned on top of a cliff, the FTT tracked a variety of aircraft and missiles at different wavelengths in 1974–75 to verify the performance of the new tracker. Test results were positive. It was these experiments that led to the fabrication of the advanced tracker AFWL used on the ALL during HEL testing in Cycle III.[24]

Although there was room for refining and upgrading the APT tracker, this did not mean it failed to meet its Cycle I tracking goals. AFWL and Hughes rated Cycle I as a complete success. In Cycle I, AFWL used a technique known as "centroid" tracking. The first step involved using the tracking gate to bracket a section of the target. Once that was done, the tracker made fine-tuning adjustments so it tracked on the center of the bracketed area. Dynamic forces—aircraft vibrations, electronic noise, and friction of the hydraulic system— caused the tracker to wander from dead center. Sometimes the jitter was so severe that the tracker broke lock. But on many other occasions, the jitter was kept to a minimum so the tracker was able to track the target. In over 150 hours of Cycle I flight testing, the IR tracker demonstrated it could track airborne targets within 12–16 microradians. (Although the goal for tracking accuracy was 10 microradians, AFWL believed 12–16 was close enough to proceed with preparations for Cycle II.) It could not do this every flight, but it did it often enough to convince AFWL that a tracker operating in a turret on a plane could accurately track airborne targets.[25]

Stabilization of the APT platform to minimize jitter was "reasonable," one AFWL official reported, "considering that the friction actually measured on the inner gimbals was higher than anticipated." Excessive friction was caused by the action of the piston and cylinder in the fine drive hydraulic actuators. Also, the AFWL/Hughes team had succeeded in isolating a number of noises generated by the signal conditioning and processing electronics in the tracker. The plan was to correct these deficiencies with the development of the advanced tracker and to have it ready for Cycle III.[26]

Meeting the objectives of the Cycle I milestone was "officially" recorded at the end of testing in November 1973. No one test was responsible for reaching this milestone. It was the entire test series, both successes and failures of individual experiments, that over time formed a pattern attesting to the tracking performance of the APT. Lamberson eagerly had watched the progress of Cycle I, keenly aware and appreciative of the strengths and weaknesses of the APT at this early stage of the game. On the positive side, he knew the APT could track an airborne target. On the negative side, he recognized work needed to go forward on development of an advanced tracker and rebuilding the hydraulic system to reduce friction during the slewing of the APT. In his and other AFWL scientists' minds, these were not insurmountable obstacles.[27]

By the end of 1973, Lamberson had reported the success of the Cycle I milestone to Systems Command. He informed the new commander, General Samuel C. Phillips, who had taken over for General Brown in August 1973, that the Weapons Lab was ready to proceed with Cycle II testing. An independent review committee set up by the Air Force, called the High-Energy Laser Milestone Assessment Board, evaluated the Cycle I data and concurred with

Lamberson that the technical objectives of Cycle I had been achieved. This led to AFSC giving AFWL the green light to move ahead with the next step to attempt to mate a low-power electrical laser with the APT to determine if the beam from an aircraft could hit an aerial target.

Low-Power Laser Goes Airborne

Cycle II: November 1973–March 1976

Although Hughes did not deliver the low-power zinc selenide window until the fall of 1974, AFWL had already started preparations for Cycle II experiments in November 1973. This portion of the ALL program consisted of two basic phases—Cycle IIA and Cycle IIB—and lasted almost $2\frac{1}{2}$ years, ending in March 1976. A total of 111 test flights occurred during Cycle II—79 in IIA and 32 in IIB.[1]

The overall goal of Cycle II was to find out if the ALL could track an airborne target and, at the same time, focus and direct a laser beam with enough precision to hold the beam steady on a very small spot on the target. In tune with Lamberson's progressive long-range plan, the Cycle II experiments were designed to use a "low-power" 150-watt CO_2 electric discharge laser as a first step to prove a beam could be propagated through the optical train to the APT and then through the atmosphere to intercept a fast-moving aerial target, in this case, a diagnostic aircraft. Lamberson certainly recognized the magnitude of this challenge. A great deal was riding on the success or failure of Cycle II. If the ALL failed to project a beam on target, then it would be very easy for the critics to advocate shutting down the program. On the other hand, if the ALL transmitted and held the laser beam exactly on the designated target spot, then this, indeed, would be a revolutionary technical breakthrough. No one had shot a laser from an airborne aircraft to hit an aerial target before. This anticipated

first-of-a-kind "proof of concept" demonstration would certainly reinvigorate the entire program, attesting to the soundness of Lamberson's approach to knock down one technical barrier at a time. It would also give reassurance that the next logical step would be to move ahead to Cycle III to determine if a "high-power" laser beam delivered from an airborne platform could hit and disable a moving target.[2]

There were many unanswered physics questions (beam propagation through the atmosphere and advanced tracking concepts) at the beginning of Cycle II, but Lamberson had good reasons to remain optimistic about the outcome. First, the tracking portion of Cycle II had already been demonstrated in Cycle I. Centroid tracking used in Cycle I worked, but it had certain limitations in terms of reliability and precision, that is, the ability to maintain lock on the same small tracking spot on target with a minimum amount of jitter. This left plenty of room for improvement. Consequently, Cycle II afforded an opportunity for AFWL to begin exploring a more promising technique—correlation tracking—to correct the deficiencies detected during Cycle I.[*] Although precision tracking had always been a nagging problem, most at AFWL believed refinements would be incorporated into a new tracker by the time Cycle III was ready to begin.[3]

Working the bugs out of the tracker was only one of the problems Lamberson had to worry about. He first had to concern himself with all of the complexities of a larger issue. What Lamberson had stressed time and again at briefings, conferences, and to his superiors at higher headquarters was the clear message that the beam control system "makes or breaks the ALL program." What he was saying was that producing a laser beam was not a particularly difficult science or technical problem. (This outlook would come back to haunt him later, as down the road it would be the complex engineering of the device that proved to be one of the most difficult problems to solve on the ALL.) The immediate issue was figuring out how to move the beam from a laser device inside the ALL and keep that beam on a predetermined horizontal course by bouncing it off a series of mirrors bolted to an optical bench on the floor of the aircraft. Maintaining the correct alignment of the beam in the presence of aircraft flexing and vibration was a tricky operation, especially at the point where one of the mirrors had to turn the beam perpendicular to its original horizontal path so the beam would move upward into the APT. Once there, the APT would focus and point the beam to hit a distant target. This first step to prove that a beam control system could operate successfully in a dynamic airborne environment was one of the fundamental reasons for Cycle II testing.[4]

SYSTEM COMPONENTS

The heart of the Cycle II experiments involved installing and testing hardware that made up the Input Laser System (ILS) located in the forward midsection of the ALL. An optical

---

[*]Correlation was a less noisy tracking process, hence reducing jitter. Also, this technique was a "frame-to-frame correlation of the total information contained in the images" to provide a clearer image using digital processing. Although correlation tracking was tested and verified at White Sands as part of the "OPTICS" experiments, it was not used on the ALL during Cycle II. AFWL used the Cycle I centroid tracker for Cycle II experiments. The correlation tracker was installed on the ALL later and used during the Cycle III experiments.

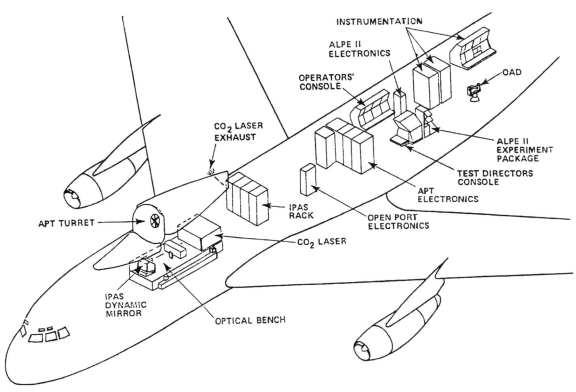

Airborne Laser Laboratory Cycle II configuration.

bench, a laser device, and an Interplatform Alignment System (IPAS) were the three major subsystems of the ILS that all had to work in harmony with one another. The foundation of the ILS was a large (5 feet × 13 feet × 20 inches high) optical bench (built by Newport Research Corporation, Fountain Valley, California) anchored to the floor of the aircraft. It essentially was a heavy (3150 pounds) rectangular metal platform designed to support the laser device and the IPAS. The bench consisted of an aluminum honeycomb structure enclosed by a steel top and bottom. To cut down on vibrations induced by the flight of the aircraft, the optical bench was attached to and supported by eight rubber shock isolators designed to absorb vibrations. The goal was to keep the bench as stable and rigid as possible so the laser beam would not bounce around in the interior of the ALL.[5]

Fixed securely on top of the bench was a small 150- to 200-watt electric discharge laser (EDL) manufactured by United Technologies Research Center in East Hartford, Connecticut. Instead of combusting fuels to generate a laser beam, electricity was injected into the United Technologies device to excite CO_2 molecules to produce a low-power laser beam at 10.6 microns. One of the main reasons for using a small electric laser was safety. Burning large amounts of volatile fuel to create high-power lasing was chancy, especially aboard a moving aircraft. The slightest leak in the system could cause the laser fuel to ignite and blow up the aircraft. To avoid this, as a first step Cycle II used the safer EDL that did not require a

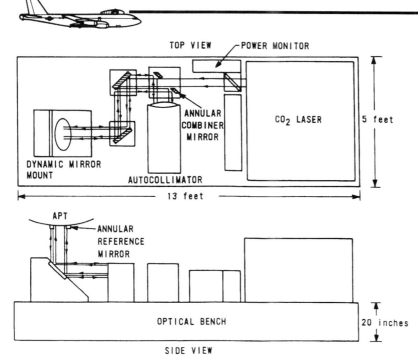

TOP VIEW — POWER MONITOR

ANNULAR COMBINER MIRROR

CO₂ LASER   5 feet

DYNAMIC MIRROR MOUNT

AUTOCOLLIMATOR

13 feet

APT

ANNULAR REFERENCE MIRROR

OPTICAL BENCH   20 inches

SIDE VIEW

Input Laser System (ILS) top and side views.

potentially dangerous combustion process. [The fuels—helium, nitrogen, and carbon dioxide— were stored in four standard "K" bottles (two bottles of helium) placed on a rack. The gases were fed into the resonator and electrically discharged to produce a 4-inch (10.1 centimeter)-diameter output beam.] Also, by first using a low-power beam, the AFWL team could afford to make mistakes in aligning the beam inside the aircraft. If the beam wandered off its desired path, such as missing a mirror and striking the side of the aircraft, no damage would be inflicted because of the weak energy level of the beam. Cycle II allowed AFWL scientists to investigate beam alignment and control system performance to build a reliable data base for directing the beam through the interior of the aircraft and then into the APT. Because they were using a low-power beam, scientists could afford to make beam alignment errors and exceed operating tolerances without experiencing catastrophic results. This would give them the data base and confidence to proceed with proper alignment of a high-power beam during the more dangerous follow-on experiments planned for Cycle III.[6]

Although the performance of the laser device and optical bench would greatly influence the success of Cycle II, the most critical issue of the ILS was how effectively the IPAS worked. Built by Perkin-Elmer Corporation (Norwalk, Connecticut), the IPAS consisted of a series of mirrors to steer the beam with enough precision from the laser device to the large primary mirror in the APT on top of the aircraft. As the 10-centimeter-wide beam exited the laser device at one end of the optical bench, it reflected off several strategically placed fixed mirrors to align it with a single steerable mirror (referred to as the dynamic mirror assembly) located at the opposite end of the optical bench from the laser device. Because the optical bench could not eliminate 100 percent of the vibrations resulting mainly from the flexing of the aircraft, the beam tended to drift off course slightly by the time it reached the steering mirror. Positioned directly under the base of the APT, this steerable mirror could be tilted in two axes to realign the path of the beam so it would be in exactly the right position to change from a horizontal to a vertical track. In simplest terms, the steerable mirror took the beam traversing the optical bench and bent it at a right angle to move it up into the APT.[7]

Shock mounting the optical bench reduced high-frequency vibrations transmitted from the aircraft to the beam, allowing the bench to move relative to the APT. With the bench and

APT moving during in-flight operations, the steerable mirror was
needed to compensate for the relative
motion between the bench and the
APT. In other words, the steerable
mirror served as the vital connecting
link between the bench and APT to
cancel out the effects of aircraft vibrations so the beam would remain
properly aligned during flight/laser
operations.[8]

Once the IPAS had delivered the
beam inside the base of the APT, still
more beam control was needed. The
beam had to reflect off six other mirrors before it could exit the turret.
This portion of the APT had not been
tested in Cycle I. The APT optical
beam train consisted of three relay
mirrors, one steering mirror, a secondary mirror, and the large primary
mirror. The purpose of the relay mirrors was to keep the beam aligned as
the APT rotated in both azimuth and
elevation. Positioning of these mirrors was critical for "dragging the
beam along" to keep up with the
movement of the APT. Also, if the

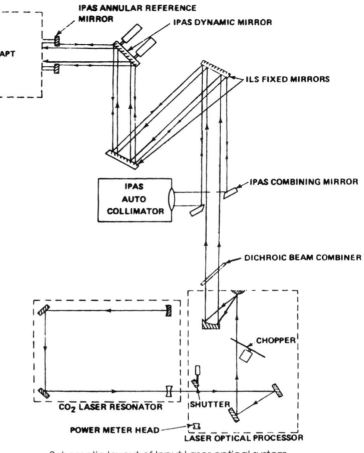

Schematic layout of Input Laser optical system.

beam failed to hit the bull's-eye of each mirror, then a portion of the beam's energy would
scatter. This meant that by the time it reached the next mirror, the power output of the beam
would be reduced, a condition weapon makers wanted to keep to a minimum.[9]

All of the mirrors originally used in the optical train for Cycle II were made of Pyrex
glass, except for the APT primary made of DURAN-50 glass. Because a low-power laser
beam had been selected, there was no danger that the beam would shatter glass mirror
surfaces. The key to Cycle II was to precisely adjust the mirrors and to verify proper beam
alignment could be maintained as the beam passed from the IPAS and through the APT. The
APT optics used during the Cycle II low-power experiments could handle beam power levels
up to 1 kilowatt. As the Cycle II laser generated only 200 watts, there was a substantial margin
of safety in the mirror train. However, for the much higher power Cycle III tests, the optics
would have to change dramatically. The glass mirrors would be replaced with more durable
water-cooled molybdenum metal mirrors (except for the primary mirror made of a glass-ceramic material) so they wouldn't be damaged when blasted with the energy deposited by a
high-power laser beam of several hundred kilowatts.[10]

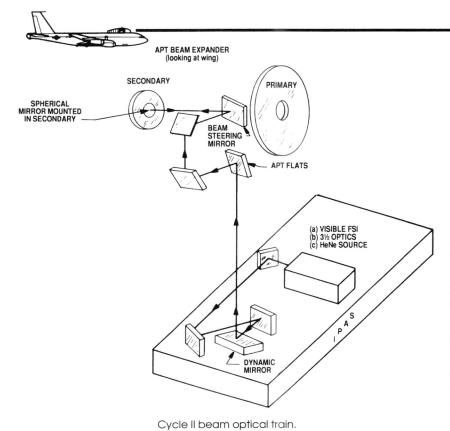

APT BEAM EXPANDER
(looking at wing)

SECONDARY

PRIMARY

SPHERICAL
MIRROR MOUNTED
IN SECONDARY

BEAM
STEERING
MIRROR

APT FLATS

(a) VISIBLE FSI
(b) 3½ OPTICS
(c) HeNe SOURCE

I P A S

DYNAMIC
MIRROR

Cycle II beam optical train.

It took 5 months—November 1973 through March 1974—to modify the ALL to accept the hardware subsystems required to conduct the Cycle II testing. General Dynamics/Albuquerque Operations, under the watchful eyes of AFWL, made all of the necessary changes while the plane was parked in Hangar 1001 adjacent to the runway on the west side of Kirtland AFB. Only minor modifications were made to the aircraft configuration to permit installation of the optical bench, the laser device, and the IPAS. This job boiled down to making sure all of these components (including the APT) and associated electronics were secured and fit in the proper locations inside the plane before proceeding with ground and flight testing. Also, an upgraded air conditioning unit was installed for cooling hardware as well as for crew comfort. Another significant modification was the removal of the F-4 radars in the nose and tail of the ALL. Cycle I had shown that the tail radar was ineffective for target acquisition. Cycle II counted on the simpler and more reliable OAD for locating targets.[11]

LABORATORY AND GROUND TESTS

Once the aircraft modifications were completed, Captain Dennis Maier led an AFWL team to conduct an extensive series of laboratory and ground tests lasting from May through December 1974. For the first set of tests, AFWL personnel removed the APT from the plane and transported it to its optics laboratory on the west side of Kirtland. The objective was to determine how well all six mirrors lined up in the APT for transmitting a laser beam under ideal conditions. This work was only a preliminary first step because of the absence of aircraft vibrations on the APT in a laboratory. Everyone recognized a progressive fine-tuning of the optics would occur during the follow-on ground testing and airborne flights of the ALL.[12]

Before any alignment experiments took place, four new mirrors—the three relay and one steerable—had to be installed in the APT. The secondary and primary mirrors were already in place during Cycle I, but they had not yet been used because there was no laser as part of the Cycle I testing. Hughes, who built the mirrors, worked with AFWL and had all of the mirrors in place by the end of the spring.[13]

After the mirrors had been installed, the next step was to adjust them so they were all aligned properly. As the EDL beam for Cycle II was invisible, it was impossible to observe where the beam hit each mirror. Plus, there were no beam sensors to pinpoint the path of the CO_2 laser beam during Cycle II. To solve this problem, AFWL introduced a low-power (5–15 milliwatts) helium–neon visible annular laser (a beam with a hole in its center that resembled a doughnut) that surrounded the EDL beam. This annular "reference" beam came out of an autocollimator assembly (also referred to as the IPAS Sensor Assembly) that was adjacent to the CO_2 device on the optical bench. The job of the autocollimator was to detect and make corrections for beam misalignment along the optical train. Both lasers were turned on at the same time, and the annular beam intercepted and surrounded the CO_2 beam. The effect was that the annular beam and CO_2 beam followed exactly the same path.[14]

Once both beams reached the base of the APT, the annular visible beam reflected off an annular reference mirror (the CO_2 beam passed through the center aperture of this mirror) and retraced its path back into the autocollimator assembly. (As the two lasers had different wavelengths, the outgoing and returning beams did not interfere with one another.) There a series of autoalignment sensors analyzed the beam to determine if it had been aligned properly from the optical bench to the base of the APT. Misalignment was measured by an error signal (recorded on the ALL analog tape system) that was then "processed and used to adjust the orientation of the dynamic mirror assembly." Continuous analysis of the error signal compensated for "all vibrations throughout the optical train." If it was determined the beam was off line, the IPAS automatically adjusted the dynamic mirror to compensate for misalignment during flight operations. Prior to flight operations, AFWL and contractor personnel made static realignments on the ground to ensure that everything was within the correction capability of the dynamic mirror. This was extremely tedious work as a large share of this labor involved repeated trial-and-error "tweaking" of the mirror to achieve the correct alignment.[15]

Getting the beam to the base of the APT solved only half of the alignment problem. The beam still had to travel through the APT. To ensure the CO_2 beam remained properly aligned on this part of its journey, a second autocollimator assembly [called the Auto-Alignment Sensor (AAS) located near the base of the APT] introduced an invisible annular reference beam (this time from a light-emitting diode) to track with the CO_2 beam up in the APT. This beam was in the short-infrared region (0.9 micron), and although invisible to humans, it could be easily seen electronically. As both beams reached the primary mirror, a second annular reference mirror peeled off the short-infrared invisible beam and redirected it back to the APT autocollimator assembly, which computed how much the APT optical train was misaligned. Sensed beam misalignments detected at the AAS were then corrected automatically by sending electronic signals to the beam-steering mirror in the APT telescope. In simplest terms, the AAS directed the beam-steering mirror to tilt to a new position so the

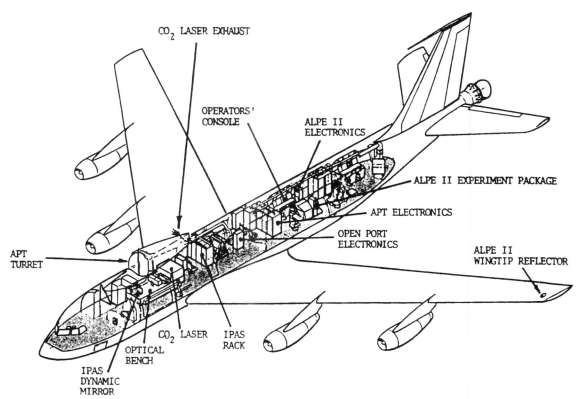

CO$_2$ LASER EXHAUST

OPERATORS'
CONSOLE

ALPE II
ELECTRONICS

ALPE II EXPERIMENT PACKAGE

APT ELECTRONICS

OPEN PORT
ELECTRONICS

ALPE II
WINGTIP REFLECTOR

APT
TURRET

CO$_2$ LASER   IPAS
RACK

OPTICAL
BENCH

IPAS
DYNAMIC
MIRROR

Airborne Laser Laboratory Cycle II configuration.

beam would be properly aligned. And once again, most of the misalignment was the result of aircraft vibrations.[16]

Both the inner and outer gimbals (that rotated the APT and its relay mirrors) and the APT steering mirror were hydraulically activated. Their functions were to stabilize the beam in spite of aircraft disturbances and to point the beam line of sight to the target. AFWL and Hughes technicians worked relentlessly looking for ways to improve the mechanics of the hydraulics of the outer and inner gimbals so the APT would slew as smoothly as possible. Hydraulic leaks were a constant problem, plus it took a long time to gain access to the place that was leaking so it could be fixed. Fine-tuning the movement of the inner gimbals, responsible for azimuth and elevation fine settings for the telescope and tracker, was accomplished by modifying the hydraulic actuator pistons to reduce friction. The mirror mounts of the steering, secondary, and primary mirrors in the APT were stiffened to compensate for vibrations and to help correct the optical misalignment between the relay mirrors and the telescope.[17]

The main benefit derived from the laboratory testing of the ILS and APT was the various measurements (such as the angular motions between the APT and optical bench, beam intensity, and beam power output) that contributed to building a baseline of scientific data.

Using this data base, AFWL was now ready to install and align the ILS and APT in the ALL aircraft and begin ground testing of the system. Loaded on the "Blue Goose" cart once again, the APT was transported along with the ILS from the 400 area to the end of Albuquerque airport's east–west runway where the ALL waited. At the time, the special hangar, or Advanced Radiation Test Facility (ARTF) designed exclusively for the ALL testing, was still under construction. Work had begun on this structure in March 1973. By November 1974, phase I of the project had been completed, which included part of the main hangar area. What remained to be done was phase II of this project—construction of clean rooms, laboratory facilities, and offices adjacent to the hangar area. By July 1975, beneficial occupancy took place—most people were moved in by January 1976.[18]

Because the ARTF hangar was not finished during Cycle IIA, ground testing of the ALL—now outfitted with the APT and optical bench—took place in Hangar 1001 belonging to AFSWC. Shortly after installation of the APT on the aircraft, an overhead crane lifted the APT off the ALL so minor repairs could be completed. Also, outside the hangar, testing involved subjecting the aircraft to different vibration levels. [Most of the ground testing of the APT (e.g., tracking, control systems, sensors, optical quality) took place outdoors on the aircraft parking area east of the ARTF.] To do this, a team from Wright-Patterson AFB attached an electrodynamic "shaker" to the underside (fuselage station 600) of the parked ALL and shook it. Several accelerometers were also put on the plane to measure all of the

vibrations and modes simulating in-flight conditions. These accelerometers recorded different vibrations on the optical bench, mirrors, and the APT to determine how these components behaved. This information again contributed to building the data base that would be used to dynamically align the mirrors so the beam would travel along the right path.[19]

Throughout the fall of 1974, the laser and optical system on board the ALL underwent extensive ground testing that in many ways repeated the earlier tests in the laboratory in the 400 area. What was differ-

Diagnostic aircraft (left) and ALL parked in front of ARTF.

ent about the testing at the ARTF test pad site was the ILS and APT were turned on inside the plane and a laser beam was aimed and transmitted through the atmosphere to hit a target.

This gave scientists an opportunity to check and calibrate all flight instrumentation on the ground in preparation for the airborne tests scheduled later on in Cycle II. (Instrumentation aboard the diagnostic aircraft was also calibrated on the ground to define the laser beam profile as it was fired from the ALL on the ground to the diagnostic aircraft parked next to it. These ground data served as a data base to compare with data collected from later in-flight tests.) Several targets were used. One was a stationary target board located 1 kilometer downrange behind the south side of the ARTF. Another was a KC-135 diagnostic aircraft belonging to AFSWC that AFWL had recently acquired. A third target was a T-39, which the ALL on the ground tried to track and hit with a laser beam as the T-39 flew by the ARTF still under construction.[20]

ACQUISITION OF DIAGNOSTIC AIRCRAFT (371)

In September 1973, Colonel Kyrazis no longer served as a consultant to AFWL. Instead, he was permanently assigned to AFWL as the Assistant for the Airborne Laser Laboratory. Officially he reported to Colonel Russ Parsons, who headed the Laser Division, but Parsons was occupied with managing a variety of programs spanning five technical branches. Consequently, Kyrazis experienced a great deal of independence as the leader for the day-to-day activities of the ALL program. Early on in Cycle II, it became clear to Kyrazis that the T-39 was inadequate to conduct the degree of diagnostics required. The T-39 target board consisted of "four infrared lamps and a circular array of 177 pyroelectric infrared detectors." Instrumentation on the target board detected where the beam hit the target and the intensity of the beam. "More sophisticated instrumentation was needed to record a large number of beam parameters," Kyrazis stated. As a result, in the spring of 1974 he began searching for a test aircraft that could be outfitted with the most advanced state-of-the-art equipment to perform the complex beam analysis. Kyrazis went to see Colonel Otis A. Prater, who was the commander of the 4900th Flight Test Group, assigned to support the nuclear readiness flight test missions of AFSWC located at Kirtland. Kyrazis knew Prater controlled three converted NC-135 test aircraft designed to detect and monitor nuclear detonations in violation of the Nuclear Test Ban Treaty.[21]

Prater told Kyrazis an NC-135 aircraft (SN 60-371) was available. Based on Kyrazis's initial inquiries, Lamberson wrote a letter to AFSWC (the agency that had been using 371) officially requesting the aircraft be turned over to AFWL. He explained the T-39 aircraft originally planned as a diagnostic platform was not equipped with the required instrumentation to support all of the Cycle II testing. "A much more comprehensive beam diagnostics effort" was needed, Lamberson wrote. The diagnostic hardware on 371 would measure beam power, quality, spot sizes, jitter, and tracking and pointing accuracies as well as defining how the beam propagated in the atmosphere between the ALL and 371.[*] He went on to state, "The

---

[*]A pyroelectric target board contained sensors to measure beam size and beam jitter. A second instrument, an IR scanner, also measured beam size and beam jitter as a cross-check to compare the accuracy and validity of data collected by both diagnostic instruments. Interferometers measured beam wavelength to compare with laboratory data collected earlier. Comparing the two sets of data defined the influence of boundary layer (turbulence) effects on the 10.6-micron beam. Aircraft 371 also took high-speed photos to document airborne tests.

NC-135 aircraft allows us the weight, power, and volume to perform many diagnostic experiments not possible with the T-39." Recognizing that 371 supported the Atomic Energy Commission's National Nuclear Test Readiness Program, Lamberson assured AFWL would return the aircraft to the AFSWC in "an agreed upon restored condition upon the resumption of atmospheric testing." While in its possession, AFWL predicted 371 would be flown on the average of seven times a month.[22]

Two weeks later, AFWL's Major Jerald N. Jensen advised AFSWC on the specific modifications

T-39 used as target during Cycle II testing. This aircraft also flew in support of the TSL program and Cycle I testing.

to 371 needed to make it ready for Cycle II experiments. Installation of optical quality germanium and fused quartz windows on the right side of 371, to allow transmission of the CO_2 laser beam from the ALL into 371, was the major modification planned. Other changes involved rearranging the aircraft's interior to accept a visible and IR vidicon to watch and define the beam profile, an IR scanner, IR lights for the ALL tracker to track, target boards, temperature sensors, and low-power CO_2 and visible lasers to send a beam from 371 to the ALL as part of the beam propagation experiments. Spatial resolution charts (circular, horizontal, and vertical markings of different thicknesses) were painted on the forward right side of the fuselage for measuring the optical quality of the APT sensors and focusing ability of the pointing and tracking system. The laser beam was directed through the two IR windows (located just above and to the left of the resolution charts). Once inside the plane, the beam was diagnosed by special instrumentation inside the aircraft.[23]

Lieutenant Colonel William E. Burke, Chief of AFSWC's Requirements Division, approved Lamberson's request for use of aircraft 371 on 9 July. AFWL would receive the aircraft, but there also was a stipulation attached. As Lamberson expected, if resumption of nuclear testing in the atmosphere occurred, then AFWL would have to "relinquish use of 371 immediately." (The Nuclear Test Ban Treaty of 1963 prohibited nuclear testing in the atmosphere, space, and underwater.) Burke's letter also authorized AFWL's plan to modify 371 in preparation for Cycle II testing. AFWL selected GD to perform the modifications. Work began on 371 in late summer and lasted only a few months.[24]

Kyrazis immediately set his engineers to start fixing the plane up to be able to do optical measurements. He described the initial work: "Basically, we built a very complete optics

Diagnostic aircraft 371.

laboratory inside that airplane, plus all the tape recording and data acquisition capabilities." Part of this modification included the installation of several special optical-quality windows along its right side. AFWL engineers and contractors designed and operated the diagnostic subsystem (an IR idicon television camera, an IR scanner, a jitter sensor, and power meters), which Kyrazis claimed cut several years off the ALL program. One of the advantages of 371's equipment was that it could reduce real-time jitter

data to allow the operators to "learn the physics immediately." Precision data reduction and analysis, however, had to wait to be performed on the ground. But the in-flight data reduction gave a rough estimate of the experimental variables. "If something went haywire," Kyrazis stated, "you knew it immediately and could alter the experiment on the spot."[25]

GROUND TESTING CONTINUES

One of the objectives of the ground testing was to calculate the aimpoint, or spot where the beam was to hit the target. Ideally, the aimpoint was on the most vulnerable part of the target (such as the cockpit canopy or guidance system in the nose of the target aircraft) so the beam could cause lethal damage. There was a fixed distance between the aimpoint and track point located on the hotter part of the target. Plus, the angle formed between the beam and track path varied as the distance from the APT to the target changed. At long distances, the tracking and laser beam line of sight almost coincided. As the target came closer to the APT, there was a greater divergence between the beam and tracker line of sight. It was for this reason that the APT needed an accurate range finder to always feed correct distance figures to the computer. The computer could then compensate by making real time calculations to ensure the beam was always pointed precisely in line with the aimpoint on target. Testing the accuracy of the range finder on the APT was an important part of the Cycle II ground testing.[26]

Ground testing of the ALL systems against T-39 flybys (approximately 40 engagements) had mixed results at best. Clearing air space with the Federal Aviation Agency was often difficult to coordinate. That agency was reluctant to allow the ALL to fire a laser—no matter how harmless it might be—while sitting at the end of the runway where commercial airliners were landing and taking off on a regular basis. Also, the very high line of sight slew rates made it difficult for the APT on the ground to keep up with the T-39 as it flew by. Occasionally, AFWL could get the beam to hit the target board near the nose of the aircraft, but in most cases, the beam failed to engage the target plane.[27]

Although a few tests were designed using the ALL to send a beam to the diagnostic plane parked approximately 700 meters away, the majority of experiments were directed from Pad 4 to the downrange target behind the new ARTF. Good results were obtained transmitting the beam from the APT to the downrange target 874 meters away. The small target facility had the most sophisticated and reliable instrumentation (an optical bench, calibration and alignment systems, control consoles, and a quick-look data reduction facility) to sample the beam and define its characteristics. Critical concerns were total power, peak intensity, jitter, dispersion, and how well the beam stayed on target.[*] In general, getting the 10-centimeter diameter beam from the optical bench to the APT was not a problem. But once the beam struck the primary mirror, things began to deteriorate. In essence, the poor quality of the primary mirror, discovered by Captain Maier and others during ground testing, prevented the 60-centimeter-wide beam reflecting off the primary mirror in the APT from being focused to a tight spot (a beam diameter of roughly 5–15 centimeters at a range of 1–3 kilometers) at the downrange site. There were optical aberrations that caused the billions of photons making up the beam to scatter instead of lining up in a precision formation. Major Denny Boesen, who ran the Beam Control and Optics Section, recalled that these early ground experiments demonstrated that "the optical quality of the telescope in the pointer/tracker was very bad." During installation of the APT on the plane in December 1974 (after lab testing), the primary mirror was damaged on the back of the Pyrex substrate. "Because we didn't have any new optics to put in," Boesen recalled, "we went ahead and flew what was called Cycle IIA with that known poor optical quality." By that time, AFWL had ordered a replacement for the defective primary mirror from GD.[28]

Denny Boesen exemplified the high quality of military personnel who worked on the ALL program. Slender and fit, he was considered the in-house expert on optics and played a major role in that area during the Cycle II and later Cycle III testing. Those on the program considered Boesen the most technically knowledgeable person on the APT. He not only drew respect for his technical competence, but he also projected the clean angular facial features of a veteran pilot, which he was, and simply looked the role of the consummate Air Force officer. One of Boesen's most striking features was the intensity of his eyes that personified a serious-minded person who could focus all of his energy on the task at hand. Others described him as low-keyed, not excitable, and a good organizer who was able to lead a diverse group of very bright optics people, who from time to time would get embroiled over petty matters. Everyone praised his leadership qualities because he didn't shoot from the hip. On the tough issues, he didn't respond instantly. Instead, he listened attentively, sifted and

---

[*]A calorimeter absorbed and measured the output power of the CO_2 laser. Jitter was measured with a jitter sensor, a pyroelectric quadrant cell, that used a small sample of the beam to measure the relative motion of the CO_2 laser beam relative to the optical bench.

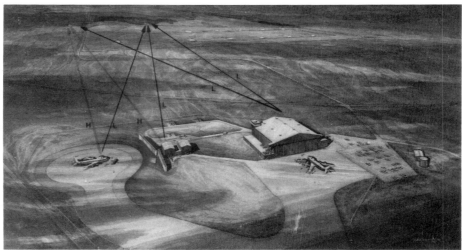

Left to right: Firing of laser beam from ALL, test cell, and hangar to downrange ground targets. L, low power; H, high power.

weighed all of the facts, and slowly and deliberately reached a decision. That type of precision in his thinking won him the confidence of all who worked for him, as well as those up the chain of command.[29]

Although Boesen and others learned firsthand the disappointment about the inferior quality of the optics from the ground testing, they managed to gather more encouraging data on the performance of the zinc selenide window installed in the APT. There were three basic concerns about the six-pane window. First, AFWL was a little nervous about the structural integrity of such a large piece of brittle zinc selenide. Second, would the pressure and thermal conditioning unit function as advertised? Finally, would the beam be unacceptably distorted as it passed through the window?[30]

Ground testing of the window, installed and fit-checked on the APT in September 1974, revealed the zinc selenide showed no signs of cracking or deforming when exposed to various weather conditions while mounted on the ALL. To keep the window at the proper temperature and pressure levels, a rather complex plumbing arrangement had to be assembled inside the relatively small turret. Part of this conditioning system consisted of tubing following the contours of the inside of the turret and leading to each of the six struts supporting the window. From there, nitrogen was fed into and circulated in the gap between the thermal panes of the window to maintain the correct temperature and pressure to ensure beam quality was not degraded as the beam passed through the window. Numerous tests on the window conditioning unit confirmed that the mechanics (e.g., pump, valves, feed lines) and flow rate of the nitrogen met all design specifications of the system.[31]

A portion of the ground testing also directed the beam through the zinc selenide window to the diagnostic airplane and the stationary targets downrange from the ARTF. The purpose of these experiments was to determine whether the quality of the beam deteriorated as it passed through the window. On the ground, the beam passed through fairly severe natural turbulence; for airborne testing, the natural turbulence was weaker, but the turbulence created by the aircraft was very strong. Measurements of the beam profile were made on both sides of the window. The evidence showed the window caused only minor deterioration of the beam quality. This was not unexpected, as earlier laboratory testing conducted when the zinc selenide was under development produced the same results. However, the nearly transparent window and its supporting struts served as an obstruction that absorbed, reflected, and scattered portions of the beam. But again, this disturbance to the laser was not that significant as 98.5 percent of the beam was able to pass through the window intact. Preliminary calculations also indicated that a single-pane window, as opposed to the double-pane window for Cycle II, would be feasible for use during Cycle III.[32]

AIRBORNE TESTS

Ground testing for Cycle IIA finished up in December 1974 and flight testing started the next month. Captain Harold S. Rhoads, assigned to AFSWC, was the ALL pilot. Flight patterns "were all over the place," one AFWL crew member recalled. Most of the ALL and 371 flying took place

Denny Boesen next to his favorite piece of hardware, the APT.

between Albuquerque and White Sands Missile Range in southern New Mexico. Extended missions took the ALL to Edwards AFB in California, over test ranges in Nevada, or to North Dakota and back again. Sixteen ALL flights, amounting to 60 hours airborne, occurred. The first flight took place on 9 January 1975 and lasted for $3\frac{1}{2}$ hours. The last was on 3 July 1975, marking the completion of Cycle IIA experiments.* Initially, the objective was to take measurements on the optical bench to check its response to in-flight vibrations and to

---

*Coinciding with the completion of Cycle IIA was the wrap up of a 6-month investigation of AFWL's laser program conducted by the Air Force Audit Agency. Although AFWL needed to improve in the administrative paperwork associated with budget and procurement actions, the Laboratory generally received high marks. The final audit report stated AFWL had "made significant progress in implementing successful policies and procedures in the management of the Air Force HEL program. The research and development . . . established by the ARTO have been effective."

determine the alignment of the CO_2 beam along the mirror train. The next step was to measure vibrations and the beam alignment in the APT and focus the beam on target. Part of the time was spent comparing how well the ground-based data predicted optical performance during airborne conditions. A variety of other tests also took place, such as airworthiness qualification of the APT window.[33]

Cycle II was a hectic time for all and in many cases involved working around the clock. Few complained as the AFWL/Hughes team was committed to overcoming the technical barricades to make the ALL work. Kyrazis recalled a typical day. "It was not unusual to work till midnight, go home to sleep, get up at 4:00 A.M. to start preflighting our equipment by 5:00 A.M., and take off at 9:00 A.M. for another 5-hour mission. Upon landing, we would debrief, decide on the details of the next day's mission, start analyzing the data, and fix or modify the equipment." Years later, on 4 May 1988, Kyrazis gave the keynote speech to honor the last flight of the ALL to its final resting place at the Air Force Museum at Wright-Patterson AFB. Reminiscing about the furious pace he had set for everyone during Cycle II, he looked out and smiled at the large crowd gathered in the ARTF hangar and said, "I suspect that there are a number of you out there who have not yet forgiven me for what I did to you in Cycle II!" Many heads in the audience instantly nodded in agreement.[34]

AFWL scientists originally were optimistic about Cycle II and believed it would take only 6 months to complete collecting and analyzing all of the flight data. They were wrong. Too many technical surprises stretched Cycle II into a $2\frac{1}{2}$-year project. Inferior mirrors in the APT, discovered during the earlier ground testing and reconfirmed during the Cycle IIA flight tests, had to be replaced in Cycle IIB and accounted for part of the delay. Other problems related to the laser operating in an airborne mode. Keith Gilbert, a captain who worked on defining aircraft boundary layer turbulence and who later became head of AFWL's Advanced Radiation Technology Office in the 1980s, was involved with trying to measure beam quality. As one of the day-to-day workers in the trenches, he conducted numerous experiments throughout the lifetime of Cycle II. Looking back on that era, he vividly remembered some of the unexpected frustrations associated with Cycle IIA.

> The first couple of flights we got up there, we couldn't even get the laser to run! It would run on the ground where nothing is moving, but then you get it up there, and something happens in takeoff—the alignment goes out, the cavity's mirrors are jiggled around a little bit, and the thing wouldn't lase. Saying it's a laboratory is a misnomer because once you get off the ground there are a very limited number of things you could do to make the experiment work. You really have to check it out on the ground completely, and then pray that when you take off that you don't hit turbulence or any or all those things that can disrupt the experiment. So what we thought was going to be a year or less extended over several years.[35]

After a few months, AFWL's prayers seemed answered. Constant fiddling around with the device had finally produced positive results, as the first flight generating a good beam occurred on 13 March 1975. From then until the beginning of July, experiments focused on continuing to build a data base on the performance of the beam control system and the characteristics of the beam as it propagated through the atmosphere from the APT to the target. Airborne testing produced mixed results. On some days no data were collected because of equipment failure in either the ALL or the diagnostic aircraft.[*] "On other days when you flew," Gilbert stated, "everything happened, the beam came out, and there was great revelry!"[36]

Once the EDL operated fairly reliably in the airborne environment, a second problem that took some time to solve was consistently directing the beam from the APT through an 18-inch germanium window on the right side of 371. The diagnostic plane flew along the left side of the ALL, usually about a kilometer away, but in some cases up to 3 kilometers from the ALL. After the console operator manually acquired the target in the APT sensor's field of view, the APT systems automatically tracked the target and pointed the laser beam. With the tracker locked on to an IR heat source (lamps) on 371, the console operator selected the aimpoint of the laser. The operator then told the system electronically to place the laser on the aimpoint, in this case, the window on the side of 371. The APT kept the beam at that point on the window regardless of the relative motion between the two planes as they flew side by side.[†37]

Captain Keith Gilbert participated in numerous Cycle II experiments to gain a fundamental understanding of the aerodynamics and behavior of the boundary layer and turbulence around the turret and aircraft fuselage.

---

[*]Generally, each test flight lasted 3 to 4 hours. Shortly after takeoff, the ALL pointing and tracking was calibrated in flight. Similarly, technicians set up and calibrated equipment aboard the diagnostic aircraft to be ready to receive and analyze the laser beam transmitted from the ALL to the diagnostic aircraft (371). Radio communication existed between the two aircraft, but it was not until Cycle III that a telemetry link was established between the ALL and diagnostic aircraft. Pilots and crew wore special goggles to protect their eyes from the range finder 1.06-micron laser beam.

[†]The diagnostic plane received the beam, and the IR scanner collected data on beam characteristics. Tape recorders usually recorded data for several minutes and then turned off. The next experiment would be set up substituting the pyroelectric target board for the IR scanner. The same experiment was repeated, and the results were compared to ensure both sets of data matched to ensure valid results were attained. Also, the APT on the ALL measured jitter which was correlated with the amount of jitter measured on board the diagnostic aircraft. This helped to determine how the major components of the system affected the beam. On returning to Kirtland AFB, the ALL propagated the beam to a downrange ground target for a postflight system calibration. The latter data were compared with preflight beam propagation data to determine how much the system performance changed during the flight. It also helped to calibrate data during flight. Once all tests were completed, ground and in-flight data tapes were removed from the ALL and taken to AFWL's computer center for data reduction. Within 8 hours, preliminary results were available. Those results were then used to plan the next flight.

Fine-tuning the placement of the beam on the target window involved the console operator lining up the cross hairs on the image of 371 appearing on the TV screen, which did not have a high degree of resolution. Consequently, this required a fair degree of dexterity on the part of the operator to get the beam perfectly aligned with the exact aimpoint on the

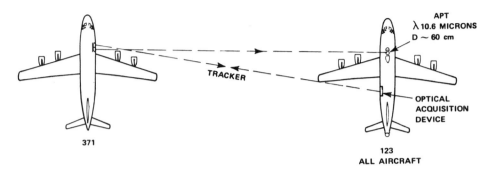

Aircraft formation for APT propagation measurements.

small window in 371. To help overcome this predicament, the console operator in the ALL was in constant communication with a member of the diagnostic team on 371. Kyrazis recalled 371 would instruct the ALL to "move it (the beam) 3 inches to the rear of the plane or bring it up 2 inches" so the beam would penetrate the center of the window and allow the 371 diagnostic equipment to analyze the beam's quality.[38]

Once the beam entered the window in 371, different optical elements of the diagnostic subsystem split the beam off to different sensors. Part of the beam went to a power meter, another portion to a jitter sensor, so four or five instruments looked at the beam simultaneously. As an example, Kyrazis explained the intricacies of the EG&G IR scanner on board 371.

> It had a wheel with a series of holes on it that would rotate at very high speed and would basically form a TV frame, and would make a frame of what the beam looked like—how wide it was and the intensity distribution and the exact shape of the beam. Because you sort of see pictures of the beam and it looks like a nice round deally-blob that smoothly comes to a peak and then slowly dies down. The real beam doesn't do that at all. In many cases it looks cruddy. It has double peaks and a bunch of stuff out in the wings. Sometimes it looks totally smeared out. Our whole purpose in Cycle II was to find out why these beams would do that and determine what was needed to correct that, because if we allowed that to happen with the high-power beam, all it was was a toaster, not a weapon.[39]

There were many more failures than successes in consistently obtaining the desired beam pointing accuracy. But on the good days when "you could move the beam and hit an 18-inch aperture (the window) at a kilometer away, then that was pretty good evidence you could point," as Kyrazis put it. "We would fly from Albuquerque to Grand Forks, North Dakota, and the beam would be on the target that whole time. I mean, we were able to maintain that pointing accuracy while 371 would move up and down, further away, closer, and that kind of thing." During the entire trip, 371's instrumentation was able to collect critical data on beam quality and the performance of the pointing and tracking system when subjected to changing mechanical and aerodynamic in-flight forces.[40]

AFWL was primarily interested in measuring beam quality in terms of power, jitter, and spot size on target. These measurements were taken under a variety of flight conditions: weather, altitude (3.9 to 8.5 kilometers), range to target (0.5 to 3 kilometers), speed (0.5 to 0.7 Mach) and climbs of the ALL, and aircraft configuration. For example, experimenters wanted to know how the beam was affected when the ALL was flown with the zinc selenide window in place and when it was removed, referred to as the "open port" configuration. Eleven open port missions were flown in Cycle II. For some of these tests AFWL installed an instrumented liner inside the APT to take pressure measurements. Other sensors recorded temperature and strain. An aerodynamic fence was placed around the opening to minimize potential resonance (noise) damage inside the turret. The data showed that the airborne turbulence was not disruptive enough to degrade or damage the internal components of the APT. Additional experiments demonstrated that the advantage of open port was that the optical quality of the beam was very good (confirming the earlier ground test data) because there was no window in the way to distort, absorb, or reflect the beam. Jitter of the beam was the same whether or not the window was installed or removed.[41]

Although there were definite advantages flying open port, there were also drawbacks. First, it was almost impossible to keep the optics clean over long periods as dust and dirt had easy access to the open APT. Dirty optics absorbed a portion of the beam, thus reducing the far-field beam intensity on target. A second disadvantage had to do with the quality of the beam during extended flight conditions. Low temperatures degraded the focusing ability of the primary mirror and resulted in the beam focused "short" of the target. Several Cycle II tests involved the ALL and 371 spending 5 hours in the air flying side by side from Kirtland to Grand Forks, North Dakota, and back again. During this entire time the laser was turned on and aimed through the diagnostic plane's window and kept there to determine if the quality of the beam changed over time. As Denny Boesen explained:

> What we found with the open port was that we initially would get very good performance, but then after it had been open for a while, the temperature started to go down. In fact, the temperature started to go down very quickly, but you could tolerate some change. But at 30,000 feet, the temperature is right nippy out there, so the temperature of the telescope would get so low that the autofocus system couldn't compensate for the temperature change and keep the beam focused properly.[42]

Although the window offered ample protection to the APT optics, a small portion of the beam energy was absorbed by the window because the zinc selenide was not 100 percent transparent. This naturally resulted in some loss of beam power. Plus, when using the low-power window, the primary mirror focused the beam "beyond" the target. But on the positive side, the beam was big enough (60 centimeters) to spread itself over the entire surface of the primary mirror so only a few watts of power per square centimeter was deposited on any point on the mirror. This was not enough power to distort the mirror or the beam as it exited the APT. Even with slight power losses, when the beam was focused, the intensity of the beam on target increased. For example, by focusing a 60-centimeter beam at the primary

mirror to 10 centimeters at a target a little more than a kilometer away would increase the beam intensity on target 36 times over the beam intensity at the primary mirror. There were trade-offs between the open and closed port configurations, but in the end more favorable conditions existed when using the APT with the low-power window installed. However, for short flights the open port gave the best performance.[43]

Once the laser beam left the APT, there were questions as to how the beam would be affected as it passed through the atmosphere. Some predicted the atmosphere would distort the beam causing it to spread out and rob it of its full power. But the airborne experimental evidence showed this was not true when the beam propagated in the relatively short 1-kilometer range. The data revealed there was almost no aberration of the laser beam caused by the atmosphere or the turbulent airflow around the aircraft. What prevented the beam from being degraded, as reported by Gilbert and Jim Davis who worked on these experiments, was its long wavelength (10.6 microns) that helped it to retain its shape and compactness. But even with the longer-wavelength beam, of the 150 watts generated in the ALL, only about 7 percent of this power was focused and deposited on target ("in the bucket") in the far field. This was the result of power losses caused by the beam snaking its way through the ALL, moving into the APT, and finally exiting the turret. Approximately 60 percent of the laser beam power was lost through the optical system on the ALL. As the beam moved from the APT through the atmosphere to the target, the beam lost about 33 percent more of its power. However, based on state-of-the-art standards, an efficiency of 7 percent was considered quite good. This translated to an average intensity ranging from 0.45 to 1.00 watt per square centimeter on target, depending on the fairing used. (For the high-power beam, the more watts per square centimeter delivered on target, the quicker the beam could destroy the target.) The full forward, closed port fairing/APT combination offered the most protection and resulted in 1.0 watts per square centimeter on target. The forward ramp, closed port resulted in only 0.45 watt per square centimeter. The reason for this was the low, forward ramp (with less frontal area than the full, forward fairing) provided less protection to the APT against the elements. Exposed to the full force of the turbulent airstream, the turret shook more, resulting in more beam jitter which, in turn, smeared the beam on the target spot and accounted for lower average intensity.[44]

What also helped to reduce beam disturbance were the fairings used during Cycle II that smoothed out the airflow next to the APT. Various fairings were tested (that had been developed during the earlier wind-tunnel experiments) using dozens of transducers to collect pressure readings from points all over each fairing. Mounted flush to the fairings and side of the fuselage were tiny semiconductor devices, each with a diaphragm with a small strain gauge attached to measure the movement of the diaphragm. As pressure fluctuations bent the diaphragm, it sent a signal that recorded how much pressure was imparted on the diaphragm. This measurement confirmed the full forward, full aft fairing combination offered the best protection. But this configuration restricted the APT's field of view to only off the left side of the plane. Testing of a low forward ramp fairing in combination with a partial aft proved most efficient for line of sight, but also had the drawback of inducing more jitter in the APT.[45]

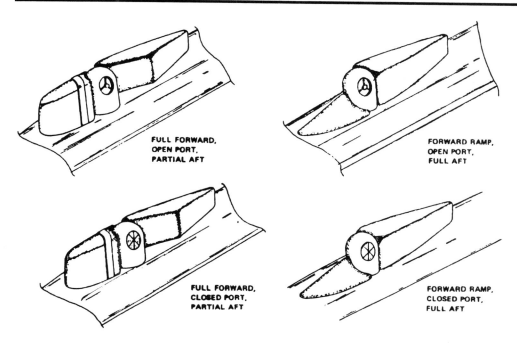

Fairing configurations used in Cycle II.

Jitter, defined as a perturbation of the beam propagation axis from the desired line of sight, proved to be the most frustrating problem with the Cycle II experiments. Vibrations induced mainly by the aircraft engines and aerodynamic forces moved the beam around (up and down and left and right creating a random motion). The tracker and autoalignment system simply could not remove all of the jitter. A small amount of jitter in the airplane translated into a large amount of beam movement on target. For example, a focused 10-centimeter beam in a quiescent environment would not move off the target spot. But jitter moved the center of the beam's diameter off the center of the target spot. This movement of the 10-centimeter beam distributed the beam's energy over a larger target area, thus reducing the beam's killing power. In effect, the moving 10-centimeter beam functioned as a weaker 20- or 25-centimeter beam in terms of killing power. For Cycle II, the goal was to establish a pointing accuracy so the beam did not move more than 20 microradians.[*] This goal was met for several ideal flight tests, but the "average" pointing accuracy for the entire Cycle II testing was 30 microradians, or about two times better than that achieved with the FTT on the ground in Project DELTA.[46]

Defining where the beam hit the target in 371, scientists could correlate this measurement to what was happening on the ALL. By working back from where the beam hit the

---

[*]The distance the beam moved off the target aimpoint is a small angular measurement expressed in microradians. A radian is an angle of just less than 60 degrees (57.295 degrees, to be more precise). A microradian is a millionth of a radian or 0.000057 degree. Twenty-five microradians is an angle of 0.001425 degree.

target in 371 to the laser device in the ALL, corrections to the alignment system could be made to move the beam to zero in on exactly the right target spot. Captains Bob Van Allen and Dennis Maier were two of the smart Air Force scientists who used this data base to build mathematical models to predict how the alignment system—mirror angles, the drive mechanism for the alignment mirrors, torque on the mirrors—would respond to a high-power laser for Cycle III. Without these essential Cycle II data, none of these models could be built; this probably would have delayed the ALL program for at least a year.[47]

Boesen and others at AFWL were bombarded with criticism from people outside the laboratory who claimed the challenging Cycle II pointing goals could not be met in the airborne environ-

Diagnostic aircraft used during Cycle II flight tests.

ment. But Cycle II proved the critics were wrong. Boesen conceded that only in the best of situations "when everything was working exactly right" were the technical performance goals met. But that was often enough to confirm that it was possible to overcome the jitter problem. "Cycle II," Boesen noted, "achieved 2 or 3 orders of magnitude better jitter stabilization than anybody had stabilized anything before."[48]

Once the Cycle IIA flight tests ended in the summer of 1975, the APT was removed from the ALL and placed back in the optics laboratory. Captain Dennis Maier, working with another optics expert, Dr. Al Saxman, had conducted a number of beam diagnostic experiments that convinced them that a new primary mirror was needed for Cycle IIB to improve focusing of the beam. For the next 4 months, work proceeded taking out the Cycle IIA primary and installing a new one manufactured by GD (Convair Division) of San Diego.[49]

The primary mirror used in Cycle IIA was made of glass; its trade name was DURAN-50. It was replaced in Cycle IIB with the new GD mirror. This improved mirror substrate was made of glass-ceramic material known as Cer-Vit (ceramic vitreous). Owens Illinois produced the Cer-Vit mirror blanks. Tinsley Laboratories of Berkeley, California, served as the major mirror fabrication subcontractor. A graphite-epoxy eggcrate structure formed the back of this mirror to provide strength and support. Bonded to this backing was the front section (faceplate) of the mirror consisting of a less than half-inch layer of Cer-Vit covered with a thin coating of silver, which served as a smooth and highly reflective surface when polished. This glass-ceramic faceplate had a very low coefficient of thermal expansion, meaning the mirror did not readily change the shape of its reflective surface in response to changes in ambient temperature or while reflecting the laser beam. Installation and alignment of the new mirror in the APT were performed by AFWL, Hughes Aircraft, and Dynalectron personnel.[50]

Of all of the mirrors in the ALL's optical train, the primary mirror was the largest (60 centimeters across). (The primary actually measured 68 centimeters across, of which 60 centimeters was used to reflect the beam. The outer 8 centimeters of the mirror was used to reflect the annular autoalignment beam.) Because of its size, it did not have to be water cooled. As mentioned earlier, the beam expanded as it moved from the secondary to the primary. Thus, the intensity of the beam's energy was not as concentrated, because it covered a larger surface area, when it struck the primary mirror. This allowed the laser to operate and be in contact with an "uncooled" mirror for several seconds. For Cycle IIB, none of the mirrors in the optical train required cooling because a low-power beam was used which did not damage or distort the mirrors. But for Cycle III, the higher laser power dictated all beam transfer mirrors in the ALL optical train be cooled, except for the primary. This meant replacing Cycle II uncooled mirrors with water-cooled mirrors in Cycle III to prevent damage by a high-power beam. The primary was not cooled for Cycle III, because its surface was large enough to withstand the intensity of even a multikilowatt laser.[51]

In between Cycle IIA and IIB, while the APT was being worked over in the optics lab, AFWL utilized the ALL still parked near the ARTF to conduct another series of beam testing known as the Airborne Laser Propagation Experiment (ALPE). The purpose was to gain more data on how the atmospheric turbulence and aircraft boundary layer affected the quality of the beam when propagating at shorter wavelengths (2- to 4-micron helium–neon laser) as opposed to the 10.6-micron ALL laser. ALPE had no direct connection with the ALL, except that the ALL was a convenient and available resource with the right diagnostic equipment to conduct the propagation experiments. ALPE experiments were conducted both on the ground and in the air. Between 7 and 16 September 1975, 12 aircraft-to-aircraft (ALL to 371) ALPE experiments produced "extremely good" data over 27 hours of flight time. The basic research data derived from ALPE, AFWL reasoned, would expand the data base and be useful in appraising shorter-wavelength lasers as potential weapons. By the mid-1970s, most believed shorter-wavelength lasers would, in the long run, be the best candidates for laser weapons.[52]

Although ALPE was an important set of experiments, its significance was secondary to other work carried out from the end of Cycle IIA flight testing in July to the resumption of airborne experiments at the start of Cycle IIB in November. During this time AFWL scientists had an opportunity to stop and analyze data collected in Cycle IIA and, based on these findings, prepare for Cycle IIB testing. Essentially, Cycle IIB was a repeat of many of the experiments conducted in Cycle IIA. The difference was that numerous changes were made to the various ALL subsystems for Cycle IIB in an attempt to obtain better performance numbers in regard to beam jitter, power, and quality, as well as tracking and pointing accuracies. Also, work continued on trying to better understand what the air boundary layer along the plane would do to degrade the beam. Measurements were made on aerodynamic turbulence and how much of that would be transmitted as jitter into the aircraft and cause the beam to wander.[53]

One of the major changes in anticipation of Cycle IIB involved removing the tracker from the APT, disassembling it, and replacing some of its key components. In an effort to reduce the high noise-to-signal ratio, which degraded tracking accuracy, AFWL and Hughes technicians worked hand in hand and replaced all of the electric circuits in the tracker. A second substantial change was the removal of the inferior detectors in the tracker, which were replaced with new detectors better suited to detect radiation off the target in only the 3- to 5-micron region. Not only was the tracker reworked, but the new primary mirror was installed and aligned in the APT, and then tested on the ground. (The new primary was assembled and polished at the contractor's facility before the end of Cycle IIA.) This was a lengthy operation, because the entire turret had to be disassembled to gain access to the area to mount the primary. Also, the secondary mirror was adjusted to reduce movement of its mirror mounts. In addition, the inner gimbals on the APT were carefully balanced to minimize vibration and inertia, and the autoalignment reference ring was modified to suppress the mechanical resonance problem identified in Cycle IIA.[54]

With these and other modifications made, AFWL flew over 20 Cycle IIB flights to collect test data on the new optics to determine the APT pointing and tracking performance. In every case, each subsystem (e.g., tracker, IPAS, autoalignment, low-power window, optical train) performed better in Cycle IIB than in IIA. Plus, IPAS was a technological breakthrough for both Cycles IIA and IIB, because it was the first in-flight system capable of compensating for optical misalignment (beam moving off its desired optical path) caused by aircraft vibrations. The airborne system hardware was still not perfect, but it was inching ahead step by step as a result of the improved empirical data gathered in Cycle IIB. Pleased with the outcome, Lamberson and his staff briefed the Cycle II findings to Dr. Hans Mark and his High-Energy Laser Review Group (HELRG) in the spring of 1976. (DDR&E created HELRG in October 1972 to serve as a special advisory and steering group for laser research and development.) Endorsement of the technical progress achieved in Cycle II by the independent HELRG was a crucial step in persuading Systems Command to give the approval to AFWL to move ahead with the high-power experiments of Cycle III.[55]

Lamberson's briefing on 8 June 1976 to the HELRG proved very convincing. His main message was that Cycle II evidence clearly demonstrated a laser beam (20–25 minutes of focused beam time was obtained on target) could be generated on board a KC-135 aircraft and directed inside the plane to the APT that then focused and pointed the beam to hit an aerial target. This was an event of major proportions, because it was the first time in history a laser had been fired from an airborne platform to engage an airborne target. However, Cycle II accomplishments were intentionally not publicized because, at the time, the ALL was a highly classified program. As spectacular as this unpublicized achievement was, more important in the long run were the extensive scientific measurements collected to create a data base that was invaluable to AFWL for understanding the physics and engineering of the optical subsystem and laser device required in predicting how the Cycle III systems would operate.[56]

There were many lessons learned from Cycle II experiments critical to the selection of improved hardware for Cycle III. The IR tracker used in Cycle II, for example, was determined to be inadequate for Cycle III requirements. This reconfirmed earlier suspicions during Cycle I that the 3- to 5-micron sensor was the wrong tracker to use for the high-power testing planned for Cycle III. Performance of the Cycle IIB tracker was good for a few tests, but in most cases inconsistency was the problem. The best single tracking experiment occurred in March 1976, when the tracker demonstrated it could track at 9.7 microradians of jitter. A few experiments showed tracking precision on the average to be about 16 microradians. In most cases, it was well above 16 microradians.[57]

A variety of trackers (visible, 3- to 5-micron, and 8- to 12-micron) tested at Oscura Peak at White Sands Missile Range as part of the OPTICS (Oscura Peak Tracker Investigation and Comparison Series) experiments, at the same time Cycle II ground and airborne experiments were under way, showed the 8- to 12-micron tracker was superior to all others in terms of precision tracking of tactical targets. (The 3–5 micron was considered better for tracking ICBM boosters, whose plume presented a hotter target.) After an extensive series of tracking tests against air-to-air missiles (AIM-9B Sidewinders and 7E Sparrows), 12 Army HAWK surface-to-air missiles, and a number of aircraft, including the F-4, F-15, F-111, and T-38, AFWL decided to design and build an "advanced tracker" to be used in Cycle III. The OPTICS tests showed the superiority of correlation over centroid tracking. In the final Cycle II technical report, AFWL scientists provided the following technical description of the proposed new tracker:

> The new tracker will have a digital correlation processor. This processor will use video information from a new highly stabilized television sensor and from a serial scan 8- to 12.5-micron infrared imager. The new tracker imager subsystem is expected to result in improved reliability as well as reduced jitter.

Also, to ensure the tracker operated at its peak performance level, improvements in the accuracy of the range finder, which fed target distance figures to the tracker, were underway.[58]

The Cycle II APT pointing accuracies for directing the beam on target measured 30 microradians on the average. The best single experiment for pointing accuracy was 10.7 microradians. Excessive jitter was the main cause of pointing inaccuracies. The inability to reduce jitter to lower levels was attributed to a number of factors.[59]

One was the IPAS mirrors needed a more rigid mounting system. However, this was considered to be only a "minor" contribution to the overall system jitter. The operation of the APT accounted for a large share of beam jitter. Problems still existed with getting the outer gimbals to smoothly rotate the APT. Jitter also increased or decreased depending on the fairing configuration used to protect the APT during flight tests. For example, twice as much jitter occurred when using the ramp forward than the full forward fairing; the ramp forward provided the least protection, causing more movement of APT components.[60]

The jitter problem was never completely resolved during Cycle II. Neither was much progress made in improving the optical quality of the mirrors. The major optical degradation was in the primary mirror which affected the mirror's focusing ability. Depressions in the

mirror surface and extremely cold temperatures were the primary factors responsible for inferior focusing. Cycle IIB did, however, demonstrate a lightweight primary mirror could be fabricated using a graphite epoxy support structure and a thin Cer-Vit faceplate. What was disappointing, though, was the new mirror performed only slightly better than the Cycle IIA mirror it replaced. In addition, problems still existed with the annular reference mirror at the base of the APT. Technicians never seemed to be able to mount it properly. As a result, the reference mirror failed to provide an accurate reference for the IPAS system, especially during the last flights of Cycle IIB. More work was needed to upgrade the mirrors for Cycle III testing.[61]

Additional work also spilled over into another phase of Cycle II identified as Cycle II.5, which actually extended to May 1977 or 14 months beyond the end of Cycle IIB in March 1976. The goal of Cycle II.5 was to better understand beam propagation phenomena as the beam passed through the airflow boundary layer surrounding the aircraft. These flight experiments measured turbulence at various altitudes and diagnosed the effects of turbulence on beam quality. Measurements were also taken on temperature variations in the atmosphere and how that influenced beam degradation. A final goal was to better define the source of jitter on the aircraft. For these flight experiments, only the diagnostic aircraft (371) was used. The ALL was not involved in this work. Captains Keith Gilbert and John Otten took the lead on these experiments.[62]

Using a very-low-power laser mounted inside 371, AFWL for the Cycle II.5 experiments transmitted a beam through the aircraft window to the outside for a distance of about 20 inches. There the beam reflected off a mirror attached to a rectangular fairing (mounted on the right side of the fuselage in line with the window) and passed through the window and into 371. Under this scheme, the beam passed through the turbulent boundary layer alongside the fuselage twice—once on the way out and once on the return route into 371. Diagnostics on 371, as Gilbert pointed out,

> would measure degradation of the laser beam and do it for a variety of aircraft Mach numbers, altitudes, and so on, so we were able to come up with a repository of data on how laser beams are affected by flow fields around airplanes. We did the same thing in wind tunnels on the turrets, trying to understand flow field and how it could affect the beam—the defocus it would cause—because when the flow comes around a large protuberance it forms, effectively, a lens, a very weak lens, and so it can change the focus of the beam. And in addition, you have turbulence there, which can spread the beam or divert it.

The basic conclusion derived from these data was that the airflow boundary layer around the aircraft would not significantly degrade the quality of the laser beam as it passed through that turbulent region. However, the shorter the laser wavelength, the more serious were effects of the aircraft boundary. This data base was the first to show that aircraft flow fields can become a major degradation source for future candidate visible- and near-visible-wavelength laser weapon systems.[63]

Cycle II clearly showed more work was needed to upgrade the tracker and optical quality of the mirrors and stabilization of the APT. One of the major bright spots in Cycle II was the very successful design, fabrication, and performance of the low-power window in the APT. Ground and flight testing consistently demonstrated the window met all reflectivity, transmissivity, attenuation (absorption and scattering), thermal, and pressure specifications. Based on these positive findings, AFWL felt confident to design and build a single-pane, 1.52-centimeter-thick zinc selenide window consisting of three sections, instead of six, for Cycle III testing. Switching to three sections reduced aperture obstructions by eliminating three of the window's support struts.[64]

In the end, Cycle II represented proof that a laser could be projected from an airplane and intercept an aerial target. This demonstration verified that the theory of airborne lasers could be applied in the real world, setting the scene for even more ambitious follow-on laser experiments. Kyrazis summed up best the value of Cycle II when he stated:

> Perhaps the most satisfying moment of my whole involvement with ALL occurred near the end of Cycle II. We had just landed from another one of our long missions. Several of us were walking back to the hangar and were reflecting on the results of the flight. We suddenly realized that we knew, with absolute certainty, that the Cycle III demonstrations would work. The results of that mission, coupled with what ALL taught us from earlier flights, gave that certain knowledge.[65]

HELRG, tasked to evaluate Cycle II accomplishments, concurred with Kyrazis's assessment. Meeting at AFWL in June 1976, HELRG, chaired by Dr. Hans Mark, conducted a 1-day review of Cycle II and concluded it had met its original objectives. Committee members were optimistic that beam power could be considerably improved during Cycle III. With installation of high-power quality mirrors with better reflecting surfaces and upgrades to the APT, beam power losses would be reduced throughout the optical system. This meant that with the 500-kilowatt CO_2 laser installed on the ALL during Cycle III, between 50 and 100 kilowatts of beam power could be placed on a target at the 1-kilometer range. This would equate to roughly 100 to 200 watts per square centimeter or enough power to damage the guidance systems of most heat seeking missiles. Based on its review, HELRG advised Dr. Robert Greenberg in DOD's Office of the Director of Defense Research and Engineering that "the Cycle II objectives have largely been achieved and that the Air Force should proceed to go ahead with Cycle III tests of the Airborne Laser Laboratory as presently planned."[66]

Completion of the collection of Cycle II experiments clearly signified a major watershed in terms of accurately pointing a laser beam in an airborne environment. Every function, except for high-power laser output, needed for a potential weapon system was incorporated into the ALL during Cycle II. However, Cycle II did not stand alone. It was one integral piece of the entire ALL mosaic. Cycle II provided the raw data to give AFWL the confidence that scaling up the power of the laser for Cycle III would result in a laser kill of a high-speed airborne target. Without the knowledge derived from Cycle II, the ALL would have stalled in place. But that was not the case. Cycle II's success forced AFWL workers to shift gears from their primary focus on the scientific "proof of concept" approach to embrace a new way of thinking. They now had to face an enormously complex engineering project to design, build, and integrate a vastly more powerful laser with new water-cooled optics. To find out if a CO_2 laser could be scaled to higher power levels and accurately pointed from an airplane to an aerial target was the formidable challenge that lay ahead for Cycle III.

First Steps to Ready
High-Power Laser

Cycle III: Phases 1 and 2

When the ALL program started, most Air Force officials confidently predicted a high-energy laser would be installed and operating on an airplane by the mid-1970s. That technological forecast turned out to be overly optimistic for two reasons.

First, laser components such as mirrors, optical alignment systems, resonators, diffusers, fuel lines, IR trackers, and combustors generally worked well under tightly controlled laboratory conditions. Data from laboratory experiments gave scientists fundamental building blocks of knowledge for gaining a better understanding of the physics of lasers. But when more powerful lasers and larger and more complicated subsystems were put inside the ALL and subjected to outside aerodynamic forces (e.g., turbulence, bad weather) and aircraft flexing and vibrations, they behaved differently than in the laboratory. The message was there was still more to learn about the basic theory and physics of lasers before improving their performance in the unpredictable real-world environment.[1]

A second stumbling block that extended the length of the ALL program was the problem of engineering many different laser subsystems into one consolidated system aboard the ALL. Lamberson referred to this process as the "science of integration" which turned out to be a major obstacle that AFWL had not anticipated at the beginning of the ALL program. Design and fabrication of numerous complex components of the ALL occurred simultane-

ously, being done by different contractors and subcontractors at different locations throughout the country, thus making it difficult to fit pieces of the system together prior to the final assembly phase. For example, General Dynamics (Fort Worth) had to design and make extensive structural modifications to the aircraft so it could accept the various laser subsystems and safely carry out the flight experiments. (The Air Force identified reconfiguring the ALL as a major "class II" aircraft modification.) Pratt & Whitney (West Palm Beach) had responsibility for designing and developing the laser device and the fluid supply system (FSS). Building the airborne pointer and advanced tracker was the job of Hughes Aircraft Company (Culver City). AFWL contracted Perkin-Elmer (Norwalk, Connecticut) to develop an extremely complex airborne dynamic alignment system (ADAS) designed to steer the beam from the laser device inside the airplane to the APT on top of the fuselage. Garrett AiResearch Company (Los Angeles), Hughes, Pratt & Whitney, and Perkin-Elmer were responsible for building high-power water-cooled mirrors for various optical subsystems. Building all of the mirrors was easy in comparison to engineering difficulties AFWL and its contractors encountered integrating the separate modules to make them work as a synchronized system. Plus, another critical factor was the few spare parts for any of the major subsystems; any modifications to hardware had to be customized on the spot.[2]

During Cycle III, many of the physics problems were solved fairly early. What delayed completion of Cycle III more than anything else was the continual engineering complications with the equipment inside the ALL. Cycles I and II, with their simple hardware relative to Cycle III, met their program objectives in rather short order—7 months and $2\frac{1}{2}$ years, respectively. Cycle III would take over 7 years, frustrating AFWL workers as well as those supporters up the chain at Air Force and DOD levels; some of them for the first time began having second thoughts about whether the ALL would ever get off the ground.

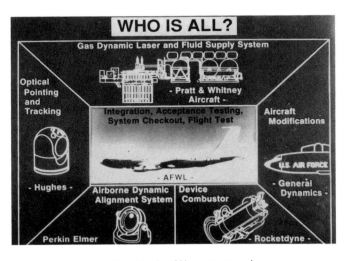

Contractors for ALL components.

Although there were many changes made to the ALL during Cycle III, the basic objectives for that portion of the program remained unaltered. In keeping with the idea the ALL was first and foremost a flying laboratory, scientists wanted to come away from Cycle III with a better understanding of the physical phenomena of lasers and optics, which influenced total system performance. The basic question centered on whether the theories of high-power lasers operating on paper and proven on the ground with the TSL at SOR in 1973 could be transferred to an airborne platform. Proving the theory required confirming that the engineering design and integration of the various laser subsystems actually worked in flight.[3]

Demonstrating the ALL could operate effectively against aerial targets was the core of Cycle III. As a first step to meet this objective, the ALL had to demonstrate its capability to disable or destroy an air-to-air missile approaching the rear of the aircraft. This was referred to as the 1B or bomber self-defense scenario. These experiments, hopefully, would disable aerial targets and, in the process, collect valuable real-world data to verify computer estimates of specific target vulnerability. A second scenario, ship-escort defense, consisted of proving an airborne laser flying at low altitude (1000 feet) could protect a ship from a low-level cruise missile skimming over the ocean. AFWL substituted a BQM-34 drone for the cruise missile. The laboratory reasoned, by the time it was ready to conduct the ship-defense scenario, it would already have collected vulnerability data on missiles from the bomber self-defense experiments. A drone's construction resembled an aircraft more than a missile. Scientists wanted aircraft vulnerability data to compare with results obtained from the laser/missile intercept. To simulate the ship-defense scenario, the ALL would fly over the ocean at low altitude. This would also allow AFWL to determine how the beam would be affected when propagating through the moist salt air.[4]

FIVE-PHASE ROAD MAP

Before any targets could be shot out of the sky, AFWL followed a rigorous milestone-driven process with many checks and balances that evolved into the operational ALL. Consistent with Lamberson's guidance, the laboratory managed the design, fabrication, and test of the various subsystems (e.g., device, pointer and tracker, FSS). Design and building of the different pieces of hardware overlapped. However, integrating and ground testing each subsystem would occur incrementally. This time-phased ground work first took place in the test cell located at the ARTF facility.[*] There each subsystem would be assembled and tested. Then all components would be connected and tested to demonstrate high-power operation of the entire laser system. Next, each subsystem was disassembled in the test cell and any design changes to improve operating performance were made before reassembling the entire system inside the aircraft on the ground. Full laser system performance first had to be demonstrated on board the aircraft. This included the ALL on the ground engaging stationary and moving targets downrange behind the ARTF building. Once the aircraft ground testing finished, the final phase would be to flight-test the laser system to demonstrate ALL lethality by the destruction of aerial targets (missiles and drones) in realistic engagement scenarios. An important feature of Cycle III, according to Colonel Dick Feaster, Director of ALL Test and Operations, was to establish an operations and technology baseline for follow-on test programs (Cycle IV) and future prototype decisions. The plan was that ALL would not die after Cycle III; more advanced versions would follow. This was to be Cycle IV that would use the same Cycle III 10.6-micron laser. Cycle IV would also incorporate an advanced system known as the High-Energy Laser Radar Acquisition and Tracking System (HEL-RATS), which AFWL planned to install on the ALL aircraft. Cycle IV experiments consisted of engaging multiple targets, in this case two AIM-9 missiles. Application studies would also

---

[*]The test cell was a laboratory with sophisticated instrumentation and control equipment for checking out and testing each component. Once that was done, components were integrated and evaluated as one complete ALL system.

evaluate more advanced shorter-wavelength airborne lasers capable of engaging multiple targets, such as low orbiting satellites and sea-launched ballistic missiles. In the long range, after the completion of Cycle IV, AFWL envisioned developing the "ALL-II" using an advanced laser and fire control system aboard a wide-body jet, such as a 747. Although this never happened as part of the ALL program, the new airborne laser (ABL) initiated by the Air Force in the 1990s used the same proposed ALL-II concept for a theater ballistic missile defense mission.[5]

To expedite the final part of the ALL program, Lamberson divided Cycle III into five distinct phases. Phase 1 involved flying the ALL aircraft to GD's facility in Fort Worth to undergo major structural, mechanical, and electrical modifications. Once this was completed, the Air Force would conduct preliminary functional flight tests in Fort Worth to make sure the plane was airworthy when flying at limited speeds. (During these airborne tests at GD, the laser, APT, and fairings were not on the ALL.) Then the ALL would be flown via Albuquerque to the Air Force Flight Test Center at Edwards AFB, California. Because the ALL was a completely reconfigured plane, it had to be recertified to verify it could fly safely within its operational envelope. Flight certification testing at Edwards was phase 2 and assessed performance and stability to include structural integrity, flutter, fairing wake turbulence, and overall handling characteristics of the airplane.[6]

Phase 3 focused on testing and evaluating the FSS, which contained the laser fuels (gases and liquids), the essential ingredients to create the right conditions for lasing to occur. Some of these fuels were toxic and explosive, demanding the crew follow strict safety procedures. After installation of the FSS in the aircraft on the ground, flight testing of the system would follow to verify pressure specifications and to detect leaks in the feed lines and the holding tanks. Assurance of FSS performance and structural integrity would give the go-ahead for the start of phase 4. This involved connecting the laser device in the ALL with the FSS to test the operation of these two major components as one system. Testing would take place on the ground (first in the test cell and later in the aircraft on the ground) and then in the air. Initially, as a safety precaution, no beam would be extracted. A beam would be extracted and directed to a target during the final phase of testing in the test cell and with the total system installed in the plane on the ground.[7]

Phase 5 consisted of two parts. Installing the mirrors would be the first step. Phase 5 also would include integration of the APT, ADAS, and optical diagnostic systems with the FSS and laser device to be tested in its full-up mode on the aircraft on the ground. A high-power beam would be extracted from the device and directed to targets downrange from the ARTF. The second part of phase 5 would be the airborne demonstrations to be conducted against tow targets pulled by F-4 aircraft at White Sands Missile Range, followed by testing the ALL's effectiveness against air-to-air missiles at the Naval Weapons Center's range at China Lake, California. For its last test, the ALL would engage Navy drones off the coast of Point Mugu, California, which was part of the Pacific Missile Range. Success with these shootdowns would signal the completion of the ALL program and either confirm or deny the feasibility of using a laser in an airborne environment.[8]

Although Lamberson had laid out the general road map for the Cycle III phases, he, Kyrazis, and others turned to and came to rely on Lieutenant Colonel Bill Dettmer as one of the key figures in the planning and implementing of the details of the ALL Cycle III ground and flight tests. Dettmer was a fortuitous addition to the ALL team and was another prime example of getting the right person at the right time to best serve the program.[9]

In early 1975, Dettmer was finishing his Ph.D. research at the Aerospace Research Laboratories at Wright-Patterson AFB when he learned of the ALL program. Knowing his flight test experience and academic background bridged the gap between the aircraft and the laser, he convinced Kyrazis and Colonel Tommy Bell, the 4950th Test Wing Commander, to bring him into the program. That summer Dettmer was transferred to the 4950th to serve as Bell's assistant for the ALL. His primary responsibility was integration of 4950th and AFWL resources to accomplish the Cycle III flight testing of the ALL. In the process, he became very familiar with the operation and laser components of the aircraft.[10]

A year later, Colonel Tommy Bell, commander of the 4950th, sent Dettmer to command Detachment 2 of the 4950th located at Kirtland AFB. Bell was highly impressed with Dettmer's capabilities and especially his attention to detail in planning long-range projects. As the Detachment 2 commander, Dettmer was in the position to head up the critical on-site team that could quickly respond to establishing flight test schedules, as well as attending to operational and maintenance requirements of the ALL. Dettmer's new assignment also coincided with Kyrazis's appointment as head of AFWL's Laser Development Division. The timing of these events and subsequent excellent work relationship forged between Kyrazis and Dettmer over the next few years turned out to be one of the main factors responsible for the success of the ALL.[11]

Dettmer was ideally suited for his job at Kirtland because of his strong operational background. He had flown B-47s, served as an Air Operations Program Development Officer in Vietnam, and later was a flight test pilot for a number of experimental programs on production aircraft. In short, he knew airplanes and especially the ALL as a result of his assignment as "Assistant for the ALL" with the 4950th. As one of his co-workers put it, "He had smarts, understood the value of progressing step by step from ground to flight testing, and knew every detail of the ALL aircraft." Consequently, he had built a credibility base that he used to convince the 4950th, whose pilots had to fly the airplane, to approve acceptance of the very extensive Cycle III modifications to the ALL aircraft. Over a short period of time the 4950th had gained a great deal of confidence and trust in Dettmer's ability and did not hesitate to accept his recommendation that the ALL would be safe to fly.[12]

A major share of Dettmer's contribution was formalized in the ALL five-phase Cycle III test plan. Dettmer, while assigned to the 4950th, was one of the primary authors of this important working document. The plan was designed to be modular, with the initial module being an overall executive description of the test program. Subsequent modules provided details of the five phases of the Cycle III tests. These modules were spread out throughout Cycle III so as to take advantage of new information as it was developed during the test program and to smooth the work load. This plan served as the blueprint providing the details for incremental testing of specific ALL subsystems during ground, flight, and scenario tests. Over the years as testing progressed, this plan was continually revised to eventually include 12 volumes/annexes to the volume I executive summary. These additional volumes covered specific test descriptions, requirements for instrumentation, communications, security,

safety, crew training, and data reduction, as well as any other support resources to accomplish the test goals.[13]

AIRCRAFT MODIFICATIONS AND FLIGHT TESTING: PHASES 1 AND 2

Shortly after completion of Cycle II in March 1976, Lamberson sent the ALL to GD's facility in Fort Worth. There the plane underwent major modifications so it would be ready to accept all of the laser subsystems to conduct Cycle III flight testing. Having modified the ALL for Cycles I and II, the GD team, headed by T. Peyton Robinson, had gained valuable hands-on experience and had become very familiar with all structural characteristics of the aircraft. Consequently, the Fort Worth contractor was in a unique position to understand and deliver the follow-on changes AFWL wanted to make on the ALL for Cycle III.

During the year and a half the ALL was down at the GD plant to undergo major modifications, the Air Force needed a capable individual to head an on-site team that could closely monitor and report on the contractor's progress. That job of Chief Modification Engineer fell to Phil Panzarella of the 4950th. One of the reasons he was selected for this critical position was his strong background in civil engineering and practical horse sense for "making things work," as one of his colleagues put it. While at the Aeronautical Systems Division at Wright-Patterson AFB, Ohio, he had gained valuable experience working on a variety of aircraft while serving in positions of aerospace engineer, project engineer, test director, and chief modification engineer. In addition, he had an impressive list of educational credentials. After obtaining a bachelor of science degree in aeronautical engineering from St. Louis University, he went on to earn a master's degree in aerospace/mechanical engineering at the Air Force Institute of Technology in 1960. Later, he was selected for the prestigious Sloan Fellowship at the Massachusetts Institute of Technology, where he received a second master's degree in industrial engineering in 1979.[14]

As Chief Modification Engineer for the ALL, Panzarella turned out to be an invaluable resource. At first, he tried commuting to his new job site in Fort Worth. He quickly realized this would not work out because he virtually had to be available on the shop floor seven days a week. As a result, he moved to Fort Worth in the late summer of 1976 and now could closely watch each step in the modification process of the laser laboratory aircraft.[15]

Panzarella had helped develop, through fault tree analysis, the probability of failure for the Cycle III modified ALL aircraft. The findings were revealing. Chances for flight failure of a normal KC-135 were one in a million. With modifications that changed the structural integrity of the ALL, chances for flight failure were drastically increased to one in 10,000. Because of this, Panzarella and his team had to be extra careful in checking and approving every single detail on the line drawings to ensure that what was on the blueprint correctly transferred to structural changes made to the aircraft. This was extremely time-consuming, but was absolutely necessary to maintain the highest safety standards for the crew and aircraft.[16]

To assist him, Panzarella had three engineers from the Aeronautical Systems Division and Sergeants Ken Vanderwall and Jim Augustine from the 4950th who would be assigned as the crew chief and flight mechanic on the ALL during Cycle III flight testing. By the time the aircraft had undergone all of the modifications, Vanderwall and Augustine, through their on-the-job training, had become intimately familiar with every aspect of the plane. Again, that knowledge would become invaluable during the Cycle III flight experiments. They also made a major safety contribution in writing many of the emergency procedures and special maintenance manuals, as well as developing the aircraft pressurization schedule and establishing safety inspection criteria for numerous on-board systems.[17]

Before any structural changes could be made to the aircraft, GD had to come up with a Cycle III engineering design plan for installing the laser and all other support equipment. This involved evaluating the advantages and disadvantages of several different design approaches. GD officials held numerous technical exchange meetings to collect input from the prime contractors who were designing the critical components for Cycle III: Hughes for the APT, Pratt & Whitney for the laser, Rocketdyne for the GDL combustor, and Perkin-Elmer for the ADAS. Most of the discussion concerned technical trade-offs and trying to reach agreement on a recommended design. The goal was to ensure the modifications would provide sufficient space on the ALL for each critical subsystem and its auxiliary equipment to function, without posing any safety hazards to the crew. What everyone wanted to avoid was any design change that might interfere with or degrade the operational efficiency of any one component. On 31 October 1974, AFWL reviewed and approved GD's preliminary design plan.[18]

That was only the first hurdle for GD. The contractor had to go back to the drawing board and develop a more detailed blueprint for every modification. Using the preliminary design as a guide, GD completed the final engineering drawings 8 months later. At the end of June 1975 GD presented its proposal to AFWL at the critical design review meeting. Over the years the lab and GD had fostered a strong working relationship based on GD's consistency in delivering quality results. So no one was surprised when AFWL quickly approved the contractor's final engineering design plan for modifying the aircraft. As the ALL would not complete its Cycle II testing until the spring of 1976, AFWL officials directed GD to continue to reevaluate and refine the final engineering drawings.[19]

In mid-May 1976, the GD team at Fort Worth completed its initial inspection, cleaning, and weighing of the ALL. For the next 14 months, the contractor performed one of the most extensive modifications ever made on any aircraft. This major overhaul ranged from cutting two holes in the belly of the plane to exhaust the hot laser gases, to rerouting flight control cables and installing airtight pressure bulkheads, as well as installing additional hardware to make the ALL as safe as possible for the crew and scientists. By the summer of 1977, the refurbished ALL resembled no other aircraft in the world.[20]

From the very start of the ALL program, Lamberson and others had made safety the number one priority. This trend received even more emphasis during the Cycle III modification because of greater potential hazards once the HEL was installed in the plane. People at Systems Command and the Pentagon became more concerned when they realized AFWL was getting ready to fly and fire two combustors inside the fuselage. Lamberson recalled, "A miniature rocket engine inside the fuselage all of a sudden started to get a lot of attention!" To deal with this concern ASD, who owned the ALL, instituted a series of Executive

Independent Review Teams (EIRTs) to periodically conduct safety inspections on the plane and its equipment. These interruptions were often time-consuming, resulting in schedule slips and additional expense in modifying equipment. But Lamberson believed these safety reviews were extremely valuable and made a significant contribution in heading off any potential dangers. He noted "there were no safety problems of any consequence on the aircraft," which he attributed in large part to the nagging EIRTs and to Panzarella's extremely capable team of Air Force workers at GD who closely monitored every aspect of the Cycle III aircraft modifications.[21]

One of the most dangerous components on the ALL was the FSS designed (by Pratt & Whitney) to deliver a variety of fluids needed to operate the laser. Five 4-inch-thick stainless-steel vacuum-jacketed (to keep the fluids cold) primary tanks (mounted on pallets) anchored to the floor behind the laser in the midsection of the aircraft made up the core of the FSS.[*] One of the two smaller spherical tanks (6.5 cubic feet) stored the liquid carbon monoxide (LCO) fuel and the other contained the oxidizer, liquid nitrous oxide (N_2O).[†] Two of the three larger tanks (20.7 cubic feet) were filled with helium (He) at 4200 pounds per square inch (psi) used to force the reactants—N_2O, LCO, and LN_2—from the storage tanks into the combustors (as gases by that point) at 2000 psi. A third large pallet tank contained the liquid nitrogen (LN_2). Overhead racks suspended from the ceiling contained methane (CH_4) and gaseous oxygen, which were combined and used as combustor igniters. Finally, two 5-cubic-foot cylindrical tanks bolted to the floor of the aircraft stored water for cooling the bench optics as well as the mirrors in the GDL resonator, ADAS, and the APT.[22]

Because some of these fluids (liquid CO and liquid nitrogen) were stored at cryogenic (very low) temperatures and high pressure, they posed catastrophic hazards. Leaks in the tanks and associated plumbing or incorrect mixing could prove deadly to personnel. CO, N_2O, He, and LN_2 were classified as either toxins or asphyxiates. For example, CO chemically bonds with the hemoglobin of blood, denying a person life-sustaining oxygen. When mixed incorrectly and ignited, CO and N_2O, or methane could detonate and cause the plane to explode in midair. Methane posed the greatest danger. Only half a pound of this gas leaking from the system could ignite and destroy the plane in flight.[23]

Several safety features were designed into the Cycle III aircraft to prevent any type of accident. One was the installation of a fire detection/extinguishing system, officially known as the Fire/Explosion Sensing/Suppression System (FESS). Fourteen ultraviolet flame

---

[*]Tanks mounted on pallets were loaded on the aircraft through the side cargo door. Pratt & Whitney subcontracted the fabrication of these high-pressure storage tanks to Beechcraft because of its experience in building similar tanks for the Apollo program. In the early design phases seven tanks were planned, but to conserve space and make the ALL more efficient five tanks were installed in the plane.

[†]These ingredients were stored in the ALL as liquids rather than gases. There was a limited amount of space inside the ALL, and more liquids, as opposed to gas, could be stuffed into the fixed amount of space available in the storage tanks. Liquid nitrogen first passed through the nozzle channel sidewalls of the laser device and diffuser to regeneratively cool both components; liquid carbon monoxide cooled the combustor. As these liquids picked up heat, they changed to gases and were eventually injected into the flow (downstream from the combustor) to improve the lasing action.

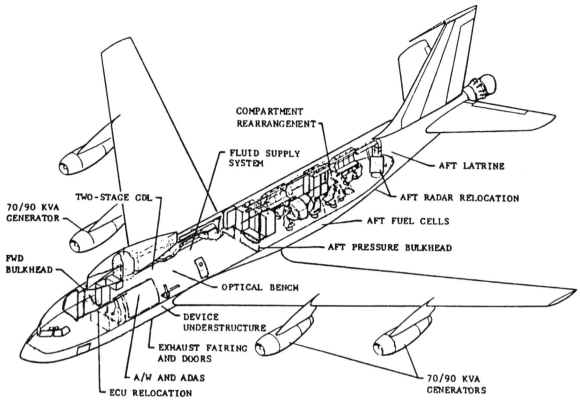

COMPARTMENT
REARRANGEMENT

FLUID SUPPLY
SYSTEM

AFT LATRINE

TWO-STAGE CDL

70/90 KVA
GENERATOR

AFT RADAR RELOCATION

AFT FUEL CELLS

FWD
BULKHEAD

AFT PRESSURE BULKHEAD

OPTICAL BENCH

DEVICE
UNDERSTRUCTURE

EXHAUST FAIRING
AND DOORS

A/W AND ADAS

ECU RELOCATION

70/90 KVA
GENERATORS

Major Cycle III changes.

as the Fire/Explosion Sensing/Suppression System (FESS). Fourteen ultraviolet flame detectors located throughout the device compartment could detect a fire or detonation in a few milliseconds. To extinguish a fire, a high-rate discharge of Halon 1301 agent (bromo-trifluoromethane) vaporized into a colorless and odorless gas in less than a second. This system worked so fast that Halon gas could stop a methane explosion in midfireball.[24]

A second major safety aspect was the design and fabrication of two airtight bulkheads separating the device compartment from the forward and aft crew compartments. (Because of structural changes made to the aircraft, the aft bulkhead could not be rigid. Consequently, a "floating" bulkhead was installed so it could flex as the aircraft flexed.) Made of aluminum and covered with a 0.032-inch stainless-steel skin, they served as pressure boundaries and firewalls. To prevent any toxic gases, especially carbon monoxide, from seeping into the personnel sections, the device compartment with the laser and FSS was pressurized 0.5 psi lower than the two adjacent compartments. In addition, each door in the bulkhead was equipped with a heat-resistant glass viewing port so crew members could visually spot any trouble when the laser was operating. Three closed-circuit color television cameras with zoom lenses installed in the device compartment kept constant watch on the laser device and FSS to help detect leaks, fires, or any abnormal operation of equipment. A fourth camera

Plumbing and fuel tanks of the fluid supply system on board ALL.

mounted in an explosion-proof housing in fuel cell number 2 (under the floor of the aft compartment) provided a TV picture of the exhaust plume. All of this could be monitored on the System Safety Operator's console in the rear of the ALL. An IR gas analysis detection system monitored any gas leaks in the device compartment. Blowout doors were also installed over the escape hatches on each side of the fuselage next to the wing so if an explosion occurred in the midsection of the ALL, the doors would blow out to relieve the overpressure and, hopefully, prevent an in-flight catastrophe. To contain a liquid spill, small cracks and openings in the device floor were sealed with URALANE, a tough two-component urethane resin to prevent any liquid from contacting and damaging flight control cables running under the floor. Finally, a specially designed emergency fuel dump mast assembly allowed for jettison of FSS liquids and gases through small openings in the right side of the ALL in case of in-flight emergencies.[25]

As mentioned, Panzarella's main job was to approve the engineering drawings of the aircraft modifications and to ensure that physical changes made to the aircraft complied with the specifications of the drawings. His and his team's first priority was always safety. Panzarella was well versed on the intent of all of the regulations governing safety. But he was also flexible enough not to be blocked by the strict letter of the regulation that might stall the ALL. His practical approach was to work within the guidelines of the regulations to keep the program moving.[26]

A good example of how Panzarella used the regulations to work in his favor concerned the requirement to secure high-pressure lines in the ALL every few inches. Because of the modified configuration of the ALL, this requirement could not be rigidly adhered to as there was not enough space available for adequate tie-down. Panzarella solved the problem by taking into account that this particular regulation was geared as a safety feature to be in place over the lifetime of an operational KC-135 aircraft. As the ALL was expected to fly only a few hundred hours, there would be little wear and tear on the high-pressure lines. Under these conditions the ALL was safe, and it was this type of decision that kept the program moving.[27]

The most critical structural change during Cycle III was cutting two 7-foot by 18-inch openings in the belly of the aircraft to exhaust the hot (approximately 1100 degrees) laser gases straight down to the outside air. With laser gases passing supersonically through a series of nozzles into the resonator (where lasing occurred), the pressure inside the resonator dropped dramatically—approximately one-tenth of normal atmospheric pressure or about 1.5 psi. As lasing took place, there needed to be some way to discharge the spent gases and heat from the laser device. Pratt & Whitney found the solution to this problem by designing and building a diffuser to bring the pressure back up to full atmosphere so the gases could be expended from the two holes cut in the bottom of the aircraft. The tapered shape of the diffuser compressed the gas flow to build the pressure up to ambient.[28]

Distribution manifold sits atop diffuser built by Pratt & Whitney. Diffuser expelled exhaust gases through two doors that opened on underside of aircraft fuselage. Pictured is one stage of laser device—two stages were flown in ALL.

No trivial amount of gases passed through the diffuser. It was about 4000 pounds of thrust. In comparison, each ALL engine exhausted from burning jet fuel approximately 11,000 pounds of thrust. Later on during the Cycle III flight testing, when the laser and diffuser were operating, the pilot had to compensate because the thrust from the downward flow of gases ejected from the underside of the fuselage would cause the nose of the aircraft to pitch upward. (Later the crew found that when the airplane was on autopilot, that system would automatically compensate for the upward pitch of the aircraft.) Intense heat generated by the exhaust thrust posed another problem. To reduce damaging effects of high temperatures from the gas flow, Pratt & Whitney lined the exhaust ducts (located directly below the

diffuser) with 0.09-inch corrugated titanium. The external duct walls were covered with fiberglass batting sandwiched between glass cloth.[29]

Two aluminum doors, hinged to the inboard and outboard sides of the exhaust slots, were opened and closed by hydraulic actuators. A layer of silicone rubber covered the inner skin of the doors for thermal protection.* The doors resembled mini bomb bay doors. It took only a second to fully extend each door to the open position; closing the doors required 5 seconds.[30]

To make room for the ALL to accept the diffuser and exhaust ducts required GD workers to cut through the floor beams, struts, stringers, and webbing that literally held the plane together. This weakened the structural integrity of the aircraft, causing GD to completely redesign and rebuild the underside of the fuselage around the two openings to ensure the plane would fly safely. Not only was the structural integrity of the plane affected by the two openings, but those changes also were responsible for causing a major rerouting of numerous aircraft control cables previously traveling along the section of the floor that the two openings now occupied.[31]

Eighteen flight control lines—four aileron, four elevator, two wing flaps, and eight main landing gear and brake cables—had to be rerouted around the two exhaust openings. (Of the 18 cables needing rerouting, 10 were on the right side and 8 were on the left side of the aircraft.) To do this, GD raised the entire device compartment floor 3 inches so there was enough room to accommodate all of the cables. This extra space was needed so cables (originally designed to run parallel to one another where the openings now existed) could be moved left or right and stacked

Flight of ALL with its two diffuser doors open—doors located on underside of aircraft forward of wings.

---

*Initially, Cycle III modification plans called for covering the fuselage downstream from the exhaust opening with a layer of silicone foam rubber as a thermal protection system. However, subsequent testing (using thermal temperature indicators, attached to the underside of the fuselage, to measure the temperature of the aircraft's skin) showed that extra protection for the underside of the fuselage was not necessary. This was because the GDL exhaust blown back by the airstream came in contact with the plane for only a few seconds, insufficient time to cause any damage.

above other cables along the portion of the floor undisturbed by the exhaust holes cut in the aircraft floor.[32]

Redirecting control cables had other important ramifications. One was the length of each control cable had to be increased to enable moving horizontally and vertically to get around the exhaust opening. AFWL decided to replace the carbon steel cables with ones made of corrosion-resistant steel (stainless steel). Grease (to prevent corrosion) and contaminants collected on carbon steel cables required more frequent inspections and maintenance; stainless steel did not need grease as a corrosion preventative. If any of the cryogenic liquids leaked and came in contact with the greased carbon cables, the grease could freeze and make those cables inoperative. Also, GD installed a new type of tensioning device consisting of an elaborate network of tension regulators and pulleys to compensate for excessive slack in the stainless-steel cables caused by temperature changes during flight operations.[33]

Concerns over temperature changes were not limited to control cables. To maintain comfort for the crew and to keep all of the electrical and mechanical systems operating at peak efficiency required reworking the environmental control unit (ECU) in the ALL. (The ECU was an evaporative, vapor-cycle air-conditioning system.) During Cycle I modifications, the basic ECU had been placed in the ALL. However, for Cycle III, to make room for the bulky FSS, laser device, and the optical bench, GD had to relocate the cooling (evaporators/fans) and heating elements of the ECU to the aft right side of the device compartment. One of the reasons for moving the ECU was that in Cycle I major hardware for Cycle III had not been finalized. During the rework, the contractor made maximum use of existing ducting. In the end, the modified ECU was able to maintain the desired temperature (70 degrees Fahrenheit) and humidity (10 percent) levels of all of the instrumentation equipment to operate and monitor the laser. Space also became a premium in the aft section as more equipment racks and consoles were added.[34]

Servicing electric power to all of the extra equipment on board for Cycle III required adding three generators. The NKC-135 aircraft came off the assembly line equipped with three 40-kVA (kilovolt-amperes) generators on engines 1, 2, and 3. These were not capable of supplying sufficient power for the Cycle III equipment. Consequently, GD replaced the older-version generators with three larger and more powerful 70/90-kVA generators (the kind used on the B-52G) on each engine nacelle. This now gave the ALL a total of four new generators, significantly upgrading its electric power supply from 120 kVA to 280 kVA. (During Cycle I modifications, GD had installed one 70/90-kVA generator on the number 4 engine.) Operating by direct drive off the engine, each generator worked at constant speed independent of the engine speed.[35]

With all of the changes made to the ALL in preparation for Cycle III, AFWL and GD officials worried about how this would affect the aerodynamic performance of the plane. Cycle III modifications added almost 40,000 pounds to the basic weight of the aircraft—from 123,112 to 162,456 pounds. Adding fuel, water for the engine injectors, and fluids for the FSS increased the ALL's takeoff weight to 240,000 pounds. The basic question remained, could the ALL take off, fly, and land safely after undergoing all of the modifications? To find out, Panzarella and his team set up its own "acceptance" flight test program in the summer of 1977 before allowing GD to formally release the ALL back to the Air Force.[36]

Modification of ALL engine to accept new generator.

Colonel Pete Odgers, who commanded the 4950th Test Wing at Wright-Patterson AFB, Ohio, was very interested and curious about these initial test flights because his organization owned the ALL. Odgers, a strong supporter of the ALL program from the start, flew his T-38 to Carswell AFB in Fort Worth to witness the first test flights. (Carswell AFB and GD facilities were on opposite sides of the runway they jointly shared.) As events unfolded, the colonel became more of an active participant than casual observer. Air Force guidance was the ALL had to be flown under visible flight conditions. But the day of the first test flight was extremely cloudy with limited visibility. Odgers wasn't about to be put off by bad weather, as Kyrazis recalled:

> Pete hopped on the T-38 and he started flying it, scouting around to find holes in the clouds so we could get through and get that airplane [ALL] up above the clouds so we could do the necessary flight tests . . . . And so off he goes touring around, radios back, "We found a hole," and comes back and we get the ALL ready to go . . . . He guides us to where the hole was and they completed the flight tests.[37]

Once Odgers had managed to make the weather cooperate, the extensively modified ALL took off to conduct its first flight test to confirm that the aircraft was safe to fly. On board were the key 4950th personnel who had been so indispensable in working closely with GD employees to make sure each step in the aircraft modification process at GD complied with Air Force specifications. Phil Panzarella and Sergeants Augustine, Vanderwall, and Bob Hoppenrath flew on this maiden flight as pilot Captain Pete Larkin put the ALL through a number of punishing maneuvers at various speeds and altitudes to determine if the plane handled properly and met airworthiness standards. Over the next few days, several more flight tests were conducted.[38]

Two design features helped to produce positive test results. One involved keeping an extra 12,000 pounds of fuel in the rear fuselage tank. Modifications made to the plane to be able to accept the FSS, laser device, and APT moved the center of gravity for the ALL forward, which affected how the airplane would handle in flight. Placing extra fuel in the rear balanced the plane to compensate for shifting the plane's center of gravity. A second aerodynamic precaution was mounting a perforated fence, or spoiler, in front of the two openings in the belly of the plane. The fence was a quarter-inch steel plate perforated with 505 three-eighth-inch holes. Its purpose was to break up the airflow to minimize the hot exhaust gases from hitting the side of the aircraft and to allow the diffuser doors to open and close efficiently. Adding the fence produced "just a little drag," according to Sergeant Augustine, but did not detract from the ALL's flight performance.[39]

Panzarella and his team were pleased with the ALL's overall performance. Although an electrical supply system had malfunctioned during the tests, that could be easily repaired prior to the next test flight. Sensing the importance of the flight test, Odgers immediately sent off a message to Headquarters, Systems Command, to inform them of this significant milestone. Odgers reported, "The Airborne Laser Laboratory successfully completed first flight on 28 July 1977. Flight time was 2.7 hours. No serious aircraft discrepancies were noted."[40]

Although results obtained from GD's flight testing were a good sign, the Air Force was required to conduct a more comprehensive and demanding series of flight certification tests. One of the Lab's chief concerns was that Boeing, manufacturer of the KC-135, had identified a flutter problem with the early model KC-135s. At certain speeds, the plane's elevator (tail) would oscillate violently, which could cause it to self-destruct in midair in only a few seconds. The Air Force wanted assurances that this was not the case with the modified ALL. Part of the flight certification, phase 2 of the Cycle III program, was to determine if flutter was a problem with the ALL.[41]

The modified ALL returned from Fort Worth to Kirtland in the summer of 1977. There the AFWL laser team made a quick inspection of the plane before it moved on for its flight certification tests at the Air Force Flight Test Center at Edwards AFB. AFWL's greatest fear was that the potential flutter problem might prevent the ALL from flying the laser. Rerouting the control cables, cutting two holes in the belly of the plane to accept the diffuser, rebuilding a substantial part of the fuselage structure, and placing a new fairing on top of the plane worried some observers. The new forward ramp/full aft fairing, built to protect the phased array HELRATS planned for Cycle IV, caused even more anxiety. HELRATS was designed to track several targets simultaneously and hand-off tracking data to the APT to allow it to engage targets as quickly as possible. For the APT to have a wider field of view (−165 to +55 degrees) to detect multiple targets from different directions, AFWL developed another version of the Cycle III fairing consisting only of a low ramp in front of the APT. This offered less protection for the blunt APT and some feared that the airflow around the APT and downstream would cause excessive buffeting on the airframe and the tail of the plane. All of these concerns turned out to be groundless. Flight testing involving numerous maneuvers at various speeds and altitudes demonstrated the ALL's flutter level was the same as a normal KC-135 and within safety standards. The new fairing (forward ramp/full aft) added about 16 percent extra drag on the plane, which affected its takeoff and climb performance. Also, wake turbulence from the fairing impinging on the vertical stabilizer was minimal, having no adverse effect on the plane's performance. Flight certification of the ALL was completed in December 1977.[42]

SPLIT STAGE DEMONSTRATOR:
PREVIEW TO THE ALL GAS DYNAMIC LASER

In one sense, the origins of Cycle III could be traced back to 6 November 1969 when AFWL let a contract (F29601-70-C-0049) for $320,000 to Pratt & Whitney. This was one of the early paper studies that first proposed a conceptual design for an airborne testbed (ATB). It identified what components were needed to make a laser work in flight and recommended where those pieces were to fit inside the ALL. AFWL's acceptance of the preliminary design plan for an ATB led to a follow-on sole source contract (F29601-71-C-0097) issued to Pratt & Whitney in May 1971 for $3.18 million to develop, build, and test a "research" GDL device known as the split stage demonstrator (SSD).* A parallel two-stage laser device was envisioned for installation in the ALL; the SSD was one-half the planned aircraft laser, hence the name "split" stage demonstrator. The SSD technology was to confirm it was feasible to build a reliable GDL that would work inside the ALL. Specific SSD goals were to verify small signal gain (population inversion) in the cavity, power extraction, flow homogeneity, and reliable operation of the diffuser. Final cost of the SSD reached almost $8 million.[43]

Pratt & Whitney's management team divided the SSD work into four basic tasks. Phase I involved designing an unstable oscillator (resonator) and the associated FSS to deliver the required mixture of gases and liquids flowing from the combustor into the resonator where lasing took place. For phase II, the contractor fabricated the resonator, the cooled mirrors and mirror mounts for inside the reso-

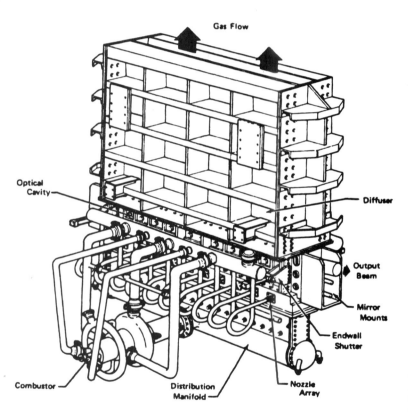

Split stage demonstrator (SSD).

*The SSD formed the foundation of AFWL's GDL Technology Confirmation Program designed to lead to the development of an airborne laser device. Concurrent with the SSD was a separate AVCO project under way to develop an electric discharge laser to compete with the GDL. In November 1972, AFWL selected the GDL, based to a large degree on the SSD test results, to be used on the ALL for Cycle III experiments.

nator, and diagnostic hardware designed to measure the output beam power and intensity profile. Assembly, installation, and alignment of the optics made up phase III. The final and most important phase (IV) consisted of operating the SSD and measuring the beam quality and small signal gain in the cavity. Because the SSD was a proof-of-concept device, AFWL directed the contractor to design and build it so it closely approximated the aerodynamic lines, operating parameters, and beam quality required for the Cycle III laser planned to go in the ALL at a future date. The laboratory, to save time and money, did not require Pratt & Whitney to lightweight the SSD. Consequently, the SSD supported on a large test stand was about five times as heavy as the GDL that eventually went in the aircraft. Dimensions of the SSD, however, were similar to the GDL on the ALL. Although some research support came from United Aircraft Research Laboratories in Hartford, Connecticut, the majority of SSD work took place at Pratt & Whitney's Florida Research and Development Center at West Palm Beach.[44]

The same physical principles used in the proof-of-concept experiments with the SSD were later transferred and used to make the GDL work in flight on the ALL. Pratt & Whitney designed and built each system so it could store fluids, deliver them (helium forced the fluids out of the storage tanks) in the correct proportions through a system of valving and metering devices to a combustor, and then control the hot flow of burning gases through a distribution manifold into a resonator where lasing occurred. As a starting point, feed lines from storage tanks injected methane and oxygen separately into a combustor. The combustor essentially was a short pipe that served as a furnace where all of the critical ingredients burned. A spark plug ignited the methane/oxygen mixture to provide a high-temperature and uniform flame front. Once this condition existed (within the span of only a few seconds), the main reactants of gaseous carbon monoxide (CO fuel) and gaseous nitrous oxide (N_2O oxidizer) were injected into the flame to burn in a stoichiometric combustion process. During this time more methane and oxygen were added and burned in the combustor to form water, which assisted the lasing action occurring downstream. Liquid nitrogen (diluent), after changing into a gas, was also fed into the secondary injectors in the downstream chamber of the combustor. Burning all of these reactants at 3000 degrees and at pressures of 800 psi produced carbon dioxide, water, and nitrogen. Composition of the hot gas flow was approximately 14 percent carbon dioxide, 85 percent nitrogen, and 1 percent water.[45]

The hot gas flow moved in a horizontal direction from the combustor into a 7-foot-long duct called a distribution manifold. Once spread throughout the manifold, the direction of the gases turned 90 degrees and moved straight upward through a bank of nozzles into the lasing cavity (7 feet long × 4.4 inches wide × 12 inches high) or resonator. The nozzle array consisted of a row of 170 individual nozzles stacked together horizontally and bolted to the inner sidewalls of the cavity.[*] In effect, the bottom (leading edge) of the SSD nozzle array

---

[*]Two holes were drilled through the length of each nozzle. Liquid nitrogen was pumped through each of these passageways to cool the nozzles and cavity walls. The entire nozzle array consisted of five interior modules (30 nozzles each) stacked and bonded together in building-block fashion with two exterior modules of 10 nozzles each flanking and connected to the interior modules.

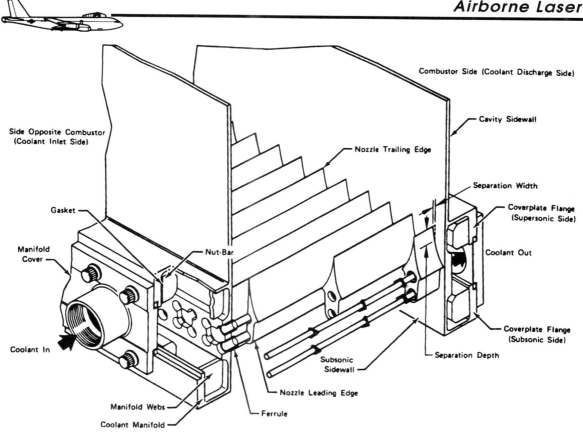

SSD nozzle/cavity assembly.

formed the floor of the distribution manifold. Two nozzles positioned next to one another formed a tiny 0.0065-inch-wide and 0.0064-inch-high throat or opening between the adjacent walls of the two nozzles. Each 4-inch-long titanium (nickel plated) nozzle was aerodynamically contoured with its tapered trailing edge (top end of nozzle) extending upward into the cavity to increase the area of uniform flow. Keeping the flow as homogeneous and smooth as possible was necessary to produce good beam quality.[46]

The main purpose of the nozzle array was to accelerate the velocity of the gas flow to produce the necessary population inversion to create efficient and high-power continuous-wave (CW) lasing action in the cavity. As the subsonic* hot gases flowed from the distribution manifold and squeezed through the small openings (throats) between the nozzle blades, tremendous pressure built up. The combination of high pressure, limited space, and the aerodynamically smooth-truncated surfaces of the nozzles caused the gases to accelerate and race through the throats of the nozzle array. When the gases exited or "popped through" the

---

*Sonic velocity, or the speed of sound in air, is 741 miles per hour at sea level. Supersonic is one to four times the speed of sound. Hypersonic is four or more times the speed of sound.

narrow throats and expanded through the last section of the nozzle, the uniform flow now expanded extremely rapidly, causing a corresponding drop in pressure (to one-tenth ambient) because more area was available for the flow to occupy. These conditions caused the flow to accelerate to hypersonic velocities (Mach 6), driving the carbon dioxide and nitrogen molecules to a highly excited state, resulting in the release of trillions and trillions of photons (or more precisely, 2.8×10^{25} photons). A small 2-kilowatt laser probe beam passing through the length of the cavity picked up the extra energy created by the population inversion of the flow. By measuring the power of the probe beam before entering the cavity, and comparing that with the increased power of the beam (after acquiring additional energy from the flow), scientists could calculate how much laser power or "gain" could be generated in the cavity.[47]

There were no mirrors at each end of the SSD resonator to allow the flow photons to bounce back and forth from each end of the resonator. Instead, photons produced by the flow latched on to the probe beam passing through the length of the resonator. Knowing how much energy was picked up by the probe beam, scientists concluded the aerodynamic flow created the right conditions for population inversion to occur in the cavity. Based on these calculations, scientists knew the cavity walls would absorb some photons. But when mirrors would be added later for the GDL resonator, enough of the remaining photons would line up and bounce back and forth the length of the cavity between the two mirrors at each end. In the case of the GDL, formation and extraction of the beam occurred perpendicular to the flow entering the cavity. The main gas stream in the GDL continued to move downward through the diffuser which compressed and decelerated the flow. This was done to raise the pressure of the flow to nearly that of the outside air pressure. Raising the pressure allowed the spent gases to be exhausted through extension ducts and out the belly of the aircraft to the atmosphere.[48]

In July 1971 AFWL accepted Pratt & Whitney's final design of the SSD. The following month the contractor started building individual components of the device at its West Palm Beach facility. At the same time, Pratt conducted analytical studies and experiments to continually reevaluate and refine various individual parts of the entire SSD system, such as the cavity sidewall and nozzle contours, and the diffuser's endwall and structural support configurations. Ten months later (May 1972), all components had been built and assembled to make up the complete SSD.[49]

Because of the relatively new technology involved in developing a device such as the SSD from scratch, the contractor ran up against a number of problems. Initial experiments showed the gas flow exiting the distribution manifold into the cavity was not a uniform temperature, which degraded the lasing process and ultimately beam quality. Another shortcoming was the difficulty in maintaining the correct spacing and contours of all nozzles for each test run of the SSD. On 31 May, the first hot firing of the SSD occurred where gases flowed throughout the system to create a beam inside the cavity. But on 13 June, after 15 firings (a total of 59 seconds of operation), small cracks were discovered in some of the trailing edges of the nozzle blades and the cavity sidewalls. As a result, the contractor had to disassemble and make repairs to the SSD over the next few months that involved coming up with a new design for and recasting the cavity sidewalls, adding external stiffening members to the distribution manifold and diffuser, and making minor repairs to the combustor. In addition, Pratt installed a heat shield in the distribution manifold to reduce thermal stress, relocated a number of fuel feed valves, and plugged several leaks in the plumbing lines carrying water to the combustor.[50]

By the end of September the contractor had completed repairing the SSD nozzle/cavity assembly and had modified other components to improve structural rigidity and durability. During the week of 23 October, Pratt began an extensive series of tests to determine how well the SSD performed as an integrated system. Interferometry measurements were made to define flow homogeneity and small signal gain.[51]

Overall, the SSD performed well. By the end of testing, aerodynamic flow data revealed "very small" shock waves in the throats and along the trailing edge of the nozzles. Colonel Rowden reported a narrow shock pattern generated by the cavity walls was corrected by "adjusting wall positions." As it turned out, shock waves were too weak to significantly degrade the operation of the SSD. Pratt also managed to measure the small signal gain in the cavity, verifying the aerodynamic flow and design of the cavity combined to create the right conditions for population inversion of CO_2 molecules. Other experimental data showed good beam quality (about $1\frac{1}{2}$ diffraction-limited) and confirmed that 70 percent (roughly 159 kilowatts) of predicted power had been achieved.[52]

The significance of the SSD was that it proved the technical feasibility of building and operating laser hardware that would serve as the model for designing and developing the follow-on laser device used in the ALL. Lamberson reported experiments with the SSD device demonstrated "excellent cavity flow homogeneity, high small signal gain, and excellent beam quality." Once and for all, success of the SSD settled the debate over the selection of one of two competing lasers—the "Humdinger" electric discharge laser (EDL) being developed at AVCO, or the GDL—for the high-power Cycle III flight experiments. Based on SSD test results, Lamberson and Rowden "judged the GDL to be approximately 18 months ahead of the technology and engineering of the EDL."[*] Consequently, Lamberson gave the green light to the AFWL contracting office in December to proceed with plans to award a contract to Pratt & Whitney to design the GDL for the ALL aircraft.[53]

---

[*]Lamberson felt the technical competition between the EDL and GDL made good management sense and was one of the strong points of the ALL program. Technical competition had worked well in the design of nuclear weapons, i.e., Los Alamos versus Lawrence Livermore, and he applied that same approach to the laser world. He believed the decision to go ahead with the GDL was the right decision based on the scientific evidence. The most limiting feature of the EDL (as attested to by experimental data) was its poor beam quality.

11

Fueling the Laser
Cycle III: Phase 3

Success of the SSD represented a pivotal milestone that Lamberson used as a catalyst to quickly get work under way on developing the laser device that would go into the ALL. Two important events initiated this process. In February 1973 AFWL awarded a contract (F29601-73-C-0071) to Pratt & Whitney to design the GDL; final cost was $5.7 million. Midway through 1973 a second contract (F29601-73-C-0117)—that eventually cost the government $32.1 million—was issued to Pratt & Whitney to build the GDL. AFWL authorized Pratt & Whitney to start building the GDL hardware prior to completion of the design effort. This was done to speed up the fabrication process on the major components of the GDL so ground and flight testing would take place as soon as possible. It would take the contractor almost 3 years at its West Palm Beach plant to fabricate and conduct preliminary testing on the device before it was delivered to AFWL in the spring of 1976.[1]

While work on the GDL moved forward, Pratt & Whitney was also busy putting together two other critical elements of the overall laser system. One was the fluid supply system (FSS) manufactured by Pratt. The second involved Rocketdyne (Canoga Park, California) building the dual combustors (essentially two rocket engines) at a cost of $1.9 million (contract

F29601-74-C-0037).[*] Integration and ground testing of the FSS, combustors, laser device, optical bench, ADAS, and APT took place first in a specially designed test cell. A tall one-story metal building (lined with thick concrete walls) situated about 50 yards east of the ARTF, the test cell served as a one-of-a-kind laboratory where each subsystem was assembled and tested separately. The test cell played a critically important role in verifying that all laser subsystems worked on the ground before similar experiments would be carried out in the airplane, first on the ground and then in the air. A variety of highly sophisticated diagnostic instruments measured performance levels of each component under rigidly controlled conditions inside the cell. Once each piece of hardware proved its own operating capability, it was integrated incrementally with the next key component in the total laser system. Those two components were then tested as a unit. For example, AFWL connected and tested the FSS and combustors as one unit. Next the laser device was joined with the FSS and combustors to determine operating compatibility of all three subsystems. This process continued until the APT (the last major element) was added to complete the full-up laser system operating in the test cell.[2]

The test cell consisted of three sections. Set up on the lower level during Cycle III were the FSS, combustors, right and left laser stages, and bench optics. The ADAS routed the beam generated at the lower level to the upper level of the test cell, which housed the APT. A 9 × 12-foot metal roll-up door on the south wall of the upper level allowed the APT to direct the beam outside the cell to downrange targets. Isolated from both of these levels was a control room resembling a blockhouse with 4-foot-thick walls. From there technicians at their consoles sent commands to activate and monitor each component of the laser system. Safety was a prime consideration in construction of the control room and the test cell. To deal with potential accidents (e.g., toxic fuel leaks, misalignment of mirrors causing the high-power beam to stray inside the cell, explosions), the test cell contained a blow-off roof and large air-handling units for rapid exchange of the air. (Fluids stored in the FSS that had to be closely monitored included gaseous oxygen, liquid nitrogen, water, liquid carbon monoxide, liquid nitrous oxide, and helium and methane gas.) The control room had its own environmental control and warning systems (Halon fire suppressant system, and CO and oxygen monitoring devices) as well as a video monitoring system to watch and record the progress of each experiment.[3]

The test cell essentially served as a mock-up of the NKC-135. All components were laid out in the test cell exactly the way they would be installed in the aircraft. There were two good reasons for this. First AFWL wanted to make sure everything fit so that there were no surprises when the equipment was moved into the airplane. Training people who would be operating and monitoring systems was another advantage. The rear crew compartment configuration in the test cell was basically identical in terms of floor space and consoles to the plane layout. This setup gave everyone an opportunity to sharpen their individual skills

---

[*]One combustor would be connected to the left stage of the laser to ignite the gas flow, while the other combustor would be combined with the right stage of the laser device. Rocketdyne was selected as the contractor mainly because of its expertise and experience over the years in building rocket engines.

and to function collectively as a cohesive team. Kyrazis felt training gained in the test cell was critical "so everyone would be familiar with what was going on next to them, who they were going to be working with, what everything was going to look like. You know, you learn certain habits. You're going to reach for a switch. That switch in the test cell was in the same position as it was in the aircraft for the most part."[4]

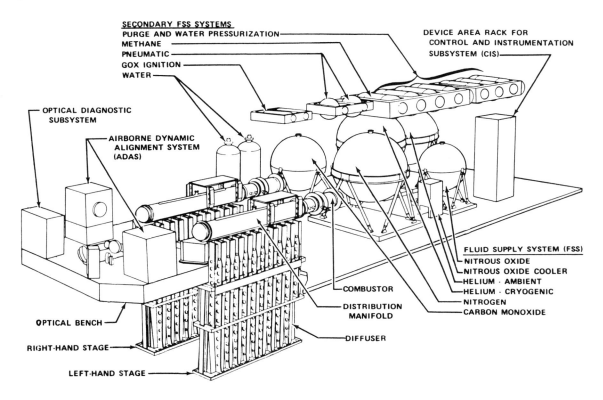

Schematic of ALL system that fit inside ALL aircraft.

The test cell was just one part of the overall ARTF compound located in a secure area with restricted access at the east end of the Albuquerque runway. A large hangar, with an APT clean room, extremely good optics laboratories, and offices, made up the heart of the ARTF. Directly behind the hangar was a test range with targets set up at the 1-, 1.5-, and 2-kilometer sites. Diagnostic equipment at each site measured the characteristics of the high-power laser beam, first sent from the test cell and later from the ALL aircraft on the ground to the downrange targets, during the Cycle III testing. At other times during this period, only a small portion of the beam was extracted and dumped in the "bucket" or calorimeter to measure beam power and quality. Also behind the ARTF was the fuel farm containing high-pressure tanks that stored the fluids to fuel the laser. An elaborate plumbing system of pipes and valves delivered the right proportion of fluids to the laser in the test cell and ALL to run experiments.[5]

ARRIVAL OF FLUID SUPPLY SYSTEM AND LASER

Once Pratt & Whitney built the FSS, its various components (e.g., tanks, plumbing) were crated and shipped in stages in air ride vans to Kirtland beginning in April 1975. Pratt & Whitney engineers worked side by side with AFWL technicians mounting the FSS in the test cell in exactly the same configuration the FSS would be installed in the ALL aircraft. Welded pipe and stainless-steel tubing carried all of the fuels from the tank farm to an interface unit on the south wall of the test cell, which served as a way station to make any final pressure adjustments for controlling the fuel flow before it was sent on to the FSS. All lines leading into and out of the FSS had to be cleaned periodically, which was a time-consuming process and one of the reasons why the FSS spent such a long time in the test cell. Cleaning involved flushing each line with Freon or gaseous nitrogen to remove residual fuels left over from previous test runs. This was followed by pulling a lint-free cloth soaked with trichloroethylene through each line. Once this was accomplished, each line was hydrostatically tested to verify it could hold its required pressure. As a safety precaution, AFWL decided each line would be certified at one and a half times its designed working pressure.[6]

Sections of transfer control module, which made up one component of the GOFSS, loaded on flatbed at SOR for transport to test cell at ARTF.

In anticipation of the arrival of the FSS, AFWL in the fall of 1974 realized they needed to build a device that they generally referred to as the ground-only fluid supply system (GOFSS). This hardware actually consisted of two main components: commercial "watermelon-shaped" fuel storage tanks and an elaborate plumbing configuration to transfer and regulate fuel flow. The plumbing unit was called the transfer control module (TCM) and was located just inside the south wall of the test cell. As fuel moved from the tank farm (behind the ARTF), it went into the TCM. The TCM monitored the fuel flow and routed each of the laser fuels (e.g., carbon monoxide, nitrogen, helium, nitrous oxide) to its designated storage tank. Those tanks for the initial testing were the tanks of the flightworthy FSS when it was installed in the test cell in the spring of 1975.[7]

The idea was that the GOFSS would match the performance characteristics of the flightworthy FSS. Once the FSS completed testing in the test cell, it would be installed on the aircraft. However, the overall Cycle III game plan called for the combustors, GDL device, diffuser, bench optics, ADAS, and APT to remain in the test cell—so further testing could take place to mate and align the optical components of the integrated laser system—after the FSS was removed from the test cell and installed on the ALL aircraft. To fuel the laser to test the remaining components in the test cell, the GOFSS watermelon-shaped tanks were installed in the test cell to substitute for the FSS tanks that now were on the aircraft. The TCM still remained in the test cell to deliver the fuel to the substitute tanks.[8]

Captain Tom Dyble headed a team made up of Captain Mike Moody, Chief Master Sergeant Chuck Edwards, Master Sergeants Joe Coyle and Chuck Carew, and civilians Frank Wells (AFWL) and Clyde Watson from McDonnell Douglas to design, build, and test a working GOFSS at SOR. Funds were not available to hire a contractor to make an expensive GOFSS. As an alternative, AFWL's Chuck Edwards scrounged old components from wherever he could to build an in-house GOFSS. Trips to Jackass Flats and Cape Canaveral located a treasure of discarded leftovers from missile programs to include big regulators, filters, valves, and gauges, which with minor modifications would be suitable for putting together the GOFSS. Once Edwards identified all of these off-the-shelf parts sitting in the scientific junkyards of the Air Force, he arranged them to be shipped by train and truck to Kirtland. AFWL then cleaned and tested the various parts, salvaging only those that could be brought up to operational standards. This involved overhauling and repairing various hardware (e.g., replacing worn-out seats and seals) before the final assembly of the GOFSS began.[9]

Clyde Watson designed the GOFSS, but assembling all of the secondhand parts collected by Edwards was like putting a massive jigsaw puzzle together. The team used a module approach in building the system, which consisted of tanks, fluid lines, instrumentation, a control computer, and operator consoles. Dyble and his crew were extremely proud of their finished product as it was lifted onto a flatbed to be transported and set up in the test cell at the ARTF in the winter of 1975. Over the next few years, the GOFSS proved to be a very useful and reliable piece of equipment.[10]

The FSS and GOFSS were similar, but there were also major differences between the two systems. On the GOFSS there were slow-acting industrial valves. Plus, the flow rate was regulated by constant pressure on the fluids as opposed to a series of finely tuned control valves on the ALL to adjust the flow. Also, fuel lines on the GOFSS were not vacuum jacketed to assure maximum cooling. Kyrazis described the final GOFSS product as "so huge and heavy that it would never leave the ground." Even though the GOFSS was not a streamlined system ever capable of going airborne, it served its purpose extremely well in fueling the laser device in the test cell, saving AFWL valuable time and money by allowing concurrent testing to be carried out on the flightworthy FSS on the plane. From the summer of 1978 through December 1979, AFWL especially got its money worth as the GOFSS served as the workhorse to determine performance of the bench optics, ADAS, and APT as a total operating system in the test cell. Testing of the lightweight, airworthy FSS in the test cell lasted until July 1978; from then until the end of 1979 the GOFSS was the primary way to fuel the laser in the test cell.[11]

AFWL began installing the FSS and its control and instrumentation subsystem (to regulate the fluids and record performance measurements of the FSS) in the test cell during the last week of April 1975. This work was completed on 27 June. Preliminary functional checkout of the FSS, involving temperature, calibration, and flow tests, started on 1 July and ended on 5 September. Seven weeks later (24 October) AFWL completed "acceptance tests" to verify that the FSS and control and instrumentation subsystem operated properly. For the next year, Pratt and AFWL personnel worked together conducting an extensive series of FSS performance tests to demonstrate stable, accurate, and responsive flow control throughout the entire flow range for both the laser fuels and the water subsystems designed to cool the mirrors in the system. The overall goal was to verify the structural integrity of the FSS and to demonstrate it was a leakproof system that could store, sequence, regulate, and deliver all fluids in the prescribed quantities from the five FSS pallet-mounted, high-pressure storage tanks to the combustors at the proper flow rates and pressures. The FSS was tested as a pressure-fed, blowdown design that used high-pressure helium gas to push liquid nitrogen, CO, and N_2O through the plumbing lines to supply the laser device with the exact amount of fuel and within predetermined time limits.[12]

Combustor testing in the test cell began at the end of October 1975 when AFWL connected the FSS to the dual combustors and began a series of cold flow tests (followed later with hot flow tests that ended in March 1976) of these two components. The purpose of the combustor testing was to verify that all of the different fluids reached the two combustors at the same time to create the right mixture to be ignited by a spark plug at just the precise moment in each of the two combustors. Another objective of the testing was to incrementally increase pressure over time (from tens of milliseconds to several seconds) in the combustors to identify weak points in the hardware. (Normally, rocket engines were tested to the point of destruction to determine operating limits. But the ALL team did not have the luxury of testing the combustors to the breaking point because there were no replacements.) Knowing the operating limitations of the combustors was an extremely important consideration, as Kyrazis pointed out:

> On the ALL we just didn't want to destroy these combustors. What we were doing, we were running at an extremely narrow range of parameters, temperatures, pressures, and so forth, just so that you knew that you're safe. And what you did is that you monitored all these parameters with the computer and, if it stepped out of bounds at all, you just shut the thing down. Then it really didn't matter in this case whether or not you shut down, because it was the safe thing to do.

Generally, this testing proceeded smoothly but not flawlessly. For example, heat buildup and flow conditions eroded some of the copper injectors, which AFWL eventually replaced; repairs took place between 30 January and 16 February 1976. In the end, results from the combustor testing showed both combustor stages started and lit at the same time, pressure rise rates were identical in each chamber, and all design operating points were verified.[13]

As the FSS and combustor experiments progressed, the two stages of the GDL device had arrived at AFWL from Pratt & Whitney in March 1976. Assembly, mounting, and preliminary flow testing (lasting less than 2 seconds for each run) of the single right stage in the test cell reached completion on 22 April. Hot exhaust gases flowed out of the bottom of the GDL through the diffuser and entered a duct in the floor of the test cell. From there the spent gases traveled through an underground tunnel to a point outside the cell and were vented up through a 40-foot stack next to the test cell. By 6 May the left stage of the laser had been set up in the test cell.* On 7 June the first round of dual-stage testing reached completion. AFWL was satisfied with these initial test results indicating the system operated properly. Only minor modifications to the laser device had to be made, such as adjusting the position of the sidewalls to maintain a uniform flow. The diffuser, originally predicted to be

a potential problem by some, started successfully and remained stable during each of the hot firings. These initial findings were encouraging, but more extensive testing was needed over the next 2 years to refine and improve reliability of operating performance of the entire laser system. (Modifying control valves, feed lines, and storage tanks, plus checking the operation of the water cooling system, took up a large share of this time.) These tests verified the three major components (FSS, combustors, and laser device with diffuser) mated properly and met technical specifications before the contractor turned the hardware over to the Air Force.[14]

One very important event took place during the first few months the GDL was in the test cell. Realizing manufacturing and testing of the optical components lagged behind the development of the GDL device, Lamberson was concerned about how to best utilize the GDL in the test cell while awaiting the installation of the bench optics, ADAS, APT, and so forth. In addition, rumors were circu-

AFWL workers adjust one stage of GDL prior to entry into test cell.

---

*An interstage beam duct connected the right and left stages of the GDL so beam buildup could occur by bouncing back and forth between the left and right stages. The right-hand stage had an aerodynamic window consisting of gaseous nitrogen to allow the beam to be extracted so it could move on and be directed by the bench optics to the ADAS and APT.

forth. In addition, rumors were circulating that Systems Command was considering reducing funding for the ALL as a way to cope with military budget decreases across the board. Lamberson wanted to head off any potential money problems to keep the momentum going on the ALL. Showing that significant technical progress was being made by the GDL was the best way to do that, especially as he had told higher headquarters back in 1973 that he would have a laser up and running in 3 years. In his mind, he knew it wasn't enough just to demonstrate correct flow patterns through the system when the GDL was integrated with the combustors and FSS. He was looking for more dramatic results, something he could pass on up the chain as a spectacular breakthrough to grab the attention of Systems Command. His motive at the time was to reassure the money holders in Washington that they indeed were getting a substantial return on their ALL investment.[15]

GDL inside test cell prior to testing.

Lamberson turned to Major Joe Zmuda, who was in charge of the GDL in the test cell, to find a better way to advertise the reliability and progress made on the GDL. Both agreed if the device could produce photons so a beam could be extracted, then that would be a major accomplishment, which headquarters could readily understand and use as hard evidence that the ALL was moving along as planned. During the initial testing of the GDL, there were purposely no mirrors at each end of the resonator to prevent photons from lining up and forming a beam. Again, the optical components outside the resonator were not yet in place to align a beam, hence, at that point there was no real need to extract a beam.[16]

To ensure funding remained on track, Lamberson called on Zmuda and his team of device experts to devote all of their efforts to extract a high-power beam from the device. The major understood the magnitude of this challenge, especially as Lamberson was eager to get results as soon as possible. "We worked day and night," Zmuda recalled, "17 hours straight was the norm." One of the first jobs was to obtain mirrors from Pratt & Whitney to place in the cavity of the device. Because time was critical, Zmuda decided to use readily available noncooled mirrors. Although the plan called for extracting a high-power beam, he reasoned by turning the laser on for only a tiny fraction of a second no damage would be inflicted on the mirror surfaces. By limiting the run time, a high-power beam could be extracted, but would not be aligned through the optical train of the entire system; the goal was to measure raw beam power and not optical performance.[17]

After a hectic few months, five uncooled mirrors were installed in the GDL and testing began. On 21 January 1977 the breakthrough Lamberson was anxiously awaiting arrived with production of the device's first photons. Eight days later AFWL announced a second low-power extraction. On 1 February Zmuda's group had extracted and "dumped" the first full-power beam into a calorimeter, which measured the beam power at 450 kilowatts. Device efficiency was measured at 4 percent, which was good by gas dynamic standards. Lamberson was jubilant. He immediately collected the data, put a briefing together, and headed off for Washington to present the latest ALL success story to General William J. Evans, Systems Command commander. Evans was excited by Lamberson's latest news and put to rest any notion that funding for the ALL would be cut. This vote of confidence no doubt relieved some of Lamberson's financial woes and allowed him to turn his full attention to ensure the laser device remained in the test cell to undergo further compatibility testing with the bench optics, ADAS, and APT. These experiments lasted through 1979 to ensure all components worked properly in the test cell before the entire system was installed on the ALL aircraft in 1980.[18]

FSS PROBLEMS AND GROUND TESTING

While in the test cell with the GDL (spring of 1976 to the summer of 1978), the flight FSS encountered a number of problems that were extremely time-consuming to fix. With several hundred valves in the entire FSS, there were several hundred locations where leaks could develop. During cold flow testing, gaseous nitrogen or helium was forced through the delivery lines to test for leaks in the system. This was an incremental process whereby the flow would be pressurized to only 10 percent of the maximum designed operating capacity of each line carrying the helium or nitrogen. Successive tests would increase pressure to 20 percent, then to 30 percent, and so on until 150 percent pressure was achieved. Liquid soap was applied to each valve and fitting to detect leaks during each test run. When a leak was discovered (indicated by the soap bubbling), it was recorded and tagged to indicate the exact location of what had to be repaired. After each run, all of the leaks were fixed before moving on to conduct tests at the next higher pressure level. Master Sergeant Dick MacCutcheon, who worked with the FSS the entire time it was in the test cell, described the tedious and time-consuming process to find leaks:

> We looked at every connection, every fitting. Just like on a valve, you'd have your line coming in and your line going out. But then you had an adaptor from that to the valve, so there were four possible leak places on just that one valve. And every little place that went down and T'd off and went to various places. You know, there were times you'd walk out in the test cell and it looked like a Christmas tree, because we had so many tags [identifying leaks] hanging on the B-nuts. Just getting it [FSS] leak tight was a long drawn-out process. And then we had one of the tanks we found—by accident—was leaking. It seems to me it was the menthane tank. One of the guys was up there leak testing the inlet fittings and outlet fittings, and he spilled some liquid soap on one of the

tanks, and the tank started bubbling when it [soap] ran down the side of the fiberglass walled tank. I was down below and I noticed fuzz—bubbling—coming out of the tank. And we had to redesign and replace that tank.[19]

Leaks were only one of many problems affecting the performance of the FSS that had to be fixed before it could be moved to the ALL aircraft for further testing to certify the airworthiness of the FSS. It took over a year to modify all of the finely tuned flight valves in the FSS to validate they performed to exacting operating specifications. The flight valves were very fast acting, opening in the range from 10 to 40 milliseconds. Tiny Teflon seats in the solenoid valves regulated the flow pressure of gaseous nitrogen to move an actuator arm back and forth. As pressure increased, the movement of the actuator opened the main $\frac{3}{4}$-inch stainless-steel fluid lines by rotating the ball flight valve seat. Extensive testing showed these valves over time would wear and not provide a good seal. Another recurring problem was the valves froze up after moderate periods of exposure to cryogenic temperatures of the liquid nitrogen and carbon monoxide. Consequently, AFWL decided to remove and take apart all of the valves from the FSS and do a complete redesign and rebuild of the seats. This work took place at a valve shop located at Manzano Base, roughly 5 miles east of the ARTF. AFWL hired Don Endicott, an engineer from McDonnell Douglas, to make these modifications, which took about 9 months before the new valves were put back together and reinstalled with the FSS in the test cell.[20]

Both stages of GDL device installed in test cell. Fuel from fluid supply system (tanks in background) flowed forward through distribution manifolds affixed to top of left and right diffusers which vented gases through floor of test cell.

Follow-on testing of the redesigned valves was tedious work requiring very precise adjustments before AFWL would declare them flight-worthy. MacCutcheon explained:

This was a critical part of testing we did a long while . . . . And you would flow and get all your pressures and temperatures and flow rates, and they did a lot of math calculations and then they'd say, go in and close it up 3/1000th of an inch. And you go in with micrometers and measure the distances, and then you turn that valve in and lock it down.

And then they would run some more testing and more testing and get a lot more data, and then they'd say, well, we need to back that out 1000th of an inch. It just took a long time. It was nothing easy to do.[21]

MacCutcheon also pointed out that no single master blueprint existed for the entire FSS. (Pratt did provide line drawings for different parts of the FSS, but there was no single drawing of the complete FSS.) Instead, Pratt & Whitney built to scale an FSS wooden mock-up at its West Palm Beach plant. Once completed, it was disassembled and shipped to the test cell at Kirtland. The mock-up, an exact duplicate of the real FSS, served as a baseline unit to keep accurate track of any alterations (e.g., addition or deletion of valves, replacement of broken lines, changes to the size and length of lines to achieve better flow characteristics) made to the FSS. This procedure allowed fabrication of lines without the costly design layouts to describe the three-dimensional line configurations. Pratt maintained meticulous records on all FSS modifications, which first had to be made on the wooden mock-up to see if new parts fit correctly and were compatible with existing parts before being installed on the FSS.[22]

AFWL traced one other problem affecting overall performance of the laser system to the FSS. This involved the mirror cooling system designed to feed water to the mirrors to cool them when the beam was turned on. When the FSS was later tested with the bench optics and ADAS in the test cell, AFWL discovered an abnormally high level of beam jitter resulting from vibrations caused by the water flow. The first time technicians opened the water tank release valves the flow created a tremendous "water hammer" effect as the water, under enormous pressure, left the storage tanks and forced its way through the 2-inch stainless-steel plumbing lines to the mirrors. This created a shock of such magnitude that it bent the $\frac{3}{4}$-inch metal plate bolted to the floor of the test cell that held the water tanks of the FSS in place. The net effect of all this was that the vibrations coupled into the optical bench and components (e.g., mirrors, mirror mounts) causing these components to move. This, in turn, not only broke some of the mirrors, but induced disturbances resulting in unacceptable beam jitter.[23]

AFWL fixed the water cooling flow-induced vibrations on the mirrors after Lieutenant Pete McQuade conducted an extensive series of tests in the test cell beginning in January 1978. As device testing was under way also at that time, the majority of tests to come up with fixes occurred from May 1978 to February 1979. One of the cures was replacing single-hole openings in the plumbing lines leading to the mirrors with two-stage multiple-hole showerhead openings to reduce shock and create a smoother water flow. By slowing the water valve openings, the flow also slowed down to help eliminate shock through the system. Another fix substituted more pliable flex hoses for a selected portion of hard metal plumbing lines. By moving the dogbone* from the bench to the tank floor and using a 4-foot flex hose to route the water from the dogbone to the bench, McQuade was able to structurally isolate the water tanks in the FSS from the bench-mounted hardware. All of these changes

---

*The "dogbone" was a manifold which split the water flow from the tanks into two 1-inch and two 5/8-inch lines to distribute the water to all of the mirrors on the optical bench. The dogbone was rigidly connected to and straddled the optical bench and the water storage tank floor.

produced dramatic results, reducing the bench-mounted mirror system beam jitter by 75 percent.[24]

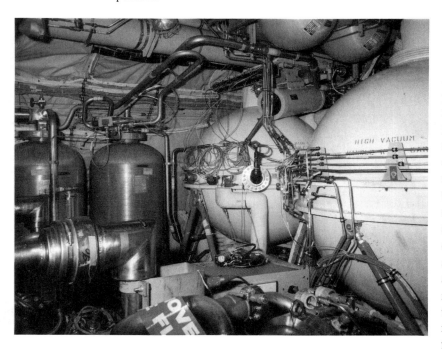

Fluid supply system (FSS) installed in ALL.

By the summer of 1978, AFWL was satisfied with the ground operation of the FSS in the test cell. To save time and to begin to get the airplane ready for Cycle III testing, the FSS was removed from the test cell and put on the ALL. Captain Don Teasdale and Dick Frosch led a team of Pratt and Laboratory technicians in July and started disassembling the FSS and installing it in the aircraft. Installation was an around-the-clock operation and took only a couple of months. During installation, the most obvious goal was to make sure the FSS physically fit in the ALL. AFWL never really doubted the FSS would fit, but that could not be ascertained until all of the tanks, fuel lines, valves, and fittings were securely integrated to make up the entire FSS in the plane. The hardware did fit, but it was a time-consuming effort (lasting from the fall of 1978 through February 1979) connecting all of the components. Once aboard the aircraft, the FSS underwent a series of tests to verify it could deliver the required fluids and gases to the laser in the ground, taxi, and flight environments.[25]

Ground testing of the FSS on the plane began on 5 March 1979 and finished 12 days later. In preparation for this phase of testing AFWL had used the same TCM that made up one of the critical components of the GOFSS. Resembling a T-shaped contraption, the boom of the TCM was lowered to rest on top of the wing of the ALL. High-pressure lines carrying fluids directly from the tank farm behind the ARTF to inside the ALL (TCM fuel lines were connected to the fuel connect ports on the right side of the aircraft) were supported by the TCM; it also regulated and monitored the flow of fluids delivered to the FSS tanks inside the plane. Part of the ground testing involved making sure the TCM servicing system operated properly and was compatible with the FSS. Once this was accomplished, AFWL pressurized the FSS to full operating pressure to demonstrate the FSS was structurally sound and to check for leaks in all storage tanks and lines. Still another important aspect of this testing was to certify all valves opened and closed at precisely the right time and to prove the Fill, Vent, Dump, and Chilldown system performed according to specifications. An overboard dump of helium was achieved using the emergency power panel. Several 16mm movie cameras photographed the ground dump which was later compared with in-flight dump data.[26]

There were no major setbacks as a result of the FSS ground testing although a few problems did surface which were quickly resolved. For example, tests revealed the helium isolation valve as well as the oxidizer recirculation valve leaked. Replacing the seals cured the problem in both cases. AFWL discov-

Transfer control module (TCM) to move fuels from ground tank farm to ALL. This was the same TCM used (without the boom) to fuel the laser system during earlier testing in the test cell.

ered other minor discrepancies, such as leaks in plumbing lines, but all were corrected prior to the start of taxi tests.[27]

Ground testing the FSS on a stationary aircraft was a logical first step for learning how the FSS would perform inside the ALL. However, AFWL wanted to find out how the system would be affected when subjected to vibrations while the aircraft was moving on the ground. Would leaks develop in fluid lines, would tanks rupture, or would the flow be disturbed because of extra stress imposed on the FSS while the plane was rolling down the runway? To find out the answers, AFWL conducted a number of taxi tests from 21 through 23 March.[28]

Pressurizing the FSS in three progressive phases—from low to medium to full operating pressures—during taxi tests provided performance data on the FSS. Teasdale explained, at each pressure level, the ALL would speed down the runway poised to take off and the pilot would suddenly hit the brakes and come to an emergency stop. One reason for doing this was to determine if any damage to the FSS occurred while the fluids were sloshing around in the tanks when the plane was moving and then came to an abrupt stop. Results revealed the FSS operated normally and showed no degrading features during the taxi tests. The final technical report concluded, "These test sequences confirmed that no elements of the FSS or associated equipment would be a hazard in the high vibration taxi environment."[29]

Ground and taxi tests were necessary first steps to assess the performance of the FSS, but the most important and revealing tests of the FSS inside the plane were reserved for five flight tests starting on 27 March and ending on 11 April. All flight tests took place over the Army's White Sands Missile Range in southern New Mexico, a short 20-minute flight from Kirtland. The main purpose of this series of tests was to verify in-flight compatibility, integrity, and performance of the integrated FSS inside the ALL to demonstrate the FSS was safe and operational so AFWL would be ready to carry out the upcoming Cycle III testing.[30]

As with the ground and taxi tests, the airborne testing incrementally pressurized the entire FSS (using inert gases and liquid nitrogen) to detect leaks in tanks, plumbing lines, and connections. A pivotal part of testing involved activating the dump system to eject the FSS fluids overboard (in case of an emergency) without damaging the aircraft or posing danger to the crew. Other objectives were to collect aerodynamic data on the dummy APT and fairing, making sure the diffuser doors opened and closed properly, and to reconfirm appropriate pressures were maintained in each of the three aircraft compartments separated by the bulkheads installed in the ALL during the Cycle III modifications.[31]

While in the test cell, testing of the FSS was limited because the airborne aerodynamic and high vibrational environment could not be reproduced. The advantage of taking the FSS to the air was the aircraft would flex and vibrate when flying at different altitudes and airspeed combinations. Exposing the FSS to takeoffs, turns and banks, sideslips, turbulence, climbs and dives, and landings was the most realistic test for determining the true performance level of the FSS. Although good data were collected on the FSS, there was a definite downside to collecting the data. It wasn't unusual for many people on board to get sick as pilot Captain Joe Hoerter put the ALL through some rather demanding and nonroutine flight maneuvers he referred to as "shake, rattle, and rolls." Kyrazis described a typical flight test:

> After pressurizing the tanks we'd go into maneuvers where we'd go over the top so that you took away the G-loads of the aircraft, then dive, and then pull out hard so that you could flex all the tubing inside to see if you could make them break. We'd go through and go almost into a stall and then into real hard 2-G maneuvers. We were bouncing around the sky like a porpoise moving through the water. Truly wild flights! Then we'd fly real low to the ground over White Sands and pick up the afternoon turbulence. The plane just shook like hell. And then we'd go in and check for leaks and eventually fix them. The main supply lines were vacuum jacketed, because the vacuum is your best insulator [to keep the fluids cold]. Well, we'd break the vacuum in those big stainless-steel lines. And we were checking other things. Was our gas detection system working? Were our diffuser doors working correctly to open to allow all the exhaust from the laser to get out?[32]

After 18 hours of rigorous airborne tests, AFWL concluded the FSS was airworthy and capable of performing its functions as part of the planned Cycle III testing. Data showed, for instance, the large gimbaled bellow FSS expansion joints had sufficient life expectancy to keep high-pressure fluid lines flexible to compensate for aircraft bending and vibration. (AFWL feared the bellows might rupture because of the constant bending and flexing of the aircraft.) There were no significant leaks in the system as all tanks and lines were able to maintain structural integrity when pressurized. In preparation for the airborne flights, it was proven that the FSS could be filled on the ground using the TCM in less than 4 hours. Airborne dump of fluids was within the limiting goal of 5 minutes as predicted. In sum, the basic goal of the flight testing was to find out if the FSS tanks and plumbing would be damaged as the ALL flew various maneuvers. "The answer," Kyrazis reported, "was the FSS came through with no real problem."[33]

MANAGEMENT CHANGES

By the summer of 1978 with a variety of activities progressing at a feverish pace, Kyrazis had set up a new management system to better focus, direct, and monitor the planning and execution of the ALL schedules for ground and flight testing. As more and more components were integrated into the aircraft, it became increasingly more difficult for Kyrazis and others to keep apprised of the status of the "big picture" as well as the numerous details of the ALL's technical progress. To deal with these issues, Kyrazis took a bold step by establishing a more informal and streamlined reporting system.[34]

This new management system was really before its time as it was one of the first working "integrated product development teams" that later became a popular management tool used by businesses in the 1980s. What Kyrazis did was to assign three individuals to his Laser Development Division office who would have total control of the three major subsystems of the ALL. Major John Otten was assigned responsibility for the laser device, Major Denny Boesen headed up the optics and APT, and First Lieutenant Jim Cooley was in charge of all matters relating to installation and integration of new equipment into the aircraft and how that affected flight safety. Although relatively young officers who were shouldering some heavy responsibilities, Kyrazis trusted them and had full confidence in their technical competence.[35]

Instead of dealing with five branch chiefs, who had other responsibilities besides the ALL, Kyrazis relied on these three individuals who he gave "carte blanche" in taking care of all of the day-to-day business of testing systems in their area of responsibility. This meant any one of the three could reach down into the division's branches and immediately "borrow" personnel and resources to help resolve any ALL problem. Branch chiefs did not particularly like this arrangement, but it was a quick and effective way to ensure the ALL progressed on schedule. As Cycle III involved the integration of numerous subsystems, Kyrazis fostered an atmosphere that encouraged Otten, Boesen, and Cooley to talk to one another daily. The benefit of this constant exchange of information was to ensure that the installation and testing of one system would not adversely affect other systems.[36]

Although Otten, Boesen, and Cooley exercised a great deal of independence, they did not have unlimited authority. Kyrazis realized he needed a strong personality to head up and coordinate the various activities of the three men leading the integrated product development team. Kyrazis, again, turned to Bill Dettmer to fill this important job.[37]

Kyrazis actively campaigned to recruit the 42-year-old Dettmer to fill the position of Director of Airborne Laser Laboratory Test and Operations. Kyrazis had convinced Major General Gerald K. Hendricks, Director of Science and Technology at Systems Command, that the timing and conditions were right to move Dettmer (commander of Det 2, 4950th) into the Director of Test and Operations slot. Colonel Dick Feaster currently held that position, but was scheduled for reassignment to the Foreign Technology Division at Wright-Patterson AFB. Knowing that Feaster would soon depart AFWL, Kyrazis kept urging Hendricks to select Dettmer to fill the anticipated vacancy. Recently promoted to full colonel, Dettmer had all of the right credentials: operational experience, flight-test experience, scientific expertise with a Ph.D. in physics (Laser/Optics), and a reputation of working well with the

ALL contractors. Hendricks approved Kyrazis's recommendation and in August 1978 Dettmer moved into his new job. In essence, he became Kyrazis's right-hand man.[38]

Dettmer's duties involved watching and coordinating technical, administrative, and safety issues related to the Cycle III progression. On the technical side, he closely monitored the decisions and recommendations of Otten, Boesen, and Cooley in carrying out the Cycle III test plan. If differences of opinion on technical matters existed among the three, Dettmer would break the tie and make a final recommendation to Kyrazis. Dettmer also worked closely with Navy range personnel at China Lake and Point Mugu in the planning of ALL flight scenarios. In sum, Kyrazis usually went to Dettmer to get the most current status on the technical and testing issues affecting the overall ALL Cycle III test plan. It was this personnel management system that proved to be extremely responsive and productive and remained in place until the end of the ALL program.[39]

One other very important change took place in 1978 that significantly altered the management of the ALL program. On 2 February 1978, Major General Hendricks presided at the promotion ceremony of Colonel Lamberson to Brigadier General at the Kirtland Officers Club. This event meant that, because the ARTO position that Lamberson held was a colonel's slot on the manning document, he could not continue to remain in that job. Although Lamberson had been the single most important influence and personality in developing a comprehensive airborne laser program, his promotion instantly qualified him as a prime candidate for the Air Force's mobility program for senior-level research and technology managers. For these reasons, the "father of the ALL" had to move on. In April 1978, Lamberson departed Albuquerque and moved to Eglin AFB, Florida, where he assumed his new duties as Deputy for Development and Acquisition of Conventional Armament at the Armament Development and Test Center.[*40]

Lamberson's departure represented a loss to the laser program at all levels. He had invested 9 uninterrupted years at AFWL working relentlessly building and establishing the credibility of the Air Force laser program. Under his leadership, the laser program at AFWL had grown from 25 people and an annual budget of $6.5 million in 1969 to 350 people and a budget of $86.8 million in 1978. By that time, laser research and development work had accounted for the single largest 6.3 program element (advanced development) within the Air Force Systems Command's laboratory structure, attesting to the rapid growth and progress of lasers.[41]

Throughout the years, Systems Command's hierarchy had always admired Lamberson for the outstanding technical direction he provided as well as his superb management abilities. General George S. Brown, Commander of Systems Command in 1971, who became Chief of Staff of the Air Force in August 1973, elevated Lamberson to a position few officers

---

[*]In January 1982 Lamberson was promoted to major general. During the 1980s he served in a variety of high-level positions to include Deputy Assistant for Directed Energy Weapons, Office of the Under Secretary of Defense for Research and Engineering (1982), Assistant Deputy Chief of Staff for Research, Development and Acquisition, Headquarters, U.S. Air Force (1983), and Assistant Deputy, Office of the Assistant Secretary of the Air Force for Acquisition (1987). He retired from active duty in 1989.

could ever hope to achieve. Brown wrote, "Colonel Lamberson's profound scientific knowledge, his superb managerial foresight and ability, and his brilliant leadership are outstanding contributions to military science today. He is, beyond doubt, the foremost military expert on high energy laser systems within the Department of Defense."[42]

Brown's glowing endorsement of Lamberson was not unique. On his departure from AFWL, Major General Hendricks presented Lamberson the Distinguished Service Medal. The narrative for this award comprised three single-spaced pages covering the numerous contributions made by Lamberson during his assignment at AFWL. Hendricks summed it up best in the last paragraph of the award write-up:

> The progress of high energy laser technology for weapon applications in the past nine years . . . has been extraordinary. These advances and the potential for this technology in the future are the direct result of the matchless dedication and effectiveness of General Lamberson as the Program Director. His superior technical expertise, exceptional management, and outstanding leadership have been the basic foundation for program development and progress; no other officer could have been more effective or successful in this position.[43]

Although he received many of the formal accolades, Lamberson realized his success was directly linked to those who worked with him. He was always generous with his praise, giving credit to the day-to-day workers in the trenches who were solving difficult technical problems to keep the ALL moving forward. It was truly a team effort, drawing on the scientific and technical talents of military, civilian, and contractor personnel.

There is no question that a diversity of people accounted for the success of the ALL. But the one person Lamberson singled out above all others was Pete Avizonis. From 1969 to 1978, Lamberson and Avizonis worked closely together combining their technical abilities and management skills to form a highly action-oriented team. Why did they work so well together? Much of the credit had to go to Avizonis, who Lamberson considered one of the top two or three laser physicists in the world. As Lamberson explained:

> While I was educated at the Ph.D. level, I recognized, relied on, and benefited from the technical superiority of Pete Avizonis. It may be true that my briefings and program management techniques breathed life into his technical genius, but make no mistake that most of the truly creative and innovative concepts of that day were formed by Pete. I doubt very much that I alone as Program Manager could have made the ALL successful, and I doubt that Pete could have, either. It was the partnership between us that overcame the monstrous hurdles encountered throughout the program.[44]

Lamberson had left behind an enduring legacy for laser work at AFWL. To many at the Lab, he bordered on being a living legend who established a reputation as a leader able to get things done in spite of the numerous technical and bureaucratic obstacles thrown in his path. People counted on Lamberson and, in most cases, he delivered. Filling Lamberson's shoes would be no easy undertaking, no matter how well qualified his successor might be.

The unenviable job of replacing Lamberson fell to Colonel John C. Rich. Rich was considered a "fast burner" in the vernacular of the Air Force and was on the right career track to make general. Promoted to colonel at the young age of 38, the soft-spoken Rich took over as Commander of the Weapons Lab on 1 July 1977. Lamberson, as Director of ARTO, had reported to Rich. When Lamberson moved on to his new assignment, Rich took over Lamberson's ARTO job. This unprecedented action was interpreted by some as a demotion for Rich, because he now reported to the new civilian Director of AFWL, Dr. William L. Lehmann. In other words, Rich had moved down in the chain of command in an organization that he had formerly commanded.[45]

Rich suspected that Secretary of the Air Force Hans Mark probably had a hand in moving him into Lamberson's old job. Mark was a staunch supporter of the ALL program, and didn't want to lose it. He knew Rich, who had earned his doctorate in astrophysics at Harvard, possessed the same level of laser technical expertise as Lamberson. Others at Systems Command agreed that Rich's experience with lasers since the early 1960s made him the best-qualified person for keeping the ALL program on track to meet its technical objectives. Rich fully realized taking over as the head of ARTO would be damaging to his career progression, or at least put on hold his chances of being promoted to general. When General Bob Mathis, Systems Command's Vice Commander, phoned Rich to advise him that the command wanted him to move over to ARTO, Rich put his personal ambitions aside and agreed to serve in that position.[46]

Rich assumed his new duties in ARTO on 1 April 1978 and relinquished command of AFWL to Dr. Lehmann on 1 July. Whereas Lamberson had deservingly received credit for success of the ALL to date, he left the program at the most critical time. Rich, by luck of the draw, took over at the most difficult part of the ALL program. Engineering integration of the high-energy laser and associated hardware into the airplane and then ensuring all of the subsystems worked in the airborne environment in perfect unison was the imposing challenge Rich faced as soon as he took over ARTO. There was no misunderstanding that tumultuous times loomed ahead. Rich was the man center stage that everyone expected to fix technical shortcomings, schedule slips, funding shortages, and most importantly, to demonstrate the ALL could shoot down AIM-9B missiles. And he had to accomplish all of this in the shadow of the reputation and legacy that Donald Lamberson had left behind. For the next 5 years, Rich struggled with these issues to get the ALL up and running. With the assistance provided by his experienced and highly competent ALL staff, to include Kyrazis, Dettmer, Boesen, Otten, and many others, Rich and the ALL team combined their talents and energies to show significant progress during Cycle III before his retirement from the Air Force in November 1982. From that point on it would be up to Colonel Dettmer, who took over from Rich, to shepherd the ALL through its final phases of the airborne demonstrations in 1983.[47]

TEST CELL WORK CONTINUES: ADAS AND APT

Before the Perkin-Elmer ADAS was built and installed in the test cell, AFWL first wanted to make sure the ADAS optical concepts could be proven using a less expensive version of the real ADAS. Hughes, who had been the primary optical consultant for the ALL since its inception, bid and won the contract (F29601-74-C-0033) to design and build a breadboard dynamic alignment system (BDAS) in April 1974. Hughes assembled the BDAS at its Culver City laboratory where it underwent integration and low-power testing starting in October 1974. Key components of the BDAS were the grating rhomb designed to divert a low-power sample of the high-power beam to determine beam alignment, and the autoalignment and beam angle sensors to keep the beam properly aligned. Although the BDAS was not airworthy and was not the same configuration as the real ADAS, the optical train in both devices was similar. Once initial tests and checkout were completed, AFWL directed Hughes to move the BDAS to Kirtland.[48]

Hughes lost no time in disassembling and loading the BDAS on an Air National Guard C-130, which flew the BDAS to Kirtland on 13 December 1974. Immediately on arrival, the equipment was trucked to AFWL's SOR—home of the TSL—and 6 days later the BDAS had been reassembled and was ready for its first laser test. The main difference between the Culver City and SOR testing in 1975 was a high-power (100 kilowatts) beam was used at SOR. This required replacing the uncooled mirrors used in the BDAS at Culver City with water-cooled mirrors to accept the high-power beam. Many problems surfaced during testing. The main one was that in the middle of the test program the TSL laser failed to consistently generate a quality beam. However, enough data had been collected to confirm timely and sufficient move-

ment of the two active steering mirrors to align the beam as it moved through the BDAS. Hughes had not accomplished 100 percent of the testing that AFWL had originally intended. However, as the final test report noted, "All of the key test objectives were at least partially attained and the test program was on the balance successful." The report went on to state, "The lessons learned on the hill at Sandia Optical Range clearly demonstrate the worth of the ADAS breadboard program."[49]

AFWL was more than satisfied with Hughes's work supporting the optical subsys-

Breadboard dynamic alignment system (BDAS) setup at Sandia Optical Range.

tems for the ALL. However, an unforeseen change of events took place as AFWL prepared the contract for design and fabrication of the lightweight airworthy ADAS in 1975. Lamberson and others at AFWL naturally thought Hughes would be the most logical choice for the ADAS as they had the expertise and hands-on experience in building the BDAS and APT. Hughes personnel were familiar with both systems and were far ahead of anyone else in

terms of understanding the physics and integrating a yet-to-be-built airworthy ADAS with the APT. Although Hughes certainly looked like the front runner, Lamberson under government guidelines could not ignore proposals submitted on the ADAS by other contractors.[50]

Perkin-Elmer took advantage of that opportunity by throwing its hat into the ring to compete for the ADAS contract. As it turned out, the Norwalk, Connecticut-based company emerged as a dark horse and came in with the best technical proposal while Hughes was still debating with AFWL over agreeing to performance specifications for the ADAS. Kyrazis recalled the situation:

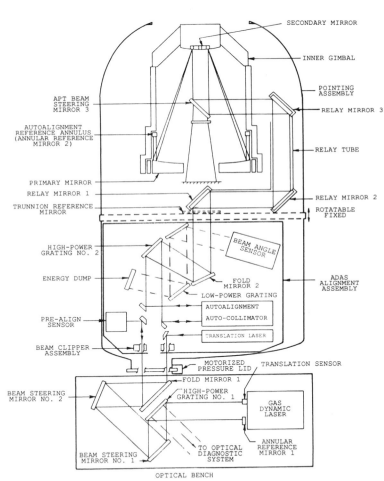

Airborne Pointing and Tracking/Airborne Dynamic Alignment System (APT/ADAS) block diagram.

Some of the performance specs that we had put down based on the Cycle II flights, Hughes would not agree to. We agreed to them as goals, but not as requirements, and Perkin-Elmer said, 'No, we accept those requirements,' and they came up with a better design. And they really did their homework. They understood the APT, they looked at it, measured things, and everything else. The guys at Hughes never bothered to go across the hall and talk to the APT folks, and that's why their proposal wasn't as good. So Perkin-Elmer got the award.

Not only did Perkin-Elmer receive the contract (F29601-76-C-0016) on 25 July 1975 to design and build the ADAS, but it delivered this critical piece of hardware weighing almost 1700 pounds to AFWL on time and under cost. Most importantly, ADAS eventually met the technical specifications as laid down by AFWL.[51]

After a year to build the ADAS, Perkin-Elmer delivered the finished product to AFWL in the fall of 1976. This one-of-a-kind device was a precision electro-optical system designed to keep the laser beam directed along the right path (dynamic translational and angular alignment) between the GDL device and the APT on top of the plane. As the beam traveled from the laser device and traversed the optical bench, the beam moved upward into the bottom section of the ADAS. Keeping the beam aligned and free from jitter was the main job of the ADAS so the beam would make its way through a very small opening in the base of the APT. If the high-power beam failed to thread its way precisely through the center of the small hole, it would damage nearby equipment in the ADAS and APT. ADAS was entirely automated requiring no operator intervention; however, the automatic operation could be manually overridden if needed. Set up first in the Optics Laboratory off the ARTF hangar, the ADAS underwent a number of rigorous acceptance tests before moving next door to the test cell building. Once in the test cell AFWL proceeded with a variety of initial performance tests from January to April 1977 to check out overall system performance.[52]

Results clearly showed there were major problems with the ADAS. As one project officer—who minced no words—put it, "The ADAS went to hell!" Many factors contributed to the system's poor performance, but three stood out. One was the water vibration problem mentioned earlier that Lieutenant McQuade eventually solved in 1978 after a lengthy investigation. A second issue, and one that was much more difficult and time-consuming to solve, was the insufficient torque (movement) exhibited by the two beam-steering mirrors located on the optical bench at the bottom of the ADAS. A third obstacle was that dust in the air ignited by the laser beam caused an unexpected sparkle effect that interfered with the sensors in the ADAS to align the beam.[53]

As the laser beam moved from mirror to mirror in the optical train, it heated the surrounding air. Tiny dust particles, close to and surrounding the beam, literally exploded and shot off in all directions when exposed to the intensely hot beam.* Sensors, whose job was to detect the beam to keep it aligned, were confused by seeing the additional light given off by flashes from the dust burning. One way to overcome this problem was through better housekeeping procedures to keep the test area as clean as possible. AFWL built a frame supporting a large cloth curtain that surrounded the bench optics and ADAS to seal out dust. MacCutcheon recalled he and others spent an inordinate amount of time taping sections of the curtain together and "using tacky mat rags trying to dust everything down." This worked fairly well, but still did not eliminate all of the dust to provide a sterile environment; remaining dust particles ignited and accelerated into the curtain burning holes and leaving scorch marks. Eventually, a better fix was redesign of circuitry in the sensors, which helped them to ignore and filter out the unwanted flashes.[54]

Although the water and sparkle issues caused a temporary setback in the test schedule, the most serious problem with the ADAS was the poor performance of the two Perkin-Elmer steering mirrors. These mirrors mounted on the optical bench were critical for keeping the beam aligned as it exited and moved from the laser device to the base of the APT. During a long and frustrating exercise in the test cell, data collected from over 100 experiments revealed a much higher vibrational level—four times greater—than AFWL had originally anticipated, which caused an abnormal amount of beam jitter. One of the frustrating problems

---

*The side of the dust particle first hit by the laser beam burned (vaporized) quickly, launching the remaining dust particle like a rocket. Propelled at high speeds, tiny dust particles crashing into highly polished mirror surfaces could cause considerable damage.

was that the optical bench could not be made stiff enough to eliminate all of the vibrations coupled into the mirrors.[55]

APT undergoing inspection in the APT lab located in ARTF hangar. Lower portion is ADAS.

To compensate for beam jitter, the steering mirrors (which also vibrated) were designed with sufficient torque to turn and tilt very quickly to keep the beam aligned on the center of the mirrors. A diagnostic system consisting of a grating rhomb on the optical bench took a sample of the beam, which was fed to the beam angle sensor (BAS) located in the ADAS alignment assembly above the optical bench.* The BAS measured angular beam misalignment. Translational jitter of the beam, that is, how much the beam moved fore and aft and left and right off its desired vertical path up into the ADAS, was calculated by the translational alignment sensor. Once these misalignment errors were computed, commands from the ADAS servo electronics sent electrical signals with just the right amount of voltage to eight actuators fitted to the backside of each steering mirror to move and minutely change the tilt angle of the mirror. Each actuator would either push against or retract from the backside of the mirror (depending on the voltage delivered to each actuator), which caused the mirror to change its angular position, thereby correcting for jitter and ensuring the beam maintained proper alignment.[56]

---

*Most of the major subsystems of the ADAS were housed in the Alignment Assembly suspended from the base of the APT in a pressure shell resembling a fat cylinder. Beneath the Alignment Assembly was the optical bench, which ran parallel to the floor of the aircraft. The Alignment Assembly hung vertically with its base over the optical bench and the top section of the Assembly attached to the base of the APT. The beam moved parallel to the floor of the aircraft on the optical bench, turned upward into the base of the ADAS Alignment Assembly, and continued on into the APT where the beam was focused and pointed to exit the aircraft.

At least that was the basic theory as to how the system was to work. The problem, as discovered in the test cell, was that the mirrors could not handle the high-frequency vibrations when the laser was turned on. In other words, the steering mirrors could not respond (accelerate) fast enough to tilt to a new position to keep the vibrating beam aligned. Precision timing and movement of the mirrors was absolutely essential for the system to work; mirror movement was so slight it had to be measured in microradian accuracies (not detectable to the eye) and had to occur almost instantaneously—within a few milliseconds. Unable to accomplish this goal of correcting beam angle and translational disturbances, AFWL expressed disappointment knowing the ALL would be further delayed while they tried to find a way to fix the mirrors. However, they could not point the finger of blame at the contractor for the unacceptable performance of the mirrors. Perkin-Elmer had kept their part of the bargain by designing and building the mirrors to technical specifications stated in the contract, but neither they nor AFWL had predicted the excessive vibrations generated by the operation of the laser.[57]

It took 2 years for AFWL to come up with a cure to fine-tune the ADAS. The first decision made by Lamberson in the spring of 1977 was to completely redesign and rebuild the steering mirrors from scratch. An improved BAS and new electronics were also to be incorporated into the upgraded ADAS. The second concern was what to do with the poor-performing Perkin-Elmer ADAS. There wasn't much choice but to continue testing and checking out the ADAS in hopes of gaining a better understanding of the physics of the system so people would be better prepared to work with the new ADAS when it arrived. Waiting for a contract to be issued and then building a new ADAS system consumed valuable time that outsiders interpreted as a major setback for the ALL. Lamberson was not happy with the state of affairs either, recognizing momentum would be slowed on the entire ALL program and undoubtedly cause anxiety at Systems Command and Headquarters Air Force.[58]

Lamberson's perception was that Perkin-Elmer had "botched up" the ADAS and he was not satisfied with their performance. Even though Perkin-Elmer had very good technical people, when Lamberson wanted answers, more often than not, the contractor's management people—who were not always the technical experts—met with AFWL personnel in trying to solve problems. This was frustrating not only to Lamberson, but also to others at AFWL who were more accustomed to dealing directly with other contractors' physicists and engineers when problems arose, rather than the Perkin-Elmer "bean counters" as one AFWL officer put it. In comparison, Hughes had always sent eminently qualified technical people to resolve issues with AFWL. AFWL over the years had been satisfied with Hughes who had forged a reputation of being extremely capable in getting optical systems to work as demonstrated by their track record with the TSL, FTT, APT, and BDAS.[59]

It was clear Lamberson wanted Hughes to take over the ADAS upgrade work to prevent any further delays. But this did not happen overnight. It took a year to formulate the new mirror specifications, prepare the contract, and finalize the selection of the contractor. In March 1978 AFWL awarded Hughes Aircraft Company the first phase of a 12-month effort to design, build, and test two high-performance steering mirrors for the ADAS optical system. The first part of the design work took place as stated in subtask statement 2-29 under the existing Hughes Airborne Pointing and Tracking contract F29601-77-C-0027. In tech-

nical parlance, Hughes proposed "adoption of a design that would increase the acceleration capacity from 80 to 800 rad/sec^2 by means of a reactionless electromagnetic drive mechanism closely coupled to a low-inertia water-cooled mirror." By August 1978, AFWL had approved Hughes's critical design review of the new mirror system.[60]

Once given the go-ahead on the critical design, Hughes's Electro-Optical and Data Systems Group began fabricating the new mirrors. Several months later the mirrors were finished. The contractor conducted initial acceptance testing on 11 and 12 April 1979 at its plant in Culver City. Hughes then delivered the mirrors to AFWL where a second series of acceptance tests took place to measure operating performance—e.g., torque disturbances, acceleration, vibrations, noise, water flow and pressure. This gave the ALL team of scientists and engineers an opportunity to check out the new system firsthand. After waiting for almost 2 years, they were more than satisfied with the upgraded ADAS. As Hughes noted in its final report, AFWL found the new ADAS mirrors "completely acceptable for the Cycle III application."[61]

Hughes's assessment was somewhat premature because to be ready for Cycle III application meant the ADAS first had to be mated to the APT and tested. This would have to be done to verify compatibility between these two critical components to ensure the ADAS and APT worked as an integrated unit in the test cell. All of the time the ADAS was being reworked, the APT had been undergoing major modifications and testing in the clean room and the test cell so that eventually it would be ready to be joined with the new ADAS. In the test cell the APT was mounted on a massive welded-steel support structure which provided the advantage of more stability than the lightweight aluminum configuration used to brace the APT in the aircraft. The APT anchored to the steel bench would encounter less vibrations than in the ALL. Protected in the test cell, the APT also would not experience the aerodynamic turbulence and buffeting the ALL would encounter in flight. AFWL recognized that the absence of dynamic airflow around the turret and the rigid test stand would produce more favorable APT experimental data than would be collected during ALL flights. On the other hand, there were some disadvantages to testing the APT in the test cell. One of the main drawbacks was the air at ground level was denser than in the airborne environment, which produced more beam thermal blooming than was expected in the airborne environment. AFWL predicted beam quality would improve considerably during the airborne experiments.[62]

During its time in the test cell, Hughes made extensive modifications to the APT. One of the most important changes was the installation and testing of the advanced tracker in the APT. (The advanced tracker arrived at AFWL in December 1977.) The advanced tracker subsystem consisted of a visible light television camera with variable fields of view (through electronic zoom and two selectable lenses), an 8- to 12-micron wavelength infrared imaging camera (often referred to as a "FLIR" for forward-looking infrared), and a processor electronics box [known as the Digital Correlation Real-Time Processor (DCRP)] which did the tracking computations. Consisting of a serial-scan sensor employing 66 HgCdTe (mercury–cadmium–telluride) elements in six lines, the IR imager was sensitive to 8- to 11.5-micron radiation. TV and IR images of targets were displayed on the APT operator's console in the rear compartment of the ALL.[63]

Mirrors built by Hughes to upgrade ADAS.

Lessons learned from Cycles I and II and the North Oscura Peak tracker tests at White Sands in 1977 and 1978 had shown the APT could track targets more accurately and reliably with an 8- to 12-micron correlation tracker than the 3- to 5-micron centroid tracker used in Cycle I. Data collected from the North Oscura Peak tests revealed that plume tracking was too unstable for precise beam control. Consequently, the APT's original 3- to 5-micron track sensor needed to be replaced with a longer-wavelength sensor that could "see" and track the cooler aircraft or missile bodies to track on after the plumes burned out. A correlation algorithm (as opposed to centroid tracking) also was developed as the preferred tracking mode, because the contrast of the target bodies was sometimes so low that centroid tracking

215

Exposure of internal components of APT.

was poor or impossible. One distinguishing feature of the new correlation tracker was it came equipped with a 10.6-micron absorption filter to cancel out any of the HEL light (in this case, the 10.6-micron GDL) reflected off the target. Finally, the new advanced tracker was less noisy and able to reject more background clutter than earlier trackers.[64]

Although the advanced tracker used a filter to absorb reflected 10.6-micron laser light to prevent interference with the correlation tracking function, the advanced tracker also incorporated a single-element staring sensor that *could* detect the 10.6-micron laser light reflected off the target. This particular sensor was used in conjunction with a circular nutation of the HEL beam to remove any boresight errors between the HEL beam path and the track sensor. This "conical scan" approach (so called because the HEL beam would trace out a spiral pattern) could also be used as a high-precision tracker. The tight spiral motion of the beam would cause a modulation of the reflected energy as the beam alternately moved closer to and farther from a reflective point on the target. This modulation could be used to steer the beam onto the laser reflective point on the target. For AIM-9B missiles, the reflective point the beam needed to remain locked on was the missile's nose.[65]

For the self-defense scenario against AIM-9B missiles, the ALL tracking system would employ centroid and correlation algorithms and conical scan tracking to place the HEL beam on the missile nose. The APT operator would acquire the missile launch aircraft manually and initiate track on the aircraft with a correlation track gate. (A "track gate" was a processor in the DCRP that would generate target position signals based on a defined region of the track sensor image. The term *track gate* was often used interchangeably to refer to the processor and to the region of the sensor field of view that it was processing. The DCRP contained five processors or gates: two correlation gates and three centroid gates.) A centroid gate would also be placed over the launch aircraft image, set to detect only light above a video threshold that was slightly brighter than the aircraft. When a missile was launched from the aircraft, the missile's bright plume would be detected by this centroid gate, which would then be given control of the APT gimbals. When the missile plume was centered and stabilized, control was passed from the centroid tracker to a correlation gate to track the missile body as the missile plume burned out. The APT's laser range finder was used during the missile track to help determine when to fire the HEL, to establish proper telescope focus, and to compute corrections for parallax errors between the track sensor and the main HEL telescope.[66]

For the future Navy escort defense scenario (also referred to as the 1N scenario) to show that a laser-equipped aircraft could protect a ship from a low-level cruise missile attack, AFWL did not plan to use the conical scan system. A BQM-34 drone was the target (to simulate a cruise missile) to be used for this scenario. Conical scan would not be used against the BQM-34 drone, because it was a large target (relative to the AIM-9B missile) and the aimpoint on the drone (such as the fuel tank) was not necessarily the brightest reflector.* For these engagements, the target drone would be acquired manually by the APT operator, who would also designate the aimpoint on his video display. The target would be tracked using a correlation gate. Once the laser was fired, a centroid gate (with a high-detecting threshold) would be used to sense the part of the drone that was being heated by the HEL beam. If this "hot spot" did not match the desired aimpoint, the system would calculate and apply an appropriate boresight correction to move the beam onto the desired target aimpoint.[67]

Changes with the optics also took place. During Cycle II the low-power beam did not require water-cooled mirrors. That all changed in preparing the APT and ADAS in the test cell for Cycle III when a high-power beam would be used. To prevent damage by the laser to optical components in the APT and ADAS, AFWL installed molybdenum water-cooled mirrors and associated plumbing to handle the high-power beam. All of the mirrors were replaced in the APT and ADAS except for the primary mirror, whose surface area (60-centimeter diameter) was large enough to accept the high-power beam for short bursts of time without any damage to the mirror. However, the primary mirror's Cer-Vit faceplate surface was recoated to improve its reflective qualities.[68]

Prior to using a high-power beam with the APT and ADAS, AFWL conducted an extensive series of experiments from December 1978 through March 1979 in the test cell to evaluate the APT. One of the goals of this work was to determine how water flow (about 7–8 gallons per minute) through the system affected mirror jitter. To find out, AFWL's Paul Merritt and Captain Robert Van Allen ran high-pressure water through the cooled optics, which was instrumented with accelerometers to measure vibrations of the mirrors. As water circulated to cool the mirrors, the two project officers bounced a low-power laser beam off the mirrors. A jitter sensor collected vibrational data to measure how much the beam moved off its desired path. Those measurements were then compared with measurements taken when the beam was turned on without water flowing through the system. In that way, one could tell what effect the circulating water had on mirror and beam movement. After analyzing the data, Merritt and Van Allen reported, "The results showed very little increase in jitter due to water flow."[69]

---

*Precision conical scan was needed to direct the laser beam to hit the small target area, the nose of the AIM-9B missile. For the Navy scenario, the laser would not be required to hit the nose of the drone head-on; instead, the laser would have a large surface area (side of drone) to intercept.

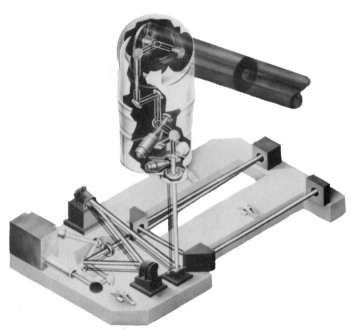

ALL beam path.

Although the water flow did not appear to present an insurmountable obstacle, AFWL was concerned with other issues that might degrade the performance of the APT. One had to do with rotating or slewing the APT left and right as well as up and down—which would happen during the future airborne scenarios to acquire a target—to determine the effect the moving turret had on beam jitter. The APT was a large and bulky device mechanically driven by a hydraulic power system and associated feed lines that generated noise and vibrations as the system operated. These disturbances and gimbal mechanical bearing wobble were very small but could be significant relative to the microradian pointing required. Data collected on slewing of the APT in the test cell, at rates from 54 to 213 microradians per second, revealed base movement of the APT was indeed a contributor to beam jitter, but Merritt and Van Allen calculated the jitter was within acceptable limits to allow the airborne testing to proceed.[70]

Before the APT could be flight tested, it first had to be mated with the ADAS, the GDL, and GOFSS in the test cell. This took place between 13 March and 30 April 1979. AFWL was fairly confident with the operation of the GDL and GOFSS based on data collected from prior testing of these two components in the test cell. In the spring of 1979, AFWL officials delivered a paper at the Third DOD High Energy Laser Conference in Monterey, California, revealing the following vital performance statistics of the GDL:

> 154 hot firings of the stages; 308 seconds of hot gas flows; and 97 seconds of photon extraction. Its nominal near-field power is 456 kilowatts. We expect to see approximately 380 kilowatts in the far field when propagating through the ADAS/APT. Most device runs have been in the 3-second range; the longest has been 7.9 seconds. After solving many problems we have a fairly reliable device which generates photons for the programmed duration approximately 70% of the time the fire button is depressed.[71]

Although pleased with the performance of the laser device, AFWL was less certain how the APT and ADAS would perform when connected to the optical bench, GDL, and GOFSS.

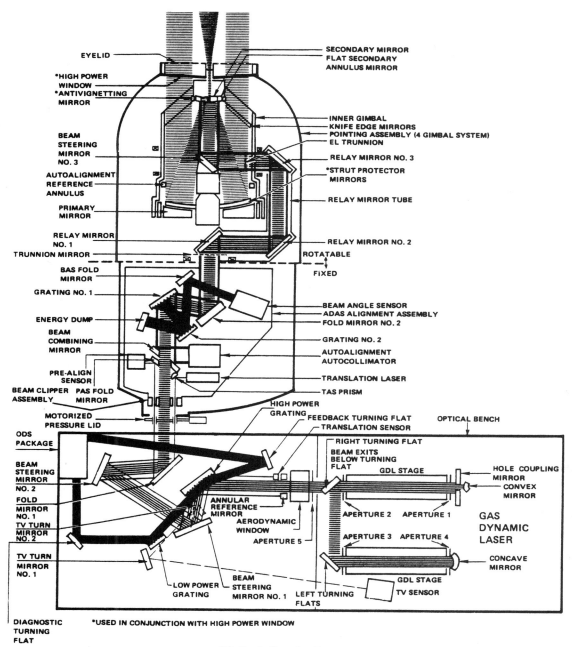

ALL Cycle III optical beam path.

Once all of these vital parts were connected, a variety of tests followed to determine how well the entire ALL system performed. Over 100 separate tests were conducted from May through mid-December 1979, with the majority taking place inside the test cell or uprange

site. Eleven tests involved sending a high-power beam from the APT in the test cell to targets downrange at the 1-kilometer site behind the test cell. A considerable amount of diagnostic equipment was set up at both the uprange and downrange locations to measure various beam characteristics. For example, instrumentation included a calorimeter to measure beam power, a jitter sensor to determine how much jitter was in the beam, accelerometers to detect the amount of vibrations in the mirrors and optical bench, and combustor vibration accelerometers to monitor points in the combustor to ensure safe operation.[72]

SENDING THE BEAM DOWNRANGE

Adhering to its policy to progress slowly in small increments, AFWL did not try to send a high-power beam downrange until it had first analyzed that beam inside the test cell. This was a two-part procedure. Placed in front of the APT exit aperture was a Hartmann plate. The side of this plate facing inside the APT was highly polished to reflect about 98 percent of the high-power beam into a calorimeter or beam dump.* Beam power then could be measured by the calorimeter, data AFWL wanted to know to determine the operating effectiveness of the laser device as well as the potential beam power that would be traveling downrange during later high-power downrange experiments. The Hartmann plate allowed about 2 percent of the high-power beam to pass through and travel downrange. In effect, the 2 percent beam was now a low-power beam propagating to the downrange target, while the remaining 98 percent of the beam was diverted to the calorimeter or beam dump inside the test cell.[73]

Besides safety considerations, the advantage of propagating a low-power beam was twofold. First, a low-power beam moving through the atmosphere was not degraded as much by thermal blooming (spreading of the beam) as a high-power beam traveling through the same air mass. Second, diagnostic instruments downrange could take samples of the low-power beam to measure beam performance such as power, intensity distribution, wavelength, and jitter. Collecting data on the low-power beam gave AFWL scientists a good idea of how the high-power beam would behave in flight, where the air was less dense.[74]

Collection of ground-test data for both the low- and high-power laser beams fired from the test cell took place at the 1-kilometer site located atop a small knoll south of the test cell. On the left end of the downrange target site was a block building (774) housing technicians who operated TV video camera controls, TV monitors, a video recorder, and a small 200-watt calibration laser. Behind the building was a high-energy experimental mobile van to record telemetry data from the target using telemetry receivers, recorders, and processing equipment. A cinder block wall extended from the experimental van to a large rotoplane anchored at the west end of the test site. Centered behind the wall was another instrumentation trailer with an IR scanner and video processor, disk recorder, and a TV camera and monitor to gather beam diagnostic information.[75]

In front of the west end of the wall was a large rotoplane target. Several meters from the large rotoplane and behind the wall was a second moving target identified as the small

---

*The 60-centimeter diameter beam reflecting off the primary mirror and then the Hartmann plate was reduced to a 10-centimeter-diameter beam to fit into the calorimeter entrance aperture.

rotoplane. The large rotoplane (obtained from a carnival ride) consisted of two 16-foot metal arms extending in opposite directions from a motor that rotated the arms at various speeds. Attached to the end of one of the arms was a target board for the APT to track and focus the laser beam. The small rotoplane sat atop a 9-foot pedestal positioned directly in front of a 20-foot-wide by 10-foot-long steel culvert resembling a large highway drainage pipe. (The small rotoplane was capable of rotating a 75-pound target board up to 2 revolutions per second on a 0.70-meter radius.) In case the beam was misaligned and missed the target or reflected off the mirrored surfaces of the rotoplane, the culvert served as a "beam catcher" to prevent the high-power beam from wandering off in all directions. If a stray beam passed through the culvert, there was no danger of a fire starting because AFWL had graded and removed all vegetation in the area behind the small rotoplane. For safety reasons, only the small rotoplane was used for high-power testing. For static tests, the small rotoplane was removed and a small gold ball affixed to the top of the pedestal served as an immovable target.[76]

Another safety precaution during high-power testing involved using six baffles to restrict and control the beam path so the laser light would be headed in the right direction downrange. Each of three steel towers, two located near and in line with the south exit door of the test cell and the other about midway through the tank farm, supported two rectangular baffles made of 10-gauge steel positioned side by side. One baffle was about the size of a commercial billboard; the other was half that size. In essence the baffles were large steel plates with cutouts in the center that were in the shape of an inverted keyhole. This allowed "APT viewing along the beam axis as well as FLIR and video viewing above the beam." The baffles were a coarse method to maintain proper alignment so the beam did not strike anything outside the designated downrange target area or move on beyond the local horizon behind the targets. Another important reason for the baffle system was to ensure the beam did not hit any of the tanks or lines in the fuel farm and set off an explosion. Tripwires were installed at the edges of the cutout of the uprange baffle. If the beam wandered from the desired path through the center of the cutout, it would burn through the wire causing a shutdown signal to be sent to the laser. This assured that a sudden slewing of the APT during firing would not endanger anyone in the vicinity of the tests. For low-power testing, the baffles were not used because the amount of radiation was not high enough to cause any damage to the surrounding area even if the beam strayed.[77]

Of the 100 low- and high-power downrange tests conducted from May to 13 December 1979, AFWL judged 37 to be successful in providing useful scientific data. The other tests had to be aborted because of safety considerations and malfunctions with individual components of the total ALL ground-configured system. For example, 12 tests had to be halted because of problems in starting up the laser device and diffuser. Another 17 tests had to be stopped because of potential problems with the fuel flow in the FSS system.[78]

Two of the most important downrange tests involved the AIM-9B Sidewinder missile and a BQM-34A drone, the same type of targets AFWL planned to use during the ALL flight testing scenarios. The purpose of the downrange tests was to find out if the high-power beam shot from the test cell (the Hartmann plate was removed for these shots) could hit the targets set up at the 1-kilometer site and deposit sufficient energy to cause damage. For the first test

AIM-9B missile target set up downrange from ARTF (center background). Test cell in right background.

conducted on 6 September 1979, AFWL set up an inert AIM-9B missile so its IR seeker dome at the tip of the missile looked back at the test cell at an angle 10 degrees left of the beam's optical path traveling from the test cell to the missile. A stationary heat gun near the test cell heated and reflected IR energy off the right side of the missile dome; the missile seeker head locked on to this local IR source to simulate in-flight conditions. The tracker in the APT also locked on to this reflected heat source so the APT could calculate the laser aimpoint on target, which was 3 degrees left of the IR heat source reflecting off the missile's dome. Simply put, the right portion of the missile dome served as the track point and the left side of the dome served as the laser aimpoint. If the beam sent downrange broke the dome, the missile's tracking mechanism would break lock.[79]

Results from the 6 September experiment demonstrated the laser failed to achieve its test objectives. When the laser was turned on, the center of the beam missed "considerably" the aimpoint on the IR missile dome. The weaker outer portion of the beam's diameter managed to cause some blistering and charring around the top edges of the IR dome, but the expected fracturing of the dome did not materialize because the hottest portion of the beam packing the most energy passed over the target.[*] Based on a frame-by-frame analysis of the color video camera, AFWL concluded the APT tracker broke lock with the heat gun and tried to lock on to the localized hot flashing caused by the beam in front of the target. Consequently, as the tracker shifted over to track the laser energy, the calculated aimpoint also moved from its original spot on the target. The next day AFWL examined the missile to verify that it had suffered no damage. As an extra precaution, the seeker was sent to China Lake for bench inspection where it was determined that the seeker experienced no internal damage as a result of the 6 September test.[80]

Two months after the first unsuccessful AIM-9B experiment, AFWL performed a second high-power laser test on the missile. This time the angle between the beam path and the IR source was increased from 3 to 10 degrees in an attempt to reduce the chances of the tracker breaking lock with the IR source. Data from this second test proved much more productive.

---

[*]Extensive studies on the vulnerability of the AIM-9B concluded there were several ways to damage the missile. One technique was to irradiate the nose of the missile with enough energy to blind the seeker, causing it to break lock with its target. Thermal fracture of the glass IR dome was the best way to permanently damage the missile. Once the beam fractured and penetrated the dome, the laser had easy access to the internal components to damage electrical circuits, optical systems, and the structural metal of the missile.

Results showed the APT in the test cell delivered and held the laser beam squarely on the aimpoint of the missile dome causing significant physical damage to the outer dome and seeker guidance and control unit inside. This was one of the most important milestones in the entire ALL program because it demonstrated the ground-based ALL housed in the test cell could operate as an integrated unit with a high degree of proficiency to disable a real-world target. With this scientific proof in hand, AFWL personnel were confident they could repeat this experiment on board the ALL aircraft against an in-flight AIM-9B as part of the aircraft self-defense scenario. But before moving ahead with the flight testing, AFWL wanted to run another series of tests with the ALL on the ground to collect data on how the ALL would perform against the BQM-34A drone planned to be used during the ship defense scenario.[81]

AIM-9 missile mounted on test stand downrange behind the ARTF.

Technicians mounted a special fuel tank on the drone for the test carried out on 9 November at the 1.5-kilometer downrange site. Inside the pressurized tank was Tonka 250—composed of 50 percent triethylamine and 50 percent xylidine—to simulate an operational Soviet rocket fuel. The beam from the test cell was aimed over the drone's fuselage to a 50-centimeter focusing mirror which redirected the beam to strike the fuel tank. As the beam struck and heated the tank, enough overpressure was created inside the pressurized tank (52.5 psi) to

Lieutenant Colonel Keith Gilbert monitoring downrange diagnostic equipment.

cause the welded-bottom seams of the tank to fail, forcing the metal surface to move downward and to cut halfway through the engine. (If the drone engines were running, AFWL believed the engine's turbine blades would have sustained substantial damage.) The laser and subsequent overpressure ignited the fuel and forced it out of the tank and into the fuselage cavity. Flames shot out of the drone, creating enough heat to buckle and flare out the bottom section of the fuselage several centimeters. Once the beam was turned off, AFWL technicians rushed to the missile to douse the flames with portable fire extinguishers.[82]

Although the drone did not blow up in spectacular fashion, clearly there was sufficient data recorded to confirm the lethality of the laser beam. Again, this was another piece of critical evidence reaffirming the safe and dependable operation of the entire laser system to disable a target on the ground at approximately the same range that the ALL would engage missiles and drones in the air. One of the most important conclusions drawn from the downrange experiments was that the intensity of the beam was more than adequate to destroy the AIM missiles and drones scheduled for the phase 5 flight scenarios. Another positive result stemming from the test cell experiments was verification that the water-cooled mirrors in the system were not degraded by the high-power beam. In a few cases, the beam caused small burns on various sections of the mirrors, but they were always able to be cleaned in preparation for the next shot. Vibrations of the mirrors caused by the flow of the water proved to be within acceptable limits.[83]

Confident with the performance of the ALL, AFWL also recognized there were still some shortcomings and unknowns associated with the operation of the laser. For example, the test cell experiments and concurrent flight mission simulations confirmed the ADAS continued to be a major contributor to beam jitter. Consequently, the ADAS had to be upgraded to cope with the expected increase in vibrations during the in-flight testing. After repeated testing, AFWL discovered laser output power decreased anywhere from 3 to 7 percent. That issue had to be studied and resolved. There were still problems with friction with the outer gimbal motion of the APT that needed to be fixed. Misalignment of the beam as it passed through the ADAS to the APT posed a major problem that would require extensive reworking of the ADAS to improve its performance. Finally, AFWL was concerned with the many environmental unknowns that would induce severe vibrations on the aircraft and the laser during flight. All of these were legitimate concerns and Kyrazis and his team redoubled their efforts to come up with solutions to produce a higher-precision ALL operating system before going airborne.[84]

High-Power Laser Goes Airborne to Encounter Air-to-Air Missiles

Cycle III: Phases 4 and 5

Once testing of the GDL device in the test cell had been completed in the fall of 1979, AFWL was ready to evaluate the device in the aircraft. The main goal of Cycle III, phase 4, was to verify the compatibility of the GDL with the ALL aircraft. This involved removing the 25,000-pound GDL and its associated hardware from the test cell and installing it on the aircraft. (The GDL contained over 60,000 detailed parts, twice as many as found in an F-100 jet engine.) Once on the aircraft, the next step was to confirm the laser functioned properly when turned on while the plane remained parked on the ground. If the laser passed that test, it would then move on to the flight testing phase.[1]

Phase 4 focused on the laser device and did not attempt to assess the entire laser system. AFWL, at this stage, only wanted to measure how the laser operated and the effects it would have on aircraft performance. That was primarily a safety issue to validate the laser would not degrade the flight characteristics of the plane as it flew the same flight patterns planned for the future phase 5 shootdown scenarios. As operating the laser device was similar to firing a rocket engine inside the ALL, AFWL's concern was to know exactly how that would affect stability, control, and structural rigidity of the aircraft. Other concerns were potential increases in vibrations that could introduce significant motions causing unacceptable disruption of the optical train. Still another salient issue was the effect of the hot laser exhaust

as it passed through the diffuser and exited the ALL. Once outside the plane, would the hot exhaust be forced by the airstream against the lower fuselage and cause any damage?[2]

Installation and testing of the FSS on the ALL had been accomplished during phase 3, prior to the laser device being placed in the plane. The FSS was a critical component that ignited the fuels to initiate supersonic gas flow into the laser device. During phase 4, the FSS and device (combined weight of approximately 11,000 pounds) had to be connected inside the aircraft before the laser could be operated and evaluated. An important aspect of the phase 4 testing was a beam was *not* extracted from the GDL. Two mirrors—one at each end of the cavity in the device—were installed and instrumented to measure how stable they were when encountering vibrations created after the laser was turned on. However, shutters placed in front of these mirrors blocked any photons produced in the cavity from exiting the cavity. This was done intentionally because the main optical subsystems, namely, ADAS and APT, were still being worked on in the test cell. Consequently, there was no need to extract a beam from the laser device, because there was no place for it to go without a complete optical train to steer the beam. AFWL planned to install the ADAS and APT during phase 5.[3]

Although the real APT was being refurbished in the optics laboratory during phase 4,

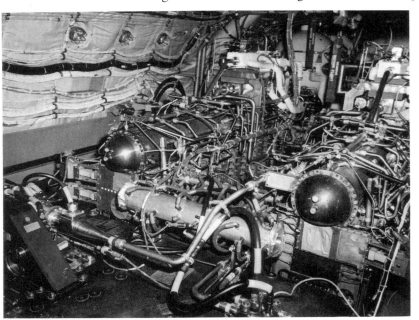

Laser device installed in ALL aircraft.

AFWL still wanted to find out how the aircraft would respond with both the laser and turret installed. To accomplish this, a dummy APT was used in place of the actual APT. The dummy APT simulated the size and mass distributions of the real APT. AFWL scientists predicted measurements of the aerodynamic loads around the dummy turret would be similar to those taken when the real APT was flown during phase 5.[4]

From January through March 1980, AFWL and Pratt & Whitney technicians devoted all of their time to installing the laser device in the plane and then connecting it to the FSS. In March, the bench optics was added. Once all of these components were in place, as a first-step safety precaution, the test plan called for cold flowing inert nitrogen in place of the combustion fuels, methane and carbon monoxide. Cold flow testing took place during April and June to check out the combustors and to measure fuel flow control systems, as well as collecting data on vibrations and pressures interacting with the system while the plane remained on the ground.[5]

ATTENTION AND VISIBILITY

At the same time work moved forward to get the laser on the ALL, New Mexico's Republican Senator Harrison "Jack" Schmitt arranged for a public hearing to take place at Albuquerque's main library on 12 January 1980. This meeting publicized the important contribution AFWL's ALL was making in advancing laser technology. Schmitt, who chaired the Senate's Subcommittee on Science, Technology and Space, which reported to the Senate Committee on Commerce, Science, and Transportation, had invited a number of prominent scientists to share their thoughts on laser applications and technology. Those who provided testimony included Dr. Hans Mark, Secretary of the Air Force; Colonel John Rich, head of AFWL's laser program; Dr. Donald Kerr, Director of Los Alamos Scientific Laboratories; and two of his premier laser scientists, Drs. Keith Boyer and Reed Jenson; Dr. Gerold Yonas from Sandia Laboratories; and Dr. Dick Begley, BDM Corporation.[6]

Senator Schmitt led off by stating that the purpose of the hearings was to provide Congress with a better understanding of ongoing efforts in directed energy programs. He singled out New Mexico for its unparalleled research capabilities and contributions in advancing laser and particle beam technologies. Part of that success, Schmitt noted with pride, was attributed to the unique synergism developed among Sandia, Los Alamos, and AFWL. The Senator's message was the nation must continue to place a "high priority" on laser and particle beam research. Failure to recognize the importance of sustaining this work, Schmitt said, "would create unacceptably high risks to international stability and our national well-being and security."[7]

Secretary of the Air Force Mark picked up where Schmitt left off. Mark's auspicious comments emphasized the potential of eventually developing lasers as weapons and praised AFWL for its substantial progress accomplished with the airborne laser. The Secretary went on to present a brief history of the development of the ALL, being careful to point out that the ALL was a proof-of-concept demonstration and not a "weapons demonstration." The ALL was a first step, Mark insisted, similar to Billy Mitchell's bombing and sinking of two deactivated German warships with aerial bombs.[8]

Mitchell had led the 1st Provisional Air Brigade in July 1921 off the Atlantic coast in an exercise to prove ships could be sunk from the air. Because of the inferior bombsights that existed at the time, Mitchell's planes could not accurately bomb from 5000 feet. Consequently, he and his air armada attacked the cruiser *Frankfurt* and battleship *Ostfriesland* at mast-height and sunk both vessels. Critics, led by former Secretary of the Navy Josephus Daniels (5 March 1913–5 March 1921), downplayed the significance of this historical first, claiming that if the ships had been armed with antiaircraft guns, then they would have easily shot down the low-flying aircraft. Mitchell dismissed such narrow thinking, pointing out technological advances would emerge in stages over time to permit aerial bombing from high altitudes where aircraft would be more survivable. He recommended in his report to Major General Charles T. Menoher, Chief of Air Service, that a Department of National Defense be established consisting of the three military services— Army, Navy, and Air Force. In that way, Mitchell argued, each service could define its combat mission and the most effective tactics and strategy to accomplish that mission. Mitchell

227

stated, "Only with such an organization would the United States be able to make correct decisions in choosing weapons for the future defense of the nation."[9]

Hans Mark was of Mitchell's persuasion. The Secretary recognized it took 20 years to turn Mitchell's experimental proof of concept into an effective military weapon. It was not until 10 December 1941 that the Japanese were the first to sink two British battleships at sea—the *Repulse* and the *Prince of Wales*—by aerial bombing.[*] Mark reminded his audience the discovery of the laser was only 20 years old, but he was confident proof of concept of the ALL in the upcoming ground and flight tests would be the first step leading to the development of HEL weapons. The nation was on the threshold, Mark believed, of using the ALL demonstration to show lasers could be developed to defend against air-to-air missiles. He felt an airborne laser patrolling the skies would be in a good position to intercept submarine- and ground-launched missiles during their early powered flight stage when they were most vulnerable to attack. An airborne laser "gun" flying between 30,000 and 40,000 feet, where the atmospheric problem was not as severe as at sea level, had a strong potential for the rapid shootdown of air-to-air and submarine-launched missiles.[10]

Hans Mark's optimistic outlook regarding the future payoffs of lasers was tempered with a hefty dose of reality. He knew the road to fielding a laser weapon would be extremely difficult and "torturous" as he put it. However, that was no reason to quit. Mark explained:

> Lasers have failed to work. They have failed to realize the design power levels. Mirrors have been broken. Mechanical systems have failed to perform as designed. Many of these setbacks have delayed us for months but every obstacle has been overcome and our progress, though often more difficult than anticipated, has been steady. This is not an unusual circumstance in an unexplored field. It has happened in virtually every major scientific endeavor with which I have been associated. Failures grossly outnumbered our successes in the early days of the Intercontinental Ballistic Missile program but today we have had over 50 launches of our Titan III vehicles without a single failure. We will have more heartbreak, more pain, and more sweat in the high energy laser development program but I am confident we will succeed.

Laser success in the near term, Mark believed, would to a large degree depend on the outcome of the ALL ground and airborne tests over the next couple of years.[†11]

[*]Aerial bombardment and damage to anchored ships moored in port occurred with the Japanese bombing of Pearl Harbor on 7 December 1941 and the British aerial attack on the Italian fleet at Taranto Harbor on 11 November 1940. In June 1942 at the Battle of Midway, U.S. dive bombers in one of the most decisive displays of airpower sank four Japanese aircraft carriers (Japan's entire carrier fleet at the time) and one heavy cruiser.

[†]Mark was also a realist about the upcoming ALL tests. He recognized laser hardware would have to become considerably smaller and more powerful before they could become practical airborne systems. The Air Force had been investigating using a shorter-wavelength laser, such as the 1.3-micron chemical oxygen–iodine laser, which was roughly one-eighth the wavelength of the 10.6-micron carbon dioxide laser used on the ALL. Mark also commented, "the Russians haven't got anything like the Airborne Laser Lab."

Many influential civilian and military leaders at the highest levels of government shared Mark's appraisal of lasers and the need to move forward with laboratory demonstrations. Dr. Ruth M. Davis, Deputy Under Secretary of Defense for Research and Advanced Technology, defended the $2 billion the Pentagon had spent over the years on lasers as a wise investment for the future. She supported infusing more money into laser research and testing at a time when rumors were circulating that the Soviets were moving ahead of the United States in laser development. The upcoming laser testing on board the ALL was a good example of the type of demonstration needed to collect scientific data to help determine how "the beam propagates and interacts with the target to cause damage and how this damage equates to target lethality." Davis knew these tests faced "technical hurdles" and would not yield all of the definitive answers. But she, like Mark, realized the ALL was certainly one logical starting point with the potential for advancing laser technology to "revolutionize strategic and tactical warfare."[12]

Dr. William J. Perry (Under Secretary of Defense for Research and Engineering, 1977–81, and later Secretary of Defense in the Clinton Administration) singled out the ALL test cell experiments completed in 1979 as a milestone because of the ability of the system to generate a beam of good quality. Based on this evidence, he predicted the ALL would be engaging air-to-air missiles in the fall of 1980. By the 1990s, Perry believed lasers would be playing an important role in the air and space. General Alton D. Slay, Commander of AFSC, agreed with Perry's assessment but was careful to point out the United States was not anywhere near fielding a bomber defense or an offensive tactical fighter laser weapon. Slay believed the real challenge that loomed ahead was translating scientific data gathered from ALL-type demonstrations to come up with "an engineering reality that is viable on the battlefield." In the general's opinion, that would not occur until the end of the century. General Richard Ellis, Commander of SAC, endorsed Slay's approach as the most realistic. Ellis thought the ALL test program would be successful, but also took the position that "there are many technology issues that must be solved before such a system could be deployed aboard a manned penetrating bomber." Others, such as Lieutenant General Thomas P. Stafford, Deputy Chief of Staff for Research and Development at Department of the Air Force, thought laser lead time could be cut dramatically. Stafford, a former astronaut, proposed that if a crash program was initiated, the United States could field a laser "prototype weapon" to use against low-level (300-mile-high) satellites by the mid-1980s.[13]

All of the "experts" were relying on educated guesses, as no one knew for sure when a laser weapon would be built. Plus, in 1980 there was no consensus as to whether a laser weapon should be designed to be deployed in space or in the atmosphere. Secretary of Defense Harold Brown felt the nation's laser weapons effort in 1980 should be redirected to space applications. He argued the technology was making progress, but still was not mature enough to make lasers a reality in the atmosphere.[*] In the vacuum of space, most of the propagation problems such as thermal blooming and scattering of the beam could be

---

[*]Because the air absorbed and defocused the laser beam, Mark believed that above an altitude of 35,000 feet the laser would be effective. At those altitudes, 99 percent of water vapor and clouds that would interfere with the beam would be below the laser aircraft.

eliminated. Hans Mark disagreed, claiming engineering of any laser system was the biggest challenge. Mark's position was, "If we're going to put a laser in space, we first had better test it on an aircraft such as the Airborne Laser Laboratory."[14]

Secretary of the Air Force Hans Mark (left) and Secretary of Defense Harold Brown.

In anticipation of the successful testing of the laser, the ALL began receiving more publicity in 1980. Previously, the details of the ALL were closely guarded, but as the end of the program drew near, the Air Force knew it would be smart to capitalize on this one-of-a-kind technology event. More and more articles appeared in the press. By September, the Air Force had set up a static display of laser progress which attracted a great deal of attention at the Farnborough air show in England. On exhibit were a seeker head of a Sidewinder missile destroyed by a laser during ground tests, a titanium plate with a hole drilled by a laser beam, and a cutaway model of the ALL. The display pointed out the significance of the ALL was it would demonstrate "the practicality of using high-energy laser components in a dynamic airborne environment" to show the effectiveness of lasers "against realistic targets." The long-range benefit of the ALL was it would provide a technology base to explore the follow-on development and application of high-energy weapons. According to one editorial, the Secretary of the Air Force wanted to give the world a glimpse of the airborne laser at Farnborough to bolster the U.S. technical image in the global marketplace. The Air Force was willing to exhibit the ALL, partly because AFWL was on the verge of extracting a beam from the plane. Since March 1980, steady progress had been made with the GDL and FSS to meet that goal.[15]

PHASE 4 PROGRESS: DECEMBER 1979–AUGUST 1980

Prior to phase 4, the mechanical integrity of the FSS had been verified during taxi and flight tests. What AFWL wanted to know now was how the FSS and laser device would operate when joined. As mentioned earlier, a combustor, located between the FSS and distribution manifold, ignited oxygen and methane to produce a high-temperature flame front. The main reactants of carbon monoxide and nitrous oxide then were injected into the flame. As that mixture burned, more methane was added to produce water. Nitrogen was injected into this burning mixture to produce the correct lasing media. The reaction products—carbon monoxide, water, and nitrogen—at 800 psi and 3000 degrees Fahrenheit moved from the combustor down a long chamber called a distribution manifold. All of the combusted gases then turned downward through the bottom of the manifold and accelerated through an array of two-dimensional minimum length nozzles extending into the lasing cavity. The Mach 6 flow of the gases through the nozzles expanded rapidly as the flow entered the cavity. This rapid expansion stimulated the colliding carbon dioxide and nitrogen molecules into a highly excited state, causing the release of photons from those molecules. As the photons lined up parallel to the resonator (cavity measuring 4 inches wide and 7 feet long) axis, bouncing back and forth between the mirrors at each end of the resonator, a laser beam formed and was extracted from the resonator. The gas flow continued to move downward through a diffuser which slowed the hypersonic flow and reduced the pressure to near ambient pressure, thus allowing the gases to be expelled from the belly of the aircraft.[16]

There were two identical combustors and laser devices which straddled the centerline of the aircraft. These two stages were coupled with an interstage beam duct that allowed passage of the beam from one stage to the other. The beam made three passes between the two stages and then exited the right-hand stage through an aerodynamic window. Once outside the laser device, the beam reflected off a series of water-cooled mirrors mounted on the optical bench. From there, the beam moved upward through the ADAS to the turret on top of the plane where the APT focused and pointed the beam to its target. Although this system

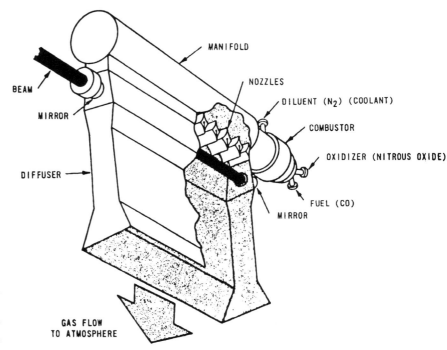

Gas dynamic laser.

was tested under laboratory conditions in the test cell during phase 3, the real test was to be conducted in phase 4 to assess the performance of the combined FSS and GDL in the airplane.[17]

The GDL device was refurbished before it was installed on the ALL. Changes in the GDL from phase 3 to 4 included redesigning the heat shields for the lower row of diffuser struts so they would not come loose as they had during the phase 3 test cell testing. The diffuser end wall ejectors and combustor injectors were thoroughly cleaned to remove deposits left by the burning gases during the test cell experiments. Also, a variable feed system, which required changes to the device plumbing lines and orifices, was installed to obtain the proper supply of nitrogen to the control valve regulating the flow of nitrogen to the aerodynamic window in the device. All of this took time, but by the first week in April the GDL had been installed in the ALL and connected to the FSS. The system was now ready to undergo phase 4 testing and evaluation.[18]

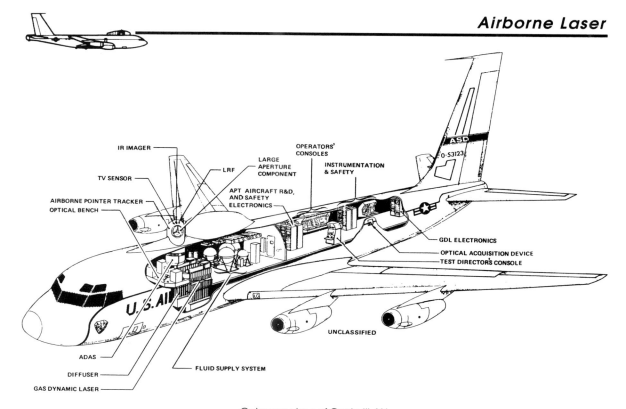

IR IMAGER

OPERATORS'
CONSOLES

LARGE
APERTURE
COMPONENT

INSTRUMENTATION
& SAFETY

LRF

TV SENSOR

APT AIRCRAFT R&D,
AND SAFETY
ELECTRONICS

AIRBORNE POINTER TRACKER
OPTICAL BENCH

GDL ELECTRONICS

OPTICAL ACQUISITION DEVICE
TEST DIRECTOR'S CONSOLE

UNCLASSIFIED

ADAS

DIFFUSER

FLUID SUPPLY SYSTEM

GAS DYNAMIC LASER

Cutaway view of Cycle III ALL.

Phase 4 consisted of both ground and flight tests. In all cases, ground testing preceded flight testing primarily as a safety precaution. If something went wrong, it was better for that to happen on the ground than in the air. As a first step, the FSS on the plane was filled with unpressurized fluids to check the structural integrity of the tanks and associated plumbing, as well as detecting leaks in the system. Pressure was increased in gradual intervals until full pressurization of the system was achieved. By March 1980 ground tests had been completed verifying the safe operation of the FSS in the ALL. On 7 April AFWL achieved the first igniter hot fire of methane and oxygen inside the ALL on the ground. Several more cold and hot fire ground tests followed over the next 2 months to verify the reliability of the combustor igniters.[19]

Convinced the combustor igniters worked satisfactorily on the ground, the next step involved airborne hot firing of the igniters over White Sands Missile Range. On 23 June the first successful airborne hot firing of the combustors occurred at 10,000 feet. During that same flight, two other igniter hot firings took place. For all three tests, ignition lasted for less than a second. The ALL returned to Kirtland, where for the next month cold flow tests of the entire GDL system were conducted on the ground. This included flowing inert gases through the resonator, distribution manifold, nozzle array, and diffuser as well as circulating water through the channels on the backside of the water-cooled mirrors mounted in the resonator and on the optical bench. None of these tests lasted more than 2 seconds as part of the overall plan to build up gradually before imposing maximum flow and pressure on the system. Once these preliminary tests demonstrated the GDL could withstand the cold flow experiments, AFWL on 25 July performed the initial hot firing of the GDL using the FSS fuels. These fuels were ignited and produced a steady-state supersonic flow through the GDL on the ground. No beam was extracted as the optical shutters in the resonator were not opened. The goal was to get all of the valves to open at exactly the right time to produce a uniform flow and to determine if the FSS and laser device would structurally hold together while the system was operating. The laser was turned off after 700 milliseconds, which was enough time to start the diffuser expelling the gases through the belly of the ALL. Time for each subsequent test increased to eventually reach full-duration runs of 7.9 seconds.[20]

The ALL returned to White Sands on 29 July and performed the first two cold flow runs of the entire GDL system while airborne. Two days later, the ALL flying at 258 knots at an altitude of 10,000 feet reached a major technical milestone by turning on the laser for 0.9 second at 10,000 feet. This first-ever airborne hot fire of the laser demonstrated the operating capability of the laser, which had to be proven before the ALL program could proceed with the Cycle III shootdown scenarios in phase 5. On 2 August the ALL repeated the hot fire flight test and increased the duration of the run time to 3 seconds. Over the next month, 19 other in-flight hot fire tests of the laser device were attempted. Half of those tests were aborted because conditions did not meet strict operating standards, e.g., temperatures, pressures, flow control. However, the other half of the tests were successful and gradually built up to 6 seconds of run time, once again confirming the operational performance and flightworthiness of the FSS and the GDL working in tandem.[21]

Interspersed with the White Sands tests were the first ALL cold flow and hot fire flight experiments conducted over Edwards AFB and the Pacific Missile Test Center. The purpose of these tests was to operate the laser device and to begin practicing the flight patterns the ALL would fly for the planned laser air-to-air missile and Navy drone intercepts to be carried out in phase 5. Another important reason for going to Edwards was to determine if the fuel farm and servicing facility for the ALL was adequate. Could the laser be safely fueled at Edwards? Could the laser device and its subsystems be worked on to make on-the-scene fixes? These were two fundamental questions that needed to be answered during phase 4.[22]

AFWL encountered mixed results during the first ALL tests in California. On 6 August there were problems with the cryogenic coolant unit and fuel storage and distribution system at Edwards. Essentially, there were difficulties in transferring the laser fuels from the tank farm to the airplane. Consequently, that flight was canceled. The planned flight to practice the Navy shootdown scenario over the Pacific near Point Mugu also had to be scrubbed the next day because of a higher priority missile test. On 19 August, when the laser was turned on in the ALL over Edwards, the mission aborted 2 seconds into a planned 4-second hot fire because of a failure with one of the combustor transducers monitoring temperature and pressure. The next day a similar failure occurred. Success was finally achieved on 21 August when the ALL completed one of two planned hot fire tests while flying the Navy scenario. That flight also collected good vibrational data around the base of the APT. A week later, the ALL repeated the practice scenario flight patterns while firing the laser for 3- and 4-second shots, before returning to Kirtland.[23]

Phase 4 flight testing ended on 29 August when the last two hot fire tests took place over White Sands. For the first test the laser was turned on for 5.7 seconds at 17,000 feet. The ALL then climbed to 30,000 feet where the final hot fire test lasted for 3 seconds. For the most part, phase 4 went smoothly and on schedule. There were aborts on a number of tests, but that was expected in both the ground and airborne experiments as part of the process to work out the bugs of the entire system. In his weekly update to Dr. William L. Lehmann, Director of AFWL, Colonel Rich, head of ARTO, reported, "All Phase IV objectives were successfully completed. Most significant were the hot gas flows at all required altitudes and the practice of the Navy low altitude scenarios."[24]

During the phase 4 test flights, AFWL collected data in three key areas. One set of data revealed how well the GDL device performed during various flight conditions. A second set of measurements dealt with aircraft response and structural loading of the aircraft during a laser firing. Firing a laser, the equivalent of lighting off a rocket engine inside an aircraft at 30,000 feet, posed some serious problems. AFWL had crunched all of the numbers on the ground and during flight cold flow tests to predict how running the laser would affect the flight performance of the ALL aircraft. Those calculations were "best estimates," but no one knew for sure if hot firing the laser would significantly degrade the airborne operating performance of the aircraft. The third category of information consisted of dynamic data used to characterize motions and disturbances around the base of the sensitive APT, which was a major contributor to beam misalignment.[25]

Analysis of the phase 4 test data showed the GDL device worked according to design specifications and within safety standards. The device repeatedly produced, aligned, and sustained a quality beam (inside the cavity), which lasted for sufficient time to support all of the anticipated phase 5 flight scenarios. Equally important was the ability of the FSS to deliver the correct mixture of fuels at a steady flow rate from the storage tanks to the device so as to maintain ideal lasing conditions. Bellows (expansion joints serving as accordionlike connections to reduce stress on the system) installed in the FSS plumbing lines provided sufficient flexibility to eliminate most of the aircraft-induced bending and vibrations on the FSS. In addition, the control data acquisition system (C/DAS) which monitored all aspects (e.g., pressure, temperature, flow) of the FSS/GDL operated successfully. During numerous flight tests, the C/DAS computer correctly sensed unsafe operating conditions and automatically aborted the laser run.[26]

On 4 September 1980, AFWL presented its phase 4 test results to the High-Energy Laser Review Group. Eleven briefings, each covering a specific aspect of the phase 4 test program, were delivered to the group of independent scientists tasked to assess the progress of the ALL. The main message was that all of the major hardware components (device and FSS) functioned properly and were compatible with the aircraft. Photons were produced but were not extracted from the resonator. Minor problems surfaced but were not serious enough to jeopardize safety considerations. For instance, at high altitudes when the laser was turned on, the thrust generated by the laser caused the nose of the ALL to pitch up slightly. However, the pilot was able to anticipate and compensate for this. Other data demonstrated the flow system designed to cool the cavity, nozzles, sidewalls, and combustors ran at a slightly higher temperature than when operating in the test cell. As a precaution for future tests, AFWL recommended laser run time be restricted at lower altitudes where the heat problems were more persistent than at higher altitudes.[27]

Phase 4 testing revealed two issues that had to be improved for phase 5 to run smoothly. One was that the carbon monoxide and nitrous oxide reservicing capability at the Edwards support facility had to be upgraded. The second issue was that phase 4 data indicated GDL firings induced high vibrational levels into the optical bench. The main question was, would the amount of motion induced on the optical bench throw off the beam alignment precision of the ADAS? That could only be answered once the ADAS was installed during phase 5.[28]

MOVING AHEAD WITH PHASE 5: MARK AND SCHMITT VISIT

Phase 5 represented the culmination of over a decade of work to perfect the technology to demonstrate an airborne laser could intercept fast-moving aerial targets in a realistic operational environment. All eyes were on this critical period of the program. Systems Command and Air Force Headquarters, who had endured and supported numerous schedule slips and funding increases of the ALL program over the years, especially were anxious for AFWL to get on with the shootdowns of AIM missiles and Navy drones planned for 1981. Although expectations were high going into 1981 to complete the ALL program, most realized achieving the goals and objectives of phase 5 would be extremely difficult.

Recognizing the enormous challenge that laid ahead for AFWL in succeeding with phase 5, Colonel Kyrazis, head of the Laser Development Division, began making preparations early to instill in his people a renewed sense of urgency. On 20 October 1980 he sent a letter to each member of the ALL team, as well as to the major contractors, reemphasizing and clarifying the goals and objectives of the upcoming phase 5 tests. He identified achieving two objectives as standards for success. The first was for the ALL laser system to kill a realistic target. Collecting accurate scientific and engineering data that would allow the Air Force to put laser weapons on aircraft in the future was the second objective.[29]

Timeliness was another underlying theme of Kyrazis's letter. He wanted to accomplish the first flight scenario of the shootdown of an AIM-9B missile by mid-January 1981. Underscoring the importance of this, Kyrazis wrote:

> If we do not do this in a timely fashion, no matter how good our work is leading up to that, the DoD's High Energy Laser Program is in jeopardy. I ask that you do anything in your power to help achieve the self-defense scenario as soon as possible ... however, this must not be done at the expense of safety, or at the expense of too high a technical risk. In other words, when we decide to go after the target, we have to be sure in our own mind that we will be able to do it. You must continually evaluate and re-evaluate what it is you are doing, and as you gain new knowledge you may need to change your mode of operation or change your short term objectives in order to minimize your total time. It is also necessary when you do this, you let appropriate people outside your working group know of your action because you will have impact on what they do. The coordination of your efforts is critical if we are to accomplish our tasks ... I also encourage you to talk to anyone else you think you might have to. On this issue I particularly wish to reiterate my policy that for technical issues, there is no chain of command. You are to go directly to anyone you please, anywhere within the Division or in the world, that will answer your questions directly and provide you with information when you feel the need for that information.[30]

Phase 5, Kyrazis pointed out, like previous stages of the ALL program, was an evolutionary process. The procedure involved installing and checking the ADAS and APT on the airplane, characterizing their performance with high-power tests inside the plane on the ground, followed by in-flight testing of the ALL to track and deposit a laser beam on towed diagnostic targets. Once these sequences were completed, the ALL would then be ready to attempt to shoot down AIM-9B missiles and Navy drones. Nothing went easy in phase 5. Numerous technical problems arose in 1981 which required additional time and money to fix the ALL system. Instead of ending the program in 1981 after the anticipated successful shootdowns, the ALL would need 2 more years to demonstrate the effectiveness of an airborne laser.[31]

The first step in getting the ALL readied for phase 5 testing involved AFWL technicians disassembling the ADAS and APT in the test cell and moving those major components to the aircraft for reassembly. By October 1980, AFWL and Hughes workers had installed the APT and ADAS in the ALL. However, all did not go smoothly as problems developed with the water cooling system. In November, water leaks discovered in the ADAS, caused by "improper connections within the plumbing of the water cooling system," required removal of both the ADAS and APT from the plane to fix the problem in the clean room in the hangar. This resulted in a 1-month delay in the schedule. By the end of December, the ADAS and APT had been reinstalled in the ALL with all of the plumbing connections made verifying the integrity of the system. In keeping with Kyrazis's policy to proceed as quickly as possible, AFWL used three shifts of workers working around the clock from 2 January to 10 January 1981 to get the first phase 5 ground tests under way.[32]

Easy does it, as workers carefully position APT and ADAS (lower section) into opening on the ALL aircraft.

Two hot fire tests of the laser occurred on 5 January. For the first test, the laser was turned on, but the system automatically shut down after less than a half second of run time because of potential safety hazards. A full duration run of the laser for 3 seconds was achieved during the second test, but, intentionally, no photons were extracted. The next day, three more tests took place. Problems with the optical system aborted the first two tests. For the third test, the laser ran for the full duration of 6 seconds—again, no photons were extracted from the laser device. However, these preliminary tests were necessary to check out the laser system to ensure it was operating properly before attempting to extract the high-power beam and direct it through the ADAS and APT.[33]

The real proof of the reliability of the ALL laser operating system, AFWL scheduled for 14 and 15 January. Plans called for extracting the high-power beam from the resonator, aligning and reflecting it off the series of mirrors on the optical bench, and then directing the beam up through the ADAS and into the APT. From there, a portion (2 percent) of the beam would be propagated to a target downrange behind the ARTF; 98 percent of the beam was "dumped" in a calorimeter located in the clean room known as the "white elephant" surrounding the APT on the airplane. Confident this first test would succeed, AFWL planned to repeat the demonstration on the 15th to coincide with a visit by Senator Schmitt and Secretary Mark. This was to be Secretary Mark's farewell to the Weapons Laboratory, as he and other officials of the Carter Administration would be leaving their jobs with the inauguration of Ronald Reagan on 20 January. Mark and Schmitt, both strong supporters of the Lab's laser program since the mid-1970s, also wanted to be on hand to witness the first series of ground tests de-signed to fire a high-power laser beam from an air-plane. Surely, that would be a historic event. Both men believed that success would serve as an extremely im-portant milestone to justify proceeding with the even more ambitious challenge of the airborne firing of the high-power laser. However, events did not turn out ex-actly as planned.[34]

APT on ALL being serviced in "white elephant" clean room on Pad 4 next to test cell.

On 14 January, the day before Schmitt and Mark were to show up to observe the full power testing of the laser, AFWL conducted a rehearsal of the test that would be repeated the next day for the two distinguished visitors. This test was identified as mission 25007. (The 2 in the mission number revealed this was a "ground" test—1 identified an airborne test; the 5 signified it was a phase 5 test and 007 represented the seventh test in phase 5.) Two separate tests actually took place and both had to be aborted because of problems traced to the combustor and incorrect flow rates of the fuels. The first test lasted only 1.3 seconds before the system automatically shut down. However, the second test ran for 2.2 seconds with the beam extracted and focused by the APT to a target downrange from the airplane. Although this run time was short, the experiment proved for the first time that a high-energy beam generated inside the ALL could be directed from the laser device, through the ADAS and APT, and propagate through the air outside the airplane. This was extremely significant because the test demonstrated all of the components of the laser system worked as one unified system to produce a beam that could be pointed to a target.[35]

Extraction of the high-power beam from the ALL on the ground on 14 January came about at great expense. In the process of directing the beam through the optical subsystems, one of the mirrors was damaged by the beam. As a result, AFWL officials were nervous because they knew they would not be able to duplicate the successful beam extraction for Mark and Schmitt the next day. Little could be done to remedy the situation and it was too late to cancel the visit. AFWL had no other choice but to proceed with the experiment on the 15th as planned. Early in the afternoon, Colonel Rich escorted the Secretary and the Senator to the ALL to witness the beam extraction. Kyrazis and his team turned the system on, but no beam was extracted. When Schmitt asked why no power came out, Rich painfully explained the difficulties the system had encountered the previous day, which would take time to fix.[36]

One AFWL project officer incorrectly characterized Schmitt as "livid" over the inability

Left to right: Senator Harrison Schmitt, Secretary of the Air Force Hans Mark, Lieutenant Colonel Thomas Moorman, Jr., and Colonel John Rich.

of the ALL to extract a beam. Later the Senator recalled, although he was "disappointed" with the demonstration, that in no way detracted from his total support of the ALL program. He recognized mistakes and delays would occur, but that was part of the price the nation would have to pay to develop the technology that eventually would lead to deployment of laser weapons in the future. Schmitt had reconciled the success of the 14th with the temporary setback of the 15th by sorting out in his mind that the system had worked on the 14th, even though he was unable to witness that event. Addressing the Albuquerque Press Club later in the day, he praised AFWL's efforts, stating, "This afternoon's test of the Airborne Laser systems being developed by the Air Force Weapons Laboratory is an important milestone in new strategic policy development."*[37]

---

*Schmitt's comments were not offhanded remarks. The Senator had briefed President-elect Reagan in December 1980 on the progress of laser technology work at Kirtland, to include the ALL. Schmitt advised Reagan the government needed to invest more heavily in laser research. The Senator recalled he came away from that meeting convinced the President was already beginning to collect information which would become the building blocks for his Strategic Defense Initiative (SDI). When Reagan announced his SDI policy on 23 March 1983, to reduce the nation's reliance on offensive nuclear weapons, Schmitt, unlike most, was not surprised. What puzzled Schmitt was why the President had waited so long before unveiling this dramatic change in strategic policy.

High-Power Laser Goes Airborne

At the same press conference, Secretary Mark also applauded AFWL for its 2-day series of ground-based laser experiments on the ALL. "I am satisfied," Mark said, "that we have, in fact, passed a very significant milestone, that the laser in the airplane is capable of firing at the full power level that we had in mind and that we are now ready to go on to the next step, which is to do that while the airplane is flying." He also recognized that recurring vibrational problems— caused by the laser generating about as much energy

Senator Schmitt and Secretary of the Air Force Mark express their satisfaction with ALL progress after tour of the aircraft.

and noise as a jet engine—still existed, which upset the delicate optical systems responsible for precise beam alignment aboard the ALL. However, the Secretary remained optimistic about the future, comparing ALL's progress to where nuclear energy technology was 30 or 40 years ago.[*] Based on the most recent AFWL tests results, Mark confidently predicted, "We can now think about shooting down the other fellow's missiles without using nuclear warheads." If that was the case, then the ALL was another potential technology card the President needed to consider adding to his future Strategic Defense Initiative (SDI) deck in formulating a new strategic defense policy.[38]

TOW TARGETS

Before anyone could think about using a laser to shoot down strategic missiles, the ALL had to first prove in the air during a series of practice runs that it could direct a high-power beam to hit an aerial tow target. Tow targets allowed the ALL aircrews to practice and refine skills in engaging airborne targets with the laser. These experiments also collected valuable information on target vulnerability and airborne far-field beam quality. Beginning on 14 February and extending over the next 3 months, AFWL tested the entire ALL system against targets towed by an F-4 aircraft over White Sands Missile Range. AFWL over the years had

---

[*] A few months later, Lieutenant General Kelly Burke, Air Force Deputy Chief of Staff for Research, Development and Acquisition, compared the potential breakthroughs with lasers on a par with the discovery of gunpowder. He predicted the United States was about 10 years away from developing a laser weapon that could change "the face of warfare."

evaluated a number of different tow targets to determine which ones were the best suited as diagnostic platforms instrumented to measure various beam parameters. A second set of tow targets was developed to simulate AIM-9 missiles and Navy BQM-34 drones. These were used to determine target vulnerability thresholds as a way to establish whether the ALL beam was powerful and intense enough to damage or destroy these targets.[39]

The diagnostic low-cost tow target (LCTT), shaped like a missile, measured approximately 1.3 feet in diameter, 16.3 feet in length, and weighed 117 pounds—when carrying instrumentation, the LCTT could weigh as much as 450 pounds. An F-4 aircraft reeled out the LCTT on a 5000-foot cable so the tow target was positioned behind and below the F-4. The ALL flew slightly ahead of and to the right of the tow target; thus, when the ALL fired the laser, the beam traveled from the APT in the direction of the left-rear of the ALL aircraft. As a safety precaution, the tow target was positioned slightly higher than the ALL to avoid any possibility of the beam striking the left wing of the ALL.[40]

Mounted on the side of the diagnostic tow target (a modified LCTT) was a high-energy target board used to evaluate the far-field performance of the ALL by measuring critical beam parameters (e.g., power, intensity, jitter). To prevent the HEL from damaging the target, the board was coated with gold. The highly reflective gold-plated surface allowed only a small portion of the beam to enter each of the 256 small openings configured in a 16 × 16 array on the target board. Behind each opening was an IR sensor that detected the 10.6-micron beam (confirming the beam had hit the target board) and recorded the intensity of the beam. By collecting and assembling all of the measurements from each sensor, beam intensity (watts per square centimeter), power, and profile of the beam over time could be calculated. Data collected by the tow target were recorded and telemetered to the Target Board Data System located on board the ALL diagnostic aircraft. Real-time video display and time history plots of beam power and

Gold-coated target board on side of tow target reeled in beneath F-4.

intensity and tracking accuracy could be observed on the diagnostic aircraft.* Once the diagnostic plane returned to Kirtland, a detailed analysis of tow target tapes processed by AFWL's computer center was conducted.[41]

Besides the highly instrumented LCTT, there were three other versions of the LCTT used in Cycle III testing. The simplest one was an LCTT fitted with a plexiglas window, which did not require beam instrumentation. AFWL workers observed burn patterns on the Plexiglas to visually assess if the beam remained steady or strayed off the tar-

Gold-coated target board mounted on low-cost tow target (LCTT).

get, as well as determining how effective the beam was in penetrating the target. (As one official put it, the Plexiglas burns were more "show and tell"-type demonstrations. Although they provided a somewhat dramatic picture of what the beam could do, the most valuable scientific data were collected by the diagnostic target board to provide a history of beam performance/characteristics over time.) There were also two other LCTTs modified, one to simulate an AIM-9B missile and the other to simulate a BQM-34 Navy drone. These targets were used to verify the ALL could engage these replicas of real-world targets under realistic conditions.[42]

An actual AIM-9B seeker head was mounted in the nose of the LCTT. The dome was instrumented with a breakwire and a pressure transducer which served as diagnostic tools.

---

*The diagnostic aircraft (371) served as a data and analysis platform that received telemetered data from the ALL and targets. These data included measurements of the ALL laser system (e.g., sensors, temperature, reactant flow, tracking parameters) and aircraft parameters (e.g., airspeed, engine power settings). This allowed 371 to monitor every aspect of the ALL's operating performance. When the ALL was loaded with the laser reactants, the risk of a catastrophic failure leading to the aircraft crashing or blowing up increased. Some referred to 371 as the "postmortem" aircraft and to some of its collected telemetry (TM) data as "postmortem TM data." The data 371 collected were vital as they would be indispensable for learning the reasons for failure if the ALL crashed. Also, reducing the ALL telemetry data to assess airborne test results as soon as possible by 371 in the air saved time and allowed AFWL workers to fix the problems in flight by making modifications to follow-on experiments during the same sortie.

LCTT with Plexiglas window mounted near center of tow target.

The breakwire was a thin wire embedded in a strip of metallic silver paint covering the inner surface of the seeker head dome. If the laser cracked the dome and the breakwire, there would be a change in the voltage passing through the breakwire. A cracked or removed dome also caused changes in pressure between the inside of the tow target and the dome, another indicator that the laser was effective in damaging the simulated missile. Also, aerodynamic forces from air entering the dome of the missile would upset the operation of a rotating reticle, which in the missile seeker collected infrared light coming off the target. In a real missile (as opposed to a tow target), this IR information in the form of error signals was processed by a "black box" behind the dome assembly to tell the missile which way to fly (up, down, left, or right) so it was always tracking the target even if the target was changing direction. If heat from the laser or increased pressures slowed down or stopped the spinning reticle, then the IR target data could not be fed to the black box to keep the missile on its proper course. (The reticle could not rotate when air was blowing on it.) If that happened, the missile would "tip over," deviate from its desired course, and eventually crash into the ground. Verification of the missile wandering off course could be obtained from ground radar tracking the missile. (Inside the missile was a transponder which gave off a beacon signal that the ground radar tracked.) In sum, the breakwire, changes in pressure inside the dome, and malfunctioning of the spinning reticle all indicated loss of track, which in effect disabled the missile and prevented it from reaching its target. In other words, the laser did not have to "blow up" the missile; all it had to do was to damage or disable the missile where it could no longer track and remain on course to its target.[43]

Designed to simulate a Soviet sea-skimming missile, the BQM-34 tow target was also a modified version of the LCTT. The fuel tank and flight control box (located in the forward section of the LCTT) served as the two target aimpoints. For this scenario, AFWL used an "edge" tracking technique for tracking on the nose of the simulated BQM. The tracking algorithm would compare a picture of the nose of a real BQM-34 with the nose of the tow target it was tracking. By matching these two pictures, the tracking algorithm could calculate the exact distance from the track point on the nose to the most vulnerable part of the target, in this case, the fuel tank or flight control box. Once the aimpoint offset had been determined,

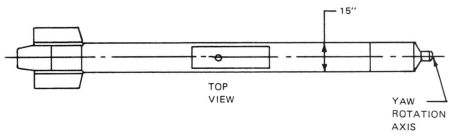

TOP
VIEW

15″

YAW
ROTATION
AXIS

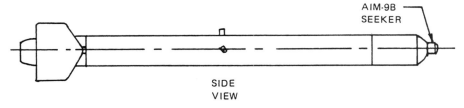

AIM-9B
SEEKER

SIDE
VIEW

LCTT simulating AIM-9 missile.

that information was fed to the APT to align the beam with the target spot. In the majority of cases, the beam usually hit close to, but outside, the exact target point. The tracking algorithm detected this initial "hot spot"—where the beam first struck the target—and directed the system to make a "one-sight boresight correction" to move the beam from the initial hot spot to the target aimpoint. All of these precision adjustments had to occur within 1–2 seconds.[44]

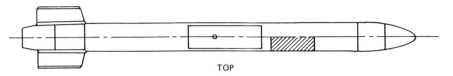

TOP

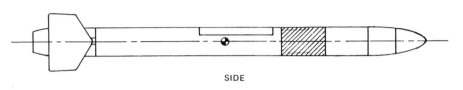

SIDE

LCTT simulating BQM.

On the back side of a metal plate located at the fuel tank position was a thermally instrumented sheet of stainless steel to record temperature changes. Recording a temperature of 460 degrees centigrade—the ignition point of the vapor in the fuel tank—meant that the laser would be able to ignite the fuel tank. A special temperature-sensitive paint applied to backside of the metal changed from blue to black when the temperature reached 454 degrees. Visible sensors installed inside at the fore and aft sections of the tow target revealed if the laser punched a hole in the target. A hole in the skin allowed sunlight to enter the dark area inside the target causing a change in sensor voltage.[45]

The first Cycle III/phase 5 flight of the ALL took place over White Sands on 14 February 1981. No photons were extracted during this 3-second hot fire test as the optical shutters were closed. The purpose of this experiment was simply to determine if the laser device could run safely and reliably. No target was shot at. Following the 14 February flight, AFWL discovered a hydraulic leak in one of the APT servovalves. The leak was in an inaccessible location and fixing it required that the entire APT be removed from the airplane. That repair operation set the program back 7 weeks as the ALL did not get airborne again until 9 April.[46]

After repairs to fix the hydraulic leak, the ALL flew to Edwards on 8 April to resume flight testing of the laser. En route, the ALL tracked the diagnostic aircraft. The reason for moving to Edwards—and the Naval Weapons Center Range at China Lake where the actual flight testing would take place—was White Sands was not available because of joint war-game exercises on the range. The next day over China Lake, on the first attempt, the ALL fired the laser for the full duration of 5.87 seconds. No photons were extracted, but it was another step forward confirming the FSS and laser device were working as predicted.[47]

On 10 April, again over China Lake, the ALL practiced the self-defense scenario with a Navy A-7 aircraft. Both planes flew a racetrack pattern, with the ALL in the lead. Part of the goal of this series of tests was to ensure the ALL and A-7 were positioned in precisely the right place and at the right time in what was called the "firing box." This was a well-defined area of restricted airspace where missiles were launched toward and engaged by the ALL. The ALL and launch aircraft were only in their respective boxes for less than 1 second. Getting the ALL and launch aircraft in their boxes at exactly the right time and at

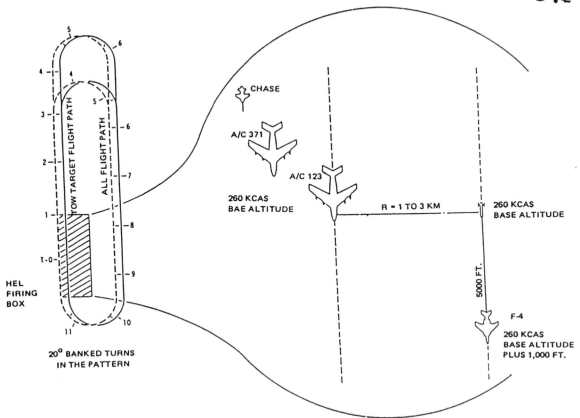

Parallel flyby profile used at White Sands Missile Range. Similar flight patterns were used at China Lake with a Navy A-7 pulling the tow target.

and pilots.[*] Once both planes were aligned correctly, the A-7 fired an AIM-9B at the ALL, yet at a distance calculated to prevent the missile from actually hitting the ALL. The AIM-9B sliced through the air at approximately 2000 miles per hour. Modifications to the AIM-9B, such as loading only a small portion of fuel that would burn for only a few seconds, ensured the missile would land on the ground within the test range perimeter. For these first experiments, the ALL's mission was not to fire the laser, but to demonstrate it could acquire and track the approaching AIM-9B.[48]

---

[*]If the ALL and A-7 aircraft were not within their narrow firing boxes at precisely the right time, then the mission was a no-go. The planes would go around their racetrack course and try to line up again. On one occasion, Kyrazis recalled it took 20 times before the planes lined up in their proper firing boxes. One of the ALL test directors—Kyrazis, Dettmer, or Otten—was in the ground control room for each of the flight tests. The test director in the control room was as familiar with the test scenario as the test director in the ALL and was able to advise both the ground controller and ALL crew during the tests.

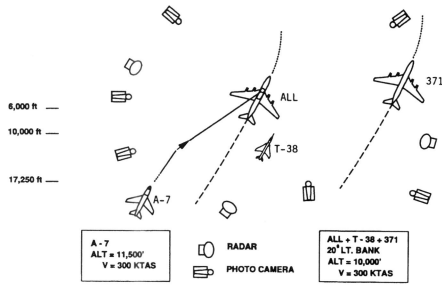

Self-defense mission.

Four AIM-9B missiles were launched (not simultaneously) at the ALL. As predicted, no missile came closer than 4500 feet to the ALL. For the first and second launches, the APT tracked the missile until motor burnout. At that point, problems developed because the launch aircraft had remained in the tracking gate with the missile. This confused the tracking logic because the APT observed two cold body targets—the missile and the plane. Consequently, the tracker broke lock with the AIM-9B after burnout and tried to shift its track to the A-7. AFWL solved this problem during the last two missile firings by the A-7. Once the missile was launched, the A-7 immediately made a hard bank and pitch-up maneuver to quickly move out of the tracker field of view. That allowed the APT to see and track only the missile until it ran out of energy and nosedived into the ground.[49]

Systems Command rated the performance of the ALL against the last two missile launches a success. The higher headquarters reported, "A lot of good tracking data was obtained in these tests." This preliminary assessment went on to point out that "after the tests, Col Kyrazis [ALL program director] was confident that they would have done the job [hit the missile with the beam] if they had fired photons out of the ALL." Kyrazis recognized two major issues had been settled by these tests. The practice scenario proved launching an AIM-9B missile at the ALL did not endanger the aircraft. Second, the APT demonstrated it could acquire and maintain track on a maneuvering Sidewinder missile. These were two important milestones that had to be accomplished before AFWL could move forward with the upcoming June airborne experiments designed to shoot down missiles with the laser.[50]

Once the ALL returned to Kirtland, AFWL completed a critical ground test on 29 April before proceeding with the next airborne experiments planned for White Sands. Parked with the nose of the ALL in the "white elephant" clean room on Pad 4 next to the test cell, the ALL demonstrated its longest duration (4.95 seconds) operation of the system where the beam was extracted, sent through the complete optical train, and directed to a target downrange behind the ARTF. That was significant, not only because the system operated smoothly for that relatively long length of time, but also because it showed the conical scan tracking and pointing techniques worked. To compensate for boresight errors between the high-power beam and the tracker, the "conscan" tracking scheme was developed. Conscan initially directed the high-power beam in a spiral pattern along the line of sight to the target. Wherever the beam hit the target, it produced a "glint" that returned a bright spot of reflected laser radiation to the APT, which was processed to define the exact point where the beam hit the target. Once that location was determined, the automatic aimpoint algorithm directed the APT to "walk" the beam forward of the target until the glint was no longer observed. Once that point was reached, the algorithm instructed the APT to stop and redirect the spiral beam "backward" until it produced a glint on the nose of the target. The APT then locked the beam on the nose, one of the most vulnerable parts of the target.[51]

A tower was located above the southeast corner of the ARTF to allow one or more safety observers to view the test range during ground tests to ensure that no one was in danger of being hit by the laser. During one of the last ground tests from the ALL, Fred Duggins was the safety observer in the tower and noticed activity on the range seconds before the laser firing. He immediately issued an abort to stop the countdown. The unaccounted activity turned out to be the Kirtland AFB commander and an entourage who were walking toward the ARTF from the southern end of the range where their automobile had broken down. This was the only incident where a test had to be aborted because of activity on the range.[52]

The first airborne propagation of a high-energy laser from the ALL aircraft to an aerial tow target occurred over White Sands on 2 May. This was a short-duration test as the laser was only turned on for 0.3 second. But in that short interval, the laser was able to strike and burn through a Plexiglas panel mounted on the side of the LCTT. On the negative side, AFWL rated the beam focus as "poor," as the spot size on target was larger than expected. But there was no denying the 2 May test clearly demonstrated that a high-energy airborne laser, for the first time ever, had intercepted an aerial target. This was a major success, but at the same time everyone realized work needed to move forward to improve beam quality and to refine the pointing and tracking system for more precise alignment of the beam on the target aimpoint.[53]

Two follow-on attempts on 13 and 18 May to engage the tow target were aborted because of malfunctions and problems with water flow in the ADAS mirrors.* On 20 May, in two

---

*AFWL flowed deionized water through the cooling system. Since April, workers had noticed the clear water had changed to a grayish-green color and its acidity level had increased. After further investigation, they discovered paramecium organisms, a type of protozoa, were thriving in the water hoses and channels running behind the mirrors. The decay products of these living organisms—literally tiny bugs—supported the spread of a bacillus-type bacteria. The combination of the organisms and bacteria produced acids which etched grooves in the mirror-cooling channels. When the "bugs" died, they formed a pasty rust-color deposit which clogged the mirror-cooling channels, hoses, and water filters. The problem was how to kill and remove the bugs from the water system. As a first step, the system was flushed with a commercial poison to try to kill the bugs. A biodispersal agent (diethylamine) was also flowed through the system to break up the coagulated residue. This was only a temporary fix, as the bugs in the water would in the long run be responsible for a $1\frac{1}{2}$-year delay in the ALL program.

Aircraft flying over White Sands Missile Range to test laser. Left to right: F-4 carrying tow target, A-7 chase aircraft, diagnostic aircraft (371), and the ALL.

separate tests, the ALL irradiated the Plexiglas with the laser. The first was a full-duration 4.0-second shot that initially hit the aimpoint, but halfway through the test the beam drifted off the nose of the LCTT. However, data collected showed—while the beam was on the target aimpoint—there was less jitter and an improvement in beam focus and intensity since the previous tests. Software problems with the automatic aimpoint designation algorithm had caused the beam to move off the aimpoint.[54]

The second flight experiment on 20 May was shut down after only 1.1 seconds of run time because of low pressure causing inadequate flow rates in the ADAS water system. The beam impacted the Plexiglas (at a different point than the first test), but hit slightly ahead of the desired target spot. Part of the beam's energy burned the outer edge of the Plexiglas as well as spilling over and scorching the aluminum skin of the tow target. Although the beam did not hit the exact aimpoint, data showed beam jitter was "nominal" and focus and intensity were good. Movement of the beam off the Plexiglas was attributed to the ground controller who was unable to position the tow target and ALL in the right flight patterns before the laser fired. The controller, as one official explained,

APT

LCTT

Aimpoint/intercept angle.

allowed the LCTT flight path to drift some 500 meters closer to the ALL flight path from the time the manual aimpoint was set until the time clearance to fire was given. As shown below, the aimpoint was set by the angle between the nose of the LCTT and the aimpoint. As the target came closer to the ALL the angle remained constant, but the intercept point changed.[55]

After the successful laser engagement of the tow target over White Sands, the ALL flew to Edwards on 27 May to first practice and then carry out the 1B self-defense scenario against air-to-air missiles. Again, the plan called for the ALL to be serviced and staged out of Edwards, but the actual airborne experiments would take place over the Navy's China Lake test range.[56]

1981 SHOOTDOWN—ALL SELF-DEFENSE SCENARIO

If the ALL succeeded in tracking and depositing a lethal dose of laser energy on the AIM-9B, then that would be the proof of demonstration needed by AFWL to confirm the

ALL program goals had been met.* Morale was up after nearly 11 years of work. AFWL scientists, engineers, and technicians were on the verge of showing that the laser and pointing and tracking system had sufficiently advanced to the point where the ALL, under realistic operating conditions, could defeat an air-to-air missile. Although expectations were high for success, events did not unfold as planned. Consequently, instead of the ALL program coming to a successful conclusion in 1981, AFWL would be forced to toil for another 2 years making extensive modifications before the system would perform as advertised.

The goal for the 1981 shootdown was to deposit sufficient energy on the nose of the AIM-9B at a range of 3 to 5 kilometers. To cause damage, the beam would have to be held steady on the target for 2 to 4 seconds. When the 10-centimeter wide beam exited the resonator (inside the ALL) the beam's power measured approximately 400 kilowatts. By the time the beam reflected off 12 mirrors in the optical train and APT and then propagated through the atmosphere, the beam would lose a significant portion of its power and its tight focus. Once it reached the target 3 to 5 kilometers away, the beam's diameter would expand to 14 centimeters and would only be able to deliver about 75 kilowatts on target. Precision in aiming and keeping the beam on target was absolutely required, considering the erratic, unpredictable, and rapid movement of the missile and the severe structural bending of the aircraft. Also, severe vibrations imposed by the aerodynamic loads on the turret of the ALL contributed to the difficulty of getting the beam to the right target spot. Again, precision was the key, as a near miss by the beam would have no effect on the target.[57]

Getting the beam from the ALL in flight to an airborne target was more difficult than directing the beam to a ground target during earlier testing. First, the imaging IR tracking system would track the missile launch aircraft. Once the missile was launched, the system would automatically transfer track from the aircraft to the missile, which had to be done within a matter of a couple of seconds. Initially, the ALL would track on the hot plume trailing the missile. After the missile motor burned out, the track would shift to the cold body of the missile. Next, based on the location of the FLIR (forward-looking IR) centroid track point on the body of the missile, the system would calculate the aimpoint offset to the nose of the missile. Feedback on the exact location of the aimpoint would then be provided to the pointing system in the turret to direct the beam to the aimpoint and to hold the beam there long enough to deposit a lethal dose of energy. Tracking accuracy to hold the beam on target would have to be approximately 25 microradians, meaning the beam would not jitter about the target aimpoint for a distance of more than 12 centimeters at a range of 4 kilometers.[58]

The day after arriving at Edwards (28 May), the ALL took off loaded with fuels in anticipation of the first firing of the beam at an AIM-9B missile. (Edwards served as a staging base to reduce the flight time to the Navy's test range at China Lake.) From the start things did not go well. Once over China Lake the weather turned bad as winds, clouds, and haze made it difficult to visually align the ALL and launch aircraft in exactly the right positions

---

*Lethality for this aircraft self-defense scenario was defined as inflicting sufficient damage to the missile to eliminate it as a threat to the ALL aircraft. This meant blinding the missile's IR guidance system or fracturing the radome.

in the firing box. The flight plan called for the Navy A-7 to launch its missile at approximately 7 kilometers from the ALL. At that distance, the AIM-9B would run out of fuel at least 2500 feet short of reaching the ALL. Flying in tight formation off the right of the ALL aircraft, a T-38 target aircraft would place one engine in afterburner to provide a strong IR signal (heat source) for the AIM-9B to home in on because as the ALL banked to the left its wings blocked out the ALL engines from the missile's view. The 20-degree left bank by the ALL also would give the APT a clear view of the missile as it departed the A-7 with a good sky background for the tracker to pick up the missile as it converged on the ALL and T-38 aircraft, which were only a few hundred feet apart. For the first eight orbits flying a racetrack pattern, the ground controller was unable to align the three aircraft (plus the diagnostic aircraft) mainly because of marginal weather conditions.[59]

Weather conditions improved slightly for orbits 9 through 19. Eventually, three missiles were launched at the ALL. The ALL was unable to acquire and track two of the three missiles. For the third missile, the ALL succeeded in tracking it until it entered a cloud and then the tracker broke lock. Clouds in the background of the missile presented less than optimum conditions for the tracker to maintain a clear image of the target, especially after motor burnout when the missile offered only a "cold" or weak signature. There were also some minor software and hardware problems with the APT. AFWL's first attempt to shoot down an AIM-9 fell short of the mark. Not only did the APT encounter difficulties with the tracker, but the ALL never fired the laser.[60]

Poor weather delayed the second airborne experiment from 30 May to 1 June. Skies were clear on 1 June and the ALL made 20 orbits over China Lake. Four AIM-9B missiles were launched and the ALL attempted to fire at three of them. For the first launch, the ALL only tracked the missile. The first laser firing (at the second missile launched) lasted for 0.9 second, but then was aborted because of misalignment of the beam in the ADAS. The AIM-9B did not travel in a straight line. The "sidewinding" action of the missile was more pronounced than anticipated, making it extremely difficult for the APT to stabilize the line of sight on the missile so that precision tracking could be accomplished. Plus, heat buildup

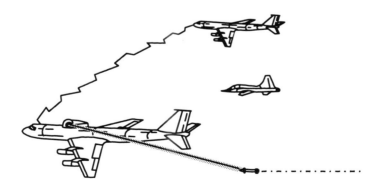

1B scenario depicting missile launch.

in the ADAS caused thermal blooming that contributed to beam misalignment. Analysis of the data showed the beam did not engage the target. On the second attempt (fired at the third missile), the beam was fired for 1.2 seconds and briefly hit the missile. This should have been an occasion for rejoicing, as this was the first time a high-energy airborne laser had ever intercepted a missile in flight. But this seemingly historic watershed was fraught with disappointment.*[61]

The problem had to do with the conical scan tracking system that spiraled the laser beam toward the target. Although the APT was able to direct the beam to the target, the beam hit several different points on the aft section of the AIM-9B, but the missile kept on flying. The beam failed to move forward to dwell on the nose of the missile, the desired and most vulnerable aimpoint. These shortcomings were attributed to the many laser radiation "glints" reflecting off other points of the missile and returning to the APT. Conscan had to see only *one* reflection point for satisfactory operation. As a result, the conscan tracking algorithm became overwhelmed trying to sort out and process all of the glints, which prevented the APT from making the desired corrections to direct the beam to move forward to the nose of the AIM-9B and maintain a good steady track on the target. However, on the positive side, the conscan sensor on the ALL received the reflected laser radiation off the target, confirming that the beam had actually hit the missile. The problem was conscan did not deliver the beam to the desired aimpoint on the target, in this case the nose of the missile.[62]

Two days later on 3 June, the ALL made three attempts to intercept AIM-9B missiles. For all three tests the ALL successfully acquired and tracked the missile launched from the A-7. On the first firing, after 0.5 second of beam time, the laser shut down because of beam misalignment in the ADAS. Insufficient pressurization in the device resulted in aborting the second firing after 0.66 second of run time; no beam was extracted. On the third shot, the beam hit the missile for 1.8 seconds. The correlation tracker remained locked on the missile for the duration of the test. When the beam was fired, the conscan precision tracker detected the laser glint return signal from the aft fin of the missile, verifying the beam had hit the target. Because of higher than normal beam jitter, the conscan tracker—designed to spiral the beam onto the missile aimpoint—broke track twice. To correct this, conscan had to go through its search mode to relock on the target. As a result, the beam dwell time on the target was insufficient to damage the missile.[63]

Although AFWL could not conclude that the ALL had the ability to shoot down or "kill" a missile, several important technical achievements had emerged from the flight experiments. First, the ALL had demonstrated it could track supersonic AIM-9B missiles from launch to ground impact. Second, the ALL succeeded in firing the laser from an aircraft moving several hundred miles per hour. And finally, the HEL beam had locked on, even though only momentarily, to an air-to-air missile zeroing in on the ALL. That was a feat of enormous consequences, which had taken over a decade to perfect and demonstrate. But meeting the original goal of the ALL, to hold the high-power beam accurately on target long enough to destroy the target, did not happen during the June 1981 China Lake flight experiments.[64]

---

*Loss of track prevented the ALL from firing at the fourth missile launched.

Some at AFWL were anxious and ready to declare the ALL a complete success, as the beam had hit the target. But cooler heads prevailed and refrained from making a pronouncement of that magnitude. After reviewing the telemetry data, Kyrazis insisted that a lethal dose of laser energy had not been deposited on the target for sufficient time to kill the missile. Telemetry data were poor and could not confirm the missile had been killed. Telemetry designed to transmit pressure changes in the missile dome was "lousy," as Kyrazis put it. He pointed out even "if we had destroyed the missile, there was no way to tell the way the telemetry was hooked up."[65]

ALL team at Edwards AFB, California, who participated in June 1981 1B self-defense scenario. Colonel Demos Kyrazis is holding balloon anticipating his upcoming retirement from the Air Force.

Kyrazis stressed more work had to be done, especially to improve beam alignment and water cooling of the mirrors in the ADAS. Also, upgrades had to be made to the conscan aimpoint designation algorithm to ensure the beam could be directed to the target with the required precision to hit and remain locked on the exact target spot. Kyrazis was not discouraged by the airborne tests, reasoning that excellent experimental data had been collected which would be used to make the necessary fixes to the ALL system so it would be ready for the next series of flight tests. He expressed the value of the June tests and his optimism for the future in a letter he sent to General Brien D. Ward, Systems Command's Director of Laboratories, on 8 June:

> In summary, the deployment to Edwards Air Force Base resulted in one of the most fruitful flight test sequences conducted with the ALL. For the first time, an air-to-air missile had been engaged by an airborne high-energy laser system. The missile was acquired, tracked and had energy deposited on it. The missile was not killed because the beam did not dwell on the target long enough. The reasons for this are known and the problems will be corrected. The information collected on the dynamics of a real engagement and on the missile imagery is extremely valuable. Much of what we saw was unpredicted in spite of prior extensive ground tests and analyses.[66]

Although Kyrazis had presented an honest and accurate appraisal of the technical results of the attempted shootdown, the press had a different assessment of the China Lake flight tests. They wanted to hear that the ALL had blasted the AIM-9B out of the sky in true Buck Rogers fashion. When that didn't happen, they reported the ALL was a "failure," neglecting to recognize the full value of all of the scientific data collected which would be used to fix the system.

13

Setbacks and Delays

A month prior to the June attempted shootdown of Sidewinder missiles by the ALL, Captain Tom Dyble, who was assigned to the Directed Energy & Nuclear Effects Division (which reported to the Director of Laboratories at Systems Command), had developed well in advance a public information release plan on the Cycle III flight test program. Questions from the news media would no doubt be persistent and hard-hitting once reporters learned of the event. Consequently, Dyble and others wanted to be prepared to present a consistent and unified Air Force position on the ALL and its technical progress. The goal was to release information in a timely and professional manner to avoid disclosure of inaccurate or classified information that might be leaked by unofficial sources.[1]

In anticipation of a successful engagement of air-to-air missiles by the ALL, Dyble had coordinated with Lieutenant Colonel John W. Duemmel of the Public Affairs Office at Systems Command and Major Lynn Gamble, Executive Officer, Advanced Radiation Technology Office at AFWL, to write a proposed press release covering all of the key highlights of the shootdown. Public Affairs prepared two write-ups to distribute to the press. One was a two page account—entitled "Airborne Laser Destroys Missile"—describing the actual shootdown with the dates left blank to be filled in later. A second and more lengthy handout (six pages) provided a broader coverage of the entire ALL flight test program addressing such issues as purpose, aircraft and equipment, past accomplishments, and future of the ALL.[2]

PRESS REACTION

In spite of all of the careful planning ahead of time, confusion prevailed immediately following the 1 June attempted shootdown. Instead of reporting the ALL firing at the AIM-9B missiles as a major step forward in the advancement of laser technology, the press reported just the opposite. Locally, the day after the 1 June test, Albuquerque's evening paper, the *Tribune*, ran a front-page banner headline announcing: "AIR FORCE LASER FLUNKS 1ST TEST." On the morning of 3 June, the *Albuquerque Journal* reported: "U.S. Laser Weapon Fails Test on Air-to-Air Missile." Similar headlines appeared in major newspapers across the country. For example, *The Washington Post* led off two articles with: "Laser Weapon Test Failure Mystifies Defense Officials" and "In Test Aloft, Air Force Laser Fails to Stop Sidewinder Missile." Several days later, *Aviation Week & Space Technology* boxed a short article introduced by "Laser Fails to Destroy Missile."[3]

Unfortunately, the press had perceived and conveyed to the public that the attempted shootdown by the ALL was a complete "failure." On the surface, it appeared the Air Force had blundered badly because of the inability of the laser to "destroy" the oncoming missiles as they approached the ALL. Instead of reporting on a major technical breakthrough designed to accelerate the process for developing laser weapons, the press interpreted the events of 1 June as a significant setback for the Air Force. What had happened?[4]

Many believed inaccurate perceptions and misinterpretation of information about the ALL could be traced directly to Colonel Bob O'Brien, who was in charge of Systems Command's Public Affairs Office. O'Brien was relatively new to the office, and although the Air Force spokesman for the ALL tests, he simply lacked the technical ex-

Laser weapons cartoon by Jack Ohman. © Tribune Media Services, Inc. All Rights Reserved. Reprinted with permission.

pertise to speak with authority on all of the various aspects of the ALL. What disturbed many Air Force people, especially those at the ALL working level, were the numerous newspaper articles quoting O'Brien as stating, "The test failed and we don't know why."[5]

O'Brien's comment was misleading because it gave the impression the Air Force botched the test and did not understand what went wrong. That was not the case. In fact, just the opposite was true as Captain Dyble tried to clarify in response to a phone call he received at 6 o'clock in the evening at his office on 2 June from George Wilson, a reporter with *The Washington Post*. Wilson asked, "Was the result of the test as negative as stated?" "No," Dyble replied. "The test was unsuccessful only in the sense that the target was not damaged as we had planned. The failure was overstated by the Public Affairs people [Colonel O'Brien]. The test was part of an ongoing test program and we obtained valuable data."[6]

The next morning in his office Dyble called Colonel Fred Holmer who was the Military Assistant to Dr. Richard Airey, Office of the Under Secretary of Defense, Research and Engineering, Director, Directed Energy Programs. Airey, at DOD level, wanted to make sure a positive message was sent to the press on the results of the ALL flight experiments. Dyble and Holmer rewrote the original news release Colonel O'Brien used and made sure the part stating "the test was not a success" was deleted. Airey's revised version read:

> The laser program has not suffered a setback. The Airborne Laser Laboratory is a *Laboratory* to evaluate and understand the physics and engineering capabilities and limitations of high power lasers on aircraft. With each test firing, scientific data is gathered and, as with all experiments in their early stages, unanticipated problems sometimes occur. There is every reason to believe the aircraft will shoot down missiles in experimental tests. It should be realized that we are not testing a weapon system, but doing *experiments* to understand the *potential* for airborne laser weapons.[7]

Efforts to clarify and to emphasize to the media that the ALL was first and foremost a feasibility demonstrator, and not a weapon prototype, began to take hold as evidenced by the press's more positive reporting of the second attempted shootdown by the ALL on 3 June. The front-page headline in the *Albuquerque Tribune* covering this event read: "Laser fails again, but Air Force says test is 'better.'" A similar article appearing in the *Aerospace Daily* noted the laser had indeed irradiated the target, but mechanical problems prevented the beam from dwelling long enough on the missile to cause any significant damage. Air Force officials had clearly recognized technical problems still existed with the pointing and tracking system, jitter, and performance of the ADAS to properly align the beam. However, Systems Command responded stating, "We understand these problems and we are working solutions to them." The critical point was that the ALL was a high-risk program and it was unrealistic to expect total success on the first-ever attempted laser shootdowns of missiles. As one official put it, "We'd rather have a success than a failure, but this is a test program. We're doing something that had never been done before."[8]

Headquarters Air Force concurred with Systems Command's evaluation of the significance of the ALL's recent flight testing at China Lake. In a memo to the Secretary of the Air Force in June, Lieutenant General Hans H. Driessnack, Assistant Vice Chief of Staff, wrote:

> We intend to press on with this program [ALL]. The information we have collected on the dynamics of a real engagement and on missile imagery is extremely valuable. We know why the beam did not dwell on the target long enough to kill it, and the problems

will be corrected. There is every reason to believe that the Airborne Laser Laboratory will shoot down missiles in upcoming experimental tests.

AFWL's next step was to take the ALL back to Kirtland, analyze the data, and conduct more ground and aerial tests to find the fixes (e.g., revising the conical scan tracking algorithms, stabilizing the optical train) to these problems.[9]

ODGERS'S TECHNICAL ASSISTANCE REVIEW TEAM

Before any fixes could be made to the ALL, AFWL and its higher headquarters first had to agree on the exact nature of each problem and what was the best way for AFWL to proceed to make the necessary corrections. To expedite this process, General Robert T. Marsh, Commander, Systems Command, commissioned Brigadier General Peter Odgers on 24 June to chair a Technical Assistance Review (TAR) team to thoroughly review the ALL test program. Marsh counted on Odgers to give him the straight information as to the ALL's probability of success of the near-term series of self-defense mission flight tests.[10]

At that point the Commander of Systems Command was somewhat frustrated because he realized the June shootdown attempts by the ALL represented still another delay in reaching the program's scientific and technical objectives. Marsh was getting impatient and wanted to know when he "would start seeing something."[*] He certainly did not want to cancel the ALL, but after 11 years of schedule readjustments and increases in funding, he was becoming more skeptical and wanted solid reassurance that the ALL would shoot down AIM-9B missiles. To get those answers, Odgers, who served as Marsh's Deputy Chief of Staff for Test and Evaluation, carefully selected from the Department of Defense a mixed team of 29 military and civilian technical experts to determine whether AFWL had developed adequate solutions to give confidence for continuation of the ALL testing.[11]

The TAR team was not given the luxury of studying the ALL issue to death. Odgers and his group started quickly in their investigation, arriving at Kirtland on 6 July. Within 2 weeks they had collected and evaluated enough data to report their initial findings back to General Marsh. Splitting into two working groups, the first team addressed nine key technical areas: beam vibration, laser vibration, beam control, conical scan, AIM-9B signature, cooling water, software, beam quality, and engineering support. The second evaluation group focused on preparation for the future attempted shootdowns: missile kill criteria/verification, AIM-9B test scenario, instrumented diagnostic targets, assessment of flight test plans, and mission scheduling. To ensure all aspects of the technical and test areas were discussed to gain an accurate perspective on the ALL status, at least two TAR members were assigned to each of the tasks mentioned above.[12]

---

[*]Marsh recalled that although he backed the ALL, he was anxious to see more dramatic results from the laser flight testing. He simply didn't have the time to devote to the ALL to monitor progress more closely because he was preoccupied with bigger and more visible programs, such as the B-1, B-2, and F-117. Consequently, he relied heavily on Odgers to keep him informed on the details and progress of the ALL.

During the first 3 days at Kirtland, the TAR team received briefings from AFWL on a variety of problems the ALL experienced during the April–June self-defense mission flight tests. An intensive series of interviews between the team members and ALL workers and on-site contractors to discuss and collect information on specific concerns and problems took place the following week. Odgers praised AFWL personnel for their positive attitude and excellent cooperation with the TAR members, which accounted for the review being completed in an efficient and rapid manner. On 16 July, Colonel Robert F. Lopina, General Odgers's TAR Deputy, briefed the team's findings and recommendations to AFWL's Advanced Radiation Technology Office.[13]

The TAR team's conclusion was that the ALL "self-defense mission is achievable." However, before this could happen AFWL needed to abide by the team's recommendations stated in each of the 14 specific task write-ups provided by the two TAR working groups. Trying to predict the future was always chancy. However, if AFWL successfully completed a series of proposed verification tests, the TAR then predicted the chances of the ALL shooting down missiles in the self-defense scenario were "extremely good."[14]

In general, the TAR team expressed concern over two key issues that AFWL should have "pursued more thoroughly" in preparation for the first missile defense scenario carried out in June. First, more tow-target tests representative of actual flight conditions (range and turret angle) should have been performed prior to the Sidewinder missile scenario to establish a more reliable data base measuring the energy deposited on the missile. A second shortcoming was that AFWL, before the attempted June shootdown, could not determine with any large degree of certainty the kill criteria and method of verification of kill. The final report stated:

> It was observed that the engagement time for each type of kill [missile blinding or break lock and a missile dome fracture] was not quantified for an engagement with an air flow over the dome representative of a missile in flight. It was also observed the far-field beam energy available was not quantified.

However, during its review at Kirtland, the TAR team and AFWL worked together to solve this by calculating "far-field beam power densities available for different engagement ranges" as well as "establishing engagement time required to achieve a dome fracture kill for various power density levels." It was this work that led the TAR team to conclude sufficient beam energy generated by the ALL was available for a missile kill in the self-defense scenario.[15]

Although these two issues were at the top of the priority list, there were numerous other related problems that AFWL was already in the process of fixing. One potential "show-stopper" was an excess of beam jitter in the autoalignment system causing the operation of the laser to automatically abort. (It was also suspected that atmospheric blooming of the beam within the telescope was another source of jitter, which resulted in too short a laser run time to get the beam to the target.) To reduce jitter (which also caused conscan to break lock) in the near term, the TAR proposed designing and installing two 100-pound tuned dampers or "vibration absorbers" on the turret. In the long run, the recommendation was to redesign and stiffen the mounts of the BAS fold mirror in the APT with the goal of raising the lowest vibrational frequency above 500 hertz. It should be noted that AFWL testing prior to the June attempted shootdown had demonstrated that the entire ALL system had functioned within acceptable levels of jitter. In fact, the TAR commended AFWL for "their accomplishment to date considering the complex system and high vibration environment in which they are operating." What accounted for the recent increase in jitter, according to the TAR, was aging of the ALL system, overload of the control system, and special requirements imposed by the self-defense scenario.[16]

Another concern was the contaminated coolant water in the lightweight mirrors of the beam control system, which clogged the in-line filters (designed to remove tiny particles in the water) and presented a corrosion and heat hazard to the mirrors. Ground water used to service the system contained bacteria and their by-products which was one cause of the filter blockage. Corrosion products in the water were attributed to rusted cadmium-plated mild steel fittings in the water system flexible hoses. The TAR recommended circulating a 2% formaldehyde solution through the very small mirror coolant passages (about 0.060 inch in diameter) to kill the bacteria, installing stainless-steel fittings to reduce corrosion, and back flushing the entire system with trisodium phosphate. These corrective steps were carried out, but there still remained some bacterial residue and corrosive products in the mirror cooling system. Although the TAR believed back flushing and frequent cleaning of individual components were only temporary solutions, they also believed enough of the contaminants had been removed to make aborts a low-risk item.[17]

There were also problems with the conscan tracking system. Glints coming off the tail of the missile were causing conscan to immediately lock on to that section of the AIM-9B missile. Ideally, AFWL wanted conscan to track on the seeker dome (one of the most vulnerable points of the target) which gave off a known steady point reflection of laser radiation. But the conscan tracker tended to lock on to the glint off the rear of the missile before the conscan tracker had a chance to illuminate the nose of the missile with the laser and reflect that laser radiation back to the ALL. In short, the conscan search pattern was always starting near the tail fin, and as a result, picked up and locked on to the glint off that part of the missile before it ever scanned the nose of the AIM-9B.[18]

One possible solution to this problem was to paint the rear section of the missile with a special absorbing coating to reduce the return signals to a level lower than what was expected from the dome return signal. Hopefully, in that way the conscan tracker would lock on to the stronger signal emitted off the missile dome. A second solution the TAR endorsed was for AFWL to continue to develop and test its offset computer algorithm* designed to:

> offset the conical scan starting point away from the FLIR tracker aimpoint [the FLIR initially tracked the target before the conical scan tracker was turned on] in the direction of the missile nose. In this manner conical scan search will start in front of the missile or at least near the nose and sweep the nose as the first likely target area. The amount and

---

*An algorithm is a procedure for solving a problem in circuit design; a rule or procedure for processing an input signal to give a desired output signal.

direction of offset is computed from inserted data on ALL location, speed, and attitude, and missile range and velocity.

The TAR strongly recommended that AFWL in the months ahead conduct more tests of the conical scan against an AIM-9B seeker head mounted on a tow target flown at White Sands Missile Range. The goal of these tests was to get conscan to consistently lock on to the seeker head so this same process could be duplicated during the planned self-defense scenario against real AIM-9B missiles.[19]

Besides addressing basic technical problems, the TAR team also offered recommendations to improve procedures for conducting the next AIM-9B test scenario to increase chances for success. One issue that arose during the first AIM-9B test scenario was that the AIM-9B missile launch range (approximately 5.17 kilometers) and therefore beam engagement range were too long for an optimum engagement. That could be corrected by repositioning the ALL and shooter aircraft so that when the laser was first turned on, the missile (launched at 4.25 kilometers) would be closer to the ALL than during the June scenario. AFWL calculated the closest the missile would get to the ALL was 1.37 kilometers (4500 feet). With the shooter aircraft and ALL closer, Kyrazis stated this would be an advantage, because both aircraft would be "working with better signal-to-noise ratio."[20]

The TAR also advised the T-38 target aircraft should fly 500 feet above the ALL to provide a better background so the AIM-9B would have a better chance to acquire and track the T-38. In addition, the TAR recommended:

> the T-38 target aircraft fly 750' to the right of the ALL aircraft so that the AIM-9B missile will approximate a pure pursuit target toward the ALL aircraft. This will reduce the aspect angle of the missile with respect to the ALL aircraft and increase the possibility of a kill. The TAR recommends that the self-defense missions not be attempted if sunlit clouds are present behind the T-38 target. To reduce the potential for the AIM-9B to break lock due to the IR return of the horizon, the TAR recommends that the self-defense mission be flown as early in the day as photo theodolite tracking camera operation will permit. The TAR concurs with the ALL procedures to insure that all AIM-9B missiles have tracking flare housings to insure consistent missile aerodynamic performance. The TAR observed that it is absolutely essential that the missile shooter crews be experienced and current in AIM-9B attacks. The shooter crews should be carefully briefed on the importance of obtaining the missile launch parameters of the self-defense mission. The TAR also observed that the AIM-9B missile is some twenty years old, and that its reliability is somewhat in question.[21]

On 21 July Odgers briefed General Marsh on the findings of the TAR team. Colonels Rich and Kyrazis and Dr. Avizonis from AFWL attended the briefing in Washington at Systems Command Headquarters. Odgers told Marsh that, although the ALL still had some challenging obstacles to overcome, the chances were good that for the next scheduled AIM-9B scenario the ALL would be able to disable a missile in flight. One of the main points was that AFWL was clearly aware of the ALL technical shortcomings and steps were already under way to make the necessary changes so that the system would work more effectively. In a nutshell, Odgers and the TAR team expressed their confidence in AFWL's ability to get the job done.[22]

Marsh's reaction to the TAR briefing was predictable in the sense that he felt reassured that the ALL was on the right track. He was not about to abandon the program at this point. Too much had been invested and too much was at stake. He recognized that the ALL was close to making a major technical contribution if it could shoot down the AIM-9B. However, he warned that he wanted the ALL program to come to its conclusion as quickly as possible. Funding for the project could not go on forever and political pressure was mounting for the Air Force to show some significant results with its airborne laser after 11 years of research and development. Consequently, Marsh approved the ALL to move forward with the caveat that timely results needed to be demonstrated in the near future. Further delay would only serve to give the critics more ammunition for proposing to do away with the ALL. Two days later Odgers presented the same briefing to Dr. Airey and several members of the HELRG who also endorsed the TAR team's recommendation that the ALL program should proceed to plan and carry out a second attempt to shoot down AIM-9B missiles.[23]

CATASTROPHE STRIKES

By the time Odgers had completed and reported his findings to General Marsh, AFWL was already at work conducting tests to come up with solutions to a variety of problems identified in the TAR report. Throughout the summer and fall AFWL continued testing to come up with the technical solutions in preparation for getting the ALL ready for the next AIM-9B scenario scheduled for December 1981. These efforts took many forms. For example, during July, AFWL technicians discovered and repaired a small water leak in one of the mirror boxes. On 4 August ground tests were conducted to collect data on beam path conditioning, which included using different nitrogen flow rates in the APT/ADAS turret to try to reduce thermal blooming of the beam in the ADAS. Downrange test data on beam focus gave the first hint of thermal bowing of the water-cooled mirrors. While this work was ongoing, others at AFWL were redesigning mirror mounts to stabilize mirrors to reduce beam jitter. In the midst of all of this, the ALL departed for Wright-Patterson AFB to undergo its required periodic phase inspections and maintenance of basic aircraft systems (laser equipment on board the ALL was not affected), which lasted until the end of the month.[24]

In September, work was under way to install passive tuned dampers on the ALL cavity mirror boxes to reduce beam jitter. More testing of the optical train revealed each mirror had a reflectivity level of 97.7 percent, lower than anticipated, but still within operating limits. However, ground testing of the high-power laser to targets downrange behind the ARTF produced disturbing results. Beam misalignment and thermal blooming were degrading beam quality in terms of producing a tight focus on the downrange targets. This was partially attributed to distortions of the mirror surfaces. In short, beam quality was getting worse instead of better and one of the main culprits appeared to be the poor performance of the water-cooled mirrors. Kyrazis, looking back on this problem years later, noted beam quality was getting worse "because the mirrors were bending from the water pressure when you were trying to cool them. It was dissolving the faceplate, and it couldn't stand the pressure."[25]

AFWL workers continued to make improvements to the ALL throughout the fall. On 9 October a series of tests at the NASA-Ames wind tunnel to improve telemetry on the AIM-9B

reached completion. That work involved installing new circuit boards on the AIM-9B missiles to transmit critical telemetry data regarding pressure and voltage changes in the missile dome as a way to verify whether the missile had been disabled by the laser. This issue had been a major concern of the TAR. NASA-Ames wind tunnel findings verified that pressure transducers and breakwires installed on the missile were able to operate within their design limits to accurately measure pressure and voltage changes caused by a laser.[26]

By the third week of October passive tuned dampers had been installed on the ADAS to reduce vibrations, but the first data collected revealed no significant decrease in the amount of vibrations. However, new mirror mounts for ADAS were tested and initial results indicated they would help in eliminating at least part of the vibrations imposed on that system. ADAS still posed a major problem in terms of getting the beam aligned properly inside the ALL. For example, during a 7 November ground test the ALL fired the laser for 1.3 seconds (scheduled for 5 seconds) before shutdown of the system caused by the ADAS. Although only turned on for 1.3 seconds before abort, the beam did cause minor smoke damage to several mirrors. In its report to higher headquarters, AFWL explained that failure of the 7 November test was related to "unknown problems in the ADAS." Later, however, AFWL traced the problem to the "misalignment of a resonator mirror which caused the input beam to ADAS to be offset. ADAS sensed the misalignment and aborted the firing."[27]

On 19 November the ALL deployed to White Sands Missile Range to practice changes in the aircraft scenario configurations recommended by the TAR report. One of the proposals was to launch the AIM-9B closer to the ALL to improve its ability to acquire and track the missile. This meant the ALL would be able to engage the missile at a closer range and thereby increase the beam intensity on the missile target and increase the chances for a kill. Once the missile was launched, the ALL would acquire the missile at a fairly close range, but then the ALL would have to speed up to prevent the missile from reaching the ALL. Five tracking tests of the diagnostic aircraft were successfully completed while the ALL flew between 260 and 340 knots. In addition, in a separate series of flight experiments during the 19 November tests, a 7.8-second closed-shutter hot fire of the laser was performed, followed by a 3.5-second firing. Reduction of the White Sands flight vibrational test data collected while the laser was firing showed that it was possible to increase the ALL airspeed to 340 knots with only slight performance penalties. AFWL also practiced its left bank turn against an LCTT pulled by an F-4 aircraft. By keeping the left turn to less than 30 degrees, AFWL found that "vectoring for the engagement [of the future AIM-9B missile] was excellent."[28]

On 30 November the ALL flew to Edwards AFB to practice the 1B (Air Force Self-Defense) and 1N (Navy Self Defense) scenarios. The next morning over the China Lake test range, the ALL flying at a speed of 350 KCAS successfully tracked two AIM-9B missiles launched at a range of 14,300 feet. Track time by the APT lasted for 20 seconds before the missiles ran out of fuel at 4200 feet from the ALL. AFWL reported all primary objectives of the 1B scenario were accomplished. However, things did not go as well on 2 December when the ALL was scheduled to practice positioning and shooter acquisition of a Navy A-7 as well as to track a BQM-34 drone during 1N scenario conditions at Point Mugu. Breakdown of the cryogenic cooler pump on the APT FLIR tracker forced cancellation of this test. (The pump cooled the sensors in the FLIR so it could detect a weak signal.) As a result, the ALL returned to Kirtland on 3 December and a replacement pump was installed on 8 December.[29]

During the next 2 weeks, AFWL conducted a number of ground experiments trying to improve the operation of a variety of subsystems. After replacing drive amplifiers for each of the beam steering mirrors, testing of the ADAS on 9 December produced encouraging results. Precision alignment of the mirrors and a low-power laser in the ADAS "was readily obtained." On 15 December the laser was turned on inside the ALL and directed downrange behind the ARTF to two stationary targets: one an AIM-9B nose cone and the other an LCTT. The purpose of this test was to demonstrate conical scan acquisition and track modes and to verify the preset focus (APT primary to secondary mirror distance). Three ground hot firings of the laser took place, but not all objectives were achieved. Although ADAS operated successfully and conical scan acquired the rear section of the targets, no stable conscan track was obtained on the nose. Part of this was the result of early shutdown of each planned 5-second hot firing caused by insufficient pressure in the fuel tanks and excessive heat in the computer system. However, against the nose cone for the third firing the laser beam did cross the target ten times, but lock-on was not achieved because of the low laser return signal— only a fraction of output power was sent downrange because of safety considerations.[30]

There were many ups and downs in trying to get the various ALL subsystems ready to work together in the fall of 1981 in anticipation of the upcoming flight tests against live air-to-air missiles. But unquestionably, the most devastating setback was the structural failure of the number 2 beam-steering mirror located on the optical bench on the floor of the ALL. What appeared at first to be an isolated event confined to one mirror, in reality, turned out to be only the tip of the iceberg signifying much deeper consequences.[31]

Disaster struck on 22 December 1981 when the ALL night crew was working on optical alignment of the mirrors in preparation for a ground test early the next morning. Part of this work involved activating a pressure system blowdown to remove trapped air from the mirror water cooling channels. Denny Boesen, who served as the optics chief, was scheduled to go to work at 4 o'clock in the morning to perform the final checks on the ALL before the actual test. Near midnight he received a panic-stricken call at his home from Captain Jim Mills who was with the ALL. Mills was obviously upset and advised Boesen that "you better come out here, we have this little problem."[32]

When he arrived on the scene, Boesen quickly assessed the state of affairs. He recalled:

> One of the mirrors had experienced some corrosion inside the mirror. When they blew the high pressure through the water lines, a little chunk had blown out of the face of the mirror and had splashed water everywhere. So then we examined all the mirrors, and every time we examined a mirror, the story got worse. And we later discovered that we had a biological contamination in the mirrors and that had changed the chemistry of the water so that these mirrors had corroded very badly.

Damage to beam-steering mirror 2 was extensive as a portion of the polished surface of the mirror literally peeled away from the front of the mirror. Bacteria (living microorganisms) in the water had eaten away sections of the 8-inch-diameter molybdenum mirror faceplate brazed to heat exchanger channels. This caused the mirror structure to weaken and pull apart

when subjected to the high pressure water flow through the heat exchanger behind the mirror faceplate. The heat exchanger was a stack of two coolant passages located behind the front surface of the mirror. The walls formed passages which were supported by posts that were machined from one wall and brazed to the other wall. Coolant water entered one side of the mirror, flowed between the support posts, and exited at the other side of the mirror. Observation of beam-steering mirror 2 revealed failure of the brazed joints and fracture of the posts.[33]

Major damage (faceplate detachment) to beam-steering mirror 2 occurred in December 1981.

AFWL immediately began a preliminary assessment of all of the mirrors to determine if their structural integrity was sound. Optical distortion measurements revealed many of the other mirrors showed signs of extensive corrosion and structural deterioration. AFWL had known from the start the rate at which molybdenum corroded in water. However, they were unaware that tiny microorganisms were living in the water and eating the metal (molybdenum), which hastened the corrosion process. When the final analysis was made, findings showed that internal corrosion of the molybdenum cooling channels was occurring at a rate as much as ten times greater than predicted.[34]

Once the extent of the mirror damage had been determined, AFWL was forced to terminate all flight testing. It was clear that the future of the entire ALL was in jeopardy. AFWL now had to devise a plan to fix the mirror problem and eventually get the program back on schedule. Coming up with options was not difficult. Convincing General Marsh to continue on with the ALL program would be a more challenging enterprise. The unenviable job of briefing the state of affairs to General Marsh fell squarely on the shoulders of Colonel Rich.[35]

On 5 January 1982, Colonel Rich (head of AFWL's ARTO) briefed Marsh in his office in Washington. Rich started by giving a brief history of the ALL, emphasizing that significant progress had been accomplished since the Odgers review team departed Kirtland in the summer of 1981. To deal with the mirror failure issue, Rich proposed three options. One option was to simply terminate the ALL program altogether to eliminate any further investment of research and development dollars. The downside of this approach was that the ALL program goals would not be accomplished and the Air Force, by not completing the ALL program, would place itself in an extremely high-risk position for acquiring future dollars for airborne and space laser weapon applications work.[36]

A second option was to proceed with limited refurbishment of mirrors and press on to complete Cycle III before the ALL was scheduled for its programmed depot maintenance

(PDM) in July 1982. This option offered possible attainment of the ALL goals, but the combination of a very tight schedule and the prospects of other mirrors failing made this a high-risk alternative. The third option proposed was to perform a major upgrade of the entire mirror system to reduce the chances of further mirror failures. While the mirrors were being reworked, the ALL would schedule its PDM as soon as possible. The main drawback was that upgrading the mirrors and then conducting the PDM would take considerable time and thereby delay the ALL program by a year to get ready for flying the missile defense scenario.[37]

Because Rich did not have all of the available data analyzed on the mirror damage on 5 January, he recommended that the final course of action be decided on 22 January. However, he was definitely opposed to option 1 because too much had been invested to abandon the ALL even though the mirror failure presented a serious problem and would cause a delay in the program. The more pressing question in Rich's mind was deciding if it was more prudent to refurbish or replace the ALL mirrors. In the end, he would choose the latter.[38]

A change in the schedule for the ALL to undergo its required PDM worked out in AFWL's favor to gain more time to solve the mirror problem. Originally, the ALL and diagnostic aircraft were scheduled for their PDM at Tinker AFB, Oklahoma, in July 1982. If the PDM date could be moved back to January 1982, then that would give AFWL time to have the mirrors fixed. Hopefully, the new mirrors would be ready to be remounted in the aircraft when the ALL returned from Oklahoma on completion of the PDM. If this could be worked out, AFWL believed it could resume high-power testing in July 1982. AFWL contacted the Air Logistics Center at Tinker to try to reschedule the ALL and diagnostic aircraft PDM to an earlier date. After AFWL explained its dilemma, Tinker agreed to reschedule the PDM on the ALL from 29 January through 6 May. New dates for the diagnostic aircraft PDM were set for 22 February through 27 May. On 28 January the ALL flew to Tinker, minus the APT, ADAS, OAD, and turret fairings which remained at Kirtland for further testing and evaluation.[39]

Because of the changes in the PDM dates, Systems Command canceled the 22 January meeting scheduled between General Marsh and AFWL to give AFSC more time to reconsider the options presented at the 5 January meeting. For the remainder of January and into February, AFWL continued to assess the undamaged water-cooled mirrors and realized, because of the corrosion, the risks were fairly high that these mirrors would eventually fail sometime in the future. To try to prevent any further damage, AFWL decided to refurbish the remaining mirrors by "chemical vapor deposition of tungsten in the cooling passages to inhibit further erosion of the molybdenum and to provide some strengthening." The idea was the tungsten vapor would be forced through and adhere to the cooling passage walls to form a barrier between the water and molybdenum.[40]

For a number of reasons the CVD technique did not work. (Tungsten coatings corroded only slightly slower than molybdenum and therefore did not offer any substantial improvement.) Consequently, General Marsh directed General Odgers to form a second Technical Assistance Review (TAR-II) team to review the progress of the ALL program since the TAR-I visit in July 1981, with special attention paid to solving the water cooling system and

controlling the mirror corrosion problems. On 18 February, AFWL briefed the TAR-II team at Kirtland. The focus of the briefing covered AFWL progress since the last TAR review, a detailed discussion of the laser mirrors, and planned corrective action. For the next few days Odgers and his team broke down into smaller groups to meet with their AFWL counterparts to address specific technical issues. After collecting all of the pertinent data, Odgers and his team published their findings on 20 February.[41]

The most significant finding from the investigation was that the TAR-II team members concurred with AFWL's decision, made on the day before the TAR team arrived, that all of the mirrors needed to be "replaced" rather than "refurbished." This meant replacing the heat exchangers and faceplates, thereby removing all of the corroded parts, essentially resulting in new mirrors. AFWL reasoned replacement was the lowest risk and the quickest way to get all of the mirrors fixed. If they opted for refurbishment, extensive metallurgical analysis of the faceplates, as well as laboratory tests of molybdenum under a variety of water conditions would have to take place. Estimated time to do all of this was 2 to 3 months, which was almost the same amount of time it would take to replace the faceplates. Plus, additional time would be lost if the analysis revealed that the faceplates and heat exchangers should be replaced. Therefore, to save time to get the mirrors operational again, the most prudent approach was to replace the mirrors. AFWL directed Hughes to manufacture all of the new mirrors, except for the resonator mirrors inside the laser device. Pratt & Whitney was to build the new resonator mirrors.[42]

During the time the mirrors were being rebuilt and the ALL was at Tinker for the PDM, AFWL seized the opportunity to investigate a number of problem areas in the optical system. AFWL could work on solving problems of the APT and ADAS beam control system in the laboratory, while these components were disassembled and removed from the aircraft. By making changes and fixes at that time, AFWL would avoid problems later when the system was reassembled and reinstalled on the ALL. For example, AFWL was concerned with fixing the BAS detector responsivity. As the TAR-II team pointed out, "The calibration and characterization of the BAS is particularly important since it is the only sensor that provides a direct measurement of the high power beam within the beam control system."[43]

Also, the TAR-II concluded AFWL needed to come up with a corrosion control inhibitor to be added to the water system. The inhibitor had to kill the bacteria in the water as well as minimize the corrosion of molybdenum. As another precaution to further reduce the risk of corrosion, the TAR team advised AFWL to flush and dry the entire water cooling system when not in use for extended periods of time. This procedure, combined with a fixed schedule to remove and replace water filters, would help prevent residue buildup in the system.[44]

Odgers reassured Marsh that AFWL had made significant progress on the ALL since the summer of 1981. Although the unexpected mirror failure represented a major setback, Odgers and his team of technical experts believed, once the mirrors were replaced, the chances of the ALL successfully completing the self-defense scenario were extremely good. The TAR-II recommendations persuaded Marsh to support the continuation of the ALL program. He became personally involved in trying to hasten the mirror recovery program by writing to Dr. Allen Puckett, Chairman of the Board of Hughes Aircraft Company, the contractor responsible for building the new mirrors. Marsh urged Puckett to give "top priority" to the mirror rebuild effort. The general went on to explain: "I consider the timely resumption of ALL testing to be critical to the Nation's high energy laser effort . . . . I ask your help in getting the necessary all-out effort from Hughes on the ALL program."[45]

Puckett wrote back to Marsh in April reassuring him that Hughes had already taken steps to expedite the mirror rebuild effort. As part of the recovery plan, Hughes established an Albuquerque Engineering Center staffed with Hughes personnel to be able to rapidly respond to AFWL's local technical needs. Puckett informed Marsh that Hughes was ahead of schedule in rebuilding the mirrors and also underscored his company's total allegiance to the ALL:

> The ALL is a critical element of the nation's high energy laser program, and it is of paramount importance that every effort be made to ensure that maximum advantage is taken of this advanced system to resolve critical issues regarding HEL weapons. This is of particular importance since for the next several years the ALL will be the nation's only HEL system test bed. Hughes has been involved since the inception of the ALL program and is committed to its success.[46]

Estimated time required to manufacture, polish, and coat the new water-cooled mirrors was 4 months. That meant the mirrors would be ready sometime in July. It would take another 5 months (July–November 1982) to install and align the optical subsystems in the turret, install the entire turret back on the aircraft, and then conduct testing to complete system checkout. In December, ground testing would restore the ALL to its status at the time of the mirror failure. All of these projected changes amounted to a 1-year delay in the ALL program schedule. An additional month (January 1983) would be required to complete ground test firings. From February through April, the plan called for pre-scenario flight tests so the ALL would be ready to conduct the self-defense scenario demonstration in May 1983.[47]

On 11 May 1982, the ALL completed its PDM and then flew to Wright-Patterson AFB to undergo its annual phase inspection. The ALL returned to Kirtland on 19 May. Two weeks later the diagnostic aircraft, having completed its PDM and annual phase inspection, joined the ALL. On 12–13 May 1982, the AFSC Vice Commander, General Robert M. Bond, visited AFWL to be briefed on the progress of the ALL by General Odgers. In this TAR-IIA brief, Odgers told Bond the ALL recovery program was on track, although not all of the technical problems had been solved. The important point was that steady progress was being made. For example, new mirror mounts were being installed on the plane and readied for testing to help reduce vibration-induced beam jitter. To better define the far-field beam energy deposited on the target and to better understand beam characteristics, AFWL was devising improved methods for instrumenting tow targets to yield more accurate data quantifying laser power on target. Also, the AIM-9B missile domes were equipped with breakwires and improved instrumentation to give more accurate data on kill criteria.[48]

Odgers was confident the ALL was moving in the right direction, but he was also up front in pointing out more work had to be accomplished in other key technical areas. Developing a revised conscan tracking algorithm remained at the top of the list. This was a potential show-stopper because, without revising and validating this complicated algorithm, the beam would never hit the target. An AFWL team, headed by Major Bob Van Allen, for

months had been working diligently to perfect the conscan algorithm, but by the summer of 1982 they had not demonstrated it would work properly during flight tests.[49]

A second major stumbling block was the corrosion problem and the failure of the water-cooled mirrors. AFWL reported on 11 May that they had discovered a compound that, when added to water, proved to be an effective corrosion inhibitor for the molybdenum mirrors. However, still at issue was the selection of the best biocide to kill the microorganisms in the system. AFWL had to ensure the biocide chosen did not degrade the corrosion inhibitor or damage nonmetals in the cooling system. Odgers also reported that Hughes was rebuilding the water-cooled mirrors at its California facility, but that work had not been completed. As those mirrors were still in the manufacturing stage in May, AFWL had no way of knowing if the mirrors would meet all of the rigid performance specifications. That would remain an unanswered question until the mirrors were delivered, installed, and tested on the ALL.[50]

A third major technical difficulty involved the problem Hughes encountered in improving the reliability of the FLIR subsystem. The purpose of the FLIR was to provide IR imagery of the target. FLIR sensors detected target heat and then converted that heat into an electrical signal. (Sensors had to be kept extremely cold to be sensitive enough to detect weak signals off the target.) The tracker/processor processed the electrical signals to project a clear IR image of the target on the tracker screen console in the ALL, allowing the ALL team to view the target from inside the aircraft. Success of this system depended to a large degree on the performance of the advanced tracker. It had to constantly make adjustments by sending electrical signals to the APT gimbals to compensate for aircraft vibration to keep the APT properly aligned on target so the image remained in the center of the screen. An analogy to the constant correcting of the position of the APT, so it was always pointed at the moving target, was a person looking through binoculars to view a house in the distance while standing on top of a moving bus. As the bus moved, the person had to move his head and the binoculars slightly every second to keep the house in the center of view.[51]

Part of Hughes's work to improve the operation of the FLIR involved rebonding the FLIR detector to the detector assembly. (The detector assembly had broken in July 1982.) First attempts at this, at Hughes's Los Angeles plant, resulted in a poor bond, which required disassembly of both components. Rebonding would take time and Hughes estimated it would not be able to deliver the rebonded FLIR detector to AFWL until mid-October. This would delay the installation of the APT on the ALL to 1 November, as the FLIR had to be installed in the APT while it was off the aircraft. Airborne tests scheduled for the beginning of January now would be pushed back to the last week of January or early February. AFWL considered this a serious setback, but could do little to rectify the situation until the FLIR detector assembly was fixed. Hughes also worked at reducing noise in the FLIR. Noise detracted from building a strong signal and a clear image. This was similar to disconnecting a TV antenna, which in turn produced noise and a "snowy" picture.[52]

KYRAZIS LEAVES THE PROGRAM

Shortly after General Odgers presented the TAR findings to General Bond, an important change in the day-to-day management of the ALL program took place. After 30 years of faithful service to the nation, Colonel Demos Kyrazis retired from the Air Force on 28 May 1982. More than a third of that time was dedicated to the ALL. He first became acquainted with the ALL while working at NASA-Ames Research Center in the early 1970s as a technical consultant to AFWL's airborne laser program. During this critical start-up phase of the ALL, he was the principal investigator on the first wind-tunnel tests on the ALL that produced invaluable data to confirm that the airplane could fly safely with a blunt body turret installed on top of the fuselage.[53]

When assigned to AFWL in 1973, he worked as the Assistant for the Airborne Laser Laboratory and then became the Chief, Laser Development Division, in July 1976. He was the lead individual responsible for the critically complex ground and flight experiments of the ALL during Cycles II and III. His success in guiding the ALL through uncharted territory was based on his commitment to three fundamental principles.

He always insisted that safety was the first priority in making any decision about the ALL. He made certain that all personnel strictly adhered to this policy, which over the years resulted in not a single incident of injury to anyone.

Second, he maintained the highest technical standards and expected others to do the same. Verifiable scientific evidence was what he exclusively relied on to allow each phase of the ALL work to proceed to the next level. He presented clear technical objectives to the ALL workers from the outset so everyone felt part of the team and so there was no misunderstanding of the direction the ALL was heading. To ensure these technical objectives were met, he established some rather unorthodox procedures by encouraging his people to operate outside the chain of command at AFWL. He challenged each of the ALL workers to go "directly" to anyone to obtain the necessary information to expedite the solution to any technical problem. This included going to industry, academia, and other government organizations for expert technical assistance.[54]

Finally, and probably most importantly, Kyrazis was willing to make the tough decisions required of strong leaders. He was the one who implemented the policy that all systems would be completely checked and tested on the ground to ensure technical objectives were met before any of those systems were flown on the ALL. He did not hesitate to make the decisions that technical objectives were not met. While others were prematurely celebrating the success of the 1981 flight scenario at China Lake, it was Kyrazis who told them that the data did not support the claim that the ALL laser had disabled the AIM-9B missile. More technical work was needed.

Kyrazis realized getting the ALL to perform as advertised was a massive undertaking. But over the years, he and his blue-suiters and contractors formed a highly effective team that persistently chipped away at solving the numerous technical and engineering problems that plagued the ALL. That work was directly responsible for the first-ever airborne laser to shoot down an air-to-air missile.

Kyrazis's decision to retire was prompted by a number of considerations. Although he would have liked to have participated in the scheduled 1983 ALL engagement of AIM-9B missiles, circumstances prevented that. At the top of the list was the large amount of time he had invested in the ALL; he logged over 200 hours of flight time on the ALL, as well as

endless hours in the test cell and hangar. Plus, he had to attend to all of the administrative details of the program back in his office.

Kyrazis later explained, "I was terribly tired. I just did not feel I could keep going. Well, now I know the reason why. Because, based on the symptoms and the fact that I really never recovered from the tiredness, I had leukemia." That, plus 30 years of service, was more than ample reason for him to retire. His loss was not taken lightly because of the very important and unique contributions he had made to the ALL.[55]

At 9 o'clock in the morning on 28 May 1982, Major General Lamberson presided at the retirement ceremony of Colonel Kyrazis in the West Theater at Kirtland AFB. As the program proceeded, several letters from senior officials were read attesting to Kyrazis's leadership and technical competence. The most revealing one was from Major General Brien D. Ward, Director of Laboratories. He wrote:

Colonel Demos Kyrazis who guided the ALL program through its critical testing.

The Airborne Laser Laboratory (ALL) is the most technically complex and challenging program ever undertaken by an Air Force laboratory . . . you have been the spearhead for this extremely difficult program. As a result of your diligence, many unknowns of the past are now known. Your accomplishments are many and include the first precision tracking of air-to-air and ground-to-air missiles from an aircraft, the flight certification and first airborne operation of a high energy laser, the first propagation of a laser beam of destructive energy from an aircraft to an aerial target, and the first engagement of a supersonic air-to-air missile with a high energy beam. Your efforts have brought the ALL program to the verge of a historical achievement—demonstrating the feasibility of airborne laser weapons. Thank you for your extraordinary contribution.[56]

General Marsh, who knew firsthand the technical challenges of the ALL, also was quick to praise Kyrazis. Marsh expressed his appreciation by writing:

You prepared yourself well for the capstone of your Air Force career—the undertaking of the Airborne Laser Laboratory program, the most important technology development program in the Air Force Systems Command. Your efforts have been a wellspring of knowledge, dedication, and discipline which nurtured this program for nearly a decade.[57]

Although Kyrazis's expertise and experience would be missed, the ALL program would move forward over the next year at a rapid pace under the capable guidance of Colonel Bill Dettmer who replaced Kyrazis.

MORE TESTING

During the summer after Kyrazis's retirement, Hughes continued to work on fixing the FLIR, while AFWL began the time-consuming process of reinstalling the basic Cycle III hardware (excluding the APT and ADAS) on the ALL, to include a modified ECS and a variety of research and development electrical subsystems. By the first week of August, the C/DAS computer was back in the plane. In preparation for testing tow targets in early September, all available target boards were assembled, calibrated, and installed into the target vehicles. The instrumented target boards were then tested on the ground with successful results.[58]

Flight testing of the tow targets from 11 to 15 September did not go as well. Each tow target was reeled out 5000 feet behind an F-4. Diagnostic aircraft 371 flew parallel to the tow target and sent a low-power laser to intercept the instrumented target board affixed to the side of the tow target. An array of IR sensors on the target board sampled the beam to measure different aspects of the beam profile (e.g., power in watts, intensity in watts per square centimeter, spot size, frequency, centroid). These data were then telemetered from the target board to the Target Board Data System on the diagnostic aircraft where the information was first given a "quick-look" analysis. Once all of the data were recorded, a more thorough analysis was conducted on the ground. This data assessment allowed the test director to determine if adjustments needed to be made to the ALL system for subsequent testing.[59]

Several problems surfaced during the September flight testing of the tow targets. One was the ground controllers had an extremely difficult time trying to vector the tow target aircraft and diagnostic aircraft into their respective firing boxes. After three passes with the first tow target, AFWL concluded not one would have been in the right position at the right time if the ALL had been used for these experiments. The third test revealed inadequate telemetry from the tow target to the diagnostic target. Telemetry improved for the final test at distances of less than 1 kilometer, but proved unsatisfactory at 3 kilometers.[60]

A subsequent set of tow target flights conducted 3 months later (13–15 December) produced better results. With more practice, ground controllers were able to align the tow-target and diagnostic aircraft into their proper firing boxes to comply with all mission profiles. Some problems still existed with transmitting data from the tow target to the diagnostic plane, but AFWL was able to determine the cause. Ultrahigh-frequency (UHF) transmissions from the diagnostic aircraft (to ground stations and other aircraft) interfered with reception of telemetry from the tow target. AFWL would study ways to eliminate this problem as well as coming up with proposed solutions for ensuring the tow target reel fully extended the target out to 5000 feet.[61]

At the same time tow-target testing was under way, others at AFWL were hard at work testing and making improvements to the APT and ADAS in the clean room. Replacing the deteriorated water-cooled mirrors on the optical bench, ADAS, and the APT with rebuilt mirrors remained the number one priority. But there also were other important modifications. For example, one of the most important changes to upgrade the performance of the ADAS involved replacing the beam steering mirrors used in previous phases with two new steering mirrors. These redesigned mirrors (with improved torque), located on the flight optical bench, were superior because they reduced the angular and translational misalignment of the beam as it moved from the optical bench up through the small opening in the ADAS and into the APT. Also, AFWL and contractors reworked the ADAS electronics to provide better performance and system stability.[62]

By mid-September AFWL workers had mated the APT and ADAS in the clean room. All of the rebuilt mirrors comprising the flight optics had been installed in the APT, ADAS, and on the optical bench. To ensure the mirrors functioned at peak efficiency, the mirror cooling water system (MCWS) underwent some changes, was reinstalled, and tested. One change was the redesign of the optical bench coolant lines, which were manufactured from flexline. Flexline reduced jitter and was more resistant to corrosion than metal lines. Two new water tanks, making the MCWS cleaner and more efficient, were also installed in the plane by the end of September. New tanks eliminated the corrosive and bacterial residue found in the older tanks. Water stored in the new tanks would be treated with anticorrosive and antibacterial compounds to maintain a clean system. In addition, two refurbished mirrors (a hole coupler and a convex mirror) delivered by United Technologies had been installed in the GDL resonator on the ALL to produce higher beam quality.[63]

Installation of the rebuilt mirrors into the APT, ADAS, and on the optical bench went hand in hand with testing the modified MCWS. The basic job of the MCWS was to cool the mirrors and mirror electronics. Because the MCWS had been modified, as a result of the mirror failure in December 1981, the mirrors and plumbing exhibited new flow and pressure characteristics. To verify the MCWS distributed water at the correct pressures and flow rates to each mirror (water pressure to each mirror differed depending on the size of the mirror) and that the opening of water valves occurred at exactly the right time, AFWL conducted a series of tests in December 1982 with the MCWS and beam control system (APT, ADAS, and optical bench) installed in the ALL aircraft.[64]

To assist in designing the new MCWS, AFWL had developed a steady-state flow analysis (SSFAN) code on its CDC 6600 computer to predict water distribution flow rates in the MCWS and to measure water pressure exerted on individual components (e.g., mirrors, plumbing lines, storage tanks, valves). In short, the SSFAN served as a useful tool to evaluate the fluid mechanics characteristics of the MCWS. If the computer code predictions matched the data collected in the upcoming December MCWS tests on the ALL, then that would be proof that the operation of the MCWS conformed to design specifications.[65]

During the initial flow testing of the MCWS, a few minor leaks were detected in some of the temporary fittings. AFWL quickly fixed those leaks. A more extensive series of tests on the MCWS began on 13 December. From the start there were problems with the instrumentation; pressure transducer response time was slow to record data. Test results showed average pressures across the individual mirrors were 10 to 15 percent higher than desired. Consequently, AFWL tried to reduce the flow rates across each mirror by adjusting the sizing of the metering orifices. A second series of tests revealed that this procedure worked well. There was good agreement between SSFAN predictions and measured MCWS data, indicating the MCWS was capable of cooling all mirrors to design specifications. AFWL viewed this as a major milestone, considering only a year earlier the failure of the MCWS was responsible for causing significant damage to the mirrors in the beam control system. In the relatively short span of a year, AFWL had managed to rebuild the MCWS and get the ALL program moving again in the right direction.[66]

Although the improved MCWS was an essential step forward, it was not the only component responsible for steering the ALL back on course. Even if the MCWS worked beyond all expectations, it would be of little value unless AFWL succeeded with the development of two other critical components. The advanced tracker had to reach the point where it consistently locked on and imaged the target. Also, AFWL still was working diligently to perfect the conical scan algorithm to direct the beam to the target.

During the fall of 1982, AFWL made progress on the tracker and the conscan subsystems. In October, Hughes delivered the upgraded FLIR, which was mated with the cryocooler. Performance testing followed. On 19 October AFWL installed the FLIR on the APT. Ground testing of the FLIR occurred on 9 November with good results. Hughes also delivered and installed the conscan detector (which sensed the return of the laser beam off the target) on the APT. By the first week of December, AFWL had mounted the APT, ADAS, turret, and operator's console on the ALL. The next step was to test all of the systems to ensure everything functioned while the aircraft was on the ground.[67]

AFWL was confident that the ground and follow-on flight testing would go well because of all of the time and effort invested over the past 12 months upgrading the beam control and tracking systems. Denny Boesen explained the value of this work:

> During this year [December 1981–December 1982] interval that we took to replace the mirrors, we not only improved the mirrors and got better optical performance on the mirrors, but we did a thousand little tweaks to the control systems and to the computer control algorithms, so that we were able to adapt for the environment and we were able to adapt for the target. And so then it was just a matter of fine tuning when we got out there and took a few dry runs. We had improved reliability, we knew where things were likely to fail and worked around that, also. So I think it was just a matter of fine tuning; that's why we were able to do it so much better the second time.[68]

During December, the ALL ground checkout consisted of numerous tests to verify hardware installation and compatibility on the aircraft. For example, major electrical power sources and interfaces had to be checked out to ensure the APT and ADAS functioned within operating specifications. This involved testing the reliability of electrical power sources and certifying the cleanliness of the hydraulic and water supply supporting the APT and ADAS. The MCWS was gradually pressurized and checked for leaks. Overall, the initial ground checkout of all of the subsystems installed on the aircraft was successful, but there were still

minor fixes that had to be made. One problem discovered was the APT FLIR boresight was not properly maintaining the correct alignment.[69]

Once the initial ground checkout of the ALL subsystems on the aircraft was completed at the end of December 1982, the next step for AFWL was to perform a series of high-power laser tests, first on the ground and then in the air. One critical part of this testing, beginning in January 1983, involved turning on the laser in the ALL, extracting a high-power beam, and directing it through the beam control system to the APT to determine if all systems operated according to design specifications. A second important part of the testing consisted of sending a sample or small portion of the high-power beam from the APT on the parked ALL to a target downrange at the 1-kilometer site behind the ARTF. The purpose of this was to determine beam quality, focus, pointing accuracy, and jitter.[70]

Downrange testing was designed to detect any flaws in the entire ALL system performance affecting beam generation, control, and propagation. One of the major concerns was to send the high-power beam through the ALL to verify no damage was inflicted on the system's optical surfaces. Other pivotal questions also had to be answered. Could the advanced tracker maintain track on static and moving downrange targets while the laser beam was turned on? Equally important was whether the conscan algorithm would work correctly to direct the laser beam to the designated aimpoint on the target. The number 1 priority was whether the system would operate safely and not present any hazards to personnel.[71]

For the most part, ALL ground tests to downrange targets took place during the 2 hours after sunrise or just prior to and during sunset. These were the times of day when conditions were optimal for propagating a beam in the atmosphere. A hot plate located downrange served as the track source for the advanced tracker. Conscan tests used a quarter-inch corner cube mounted on an oblique Mylar reflecting surface. When the beam was not directed downrange, it was not uncommon for testing to last up to 8 hours during the day. Many of these experiments involved taking near-field measurements of the high-power beam. A sample of the beam was projected into a large integrated telescope (LITE) assembly—located in the auxiliary environmental control unit—to collect data on beam characteristics.[72]

Thirteen ground tests took place from 7 January through 19 March. Generally, these tests produced good data on beam characteristics and were successful in correcting FLIR boresight errors. Also, test results showed the airborne beam control system operated normally in aligning the beam throughout the ALL. However, AFWL still encountered difficulties in getting the conscan to work. Attempts to track the quarter-inch corner cube mounted on the downrange target board were not successful. Bright glints produced from the laser hitting other points of the target (other than the corner cube) upset the logic sequence of the conscan algorithm.* This confusion also occurred when the conscan tried to track the stationary nose cone of the AIM-9B missile. Part of the problem, AFWL reported, was the complication of trying to track the nose cone "against the sky reflector background." This problem would be eliminated when testing took place in the airborne

---

*Conscan was designed to offset the beam toward the nose to avoid unwanted glint returns from other points on the target.

environment. In the meantime, AFWL scientists used the ground test data collected and computer studies to make modifications to the conscan algorithm in hopes of perfecting the system.[73]

Although there was some disappointment with the speed of upgrading the conscan algorithm, most at AFWL believed the problem would eventually be solved. At any rate, AFWL did not judge the conscan situation to be so severe to prevent moving ahead with scheduled flight tests. The reasoning to proceed was that the realistic airborne environment would be more receptive to the conscan operation. Plus, other subsystems of the ALL had already checked out on the ground fairly well and AFWL scientists were confident that once in the air they could get conscan to operate successfully against missile nose cones on tow targets. In short, scientists were willing to take the calculated risk of perfecting conscan against realistic airborne targets (as opposed to trying to perfect the system against ground targets) so as to keep the ALL program on schedule to meet the upcoming missile defense scenario in May.[74]

In preparation for the AIM-9B and Navy drone scenarios, 16 airborne tests took place from 7 January through 7 May 1983. Most of these tests were against tow targets to gather data on beam quality, tracking accuracies, and the ability of conscan to direct the beam to the designated aimpoint or nose of the target. The AIM-9B tests were conducted over White Sands Missile Range, and the

ALL flight mechanic Jim "Augie" Augustine in center foreground takes break in the action with members of Hughes team (left to right: K. C. McGrail, Fred Ziegler, and Dewit Weir) prior to ALL takeoff for testing at White Sands.

BQM-34 experiments took place at Pacific Missile Test Center near Point Mugu. To characterize the operation of the laser system at various altitudes, the ALL was flown at low (3000 feet), medium (10,000 feet), and high (30,000 feet) altitudes for a variety of experiments.[75]

Airborne test results were mixed, depending on what subsystem was experimentally assessed. For example, new IR detectors that had been installed in the FLIR assembly worked extremely well for the majority of the airborne tests. Data collected showed the new detectors had much greater sensitivity than the original detectors. Another airborne success story was hardware changes that reduced vibration and jitter levels. Replacing flexure mirror mounts with newly designed ball-in-cylinder mounts proved effective in reducing jitter. At that point, AFWL officials considered jitter to be sufficiently low so as not to pose a major threat to mission success for the upcoming 1B (air-to-air) and 1N (fleet defense) scenarios.[76]

Demonstrating vulnerability thresholds of tow targets to laser radiation turned out to be a much more difficult undertaking. Again, the culprit was conscan's performance. An AIM-9B seeker head installed in the nose of the tow target (to simulate an AIM-9B missile) served as the aimpoint for the conscan-directed laser beam. Inside the nose (dome) of the tow target was an instrumentation package consisting of a breakwire, interior pressure sensors, and 10.6-micron sensors. These components could measure beam radiation inside the nose cone and transmit those data to the ALL diagnostic aircraft for analysis. In short, based on the measurements obtained by the various instruments, AFWL could verify whether the laser beam deposited enough energy to damage the AIM-9B.[77]

Damage inflicted by ALL on AIM-9B nose cone installed on low-cost tow target. Test took place on 20 April 1983 over White Sands Missile Range.

The ALL was to engage the simulated AIM-9B tow target at a minus 140-degree azimuth measured from the nose to the left side of the aircraft. Keeping the ALL, the tow-target aircraft, and tow target properly aligned within a range safety-constrained firing configuration turned out to be much more difficult than originally anticipated. Also, for most flight tests numerous laser glints reflecting off the tow target body were seen by the conscan system which diverted the beam from locking on to the nose cone. Major Bob Van Allen, the lead architect for designing the conscan system, explained the glint problem:

> This [glint problem] was finally traced to a boresight error left and high which started the beam acquisition on the target body rather than ahead of the dome. This bias was electronic in nature and removed in software. One flight was subsequently made which verified the problem was solved—dome was immediately acquired with no evidence of glint tracking.[78]

The flight Major Van Allen was referring to occurred on 20 April over White Sands. After many previous tries, this test was the ALL's first successful airborne engagement of an AIM-9B nose cone mounted on the front of a tow target. The ALL flew at 14,000 feet with the APT looking rearward (−140 degrees) and slightly upward at the tow target positioned at 14,500 feet and a distance of several kilometers. Almost immediately after the beam was turned on, conscan directed and locked the beam on to the nose cone. Telemetry from the nose cone verified the IR dome fractured after only about 2 seconds of irradiation.[79]

Success of the 20 April test over the remote desert of New Mexico was not just another routine experiment. Those who had labored long and hard to get the system to work knew without a doubt that this was one of the most significant turning points in the entire ALL program. It was this momentous event of 20 April that conclusively proved all of the subsystems of the ALL worked effectively as a single laser unit. Interception of the AIM-9B nose cone by a laser beam fired from an airborne platform served as the main catalyst to propel the ALL into its most critical next stage. Now the ALL would have to move one more dramatic step further by attempting to shoot down a real AIM-9B missile porpoising through the sky and zeroing in on AFWL's laser aircraft.

General Brien D. Ward, Director of Laboratories at Systems Command, recognized the importance of the White Sands tow-target tests. He emphasized these data "verified the ALL laser power, beam quality, and pointing and tracking jitter all are significantly improved following the FY82 refurbishment of the ALL beam control system." He went on to point out that this information has "given us the confidence to proceed into the missile engagements."[80]

Before General Marsh would allow the ALL to try to engage the air-to-air missile launched from a fighter aircraft, he wanted reassurance that the planned upcoming test would not be a repeat of the June 1981 attempted shootdown 2 years ago. Once again, he called on Major General Odgers to lead another technical review team to thoroughly investigate the progress AFWL had made toward the goal of being ready to embark on a second bid to shoot down a missile as part of the 1B self-defense scenario.* Odgers and his 12-man team arrived at Kirtland the second week of May to begin assessing the operational readiness of the ALL to perform the self-defense scenario.[81]

Odgers's group concentrated exclusively on actions taken by AFWL to correct deficiencies since the last two TAR team (February and May 1982) visits. A sufficient amount of new data were available to evaluate. Since high-power testing resumed in November 1982, the laser had been fired 39 times producing a total of 169 seconds of beam time. To come up with a realistic recommendation for General Marsh, the team conducted a risk assessment of how each critical component of the ALL performed regarding beam energy, conscan, missile signature, corrosion, cooling system, vibration, software, kill criteria, and autoalignment.[82]

General Odgers and his team presented a glowing report on the progress of the ALL. Conscan and the mirror cooling system, problems that had plagued the ALL for the past year and a half, were now upgraded from a yellow to green status. The 20 April conscan test was highly successful and based on these results the TAR rated the conscan probability of success

---

*Marsh had put a great deal of faith and stock in Odgers's scientific judgment and previous knowledge of the ALL. In September 1982, Odgers had left Marsh's staff to assume command of the Air Force Flight Test Center at Edwards AFB. However, Marsh recalled Odgers to head up the ALL technical review team. Marsh wanted straight answers on the status of the ALL operating systems and felt that Odgers was the man to provide an objective assessment.

for the self-defense scenario as high. Major Van Allen and his team concurred with the TAR's assessment, although Van Allen admitted he would have been more comfortable with another successful conscan tow-target test as extra insurance before proceeding with the actual self-defense scenario.[*] However, AFWL's tight schedule would not permit this.[83]

A second major concern that needed to be resolved was corrosion of the mirror heat exchangers that had caused mirror failure. Corrosion, plus biological infestation in the mirror cooling system, blocked filters and caused system aborts related to loss of coolant flow. The TAR's Majors Robert Hemm and John O'Pray reported on corrective action taken by AFWL in this area:

> A superb effort has been made in correcting all problems. 100% refurbishment of the mirrors has eliminated all bad optics and essentially made all the mirrors like new. Addition of a corrosion inhibitor to the water has reduced corrosion to a very acceptable level. Corrosion monitoring procedures both confirm the low corrosion rate and spot incipient problems. The use of a biocide and improved water handling procedures has eliminated the biological attack problem. All filters have been cleaned. Cracked rubber liners in the water and nitrogen tanks have been replaced by Inconel liners. The mirror cooling water system should easily meet all ALL lifetime goals.[84]

With the conscan and corrosion problems fixed, this meant all systems of the ALL were given a green status by the TAR investigative team. This prompted General Odgers to inform General Marsh that "the overall unanimous opinion of the TAR team members was that the AFWL has adequately corrected all previously noted deficiencies and that ARTO has a high probability of successfully executing the self-defense mission scenario." That was the response Marsh wanted to hear. With Odgers's reassurance, Marsh directed AFWL to devote all of its efforts in preparation to carry out the self-defense scenario scheduled for the last week of May. By early June, the Air Force and DOD would know only too clearly if all of the years of investment in the ALL had finally paid off.[85]

---

[*]Not all agreed that substantially more useful data would be acquired by additional airborne tow-target tests. John Otten, Test Director for the 1B scenario, pointed out that the airborne tests were slightly flawed because the tow target head-on configuration was not the same as the real AIM-9B missile. For one thing, the tow target had a different track signature than the real missile. A target pulled slowly behind an aircraft did not heat up as much as a real missile traveling supersonically. Hence, the tow target presented a cold signature to track on, which was less ideal than tracking on a hot signature. A second problem with the tow target was the straight path it followed; the real missile would "meander around" making tracking much more difficult. Still a third problem was, even though a real AIM-9B nose cone was fitted on the tow target, the configuration of the tow-target nose and real AIM-9B was different and presented different pictures to the APT. The nose on the real AIM-9B had a clean and symmetrical shape approximately 5 inches across. On the tow target, a protruding "skirt" of metal flared out on the side of the nose cone that transitioned the nose cone onto the body of the tow target. This enlarged the front of the nose cone to 15 inches across. In sum, Otten concluded that the tow target data were useful but not perfect. The tow target, because of its cold signature, straight flight path, and geometry of its nose cone area, could not replicate the real AIM-9B missile. Therefore, Otten and others were anxious to move ahead with testing the ALL against the airborne AIM-9B missiles in the 1B scenario.

14

Success: Two Shootdowns

Once General Marsh gave his approval for AFWL to commence preparations for the 1B air-to-air missile defense scenario, he began monitoring the ALL program more closely by requiring periodic briefings from AFWL personnel. The stakes were enormously high at that point. After setbacks, delays, technical fixes, and seemingly endless rescheduling, Marsh wanted what he had always desired most from the ALL—results!

Colonel Jerry Janicke, who had taken over as chief of AFWL's Laser Development Division in January 1983 from Colonel Bill Dettmer, who replaced Colonel Rich as head of the ARTO, vividly recalled the tension at Systems Command just prior to the ALL's departure to Edwards in the middle of May. Janicke had traveled to Washington, D.C., to advise Marsh and his deputy, General Robert M. Bond, on the most current status of the ALL to raise their confidence level that the ALL would be successful. At the end of the briefing Marsh in no uncertain terms reminded Janicke, "Don't embarrass us, and get it done." To underscore the importance of that message, General Bond looked Janicke squarely in the eyes and, slowly in a very deliberate tone, repeated the essence of his boss's guidance: "In case you didn't hear the general, don't embarrass us—get it done!" Janicke's reply was a swift, "Yes, sir."[1]

When the ALL arrived at Edwards on 15 May, the AFWL flight crew was not billeted on base or at the nearby town of Lancaster. Instead, they drove 90 miles west to the remote Silver Lake Lodge located near Barstow. There were a couple of reasons for this. First, AFWL did not want to tip off the press ahead of time that the ALL shootdowns would be under way shortly. Isolating the crew, hopefully, would prevent any inadvertent leaks to the press. In that way, in the event of a failure with the AIM-9B shootdown, the embarrassment issue Marsh was so acutely aware of would be avoided.[2]

A second reason for staying away from Edwards was the potential of the Soviets tapping the phones to acquire information on the laser tests that might be useful to them. During the 1981 attempted shootdown, Kyrazis felt fairly certain that someone was tapping his phone as he passed on the results of the flight experiments to higher headquarters. To his amazement, the exact words he used on the phone appeared in the papers the next day. Dettmer concurred. He recalled, "I don't know where they [Lancaster radio station and newspaper] were getting their information, but it was certain there was a leak of sorts and information was given." The source of the leak was never discovered. But because of this experience and to remain on the safe side, the extra precaution was taken to billet the Cycle III ALL crew at a site few knew about.[3]

Although the billeting location of the crew was not publicized to outsiders, everyone connected with the ALL knew airborne testing would take place over the Naval Weapons Test Range at China Lake. There were several reasons why China Lake was selected. AFWL had investigated using White Sands Missile Range, but the Army was somewhat skeptical about sending a high-power laser beam against real missiles flying across the skies of New Mexico. Other ranges were looked at, but all turned the Air Force down. Edwards was not set up to fire live missiles, as its charter was to flight-test airplanes. However, the Navy was willing to accommodate the ALL as China Lake had the right geometry and safety parameters to conduct airborne experiments against air-to-air missiles. Plus, AFWL rated the Navy ground controllers, a critical ingredient in aligning all of the in-flight aircraft, as highly dedicated and competent. Another advantage of China Lake was that it was close-by as it was adjacent to Edwards's range. Finally, the Navy wanted to be cooperative for purely parochial reasons. They knew later in the summer the Air Force would support one of their programs by using the ALL to engage a drone over the Pacific to determine the potential of lasers as a ship defense weapon.[4]

1B SELF-DEFENSE SCENARIO

Lieutenant Colonel John Otten served as Test Director for the 1B scenario, designed to simulate a bomber defending itself against air-to-air missiles while flying to its target. The test objective was to operate the total laser system in an all-up airborne environment to demonstrate the beam could be kept on target for sufficient time to destroy or disable the missile's radome. Otten had complete control of the flight crew and the operation of all subsystems of the laser. Safety and timing of the start-up and operation of the laser system were his prime responsibilities. One of his other important obligations was deciding the exact time the laser should be fired. That was a long and painstaking process because extensive preparation of all systems, first on the ground and then in the air, were required before the firing command could be given.[*][5]

Ground preparation and servicing of the ALL started at noon on the day before the scheduled flight mission. About 8 o'clock in the evening, technicians began checking valves regulating the gas flow, aligning mirrors in the resonator, on the bench, and in the ADAS, and flowing water through the system to verify mirror alignment remained in place. By midnight, liquid nitrogen

Lieutenant Colonel John Otten, left and back to camera, conducts preflight briefing at Edwards AFB for upcoming ALL test against AIM-9B missiles.

was pumped on board to begin cooling the entire laser fuel supply system so that when the cryogenic laser fuels were added later they would not boil off. Around 3:00 in the morning, the fuels (liquid carbon monoxide and liquid nitrous oxide) and helium were loaded, which took approximately 3 hours. This was happening at the same time the flight crew at Silver Lake Lodge was getting up and then making the 90-minute drive to Edwards. They arrived at the ALL between 4:30 and 5:00 A.M., or about 2 hours prior to takeoff.[6]

An hour before flight, the fuel and coolant lines were disconnected from the ALL. Thirty minutes out the ALL was ready to go. Once the ALL was airborne, it made several orbits around its racetrack course. The purpose of this was to establish the weather background for firing that particular day, to bring the laser up to the point where it was ready to fire, and to work with the ground controllers to initiate the proper vectoring of the ALL and 371, the diagnostic airplane. Generally, 371 had a crew of up to 35, whereas the ALL crew numbered 15.[7]

The T-38 and the shooter aircraft, the Vought Navy A-7 Corsair, then came up and joined the ALL and 371. All aircraft flew a racetrack course that took between 10 and 15 minutes to complete. (It wasn't unusual for all aircraft to make ten or more racetrack orbits before they were properly aligned.) Ground controllers devoted all of their efforts to vectoring four

---

[*]Prior to the first 1B scenario flight, John Otten and Denny Boesen went to see the movie "Star Wars." On their return, Otten offhandedly mentioned to Master Sergeant James "Augie" Augustine, flight mechanic for the ALL, how impressed he was with Han Solo's spaceship named Millenium Falcon. The next morning, Otten boarded the ALL and noticed the words "Millenium Falcon" painted on the left forward fuselage of the ALL aircraft.

ALL team at Edwards AFB, California, preparing for 1B self-defense scenario to engage AIM-9B missiles over China Lake in May and June 1983.

aircraft (ALL, 371, A-7, and T-38) into their firing boxes at exactly the right time and distance from one another. The Navy A-7 was positioned several miles away. Once each aircraft was in its box, and at the correct heading, orientation, and speed, the laser was ready to fire, if all operating parameters had been met. The job of the T-38 was to put out a bright heat signature for the shooter aircraft to lock on. (This was required because the geometry of the engagement prevented the AIM-9B from receiving a clear signal from the ALL aircraft at the launch range dictated by safety considerations.) To accomplish this, the T-38 pilot would light one afterburner while using the plane's speed brakes and lowering its landing gear and flaps to hold down its speed. That allowed the T-38 to provide a good target signature (better than the ALL engines) and at the same time to "stay around" the ALL rather than accelerating out of the area.* But because it burned large amounts of fuel in the afterburner stage, the T-38 could only remain airborne with the ALL for about 60 minutes. Usually, a second T-38 was on call to support the ALL or the first T-38 would have to go back to refuel and rejoin the ALL to get ready for a second laser firing.[8]

The shooter aircraft carried two Raytheon/Ford Aerospace AIM-9B air-to-air missiles. Because the 9Bs were older-model AIMs (AIM-9Bs had been around for 20 years), the Navy

---

*Major Ted Wierzbanowski, one of the T-38 pilots, recalled flying alongside the ALL was one of the most unusual missions he had ever flown. As his plane was the target of the Sidewinder, he was understandably nervous. "If I ever saw a missile getting close, I knew what I was going to do," he recalled, "I was going to go straight up. What would happen to the laser lab . . . they'd have to figure that out." In fact, what Wierzbanowski described was the escape maneuver for the T-38 in the highly improbable event that the missile might intercept the T-38. The ALL escape maneuver was to turn and dive. Of course, the slow-moving ALL could never really get out of the way of a supersonic missile. Otten recalled the ALL escape maneuver was actually executed once when the A-7 pilot inadvertently released the AIM-9B at a slight nose-up attitude—the only possible intercept launch position. "Obviously, we were worried," Otten remembered, "and the ALL crew sure was quiet for a while until the danger passed!"

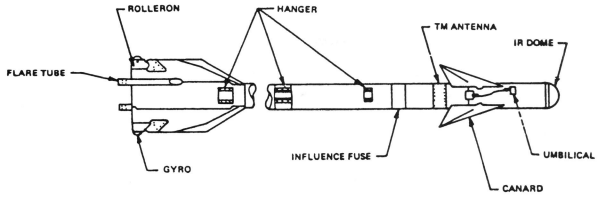

AIM-9B Sidewinder air-to-air missile.

A-7 had to be modified to carry these vintage missiles. For obvious safety reasons, the high-explosive warheads of these missiles were removed and replaced with a spotting charge. One of the advantages of the A-7 was it was not severely restricted by fuel capacity; it could remain with the ALL for two firings of the laser, one against each of the two AIM-9Bs launched. Generally, the earliest the A-7 would launch was the third orbit of the ALL.[9]

Originally developed by the U.S. Naval Weapons Center at China Lake, the AIM-9 Sidewinder (first successfully tested in 1953) was a slim heat-seeking guided missile designed for air-to-air combat. Measuring 111 inches long and weighing only 155 pounds, it had a range of 3.2 kilometers (2 miles). Its main parts included a seeker head (radome/dome/nose), a control system, a warhead, and an engine. Once the pilot placed the target in his sight, he fired the missile when he heard a tone or "growl" in his headset. This fire-and-forget feature and its relatively low cost ($3000 in 1960) and simplicity (less than 24 moving parts) made the AIM-9 a highly attractive weapon. To improve its performance, 14 versions of the Sidewinder have been developed. One of the advantages of using the AIM-9B was that it injected a strong flavor of realism to the 1B self-defense scenario. The Soviet AA-2 "Atoll" air-to-air missile was a carbon copy of the AIM-9B. Performance of the airborne laser against the AIM-9B would produce the same effects against the AA-2 Atoll, thereby allowing the Air Force to better evaluate the Soviet threat.

In a series of seven airborne tests from 17 May through 1 June, the Navy A-7 launched a total of 24 missiles at the ALL. (For these types of tests, it was not unusual for the ALL to spend 4 hours or more in the air.) The Sidewinder traveled toward the ALL at a speed of about 2000 miles per hour. Acquisition of the target and precision tracking was the purpose of the first group of missiles fired at the ALL. To do this required that the personnel and equipment on the ALL follow a rigid set of operating procedures. Basically, the first step was for the operator of the OAD looking out the left side of the ALL to acquire the shooter aircraft and electronically hand off that information (e.g., coordinates, azimuth, distance) to the APT. Once the target was framed by the APT tracking gates, the APT began automatically tracking the shooter aircraft. The T-38 aircraft (flying in close formation off the right rear of the ALL) served as the heat source and target for the missile because during its left turn as

Loading AIM-9B missiles in preparation for 1B self-defense scenario.

part of the 1B scenario, the ALL's wings shielded the engines from the missile's view. In other words, the missile's guidance could not see the engines' signature to track on once the ALL went into its turn at the long range required for safety. Plus, the 20-degree left turn made by the ALL forced the AIM-9B (approaching from the left rear and above) to follow a course so its nose was always in a direct line of sight to the APT. This was vitally important, because when the beam was turned on it wanted to hit the nose rather than the side of the missile.[10]

As all of the aircraft hit their designated firing boxes (and remained there for only a few seconds), the A-7 would lock the missile onto the afterburner signature of the T-38 aircraft. The T-38 pilot would receive a tone in his headset to confirm lock-on. Ground controllers and the ALL test director could also hear the tone. However, because of safety considerations, only the ground controller could authorize launch of the missile. He was the one who had to make the final decision to launch after verifying that all aircraft were in their proper firing boxes. Colonel Otten, test director on the ALL, also had authority to cancel any launch if he felt that conditions for firing the laser were not quite right.[11]

Once the ground controller gave the launch command, the A-7 pilot released the missile with a call of "Fox" away (indicating launch), and then made a hard right climb to depart the area. For the first second or two at and after launch, the missile produced an extremely bright flash caused by its engine burning large amounts of fuel. It was this plume that the ALL's centroid tracker, operating in "missile launch detection mode," automatically locked on to initially track the missile. (Manual track of the missile was not possible because the speed of the missile was much too high.) At this

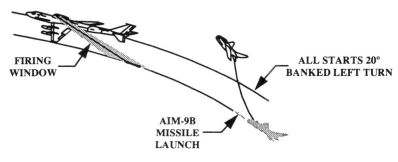

FIRING WINDOW

ALL STARTS 20° BANKED LEFT TURN

AIM-9B MISSILE LAUNCH

1B scenario: nose-on, aft-look profile.

time the tracker ignored the A-7 which climbed to a higher altitude and disappeared from the track gate. Once the plume disappeared, the ALL shifted to correlation tracking the body of the missile.[*] At first, because the missile was relatively cold a few seconds after launch, the image of the AIM-9B missile body was not visible on the screen in the ALL, even though the tracker knew it was tracking the hot body and residue plume of the missile. But as the missile accelerated to the target, it aerodynamically heated up and provided a clearer image on the screen.[12]

When the missile reached a predetermined distance along its flight path to the ALL (measured by the ALL range finder), the computer on the ALL gave the command that fired the laser. The beam spiraled out toward the target with the goal of intercepting the target. What often happened was the beam would hit the missile, but not on the nose, which was the desired aimpoint. A glint off the body of the missile would return the 10.6-micron beam back to the APT. With that information in hand (coordinates of the glint), the ALL began tracking the missile using the 10.6-micron laser beam return. Then the conical scan algorithm would walk the beam forward

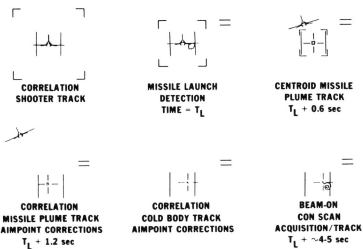

ALL tracker missile launch detection sequence.

of the initial glint location on the missile until the tracker no longer could detect a return. (The computer on the ALL knew the coordinates of the glint on the missile as well as speed and trajectory of the missile. Therefore, it could tell the beam the rate and direction the beam should move forward along the velocity vector.) This meant the beam was now positioned in front of the missile's nose. At that point, the conical scan directed the beam to back up until it produced a glint off the nose. Once the APT confirmed the beam had hit the nose, it held the beam on the nose to inflict damage. The conical scan track point and aimpoint then became the same. This entire process, from the time the beam was turned on through the beam dwell time on target, lasted only a few seconds.[13]

The first 11 of 24 missiles were fired to give the ALL ample opportunity to perfect its tracking for the 1B scenario. In most instances, the A-7 launched each missile at an altitude of 10,000–15,000 feet and about 7 kilometers from the ALL.[†] It immediately became clear

---

[*]Correlation tracking of the missile continued for the remainder of the AIM-9B flight, even after conscan tracking was initiated. This meant two trackers were used simultaneously—correlation to track the body of the missile and conscan to track the nose. If conscan failed, the ALL would not lose the missile because it would still be tracked using correlation tracking.

[†]The ALL flew at a relatively low altitude because it was safer as there was more drag on the AIM-9B missile. More drag prevented the missile from operating at peak efficiency and reduced the chance that the AIM-9B would ever reach the ALL.

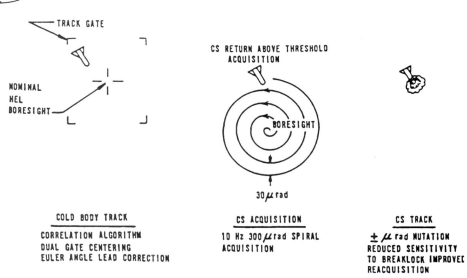

Conical scan sequence.

that AFWL had underestimated the difficulty in tracking each missile. The problem was that each missile behaved and looked differently to the tracker. Some veered left whereas others veered right as they came off the A-7 launch rail. Plus, the missiles were unstable and unpredictable in terms of the erratic trajectory they followed to the target. Indeed, the missile behaved in accordance with its namesake "Sidewinder." Signature of each missile also varied each day depending on changing weather conditions. For these reasons the ALL team needed a number of practice runs over a dynamic range of possible missile responses to make minor modifications to tracking algorithms to consistently obtain a good tracking lock on each missile. That had to be accomplished before the laser could be turned on to attempt to intercept an AIM-9B.[14]

After practicing against the first 11 missiles launched, AFWL was confident it could acquire the shooter aircraft, pick up the missile at launch, and continue to track the missile as it moved toward the ALL. For the next 8 missiles, the goal was to try to conscan track each missile as it came closer to the ALL and attempt to direct the laser beam to the nose. This was a critical stage because this was the time the laser was turned on. In general, these tests were not completely successful for a number of reasons (e.g., safety, incorrect flow settings, mirror misalignment). Consequently, the computer on board the ALL automatically shut down the laser device early in its run time. Conditions had to be near perfect before the laser could be turned on to use conscan to try to track and then destroy the missile. Many of the abort parameters intentionally were not tight at first. As the system matured, the ALL could reliably engage a target at a preselected time and place.[15]

The practice of aligning all of the aircraft in their proper firing boxes, acquisition and initial track of the AIM-9B, and going through all of the steps to get the laser ready to fire, proved to be an extremely valuable experience. But the next step of firing the beam and directing it to the target became extremely difficult and frustrating. For example, tests on 17 and 21 May showed the beam consistently hit the aft section (fins and rollerons) of the missile. Even though the beam would sporadically bounce off the nose, the laser was unable to remain on the dome long enough to cause any damage. With each subsequent missile launch, the ALL team gained more confidence by making incremental refinements to the conscan tracking algorithms, which led to improvements in the operation of the entire system. But up to that point, although significant progress had been made in fine-tuning and calibrating the system in the real-world environment, no missile had been shot down.[16]

That would change dramatically over the next few days as the ALL intercepted the next five missiles launched. On 26 May, the first successful engagement (mission 15043) of the ALL beam destroying an AIM-9B missile occurred over China Lake.* Colonel Otten described that historic first shootdown as "if someone wrote the textbook." Everything happened just as it was planned. Tracking was rock solid during the duration of the experiment, which lasted 4 to 5 seconds while the beam was turned on. Conscan initially picked up the missile at a range of 3 kilometers and directed the beam to the nose of the missile as evident by the bright glint shown on the tracking screen in the ALL. More importantly, the beam remained on the nose long enough to heat up and damage the sensitive components of the guidance system, causing the missile to break lock and veer off course. Otten realized immediately the lasting significance of this momentous event when he recalled, "That was the end of the story. Nobody had to show us anything. We didn't even need to see any data!"[17]

In fact, more than visual evidence was needed to prove conclusively a kill had occurred. Observation from pictures provided by the TV camera and tracker on the APT was only one form of proof that the AIM-9B had made a hard turn away from the T-38 aircraft the missile was trying to home on. The critical scientific evidence came from telemetry data transmitted (28 separate signals) from the missile's instrumentation package to 371. Plus, radar tracks plotted by radar ground stations tracking the AIM-9B showed the missile made an abrupt change in its course trajectory the instant it was engaged by the laser. The result was the missile dive-bombed, crashed, and buried itself several feet into the ground.† All of these factors corroborated Otten's jubilant announcement that the beam had clearly hit the nose of the target to cause sufficient damage to the missile's guidance system, indicated by the missile breaking lock with its target.[18]

There were two verifiable kill mechanisms substantiated by telemetry data. One was thermal. An electric current flowed through a silver strip (Dupont Silver Preparation 4666) or "breakwire" painted on the inside of the dome. If the beam burned through the breakwire or the nose dome shattered, the flow of electric current would be interrupted. Instrumentation

---

*The first AIM-9B disabled on 26 May was actually the second of three missiles launched that day. When the first missile was launched, the laser sequence had to be aborted because of safety considerations associated with reappearance of the shooter aircraft on the tracking screen after the A-7 launched the missile. There was a remote possibility that the laser beam might be mistakenly directed at the A-7 instead of the missile.

†Postflight analysis of the effects of the beam could not be determined through recovery of the missile because of extensive ground impact damage.

inside the dome electronically measured that interruption (loss of voltage) to confirm the electric circuit had been broken. This decrease in voltage would be telemetered to the

AIM-9B tracker head general construction.

diagnostic aircraft. This was one way of showing the beam had penetrated the dome. Once inside the missile's nose or dome, thermal energy of the beam burned the seeker sensor and reticle (focusing mechanism) inside, which caused the missile to lose altitude and break track. Damage to the reticle caused it to spin down from about 70 cycles per second to zero. This was a catastrophic failure and meant it no longer had the ability to track anything. In effect, the hot beam had "poked the missile in the eye" causing the AIM-9B to lose its ability to follow its computed trajectory and guide to its target. This "blinding" of the missile took less than a second, but was the principal reason the missile was unable to complete its mission. Again, instrumentation in the nose detected damage to the internal components of the dome by providing data that proved the missile's guidance system had malfunctioned.[19]

An increase of pressure inside the nose of the missile was the second kill mechanism. As the beam penetrated the nose, it caused the glass dome to fracture or break off sections of the dome. Those fractures and breaks provided passages for entry of increased pressure (ram air pressure originating from the atmosphere outside the missile) into the AIM-9B dome. Pressure and turbulence, once inside the dome, caused structural damage to the tracking components. Instrumentation inside the dome measured pressure changes to provide empirical data confirming that pressure had increased.[20]

Bolstered by the near-flawless shootdown of the first AIM-9B, the ALL team made several additional orbits for alignment into its proper firing box in preparation for engaging a second AIM-9B. Once again, the A-7 launched the Sidewinder and the ALL was able to lock on and track the missile. When the conical scan was activated with the turning on of the laser, the beam for a second time that day was able to track and remain on the nose to establish a confirmed kill. Telemetry data revealed thermal energy of the beam and structural damage to the dome combined to divert the missile off course. This second missile shootdown sequence was almost identical to the first. The main difference between the two shots was that the beam dwelled on the nose of the second missile for 3.8 seconds or 1 second less than the dwell time of 4.8 seconds of the first missile. Although the beam remained on target for less time on the second missile, results were the same. The second AIM-9B suffered substantial damage to its guidance system causing it to make an abrupt hard turn off course and eventually to fly out of control until it crashed on the desert floor below.[21]

When the ALL landed at Edwards, Otten and his crew had much to celebrate over the next few days. Everyone knew that two missiles had been fired and two missiles had been successfully disabled by the high-power laser firing from an airborne platform. This single mission fully completed and met all technical objectives of the 1B scenario milestone demonstration. To commemorate this extraordinary technical breakthrough, beer and war stories flowed freely over the Memorial Day weekend. Janicke recalled

First Lieutenant Frank Zoltowski inspects three-pane, high-power window after first AIM-9B engagement.

fondly, ". . . made it to the bottom of the pool eight times that night with beer in hand!" But the partying was to be relatively short lived. By Monday the crew was turning to more serious matters in getting the ALL readied for its next mission on 31 May. The question facing everyone was, could the extraordinary feat of 26 May be repeated to collect additional scientific data?[22]

The 31 May shootdown (mission 15044) was radically different from any of the other laser firings. One of the problems with the first two shootdowns was that there was no spectacular explosion of the missile when engaged by the laser. The beam was invisible and damage to the missile was verified by telemetry data. These data were recorded on tapes on the diagnostic aircraft and then carefully analyzed on the ground after the flight. What AFWL needed to do was to add some exciting visual evidence that the laser beam had a devastating effect on the missile. As Otten explained:

> There's nothing to see. The problem was that when you hit these things you get tracks careening to the right. You've got a lot of wiggly lines that do a lot of great things. There is a lot of telemetry data that does all this wonderful stuff. Super for the scientists and engineers. But doesn't have much press value. Nothing explodes.[23]

To convey to the press, government officials, and the general public the extent of the laser's killing power, AFWL intentionally rigged the nose of the third AIM-9B with a small

spotting charge. The breakwire in the nose was connected to the charge. If the laser burned through the breakwire, the spotting charge would detonate. That small explosion would, in turn, put out a bright trail of smoke to give visual proof that the missile had been intercepted by the laser.[24]

It took numerous orbits to position all of the aircraft for the 31 May shootdown. Once in place, however, tracking of the missile, conscan, and hitting the nose with the beam worked to perfection. The only problem was that the beam dwell time on the dome was 2.4 seconds—about half the laser time on the first AIM-9B damaged on 26 May—before the laser shut down because of alignment difficulties in the APT. However, that was plenty of time for the beam to burn through the breakwire that set off the spotting charge. The result was a bright plume of white smoke engulfing the nose to simulate a dramatic explosion for all to see. The missile itself did not explode because for all tests its live warhead had been removed. But the sudden puff of smoke and bright light gave the appearance that the missile had exploded.[25]

Otten realized the breakwire explosion was designed to accommodate the plans and desires of higher headquarters to publicly advertise the first successful use of an airborne laser against real missiles. Although he knew all of these theatrics would not contribute in any significant way to the body of scientific knowledge, he also clearly understood that an explosion would appeal to the nontechnical audience as proof of the highly deadly nature of the laser. As he pointed out later:

> The blow-up was staged. The kill was real. And the reason was that—the visual evidence of the kill—was not obvious [for the nonexplosive breakwire shootdowns]. You knew it on the airplane that you had engaged this target, you obviously had deposited energy on it. If you ever got a few seconds worth of beam on the nose of the missile, it was history. This one [third shootdown] went way beyond the threshold of the kill factor. It wasn't any problem that we killed them. We knew that and we could show that to a technical audience. But we also got the data [explosion] on CNN.[26]

The 31 May shootdown was the only one where the missile contained the spotting charge connected to the breakwire. On the next day, the ALL tried and was successful in shooting down its fourth and fifth consecutive missiles minus the spectacular burst of smoke and light. Once again, this was rather an astounding record of reliability considering the complexity of getting all components of the laser system to work together and to maintain an extremely high degree of precision to aim and keep the high-energy beam on the same spot on a moving target.* Beam time on target for the fourth shootdown was 3.6 seconds and 3.1 seconds for the fifth missile shot down. Although in both cases this was sufficient run time to disable each missile, neither shot could rival the first shootdown, for which the beam engaged the target for 4.8 seconds.[27]

---

*Colonel Janicke later commented on the robust nature of the ALL laser. Designed for roughly 500 firings, the laser had fired over 1900 times by the time the ALL program had shut down at the end of 1983. This was an impressive record and conclusively showed that the laser could produce photons on demand.

Telemetry data from the fourth and fifth shootdowns (mission 15045) revealed in each case the breakwire had been burned through. In addition, both missiles showed loss of sensor signal, reticle spindown or complete stopping, seeker gimbals malfunctioning, and distinctive changes in the programmed flight path. The data from the last two missile shootdowns—combined with the evidence gathered from the first three missile engagements—marked a major milestone in terms of building a strong scientific and engineering case that

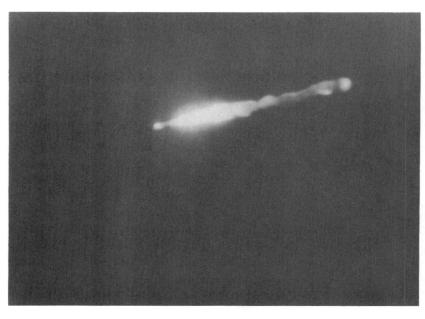

Engagement of AIM-9B missile by laser on 31 May 1983. Laser burn-through set off a small spotting charge that resulted in visual effects of bright light and smoke.

an airborne laser indeed did have the potential to be developed into an operational weapon. The five shootdowns clearly had moved the HEL out of the lofty theoretical physics arena and into the real-world environment of vibrating aircraft and deadly air-to-air missiles.[28]

There were many encouraging signs drawn from the five shootdowns. One was the laser device was reliable and ran almost routinely when the fire button was depressed. However, there were occasions during the laser run time that the device would automatically shut down because of built-in safety considerations. With 150 transducers monitoring every aspect of the complicated gas flow/pressure regulation system, there were bound to be some aborts to keep the system operating within strict safety guidelines. Other instruments measured outside air temperature, airspeed, engine power settings, control and error signals of the tracker, and more. These data were telemetered to 371, which served as a portable receiving station. Aircraft 371 then was able to use these data to closely watch the performance of all systems on board the ALL from takeoff to landing. In the worst-case scenario—the ALL blowing up—the 371 data would be used to conduct the postmortem analysis. Aircraft 371 also was needed as there was no telemetry tracking to follow the ALL as it flew to staging points fully loaded. Plus, 371 allowed the "gerbils," the data reduction specialists in 371, to work the data, freeing the ALL crew to concentrate exclusively on the operation. For example, a low level of diluent caused early shutdown of the laser during the second shootdown.[29]

ADAS posed a second problem. Over the lifetime of the ALL, the ADAS remained a sensitive piece of hardware. Designed to thread the beam from the floor of the ALL up into

the APT, the ADAS experienced some fundamental difficulties. As one final technical report pointed out:

> The ADAS aborts were caused by control instabilities due to high beam jitter and pathological events due to HEL beam effects. For example, beam path sparkle was caused by the laser heating particulates in the beam path. The heated particles glowed brightly, and their emitted light was sensed by the ADAS, caused rapid transients, and led to an abort condition.[30]

But although there were weak links in the overall system, the fact remained that the ALL had achieved an extraordinary technical accomplishment and level of reliability. This highly complicated system had succeeded in generating 400 kilowatts of laser power inside an

aircraft and had managed to weave the high-power beam through an intricate optical system to be focused on a target several kilometers away. During the process, the beam did lose some of its power as it exited the APT with about 250 kilowatts. Additional power losses were experienced by the beam as it moved through the atmosphere. Between 75 and 100 kilowatts of beam power reached the target. That was more than sufficient, as only about 20 kilowatts was needed to kill the target.[31]

It was Janicke's job to inform higher headquarters after each successful shoot-

Diagnostic aircraft 371 team that supported ALL during shootdown of AIM-9B missiles in May and June 1983.

down. That turned out to be somewhat of a delicate balancing act because each general wanted to be the first to receive the much-awaited news. General Odgers, who had headed all of the TAR teams and now resided at Edwards as the Flight Test Center Commander, was the first to be informed of the ALL's success. Janicke burst in and interrupted Odgers's staff meeting on 26 May to advise the general of the ALL's astonishing accomplishments. Later he called General Brien Ward, Director of Laboratories at Systems Command, to pass on the good news. General Marsh was not immediately contacted because he was attending the Paris Air Show. Also, General Forrest McCartney, Commander of Space Division, AFWL's immediate higher headquarters, was notified of the ALL's success.[32]

When Marsh later heard the news, he was "very pleased." The general, after more than a decade of work, had something very special and tangible to advertise on behalf of the command. The science and technology were outstanding and had met every ALL program goal. The Air Force had done a superb job as the integrating contractor. And as pointed out by Janicke, perhaps most important was "the ALL was a good show-and-tell item from the point of view of demonstrating you could hit a 2000-mile-per-hour target with a laser and keep the beam on it. That's a fairly difficult challenge!" Marsh proudly put his sentiments into words and dispatched a congratulatory memo to the ALL team.[33]

Later, Janicke went to the Pentagon to brief General Lamberson, Assistant for Directed Energy Weapons to the Under Secretary of Defense for Research and Development, on the details of the ALL's success story. Lamberson, who had guided the ALL through its early turbulent days, realized more than anyone the scope and difficulty of the ALL's technical progress over the years. The general was truly one of the first who had set and sold the vision of airborne lasers. He had fought hard to keep the ALL program afloat while others were attacking the program as futile and a waste of resources. Indeed, on hearing of the five AIM-9B shootdowns, Lamberson must have leaned back in his chair with a great deal of satisfaction and pride. Quoted in the September 1983 issue of *Lasers & Applications*, Lamberson stated the ALL experiments at China Lake marked "the highest level of system performance yet achieved."[34]

Lamberson and others were especially grateful that there were no accidents during the five shootdowns. Safety had always been the first priority for any mission. However, that record was placed in jeopardy on 2 June, the last day of testing for the 1B scenario.[35]

On 2 June the ALL laser ran for 6 seconds but failed to engage the AIM-9B. Because of problems with the tracking—conscan never acquired the target—the beam missed the missile. Conscan return from the unusually high dust and moisture levels caused conscan to lock on to the clear sky rather than the missile. Two subsequent firings did collect aerosol scatter data to be used to establish conscan track thresholds demonstrating the behavior of the beam return off particles in the atmosphere. These last two tests were aborted early into the laser run time because of tracking problems and malfunctioning of transducers that monitored the performance of the entire ALL system. Although there was some disappointment in not intercepting a missile during the 2 June tests, the ALL crew's spirits were still high. After all, the previous five tests were a huge success in meeting all of the objectives of the 1B scenario and the euphoria from that experience had not worn off yet.[36]

Following the last airborne laser run on 2 June, instead of returning to Edwards, the ALL, as planned, headed home to Kirtland. En route, Otten recalled, "We almost blew up the aircraft," while the ALL crew worked on venting the system down after the last laser run. That involved dumping all of the liquid fuels and gases overboard by pumping gases through the lines.* Another objective of the venting was to reduce the high pressure (about 6000 psi) in the fuel tanks and lines to near ambient pressure as a safety precaution. Toward the end of this process, Otten and others heard a "pop" in the device compartment. What happened

---

*Fuels were dumped out the side of the aircraft through tubular-looking vents in the shape of ram horns. They extended about 15 inches off each side of the aircraft and could be angled in such a way that the dumped fuels would not splash up against the aluminum side of the aircraft.

was one of the FSS bellows—the vacuum-jacketed diluent pressurization line designed to provide system flexibility to reduce bending and thermal stresses—had ruptured and injected pressurized gas into the device compartment. In the process, bits of insulationlike substance (officially identified as "getter" material) exited the vacuum jacket and were wildly thrown about the device compartment. Luckily, because of the timing of the accident, a major catastrophe was avoided. Otten explained:

> If that [bellows rupture] had occurred 10 to 15 seconds earlier in the venting process, because it vented pretty quickly, say, 5 or 6 seconds before that when there was a lot of pressure laying in those lines, that would have been an exciting time. That's probably one of the closest problems to a major in-flight incident that we had. Everyone was on such a high, however. We really didn't realize the magnitude until after we got home. As soon as you land, you went in and you turned off the fire suppression systems, put caps on the halon bottles. They were very dangerous. And so we'd immediately go in, and there were a bunch of safety things that you had. And we opened up the bulkhead door and went in there and said—in fact, I think it was Patty Norris, Chief Norris, who came and said, 'You had a serious problem.' And we had a serious problem as it turned out.[37]

The bellows failure did not detract from the successful shootdowns. Few knew about this near disaster and AFWL was not anxious to advertise the mishap. However, they did fix the problem by going back to Beech Aircraft to manufacture a new bellows. This tasking was buried in the schedule used in preparation for the upcoming Navy scenario. AFWL completed a redesign analysis of the bellows and gave Beech the new specifications. Janicke recollected, "Beech bent over backwards to fix that thing." By the end of the summer, Beech had delivered the new bellows, which was then installed on the ALL and flight-tested successfully at White Sands.[38]

Although the ALL had avoided a serious in-flight accident on the way back to New Mexico, that was overshadowed when the ALL landed at Kirtland. There, an elated Colonel Tony Johnson, AFWL's commander, headed an enthusiastic welcoming party eagerly waiting to deliver their congratulations to the victorious ALL team. As the plane taxied to the front of the ARTF hangar, loud cheers and applause from the large and partisan AFWL crowd greeted the returning ALL crew. One by one, Otten and his team departed the aircraft and tried to navigate their way to the safety of the hangar, only to be intercepted and ceremoniously hosed down by the fire department in recognition of their highly successful mission over the past 2 weeks. No one escaped the deluge, and none of the soaked victims' spirits were dampened as the festivities began.[39]

Inside the spacious ARTF hangar, an informal ceremony took place where the Lab commander praised the laser team for a job well done. Colonel Johnson's comments profoundly captured the spirit of the moment and left a lasting impression that something extra special had transpired recently over the skies of China Lake. "This is a proud moment for the Air Force and the nation," Johnson informed the assembly of attentive onlookers. "I'm proud to be part of an organization that was given a tough job and tackled it with vigor." He was careful to point out to his audience that the ALL was a flying laboratory and not a prototype weapon system. However, this in no way detracted from the scientific and technical accomplishments of the ALL. Indeed, the ALL was a milestone of major proportions. As

Johnson reminded those congregated in the hangar, the ALL's most significant contribution was it furthered the nation's understanding of the technical feasibility of laser weapons. The China Lake airborne experiments, the AFWL commander emphasized, established an invaluable technology data base that will be applied to future laser development efforts.[40]

Johnson was not the only one who praised the outstanding contribution of the ALL and its crew. In contrast to the perceived failure of the June 1981 shootdown, press coverage of the 1983 shootdowns was positive and highly complimentary of the Air Force's progress in the world of lasers. AFWL had learned an important lesson from 1981 in terms of timing the release of information. A complete analysis of all of the data was conducted and rechecked before the Air Force decided on 26 July 1983 to announce its most recent success. AFWL wanted ample time to verify that the data showed conclusively that the five missiles could not complete their mission. Delaying the release of information to the press was done to dispel any premature criticism that the ALL was a failure.[41]

Negative headlines run by the Albuquerque newspapers depicting the ALL as a "failure" in 1981 after the first attempt to shoot down missiles changed dramatically for the good in 1983. The *Albuquerque Journal* reported, "Plane's Laser Hits Missile," while the evening edition ran, "Laser Test: 'Star Wars' air defense system cripples 5 missiles."[42]

There was a similar reaction by the national media in terms of favorable press coverage. Across the country, city newspapers responded in a highly positive fashion to the ALL's most recent success. On 26 July, *The Washington Post* announced, "Airborne laser beam used successfully by Air Force." On the same day, *The Los Angeles Times* informed its readers, "AF Laser 'Defeats' Speedy Missiles in Test: Airborne Result Called 'Major Milestone' in Weapon Development." "Laser Passes Key Test—Missiles KOd," appeared on the front page of the *San Francisco Chronicle*. Success of the ALL even reached across the Pa-

Celebration in ARTF hangar on successful ALL shootdown of AIM-9B missiles. Colonel Otten in foreground and Colonel Bill Dettmer, ARTO director, hands clasped in first row, on hand to share the celebration.

cific to the Japanese newspaper *Asahi Shinbun* that proclaimed "Laser Beam Destroys Missile." *Aviation Week & Space Technology* used two photographs showing a smoking

AIM-9B careening out of control with the description, "Sidewinder Destroyed in Airborne Laser Test."[43]

Bolstered by all of the favorable publicity, the ALL team was ready to pursue its next experimental challenge to try to disable Navy drones with a laser beam. Preparations for this final test began in June. The mood was one of optimism. With the success of the shootdown of AIM-9Bs under their belt, most believed it would be easier to use the airborne laser to disable slow-moving drones.

SHOOTDOWN OF NAVY DRONES—1N SCENARIO

Since the early 1970s, the Navy had been interested in the potential of lasers. As one of the key participants in the first tri-service ground-based laser program, the Navy recognized the development of the laser as a weapon could significantly contribute to upgrading the fleet's antiship missile system. Never considered a stand-alone weapon, the Navy looked to the laser to supplement its arsenal of antiship weapons, to include the Phoenix air-to-air missile system, Aegis/standard surface-to-air missile system, and NATO Sea Sparrows.[44]

Ships are extremely vulnerable to missile attack. Practical experience clearly showed that to be the case. For example, in 1967 an Egyptian SS-N-2 missile sank the Israeli ship *Eilat*. During the battle for the Falklands in 1982, a sea-skimming Argentine Exocet missile slammed into the side of the British ship *Sheffield*, sending her to the bottom. To cope with these very real threats, the Navy depended on its use of conventional missiles to defend its battle groups from incoming missiles. But intercepting a missile with a missile was no easy task. Consequently, looking to the future, the Navy felt lasers offered another viable option to pursue to deal with the dangerous missile threat posed by any adversary.[45]

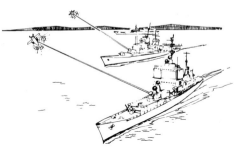
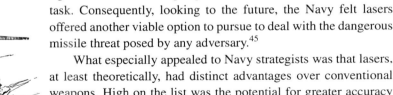
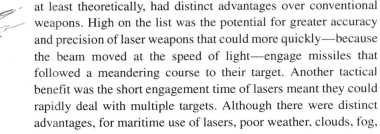

Ship defense against air attack.

What especially appealed to Navy strategists was that lasers, at least theoretically, had distinct advantages over conventional weapons. High on the list was the potential for greater accuracy and precision of laser weapons that could more quickly—because the beam moved at the speed of light—engage missiles that followed a meandering course to their target. Another tactical benefit was the short engagement time of lasers meant they could rapidly deal with multiple targets. Although there were distinct advantages, for maritime use of lasers, poor weather, clouds, fog, and haze all presented major problems for the beam moving through the air. However, this was not as severe a problem for a laser attempting to intercept low-flying cruise missiles (a few feet above the water and below the cloud cover) as it was for a laser trying to counter high-altitude missiles diving at ships.[46]

Although in many cases a shipborne laser would operate below the cloud cover, unique issues had to be addressed with a laser operating in a marine environment. The continual rocking movement of a ship plowing through the sea provides an unstable platform that adversely affects the laser's precision pointing and tracking system. Turbulence over water is less than over land, but the constant motion of a ship kicks up water and sea salt that degrades the quality of the beam as it propagates through the air. To learn more about these shortcomings and overall performance of the laser in the marine environment was the main reason the Navy teamed up with the Air Force's ALL to carry out the 1N or fleet-defense scenario over the Pacific in the fall of 1983. The more immediate objective of the 1N mission was to show a laser aircraft could protect a ship from a low-level cruise missile attack.[47]

In preparation for the 1N scenario, AFWL's leadership assigned a new ALL team to replace the one that had completed the bomber defense (1B) testing at China Lake. Lieutenant Colonel Denny Boesen became the 1N Test Director supervising the activities of approximately 14 scientists, engineers, and technicians aboard the ALL. There were two reasons for this organizational change. First, with the 1B and 1N scenarios being radically different operations in terms of targets and how they were launched, target speed and trajectories, and maneuvering of the ALL, it made sense to put together two highly trained crews to meet the specific requirements of each mission. Second, there was a strong desire by Colonel Bill Dettmer, who headed the laser program at AFWL, and others in the ARTO, to give more people who had worked on the ALL over the years an opportunity to participate in a very important and historic flight test. In short, it was good for morale to spread the work load around.[48]

As with the 1B scenario, numerous airborne practice tests took place over White Sands Missile Range. The objective was for the ALL to rehearse against tow targets mirroring the same flight geometries the ALL crew expected to see in the 1N scenario. Acquiring and tracking the tow target, verifying operation of the laser device and APT, and measuring beam quality and target vulnerability levels were important steps that had to be accomplished before the ALL attempted to shoot down the Navy's self-propelled missile target drone designated the BQM-34A and nicknamed "Firebee." Categorized as an unmanned aerial vehicle, the BQM-34 had been developed in the late 1950s and used as a research target. Manufactured by Teledyne Ryan Aeronautical Company, the BQM-34 weighed 2500 pounds and was powered by a General Electric J85 turbojet engine.[49]

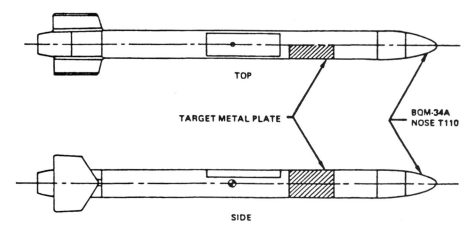

TOP

TARGET METAL PLATE

BQM-34A
NOSE T110

SIDE

BQM-simulated low-cost tow target.

No drones were used at White Sands. Instead, LCTT outfitted with the nose of a BQM-34A were flown to simulate the drone. Also, installed on the forward right side of the tow target's fuselage was a target metal plate made of the same material (0.063-inch-thick aluminum) as the drone. This plate was used to determine if the ALL could deliver the beam to the designated target spot and to calculate burn-through times and vulnerability thresholds to determine the ALL's lethality capability. Behind the outer aluminum plate was another plate made of stainless steel (0.125 inch thick) to duplicate the inner wall of a special fuel tank mounted on the BQM-34A. By measuring the temperature of this inner plate heated by the laser, AFWL workers could determine if the beam produced enough heat to detonate fuel vapors inside the tank. Ground tests indicated that if the temperature of the inside wall of the fuel tank reached 460 degrees centigrade, then the fuel inside the tank would ignite. As the fuel burned, pressure inside the tank increased—this varied depending on a rich or lean fuel mixture. Two Navy ground experiments showed one fuel tank ruptured when it reached a pressure of 318 psi; the second tank burst at 349 psi. AFWL also had conducted a series of vulnerability tests directing the laser beam to fuel tank targets located downrange behind the ARTF hangar. These experiments produced results similar to the Navy tests, showing the laser was capable of detonating fuel tanks.[50]

The series of rehearsal tests in preparation for the 1N scenario took place from 13 August through 9 September 1983 at White Sands. For the first few tests, the OAD operator, Captain Steve Coulombe, had a hard time acquiring the tow target as it approached from the left rear and eventually caught up with and flew parallel to the ALL. Coulombe's job was to look through the OAD (equipped with a zoom lens) and place the cross hairs on the tow target. After he initiated tracking, signals would be sent from the OAD to the APT to drive the turret to the same line of sight as the OAD. Once that happened, the beam control operator would begin tracking the target using the TV tracker in the APT. He next would see the target on his IR video display, at which time he would place the tracking gates on the target. With that accomplished, the APT would automatically track the target.[*][51]

As Coulombe strained to pick up the tow target through the OAD, he could see only a barely visible "washed out" target. The sunshine at White Sands was intensely bright as it entered the small OAD (14 × 15 inches) window on the left side of the aircraft. Once inside the plane, the light reflected off the pale green insulator material fitted up against the inside of the fuselage. The light then reflected onto the inside of the OAD window, which resulted in the washing out of the target image. To cure this problem on subsequent test flights, he draped black cloth over the green insulation around the window. The black fabric absorbed the incoming sunlight and eliminated the false image on the inside of the window that obscured the target. Target detections with the OAD became much more reliable after that.[52]

For the most part, the tow target testing at White Sands went as expected and confirmed the laser and all of its subsystems worked according to plan. For 7 of 14 attempted shots at the target, the laser system performed for the full duration of the experiment. Half of the

---

[*]Even after tracking was turned over to the APT, the OAD continued to follow the target in case the APT broke lock. This was done so the OAD would always be ready to feed target coordinates to the APT for it to quickly reacquire the target.

experiments were aborted because of malfunctions with the ADAS, APT, and computer software. Alignment of the ALL and tow targets flying parallel to and at the same altitude as the ALL made it relatively easy for the ALL to engage its targets. Mainly because of safety reasons, the tow targets were not flown at the low altitude (100 feet) planned for the 1N scenario over the Pacific. Consequently, with the ALL flying at a lower altitude for the 1N demonstration, it would be a much more challenging assignment for the ALL to look down and intercept a self-propelled drone flying 100 feet above the water.[53]

The first two full-duration tow target experiments conducted on 13 August 1983 revealed the airborne laser hit and penetrated the aluminum target plate near the center of the aimpoint. Sufficient energy was deposited to turn the temperature-sensitive paint on the plate black, indicating the temperature had reached 460 degrees. Similar results were obtained on 1 September when the ALL flew for $2\frac{1}{2}$ hours and successfully intercepted two tow targets. Again, the beam bored through the target plate and burned several Teflon-coated wires making up the sensor instrumentation package. The burning wiring produced a drop to near zero volts from the telemetry system, which verified burn-through. Both targets were recovered and later examined on the ground. Inspection provided visual evidence of the hole punched in the target by the laser, which matched airborne telemetry data collected by the diagnostic aircraft (371).[54]

In sum, the White Sands tests were important because they showed target heating and surface burn-through times were of sufficient magnitude to verify the ALL's ability to kill its target. Operation of the laser device and beam control system, beam quality, and energy on target were well within the ALL's desired performance parameters. Boesen's team, with the completion of the White Sands flight experiments, had attained positive results and were reinvigorated with a renewed sense of confidence as they headed off for Edwards AFB to prepare to shoot Navy drones out of the sky off the coast of California.[55]

The ALL landed at Edwards AFB on 16 September. Edwards, as had been the case with the 1B testing, served as the staging area for the ALL during the 1N scenario. As the 1N testing was a cooperative effort between the Air Force and Navy, personnel from the Naval Weapons Evaluation Facility (Kirtland AFB) worked closely with AFWL coordinating the details of the launch and flight control of the drone with Navy range controllers at Point Mugu. (The Navy maintained an inventory of drones to use as targets for gunners on ships and planes to shoot at.) Besides range control support, the Navy modified each drone by installing instrumentation to measure the effects of the laser beam.[56]

The overall objective for the 1B and 1N scenarios was the same: destroy aerial targets with an airborne laser. However, there were major differences in the targets. The 1B engagements involved directing the beam to intercept high-speed air-to-air missiles. Launched from a fighter aircraft, each AIM-9B Sidewinder missile was a relatively small target that did not travel in a straight line. This made it extremely difficult to acquire, track, and precisely point the beam so it hit exactly on the desired target spot. The ALL had only one chance to fire at the missile. Plus, this was a more dangerous mission because the missile was aimed at the ALL aircraft.[57]

For the 1N fleet-defense scenario, the complexity of operating the laser device and directing the beam from the ALL was identical to the 1B scenario. However, the drone target

was radically different from the AIM-9B missile and its flight pattern. The drone was ground launched (with rocket assist) off a rail at Point Mugu. A ground controller tracked the drone on radar and used a joystick to fly it. It was similar to flying a model airplane.

The large (23 feet long, 3 feet in diameter, with a wing span of 13 feet) and bright orange, slow-moving (subsonic with a maximum speed of about 690 miles per hour at 6500 feet) drone was a relatively easy target to acquire, track, and hit with a laser beam. Once 20 miles out at sea, the drone flew a racetrack pattern. Unlike the AIM-9B missile, the drone had enough fuel to remain airborne for about 25 minutes. This gave the ALL three or four chances to track and fire its laser at the drone. If something went wrong on the first pass, the drone would continue on its racetrack course and come around for a second pass so the ALL (also flying a matching racetrack pattern) would have another opportunity to acquire, track, and engage the target. This was an added luxury that was not available during the 1B scenario.[58]

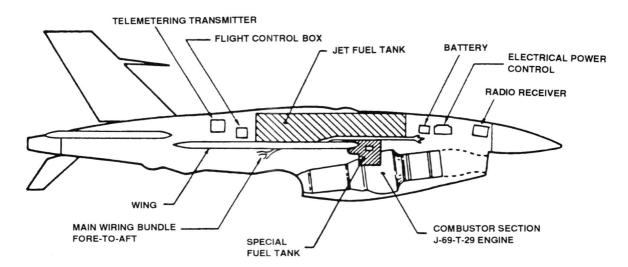

BQM-34A drone cutaway.

AFWL personnel followed the same step-by-step ground procedures used for the 1B scenario to get the ALL ready for each of the 1N airborne experiments. A few hours before midnight, technicians began checking valves controlling the gas and water flow, as well as realigning mirrors in the beam control system. By 1:00 A.M., liquid nitrogen was being circulated through the laser system. That was done to cool the system so when the laser fuels were pumped on board over the next 3 hours, those fuels would not evaporate. By 5:00 A.M., the ALL technical crew began arriving to conduct their preflight checks and preparations. Coolant lines were disconnected an hour before takeoff. Usually by 7:00 A.M. the ALL was airborne and heading out to take its position to get ready to rendezvous with the drone.[59]

For the first 3 days of rehearsals (16, 19, and 22 September), the ALL took off from Edwards to practice getting into its racetrack formation and entering its designated "firing box" at the right time. The ALL team turned on the laser as well to test the reliability of the system, although no beam was directed out of the aircraft. For a couple of practice runs, a drone was launched from shore to determine flight times to its position and for the ALL crew to observe its flight behavior. In addition, the ALL used this time to practice acquiring and tracking the drone as it made several loops around its aerial racetrack. By the end of these 3 days of practice runs, Boesen and his team felt well prepared and were anxious to get on with the first laser shot against the drone.[60]

On 23 September, after several practice orbits to check all systems, Boesen's team was on station eagerly awaiting the arrival of the drone to proceed with ALL mission 15063. The OAD operator knew exactly the direction the drone would be coming from, and in less than 5 minutes after launch, he had the drone in his sights. Handover of tracking from the OAD to the APT occurred on schedule.[*] A few seconds after turning on the laser, problems with the autoalignment system that steered the beam through the aircraft forced shutdown of the ALL system. The drone eventually ran out of fuel and dropped into the water. Although the ALL never got a shot off for its first attempted drone shootdown, technicians on board were able to quickly make minor adjustments to the beam autoalignment system to get the ALL ready to engage the next drone.[61]

With the second drone launch of the day, the ALL again was able to acquire and track on the nose of the target in relatively easy fashion. Flying at 100 feet above the water at a speed of about 550 miles per hour, the drone was approximately 900 to 1000 feet below and parallel to the ALL at a distance of about 2 kilometers off the aircraft's left wing. At a predetermined point in its flight path, the ALL banked slightly (10 degrees) to the left so the laser had a clean shot at

BQM-34A drone prior to launch.

the target. When the laser fired, the beam immediately hit its target spot, a special stainless-steel fuel tank (to simulate the fuel mixture of a Soviet cruise missile) mounted in the fuselage

---

[*]For the 1N scenario, centroid tracking was used. Once the track point was established (either on the nose or body of the drone), the tracking algorithm calculated the location of the aimpoint (a fixed distance from the track point) for the laser to shoot at. The aimpoint was either the special fuel tank or flight control box.

just forward of the wing root. The wing root was the area where the wing attached to the fuselage.[62]

When the beam hit slightly high and to the left of the fuel tank aimpoint, a brief flash was observed. The beam remained on the wall of the tank for several seconds, long enough to raise the temperature to ignite the fuel vapor inside the tank. This caused the fuel to detonate, but the explosion was not of sufficient force to rupture the tank.[63]

The most plausible explanation of why the tank failed to split open was traced to the composition of the fuel mixture in the special fuel tank. Understanding the combustion process of a rich level mixture was a complex and tricky business. Temperature and pressure instruments on the drone provided data that indicated the temperature and pressure inside the tank were not high enough to cause one big explosion. Instead, there were a series of intermittent miniexplosions that prevented the pressure from building up to one big bang that would destroy the fuel tank. Earlier ground tests had shown that a fuel tank containing an overly rich mixture would have to reach the vicinity of 320 psi and higher to produce a catastrophic explosion. Instrumentation on the drone revealed peak pressure inside the special fuel tank reached only about 300 psi.[64]

From an experimental point of view, this first drone engagement was considered at least partially successful. The ALL did generate a laser beam on board the aircraft and was able to direct it to intercept the drone at the selected target spot. There was enough energy deposited on the target to cause the vapors inside the fuel tank to explode, but the force was not great enough to rupture the tank. Consequently, the drone continued to fly and showed no signs of deviating from its designated flight plan, meaning it would be able to complete its mission. The drone eventually ran out of gas and landed in the water and sank because of a malfunction with its watertight flotation compartment. It was never recovered. In the final analysis, even though the laser intercepted the target, this first engagement was not rated as an official "kill" because the laser failed to "destroy" the drone.[65]

Boesen's ALL team certainly did not interpret the results of the first attempted drone shootdown as a failure. From a scientific point of view, the ALL had managed to accomplish what the system was designed to do—accurately project a beam from a moving aircraft to hit an aerial target. The fact that the laser did not meet the goal of destroying the drone, because of the overly rich fuel mixture, did not detract from the overall operating performance of the airborne laser system. In any case, ALL scientists and technicians were eager to get on with the next test to demonstrate the laser's lethality against a second drone.[66]

Three days later on 26 September, the ALL (mission 15064-1) again attempted to place the beam on the drone's special fuel tank. This time the tank had a leaner fuel mixture to improve the chances of a large explosion that would rupture the tank. However, the anticipated explosion never occurred because the ALL was late in firing the laser at precisely the predetermined time because of a software glitch.[67]

Instead of striking the designated aimpoint on the center of the outside wall of the special fuel tank, the beam hit the aft section of the tank. But the beam almost immediately shifted off the tank. A Navy ship recovered this second BQM-34A from the water. Based on the physical evidence, the drone showed that the beam had mostly dwelled on and had torn a hole in the wing root. Most of the beam's energy caused damage to the front edge of the wing. A portion of the beam had spilled over from the wing to the fuselage and for a brief instant penetrated the outer skin of the fuselage. But the beam did not remain on the fuselage for very long, as the heat of the beam barely scorched the foam insulation on the inside wall of the drone.[68]

Although the beam had missed the exact target point, the laser did cause structural damage to the drone. However, the damage (similar to several small-arm

Beam striking wing root of drone on 26 September 1983.

rounds penetrating the drone) to the wing root and fuselage was not significant enough to cause the drone to fly out of control and crash into the sea. The BQM-34A's flight control system was not affected and was able to make adjustments (through its aileron control movement) to compensate for the damage to the drone. So for a second time, the ALL had managed to hit the drone, but had been unable to inflict sufficient damage to prevent the drone from accomplishing its mission. Therefore, this airborne experiment could not be categorized an official kill, even though the ALL was successful in hitting the drone with a beam.[69]

Colonel Janicke recalled that the recovered drone with damage to the wing became "a beautiful show and tell." He also lamented, "We really destroyed the heck out of that BQM, but it was labeled a failure because it didn't follow any of our success charts."[70]

Up to this point, the ALL was zero for two in terms of killing the drone. But for the second shot (mission 15064-2) at a drone on 26 September, the ALL team had one more opportunity to prove the ALL was capable of destroying a drone. For this last shot in the series, the ALL tracked on the body of the drone (instead of the nose) and aimed the beam at the flight control box (instead of the fuel tank). The flight control box, located on top of the fuselage two-thirds to the rear of the nose, contained the wires and circuitry that controlled the flight pattern of the drone.[71]

Recovery of drone from ocean with damage to wing root.

As the drone came into view, the ALL began tracking on the fuselage and calculated the aimpoint offset based on the tracking location.[*] This time the ALL fired at precisely the right moment at a range of about 1.8 kilometers and placed the beam on the flight control box. Almost immediately, the flight control signals telemetered from the drone to the diagnostic aircraft (371) monitoring the drone were severely disrupted. These perturbations in the telemetry signals were caused by the laser burning through numerous wires in the flight control box, which caused multiple circuit failures.[72]

Burn-through of the fuselage skin into the flight control box's sensitive electrical components by the laser immediately caused the drone to lose control. From the time the beam hit, telemetry data revealed the drone began veering off its predetermined course. Those on board the ALL observed (also captured on film) that as soon as the laser punctured the flight control box the drone began rapidly rocking back and forth. After a few seconds, the drone made a hard 90-degree roll to the right and abruptly made a sharp pitch down and crashed into the sea.[73]

As the drone nose-dived out of control and splashed into the ocean, there was a great deal of excitement and satisfaction on the part of Boesen's ALL team. Aboard the ALL, the excitement of that once-in-a-lifetime moment was captured on the audiotransmission tape. As the beam hit the drone, one ecstatic voice rose in a crescendo above all others, "All right! I think that was it. Boy, if that didn't do it, nothing is going to!" There was absolutely no doubt this third and final laser shot at the drone resulted in a "kill." Both visual evidence from those aboard the ALL and film taken of this event, as well as telemetry data confirmed the lethality of the beam. In addition, the operator on the ground who controlled the flight of the drone using a joystick reported he lost total control of the drone once it was struck by

---

[*]For the first two laser shots at the special fuel tank, the target aimpoint was calculated by measuring a fixed distance back from the nose of the drone. For the shot at the flight control box, the ALL tracked on the fuselage of the drone and computed the distance from that point back to the control box.

the laser. It was clear, a few seconds after the beam arrived at the control box, the drone abruptly changed course and was unable to continue to follow its intended flight path to complete its mission.[74]

Depending on how one kept score, the 1N scenario was either a complete success or at least a partial success. From a technology standpoint, the ALL team batted 1.000 for delivering a laser beam from an airborne platform to an aerial target. For two of the three shots, the beam hit the designated target spot on the drone attesting to the accuracy and precision of the ALL operating system. On the other shot, the beam missed the target aimpoint and hit the wing root.[75]

But getting the beam to the target only achieved half the 1N scenario objective. The ultimate goal was to disable or destroy the drone. For only the third shot was the ALL's beam able to cause substantial structural and electrical damage to the drone to prevent it from completing its mission. That made the ALL one for three in terms of drone kills. Three for three would have been better to demonstrate consistency and effectiveness, but the final shootdown still was an extremely important first step in proving that it was scientifically feasible to engineer and use a complex laser system to destroy aerial targets.[76]

On learning of the completion of the 1N scenario, General Marsh was relieved because he no longer had to fight to get funding each year to sustain the program. It was hard to argue for money in recent years in light of recurring ALL technical problems and schedule delays. Plus, some critics in Congress, the Office of the Secretary of Defense, the Office of the Secretary of Air Force and Air Staff, and technical advisory groups began to more closely question the wisdom of investing more dollars—especially since 1981—in a program that still had not proven itself. Scattered throughout these organizations, as Marsh recalled, people were getting "itchy" to bring the ALL to a conclusion quickly. In spite of growing resistance to the ALL in some quarters, Marsh noted he didn't think "anybody of importance ever came out and said let's kill the program, like the Chief of Staff or the Secretary of the Air Force. Nobody that I know of said 'let's kill it.'"[77]

Over the last few years of the ALL program, Marsh had managed to fund the ALL at an annual rate of about $30 million. However, with the end of the 1N experiments, the constant burden of fighting for dollars was lifted from Marsh's shoulders. In his mind, achieving the goals of intercepting the Navy drones signaled the successful end of the ALL program.[78]

Marsh, ever the realist, recognized the ALL did not go out with a big bang. The 1N scenario was anticlimactic in comparison with the earlier AIM-9B shootdowns. Denny Boesen, the 1N Test Director, perhaps best summed up the sentiment at the time when he recalled, "Nobody was as excited about the Navy scenario." Maximum energy and attention had been expended by the Air Force to advertise the success of the more difficult 1B scenarios because those experiments represented the first time an airborne laser had destroyed real-world missiles. Those tests were extremely demanding in demonstrating advanced pointing and tracking techniques against supersonic targets. Against large subsonic drones *simulating* cruise missiles, everyone came to expect the ALL would have little trouble intercepting the BQM-34A targets. As Marsh put it, "If you expect something to happen for a long, long time and talk about it and publicize it well, then when it happens it's not nearly as 'grabby.'"[79]

Marsh's assessment was correct. In comparison with the 1B shootdowns, the 1N accomplishments received little press coverage at the local or national level. To make certain it had all of its facts straight and data analyzed, the Air Force did not release information about the 1N flights to the press until 28 November. Only a few papers provided any in-depth coverage.

The response was more positive at the Air Force level. General Marsh sent a congratulatory message to the ALL team for their recent success marking the completion of the ALL Cycle III test program. He praised all who participated in the 1N work effort, stating, "The technical achievements obtained during this program will significantly benefit our high-energy laser technology program. The dedication and ability of all those involved with the ALL reflect very favorably upon the USAF, AFSC, and the Air Force Weapons Laboratory. Please thank everyone for a job 'well done.'" Colonel Johnson, AFWL commander, ensured Marsh's message reached every member of the ALL team.[80]

Colonel Janicke endorsed Marsh's interpretation of the value of the 1N experiments. Janicke remarked on the 1N events in a local newspaper article, "I'm pleased with this accomplishment. This test culminates 12 years of hard work on this project. During this past year we've seen all the necessary parts come together to let us perform two major proof-of-concept demonstrations."[81]

With the end of the 1N demonstrations, General Marsh wanted AFWL to begin taking steps to shut down the ALL program as soon as possible. The ALL had served its purpose as it had convincingly proved that an airborne laser could shoot down aerial targets confirming lasers had the potential to be turned into weapons. However, Marsh recognized there was no chance for a follow-on ALL program. The political climate would not permit it. With the establishment of President Reagan's Strategic Defense Initiative Organization (SDIO) in March 1983, there was a shift in strategic emphasis for research and development efforts to be directed at the potential for space-based lasers. Plus, Marsh wanted to put the ALL to bed because he was consumed with higher priorities, such as the B-1, B-2, and F-117 programs. Consequently, Marsh and others knew there was no groundswell of support to move ahead with a second-generation airborne laser, mainly because it would be unrealistic to expect the funding would be available. The general feeling was the airborne laser had run its course, at least for the time being.[82]

ORDERLY TERMINATION

It was clear by the early summer of 1983 that the ALL program would come to an "orderly termination" on completion of the 1B and 1N flight demonstrations. The House and Senate Armed Services committees had recommended funding for the ALL would cease in FY84, although $10 million would be included in the budget to cover costs to bring the ALL to a conclusion. This money would be used, in part, to pay for a series of limited aero-optics flight experiments (costing $5.3 million) to collect additional data on flow field characteristics around the turret. Costs to use the White Sands Missile Range, TDY and per diem for supporting contractors, and charges for flying and maintaining the ALL by the 4950th, all figured into the $5.3 million. After the aero-optics tests, other funds would be needed to remove the APT and other equipment from the plane, as well as making arrangements to store the airplane, and working with the contractors in preparing the various technical reports to document the final phases of the ALL flight experiments.[83]

One portion of the "wrap-up" tests involved propagating a low-power laser from the ALL to collect additional data to better define beam characteristics to include power, size, and quality. The GDL high-power laser was not used for any of these experiments. In December, the APT was removed from the ALL and replaced with the dummy turret to be used during the aero-optics testing. This effort conducted at White Sands Missile Range, under the direction of Steve Coulombe, consisted of designing four different test configurations to map the flow field around the turret.[84]

For this flow field testing, Coulombe explained "wing structures" were installed next to the dummy turret on top of the airplane. These wing structures contained motorized transport mechanisms that moved anemometer probes that measured the speed and turbulence of the airflow at various locations around the turret. For each of the four test configurations, the plane was flown at different speeds and altitudes in an attempt to get a comprehensive mapping of the airflow under a variety of flight conditions. There were no spectacular breakthroughs in the nature of the data collected. Essentially, the

Probe drive assembly position in front of APT.

data served to verify and refine aerodynamic scaling laws developed over the last 10 years. By March, these tests reached completion and signaled the close of the ALL flight experiments.[85]

Authorization for the wrap-up testing was approved by Systems Command as part of the second of two options for the ALL phase-down plan. Lieutenant Colonels Otten and Alexander Halber, Chief of the Laser Device Branch, had briefed AFSC in September on two choices outlining what should be done with the ALL. The first option involved "demodifying" the aircraft by removing and salvaging all of the systems on board (e.g., APT, GDL, FSS, ADAS). Once that was completed, the plan called for returning the ALL and diagnostic aircraft either to AFSC's research and development fleet or to the Air Force's airplane "boneyard" at Davis-Monthan AFB, Arizona.[86]

The first option was discarded in favor of the second proposal for the ALL phase-down, known as the "caretaker status" option. This approach called for carrying out the final aero-optics testing. Once that was finished, the ALL would be stored in its hangar for 3 years and be maintained by the 4950th Test Wing. Key components, such as the airborne beam control system (ABCS), would be removed from the ALL and used for experiments to support other programs. When not being used for experiments, the ABCS and its specialized support equipment were to be kept in the APT laboratory in a "controlled storage" status.[87]

By preserving the ALL in a caretaker status, the Air Force would have the flexibility to resurrect the ALL within 1 year to its Cycle III flying configuration in response to future national policy requirements. For instance, with the creation of SDIO in 1983, there was the possibility that the ALL might be reactivated to participate in flight demonstrations to collect beam control data that could be used in support of advancing laser technology in the space arena.[88]

In February 1985, a memo from the Office of the Under Secretary of Defense for Research and Development (OUSDR&E) urged Air Force Systems Command to contact all of the military services, Defense Nuclear Agency, Defense Advanced Research Projects Agency, and SDIO to determine if there were any research requirements they might have that could be satisfied by the ALL in its current or modified condition. The OUSDR&E memo emphasized they wanted to pulse the system to "re-examine our requirements and future plans before we move past the point where this resource [ALL] is lost forever."[89]

SDIO had briefly flirted with the idea of bringing back the ALL to collect near-term data relating to the general categories of beam control, tracking, and target imaging. However, none of these plans to reactivate the ALL materialized. Part of the reason was the high costs involved. Estimates ranged from $85 to $148 million to get the ALL to acceptable operating standards and to fund follow-on testing. Plus, time was a factor. AFWL estimated it would take about a year to restore the ALL once the decision was made to go forward. In the end, SDIO decided against resurrecting the ALL, as did the military services and other government research agencies.[90]

With little interest expressed to use the ALL to support other programs, at least in the near term, AFWL parked the aircraft in the hangar (building 760) at Kirtland. During its time in storage, Detachment 2 of the 4950th Wing maintained the aircraft in a "semi-operational" status. The ALL was washed and lubricated before it went into storage. Covers were installed on aircraft engine and air-conditioning openings to prevent birds and small animals from nesting in those areas. Fuel remained on the aircraft to prevent fuel cell deterioration and for periodic engine runs every 45 days. Every fourth engine run, the water injection system was serviced and operated. At regular intervals, the air-conditioning system was turned on and checked and the tires were rotated.[91]

These regularly scheduled storage procedures remained in place over the next several years until the spring of 1988 when the ALL underwent a complete preflight checkout in preparation for its last flight. At the urging of Major General Lamberson and others who had been associated with aircraft over the years, Systems Command decided the ALL's final and most appropriate resting place would be the Air Force Museum at Wright-Patterson AFB located in Dayton, Ohio. There the ALL would be put on permanent display for the public to view and admire. Although the ALL was a lasting testimony to advances in aviation and science, those who would pass by and read the brief description of the ALL's accomplishments could only begin to understand the enduring contribution of one of the most unusual aircraft ever flown by the Air Force.[92]

The "flyaway" ceremony for the ALL took place on 4 May 1988. At 10 o'clock in the morning, a crowd of several hundred jammed into the ALL hangar to witness the final farewell to the ALL. Many familiar faces from the past showed up for the festivities, including General Lamberson, Colonels Otten, Dettmer, Boesen, Janicke, and a vast array of other military, civilians, and contractors who were so instrumental in ushering the ALL through its lifetime. Colonel Otten served as master of ceremony. After introducing the distinguished guests in the audience, he presented the keynote speaker, Demos Kyrazis, one of the most inspirational leaders in the ALL program.[93]

With the large hangar doors open, the ALL was parked just outside for all of the audience to see and provided a fitting backdrop as Kyrazis approached the podium. Kyrazis, now a civilian, gave a stirring speech reflecting on the truly unique contributions of the ALL. After giving a brief history of the ALL spiced with humorous anecdotes, the retired colonel turned more serious in looking back and trying to come up with reasons to explain the success of the ALL. He posed the question:

> What, then, brought about our success? The answer is simple. It was the commitment of the people, both in Government and Industry, who made this program succeed. But before people will commit, they must have a vision of what must be accomplished and the value of that accomplishment. The man who gave us that vision and taught us why we must achieve that vision was Don Lamberson, and the people on this program responded to this vision.[94]

Although Lamberson set the course and tone for the ALL program, Kyrazis cited other reasons responsible for the success of the ALL. Personal integrity of the work force was at the top of the list. Throughout the entire program, scientists, engineers, and technicians were willing to admit mistakes and correct them and to stop any part of the operation that posed a safety hazard. Highly pressurized toxic fuels and the operation of a laser on board an aircraft created the potential for an in-flight explosion. Because of these conditions, safety was al-

Flyaway ceremony conducted in ARTF hangar with ALL in background.

ways rated the number one priority. That concern over the years resulted in an incredible safety record—no one was killed or seriously injured.[95]

Kyrazis also pointed to the chain of command as another factor contributing to the success of the ALL. From the highest level of government starting with the Office of the

Secretary of Defense down through the Secretary of the Air Force, Headquarters Air Force, and Systems Command, AFWL received the support it needed to complete the ALL program. Although there were disagreements from time to time, the higher headquarters supported the decisions of the AFWL technical people and worked hard to ensure the necessary funding was forthcoming. Plus, the close and trusted relationship and teamwork established between the AFWL team and contractor personnel proved indispensable for getting all of the diversified subsystems to operate as one unit.[96]

An enthusiastic crowd of well-wishers salute the ALL as it departs on its final flight to the Air Force Museum at Wright-Patterson AFB in Dayton, Ohio.

On conclusion of Kyrazis's remarks, the crowd reassembled outside the hangar and observed the ALL as it started up its four jet engines and taxied to the end of the runway poised for takeoff. Moments later the NKC-135 rumbled down the sun-drenched New Mexico runway and slowly lifted off belching black smoke from its water-injected engines. The ALL circled to make one final pass before the cheering crowd as its final salute, symbolizing the end of an era. The excited congregation of bystanders continued to cheer as the ALL gained altitude and disappeared over the horizon en route to Dayton for its final resting place at the Air Force Museum.[97]

Epilogue

Major General Donald Leslie Lamberson strained and squinted to fight off the bright New Mexico sun as he observed, for the last time, the ALL aircraft slowly fading into a distant speck and then vanishing in an instant over the deep-blue skies of Albuquerque. At that moment, a local TV reporter walked up and asked the general what all of the commotion was about. Lamberson smiled. Very patiently in his deep, husky voice, the father of the ALL responded to the young reporter in the same understanding tone an eminent college professor would use in trying to get a not so obvious point across to one of his most promising graduate students. "The ALL," Lamberson proceeded, "has done all the missions that we asked it to do. It proved we could take a high-energy laser, put it in an aircraft, could fly it, could direct the beam against targets and destroy those targets. The fact that this was successful has led now to many other applications of this technology." The reporter, although unfamiliar with the science and technology of the airborne laser, clearly sensed the ALL represented a major watershed in the history of aviation and science.[1]

Lieutenant General James A. Abrahamson, Director of the Strategic Defense Initiative Organization in the Pentagon, echoed Lamberson's sentiments. In a congratulatory letter praising everyone who worked on the ALL program, Abrahamson wrote:

> The delivery of the Airborne Laser Laboratory to the Air Force Museum is a fitting tribute to the men and women who made it possible. The technical accomplishments you demonstrated still serve as a benchmark to judge the integration of modern technologies into successful test products. Even after five years the ALL has been in storage, no other high energy laser system has come close to being such a complete weapon.

Abrahamson was perhaps somewhat premature in suggesting the ALL was close to a "complete weapon." However, the ability of the ALL to shoot down missiles had a profound influence on the Air Force to move forward with the conception and initial planning and preliminary development of a second-generation airborne laser in 1991.[2]

This second-generation airborne laser (ABL) grew out of conditions that the military had faced in the Gulf War. One of the lessons from that conflict was that the United States needed to find a better way to locate and intercept theater ballistic Scud missiles. Although the Army's Patriot antimissile system was capable of engaging Scud reentry vehicles in their terminal phase of flight, the number of Scuds hit by Patriots was much lower than originally reported. What the Air Force proposed, to complement the Patriots, was the development of a short-wavelength chemical oxygen–iodine laser, with a nose-mounted pointer and tracker, designed to fly on a 747 aircraft. This new precision ABL weapon system would be capable of locating and taking out Scud missiles in the boost phase, the most vulnerable segment of the flight trajectory.[3]

The second-generation airborne laser (ABL).

Theater ballistic missiles will continue to grow in popularity, especially for developing countries that desire to wield influence over neighboring states, yet are unwilling or unable to maintain the economic burden of supporting a modern air force. An increased threat of theater ballistic missiles is very likely if no credible defense exists against them.[4]

Air Combat Command played a major role in urging the Air Force to develop and test the next-generation ABL. Space & Missile Systems Center, Los Angeles AFB, and Phillips

Laboratory at Kirtland AFB, New Mexico, are leading this effort. Since 1994, two competing defense contractor teams—Boeing and Rockwell International—developed separate proposals defining engineering design concepts for integrating a high-energy laser with precision pointing and beam control options on an aircraft. On 12 November 1996 General Ronald R. Fogleman, the Air Force Chief of Staff, announced at a ceremony held at the Pentagon that the Boeing Defense and Space Group, Seattle, Washington, was selected as the winning contractor to proceed with building an airborne demonstrator that will be flight-tested to evaluate its ability to acquire, track, and kill a boosting theater ballistic missile. TRW (Redondo Beach, California) and Lockheed Martin (Sunnyvale, California) were the other two key members of the Boeing team.[5]

Award of the $1.1 billion contract requires Boeing to produce a newly designed YAL-1A, a prototype attack laser-aircraft using a commercial 747-400F airframe. The schedule calls for Boeing to deliver the aircraft to Wichita, Kansas, in the spring of 1999 so it can be modified to accept the laser and associated equipment. Flight testing of the new aircraft will follow with expectations that the ABL will destroy a boosting theater ballistic missile by 2002. Once that occurs, the Air Force will issue a follow-on contract to build the first three operational aircraft by 2006. By 2008, the plan calls for four additional ABL aircraft to enter the Air Force inventory.[6]

Laser leaders reunite at the 13th annual ALL reunion held in Albuquerque on 23 November 1996. Left to right: Don Lamberson, Demos Kyrazis, and Bill Dettmer who led the ALL program through both good and bad times. At far right is Colonel Dick Tebay, the heir apparent of the ALL, who heads the second-generation airborne laser (ABL) currently under development.

One of the most important reasons why the Air Force moved forward with such conviction and confidence in deciding to invest in the second ABL was the ability to draw on the invaluable scientific and technical data base derived from the first airborne laser demonstrations. Using the ALL data as a starting point, the creators of the ABL at Phillips Laboratory are in a much better position to assess the best direction to follow for developing the most advanced type of laser, a sophisticated pointing and tracking system, engineering integration designs, and the best aircraft suited for the second-generation ABL.

Colonel Richard D. Tebay, System Program Director for the Airborne Laser Program at Kirtland AFB, consistently has praised the ALL team for their invaluable contributions in advancing airborne lasers. Addressing the large crowd assembled for the 13th Annual Airborne Laser Laboratory Reunion held in Albuquerque on 23 November 1996, Tebay underscored the unusually strong ties existing between the ALL and ABL programs. In paying tribute to the people who made the ALL work and in recognition of the historical importance of that program, Tebay proudly proclaimed that "without the ALL it would have been impossible to have gained the high-level support to move forward with the ABL."[7]

Success of the ABL offers the prospect of significantly reducing casualties and death during war. An effective ABL system catching Scuds in their boost phase will prevent their deployment of warheads. If the laser intercepts the Scud shortly after it lifts off, the missile and its warheads will not be able to complete its mission and land at the desired target location. (Because the missile debris would most likely fall back on or near the launch site, the enemy is forced to consider the risk of mass fratricide if they employ chemical, biological, or nuclear payloads.) Warheads, disabled during the first part of the missile's flight, would never arrive on target and thus the loss of human life would be significantly reduced. Plus, the long operating range of the ABL will allow it to perform its mission from friendly air space to also reduce the chances of friendly casualties.

There is no doubt that the scientific and technical bricks and mortar of the ALL laid the foundation for any follow-on airborne laser. The most important legacy of the ALL was that it represented a vital and irreversible first step. As with the evolution of any revolutionary system, such as the airplane and ICBM, transition of the ABL to an operational weapon system sometime in the next century will be determined by small incremental gains and refinements to each succeeding airborne system. When that threshold is crossed, the airborne laser of this study will rank on the same order of magnitude, technical merit, and historical significance as the Wright flyer.

Robert H. Goddard, who pioneered U.S. rocketry, reminded his advisors from the Carnegie Institution of Washington and the David Guggenheim Fund in 1931 that venturing into unknown scientific territory to test theories through experiments and tests was often a frustrating and time-consuming process. Results did not come quickly. Problems arose because of all of the uncertainties associated with breaking new ground. In trying to explain why progress was slow with his work, Goddard stated, "The liquid-propellant rocket is a new problem in research and is entirely unlike designing and constructing a special engine of a type in common use."[8]

The reality was that Goddard was a very long way from providing a reliable operational rocket in 1931, but by the late 1950s the United States was routinely mass-producing ICBMs.

Similarly, the ALL, like Goddard's early test rockets, made slow but steady progress. One by one, scientific and technical barriers were knocked down. These essential first steps eventually led to the development and successful flight testing of a more proficient airborne laser.

For any future system, there are valuable lessons learned from the unique ALL experience. First, the advancement of the science and technology in the ALL program provided the basis for believing that it would be possible to eventually field an ABL weapon. The evidence supports that the ALL began that process. Expanding the technology base in the areas of laser devices, a reliable pointing and tracking system, optics, fluid supply systems, engineering integration to fit the system on board the aircraft, safety, turbulent flow studies, data collection and analysis, and more all were major contributions.

Advancing the technology and ensuring that all subsystems worked as one harmonious unit resulted from the combined efforts of the ALL team made up of highly motivated and qualified military and contractor scientists and engineers. Not only were they able to get each subsystem to perform to operating specifications, but they also succeeded in designing and engineering the total system into the relatively cramped space of an NKC-135 aircraft.

All of these accomplishments showed the United States was clearly the leader in advancing the physics and technology of HEL systems. More specifically, the Air Force laboratory system was in the forefront of this effort and was responsible for demonstrating the first proof of concept for airborne lasers. The evidence of that was the first-ever demonstration of the successful interception and destruction of five AIM-9B Sidewinder missiles by a HEL on an aircraft and the first-ever interception and destruction of a simulated antiship missile (Navy BQM-34 drone) flying at low altitude over the ocean. These events unequivocally showed the potential for developing and using HELs as airborne weapons to acquire and destroy a variety of aerial targets, to include theater ballistic and air-to-air/air-to-ship missiles and aircraft.[9]

It is important to note that the ALL process reinforced the position that after World War II government laboratories definitely shaped and defined scientific research designed to lead to military applications. There is no doubt that without the military's influence to explore the feasibility of harnessing directed energy as a new class of weapons, there would have been no ABL program. But one of the distinguishing features of the ALL experience was that it deviated from the traditional view that civilian scientists and engineers were mainly responsible for breakthroughs of major proportions. In the case of the ALL, a talented and highly educated military work force accounted for a large share of the scientific and technical success of the program, especially in the areas of beam quality and beam control, gas dynamic flow, turbulent airflow analysis, optics, conical scan algorithms, fluid dynamics, and system integration.

In sum, the ALL left an indelible mark on modern science and technology. Substantial risks were inevitable in a program of this type. But by definition, taking risks and betting on the long shots to come in was the business of laboratories. Before celebrating a single success, scientists and engineers had to endure numerous experimental failures. As Lamberson in the early years of the ALL was fond of saying, "No one knew where the research would take us."

By 1983 it was clear where the research had ended for the ALL. Although it took longer than expected, the AFWL team had advanced the science and technology of lasers to the point where a complex airborne system conclusively demonstrated a laser could shoot down aerial targets. In the process, those military, civilian, and contractor workers gained the expertise and knowledge to become the leaders in the laser community. It was this cadre of people who went on to apply their knowledge to support numerous Strategic Defense Initiative programs and the second-generation ABL born at Phillips Laboratory. Today, a large fraction of the leading scientists and engineers in the ABL and other high-energy systems are "graduates" of the ALL design school.

Success in any endeavor depends on the vision, commitment, and strong work ethic practiced by an individual or group of people. Certainly that was true with AFWL and contractor scientists and engineers who formed a closely knit team that was not afraid to articulate its vision by taking the initiative to strike out into unknown territory. They refused to be paralyzed by fear that the ALL would not work. None of them were Nobel scientists, but they demonstrated unusual courage and an unrelenting desire and persistence to move forward in the face of incredible odds. In the end, they beat those odds, showing that a light ray might truly turn into the bullet of the next century.

Notes and References

INTRODUCTION

1. USAF News Release (AFSC #83-79),"Airborne Laser Laboratory Experiment," 25 July 1983; USAF News Release (AFSC #83-130), "Air Force/Navy ALL Tests," 29 November 1983; Jeff Hecht, *Beam Weapons: The Next Arms Race* (New York: Plenum Press, 1984), p. 30.

2. Maiman was only 32 when he invented the laser. He earned a B.S. in electrical engineering at the University of Colorado and a Ph.D. in physics from Stanford University in 1955. Before joining Hughes, he worked for Lockheed Aircraft on guided missile communications. After discovery of the laser, Maiman left Hughes in 1961 to become the founder and president of Korad, Inc. in Santa Monica. In 1968, he started Maiman Associates in Los Angeles and then formed his third company, Laser Video Corporation, in 1972 in Los Angeles. Three years later, he went to TRW Electronics as one of its vice presidents in charge of exploring new product lines. T. H. Maiman, "Stimulated Optical Radiation in Ruby," *Nature*, 6 August 1960, pp. 493–494; T. H. Maiman, "Stimulated Optical Emission in Fluorescent Solids. I. Theoretical Considerations," *Physical Review*, 15 August 1961, pp. 1145–1150; T. H. Maiman *et al.*, "Stimulated Optical Emission in Fluorescent Solids. II. Spectroscopy and Stimulated Emission in Ruby," *Physical Review*, 15 August 1961, pp. 1151–1157; George F. Smith, "The Early Laser Years at Hughes Aircraft Company," *IEEE Journal of Quantum Electronics*, June 1984, pp. 577–584; John A. Osmundsen, "Light Amplification Claimed by Scientist," *New York Times*, 8 July 1960, pp. 1, 7; "Laser Research and Applications," prepared at the request of Honorable Howard W. Cannon, Chairman,

Committee on Commerce, Science, and Transportation, U.S. Senate (Washington, D.C.: U.S. Government Printing Office, 1980); A. Javan *et al.*, "Population Inversion and Continuous Optical Maser Oscillation in a Gas Discharge Containing a He-Ne Mixture," *Physical Review Letters*, 1 February 1961, pp. 106–110; Barry Miller, "Optical Maser May Aid Space Avionics," *Aviation Week*, 18 July 1960, pp. 96–99; Jeff Hecht, *The Laser Guidebook* (New York: McGraw–Hill, Inc., 1986), p. 12; Hecht, *Beam Weapons*, pp. 23–24; Harland Manchester, "Light of Hope—Or Terror," *Reader's Digest*, February 1963, pp. 97–100; "Light Ray—Fantastic Weapon of the Future," *U.S. News & World Report*, 2 April 1962, pp. 47–48; "First Working Optical Maser," *Aviation Week*, 11 July 1960, p. 38; F. Clifton Barry, Jr., *Inventing the Future* (New York: Brassey's Inc., 1993), pp. 46–51.

3. Five months after Maiman's historic breakthrough, Ali Javan and his colleagues at Bell Telephone Laboratories succeeded in operating a helium–neon electric discharge laser, the first gas laser. (See A. Javan, W. R. Bennett, and D. R. Herriott, *Physical Review Letters*, Vol. 6, 1961, p. 106.) Over the next few years, the development of gas lasers occurred very rapidly. Gas lasers extended from the near ultraviolet to the millimeter wavelengths; power increased from milliwatts to hundreds of watts. Many in the scientific community rated lasers on the same level as the discovery of the transistor in 1948 by another team from Bell Laboratories (John Bardeen, Walter H. Brattain, and William Shockley) and the first modern computer invented by Howard Aiken in 1937. Donald R. Herriott, "Applications of Laser Light," in *Lasers and Light*, ed. Arthur L. Schawlow (San Francisco: W. H. Freeman & Company, 1969), pp. 313–321; C. Breck Hitz, *Understanding Laser Technology: An Intuitive Introduction to Basic and Advanced Laser Concepts* (Tulsa, Oklahoma: Pennwell Publishing Co., 1985), pp. 1–6; Hecht, *Beam Weapons*, p. 40.

4. Laser bar code scanners not only speed up customers moving through checkout lines, but also allow the store owner to keep accurate track of the inventory to determine which products are selling well. The Marsh Supermarket in Troy, Ohio, was the first to install and use laser bar code scanners in 1974. Herriott, pp. 313–321; "Retailers Want More Efficient Bar Codes," *Albuquerque Tribune*, Business Outlook Section, 30 March 1992, p. 3; Jeff Hecht, "30 Years Reflections: Lasers as a Commercial Technology," *Laser Focus World*, May 1994, p. 83; Barney Finn, Draft for Laser Exhibit at National Museum of American History, 5 April 1985.

5. Stewart E. Miller, "Communication by Laser," in *Lasers and Light*, pp. 323–331; Hitz, pp. 1–6.

6. U.S. Senate, "Laser Research and Applications"; Barnaby J. Feder, "Lasers Finally," *New York Times*, 2 August 1981, p. C-9.

7. Dennis Gabor invented holography in 1947 using ordinary light to illuminate photographic plates to produce a hologram. He taught at London's Imperial College, pioneering experiments in the physics of holography for which he won the Nobel prize in physics in 1971. Samuel L. Marshall, *Laser Technology and Applications* (New York: McGraw–Hill, Inc., 1968), pp. 243–244; Emmett N. Leith and Juris Upatnieks, "Photography by Lasers," in *Lasers and Light*, pp. 339–350; Keith S. Pennington, "Advances in Holography," in *Lasers and Light*, pp. 351–359.

8. In 1981, Spectra-Physics, Incorporated (Mountain View, California) and Coherent, Incorporated were the two largest firms providing lasers and related equipment for the commercial market. Hughes Aircraft Company led the field in support of laser research and development for the military. Schawlow, "Lasers: The Practical and the Possible," *The Stanford Magazine*, Spring/Summer 1979, pp. 24–29; "Already Dozens of Civilian Uses for the Laser," *U.S. News & World Report*, 18 October 1971, pp. 86–87; Maj Walter M. Breen, "The Laser: Its Function and Its Future," *Air University Review*, May/June 1975, pp. 59–67; Feder, p. C-9.

9. Interestingly, Maiman never envisioned that his laser would ever evolve into the ultimate "death ray" weapon to replace machine guns, rifles, tanks, and the like. He believed lasers would have supplementary military applications, especially as range finders and laser-guided bombs. He anticipated lasers would play a much larger role as a precision scientific tool used in communications, TV systems, medicine, microwelding, and so forth. Hecht, *Beam Weapons*, pp. 1–13;

Interim Report of the USAF Scientific Advisory Board Ad Hoc Committee on Laser Technology, May 1974, pp. 2–3; Richard Burt, "New Laser Weaponry is Expected to Change Warfare in the 1980s," *New York Times*, 10 February 1980, pp. 1, 54; Barry L. Thompson, "'Directed Energy' Weapons and the Strategic Balance," *Orbis*, Fall 1979, pp. 697–709; Transcript, Tony Edwards, "NOVA: Light of the 21st Century," BBC-TV, Boston, 1978, pp. 6–9.

CHAPTER 1. MILITARY'S EARLY INVOLVEMENT IN LASERS

1. There is an ongoing debate as to the authenticity of whether or not the fleet was set on fire using mirrors during the Second Punic War. See D. L. Simms, "Archimedes and the Burning Mirrors of Syracuse," *Technology and Culture*, Vol. 18, 1977, pp. 1–24; Jeff Hecht, *Beam Weapons: The Next Arms Race* (New York: Plenum Press, 1984), pp. 16–18; A. C. Claus, "Archimedes' Burning Mirrors," *Applied Optics*, October 1973, p. A14; D. L. Simms, "More On That Burning Glass of Archimedes," *Applied Optics*, May 1974, pp. A15–A16; Klaus D. Mielenz, "That Burning Glass," *Applied Optics*, February 1974, pp. A14, A16; "Archimedes' Weapon," *Time*, 26 November 1973, p. 60.

2. H. G. Wells, *The War of the Worlds* (London: Octopus Books Limited, 1898, 1983 ed.), p. 80.

3. Hughes Research Laboratories developed operational range finders by late 1961. By the mid-1960s, the Army was using laser range finders that could calculate distances within a foot of the target. During the Vietnam War, after repeated attempts (6 years and 872 bombing attacks) using conventional bombs to destroy the Thanh Hoa Bridge in North Vietnam, the Air Force on 13 May 1972 finally collapsed the bridge with a laser-guided bomb. An electro-optical seeker head fitted to the bomb sensed reflections of a laser beam illuminating the target. The seeker head then sent signals to the bomb's movable control surfaces (movable fins) to readjust the position of the fins so the bomb would home in on the laser energy reflected off the target. Laser or "smart" bombs were extremely accurate, typically hitting within 10 feet of the target. Iron bombs usually impacted within 100 yards of the target. Although progress had been made with range finders and smart bombs, no hand-held operational laser weapon had been developed. From time to time articles appeared in the press hinting of an imminent ray-gun breakthrough. For example, the 5 November 1990 issue of *Defense News* (p. 37) reported the Army had tested a hand-held laser built by Allied-Signal, Inc., Morristown, New Jersey, and McDonnell Douglas of St. Louis. U.S. Senate, "Laser Research and Applications," 1980; DOD Reg 5210.61 (Encl 1), "Specific Security Classification Guidance on High Energy Lasers Information for DOD Activities," 7 April 1977, p. 1; Barry Miller, "U.S. Plan to Accelerate Laser Development Spurs Market," *Aviation Week & Space Technology*, 21 August 1967, pp. 92–93; Barry Miller, "U.S. Begins Laser Weapons Programs," *Aviation Week & Space Technology*, 26 March 1962, pp. 41–45; Jeff Hecht, *The Laser Guidebook* (New York: McGraw–Hill, Inc., 1986), p. 8; C. Breck Hitz, *Understanding Laser Technology: An Intuitive Introduction to Basic and Advanced Laser Concepts* (Tulsa, Oklahoma: Pennwell Publishing Co., 1985), pp. 5–6; Drew Middleton, "Mass-Produced Precision Guided Weapons are Said to be Revolutionizing Military Doctrine and Tactics," *New York Times*, 23 February 1976, p. 15; General P. F. Gorman, USA, "What the High Technology Edge Means," *Defense 83*, June 1983, p. 25; Charles D. Bright, ed., *Historical Dictionary of the U.S. Air Force* (New York: Greenwood Press, 1992), pp. 563–564.

4. Program, First Conference on Laser Technology, U.S. Marine Corps Recruit Depot, San Diego, California, 12–14 November 1963.

5. U.S. Senate, "Laser Research and Applications," 1980; Rpt, Colonel John C. Scholtz, Jr., "The Air Force High Energy Laser Program" (Air War College Professional Study No. 4229), November 1970, pp. 1–6; "Light Ray—Fantastic Weapon of the Future?," *U.S. News & World Report*, 2 April 1962, pp. 47–50; Hitz, pp. 1–2; Arnold L. Bloom, "Optical Pumping," in *Lasers and Light,* ed. Arthur L. Schawlow (San Francisco: W. H. Freeman & Co., 1969), pp. 46–54; Rpt,

"Laser/Optics: Short Course," Physics Department, Air Force Institute of Technology, Wright-Patterson AFB, Ohio, June 1979.

6. The widely respected German physicist Max Planck in 1900 proposed that radiation consisted of small packets of light called *quanta*. The shorter the wavelength of light, the more energy is packed into each quantum. A quantum is similar to an atom, which combined with other atoms, are the building blocks of matter. (For a good layman's discussion of Planck's work, see Gary Zukav's *The Dancing Wu Li Masters: An Overview of the New Physics*.) Later, Albert Einstein postulated energy levels can only change in increments equal to a quantum or photon, as he called it. Photons have a dual nature and can be thought of as a bundle of waves in particle form. Arthur L. Schawlow, "Optical Masers," in *Lasers and Light*, ed. Arthur L. Schawlow (San Francisco: W.H. Freeman & Co., 1969), pp. 224–233; C. Kumar N. Patel, "High-Power Carbon Dioxide Lasers," in *Lasers and Light*, ed. Arthur L. Schawlow (San Francisco: W.H. Freeman & Co., 1969), pp. 265–275; Bela A. Lengyel, "Evaluation of Masers and Lasers," *American Journal of Physics*, Vol. 34, March 1966, pp. 903–911; Edgar F. Ulsamer, "Laser: A Weapon Whose Time is Near," *Air Force Magazine*, December 1970, pp. 28–34.

7. Arthur L. Schawlow, "Advances in Optical Masers," in *Lasers and Light*, ed. Arthur L. Schawlow (San Francisco: W.H. Freeman & Co., 1969), pp. 234–235; Lee Edson, "The Advent of the Laser Age," *New York Times*, 26 March 1978, Section 6, pp. 35–46.

8. Arthur L. Schawlow, "Laser Light," in *Lasers and Light*, ed. Arthur L. Schawlow (San Francisco: W.H. Freeman & Co., 1969), pp. 282–290.

9. There is not a one-to-one relationship for energy pumped in and converted to a laser beam. Generally, if 5–7 percent of the input energy is transformed into a beam, then that is considered a good efficiency rate. (An automobile's internal combustion engine's efficiency is about 16 percent.) Efficiency is defined as the emitted energy (photon) divided by the amount of energy required to stimulate an atom to its excited (upper energy) state. If it takes 100 units of energy to move an electron from its normal orbit to an excited state and the photon released represents 5 units of energy, then the efficiency would be 5/100 or 5 percent. Calculation of efficiency in this way is the absolute or quantum efficiency of the laser. In practice, however, the atoms in a gas have different kinetic energy levels, which means that some electrons require more energy than others to move up into higher energy levels—some electrons will jump out of their normal state but will not make it to the lasing energy level, thereby wasting input energy—input energy is expended in trying to stimulate these electrons. The result is only a fraction of the input power source is used to excite electrons to the upper lasing level. Therefore, the "working efficiency" of the laser—defined as output power divided by input power—will always be much lower than quantum efficiency. Interview with Leonard J. Otten III, Colonel, USAF, Phillips Laboratory, Albuquerque, New Mexico, 30 August 1991; Alexander Lempicki and Harold Samelson, "Liquid Lasers," in *Lasers and Light*, pp. 255–263; Arthur L. Schawlow, "Lasers: The Practical and the Possible," *The Stanford Magazine*, Spring/Summer 1979, pp. 26–29.

10. Lasers are often referred to as "man-made light," because they do not exist in the natural order of the universe. However, some scientists have theorized a "natural laser" exists in the upper atmosphere of Mars. Interview with Darrell Spreen, Phillips Laboratory, Albuquerque, New Mexico, 31 August 1994; Schawlow, "Laser Light," pp. 282–290; U.S. Senate, "Laser Research and Applications," 1980; "Light Ray—Fantastic Weapons of the Future?," *U.S. News & World Report*, 2 April 1962, p. 47; Hitz, pp. 1–2; Philip J. Klass, "Special Report: Laser Weapons," *Aviation Week & Space Technology*, 18 August 1975, pp. 34–39.

11. Schawlow, pp. 282–290; U.S. Senate, "Laser Research and Applications," 1980; Miller, pp. 41–44; Hitz, pp. 1–6.

12. All light, regardless of its frequency and wavelength, moves at exactly the same velocity of 186,230 miles per second in a vacuum or approximately Mach 1 million! Frequency times wavelength *always* equals the constant speed of light. Therefore, if frequency is known, wavelength can be easily calculated or vice versa. For example, a light wave with a frequency of 2

would have a wavelength of 93,115—186,230 divided by 2. Similarly, the higher the frequency, the shorter the wavelength; the lower the frequency, the longer the wavelength.

Laser beams travel in a straight line rather than in an arched trajectory like a bullet fired from a rifle. Because lasers propagate in a straight line and do not follow the curvature of the earth, this was considered to be a limitation. Also, dust, humidity, clouds, fog, and snow tend to defocus and absorb the beam. However, these obstacles could be overcome by basing a laser in the vacuum of space, which would also eliminate the problem of the target being blocked out by the curvature of the earth.

Because laser beams are orders of magnitude faster than the high-acceleration antimissile weapons, such as the Martin Sprint missile developed in the 1960s, more time would be available for laser operators to select and track targets. This is a definite advantage when trying to engage maneuvering multiple independently targeted reentry vehicles.

Scientists at Massachusetts Institute of Technology's Lincoln Laboratory at Lexington, Massachusetts, in May 1962 fired a laser beam from the roof of a building to the moon 250,000 miles away. It showed the focusability of lasers, as the beam spot size on the moon was only 2 miles in diameter. Statement by Hans Mark, Secretary of the Air Force, Before the Subcommittee on Science, Technology, and Space, Albuquerque, New Mexico, 12 January 1980, p. 222; U.S. Senate, "Laser Research and Applications," 1980; Barry Miller, "Optical Maser May Aid Space Avionics," *Aviation Week & Space Technology*, 18 July 1960, pp. 96–97; "Laser Space Weapons in R&D," *Electronics*, 10 November 1961, p. 81.

13. Interview with Demos T. Kyrazis, Colonel, USAF, Retired, Albuquerque, New Mexico, 8 August 1985; Interview with Russ Parsons, Colonel, USAF, Retired, Albuquerque, New Mexico, 15 December 1988; "Laser Weaponry Seen Advancing," *Aviation Week & Space Technology*, 12 January 1970, pp. 16–17; Anthony Ripley, "Laser is Stirring Imagination of Weapon Scientists," *New York Times*, 16 October 1971, p. 15.

14. Interview with Petras V. Avizonis, Air Force Weapons Laboratory, Albuquerque, New Mexico, 16 September 1988; Rpt, ARTO-73-1, Colonel Donald L. Lamberson, "Program Management Plan: Advanced Radiation Technology Office," June 1973, p. 49; Phillip J. Klass, "Research Nears Application Level," *Aviation Week & Space Technology*, 14 August 1972, p. 13; Phillip J. Klass, "Major Hurdles for Laser Weapons Cited," *Aviation Week & Space Technology*, 9 July 1973, pp. 38–42; Kostas Tsipis, "Shortcomings of Laser Weapons," *Laser Focus*, July 1983, p. 8.

15. Heating of the air (dust and other particulate matter) and the target can produce ionized plasmas (gases) which absorb and block the beam's passage. However, on an airborne platform, the laser beam is constantly changing its path to the target, since the plane is moving. Therefore, the beam is boring through a channel of cooler air which reduces but does not eliminate thermal blooming.

Shining a flashlight through foggy or misty air causes the light to diffuse or spread out. Similarly, atmospheric conditions (e.g., fog, rain, snow, clouds) cause a pencil-wide laser beam to expand, preventing the beam from delivering maximum energy on target. Interview with Avizonis, 16 September 1988.

16. Perfecting countermeasures would be difficult. For example, spinning a missile would require development of an extremely advanced guidance system. Adding hardening material would add substantially to the weight of the missile, thereby compromising the completion of its mission. Ultimately, countermeasures would only delay rather than prevent a kill with a highly effective laser. Interview with Avizonis, 16 September 1988.

17. Transcript of "Laser Flight" film, Office of the Assistant Secretary of Defense, 27 March 1986; USAF Fact Sheet, "Department of Defense's High Energy Laser Program," May 1979.

18. U.S. Senate, "Laser Research and Applications," 1980; Barry Miller, "Government Funds Boost Laser Research," *Aviation Week & Space Technology*, 12 March 1963, pp. 229–232; Bonner Day, "Progress on Energy Beam Weapons," *Air Force Magazine*, July 1979, pp. 62–65; Walter Pincus, "Soviets Said Choosing Lasers as Next Lap in Arms Race," *Washington Post*, 23 August 1979, p. A2; "DOD Funds $15 Million for Laser R&D," *Aviation Week & Space Technology*, 8 April

1963, p. 75; James Canan, *War in Space* (New York: Harper & Row, 1982), p. 147; Barry J. Smernoff, "Strategic and Arms Control Implications of Laser Weapons," *Air University Review*, January/February 1978, pp. 38–50.

19. In 1954, Townes operated the first microwave amplifier or ammonia maser. His findings on the maser appeared in the July 1954 issue of *Physical Review*. While a consultant to Bell Telephone Laboratories in 1958, Townes worked with his brother-in-law, Arthur Schawlow, to develop the optical maser concept. Devices that generate or amplify visible or nearly visible radiation are called optical masers. Both men received a patent in 1961 for a maser device, with Bell Laboratories retaining all rights. Townes also shared the 1964 Nobel prize in physics with two Soviet scientists, Nikolai G. Basov and Aleksander M. Procharov, for discovery of the maser principle. Arthur L. Schawlow and Charles H. Townes, "Infrared and Optical Masers," *Physical Review*, Vol. 112, No. 6, December 1958, pp. 1940–1949; Barry Miller, "Optical Systems Have Space Potential," *Aviation Week & Space Technology*, 14 December 1959, pp. 87–91; "Industry Observer," *Aviation Week & Space Technology*, 23 October 1961, p. 19; Barry Miller, "Aerospace, Military Laser Uses Explored," 22 April 1963, pp. 54–69.

20. The Army speculated lasers could be used as antipersonnel or antitank weapons. The Navy looked to lasers as potential ship defense weapons. The Air Force believed lasers might be effective against ICBMs or aerial targets in the atmosphere and be used for surveillance radars, guidance systems, and communications. An extremely narrow laser beam was attractive because it would be almost invulnerable to jamming. The Department of Energy's Lawrence-Livermore Laboratory explored the possibility of using lasers as a source of concentrated energy to initiate thermonuclear fusion (combining atoms under great pressure) as the triggering device for nuclear weapons. Miller, "U.S. Begins Laser Weapons Programs," pp. 41–44; "Optical Maser R&D," *Aviation Week & Space Technology*, 12 February 1962, p. 32; Klass, "Research Nears Application Level." pp. 12–15.

21. Most of ARPA's money earmarked for laser research in the early 1960s went to the Office of Naval Research and the Air Force Special Weapons Center. By the end of the decade, as laser technologies matured, responsibility for advancing the state of the art shifted from ARPA to the individual military services. The Air Force took the lead in investing money for high-energy laser research, followed by the Navy and later the Army. In 1972, ARPA's name was changed to the Defense Advanced Research Projects Agency (DARPA). In 1993, DARPA reverted back to its original name, ARPA, as a result of the deemphasis on "Defense" driven by the near elimination of the Soviet threat. George Heilmeir, "Contributions to Science and Technology by the Advanced Research Projects Agency (ARPA) 1957–1970," *Commander's Digest*, 7 October 1976, Vol. 19, No. 21, p. 2; Barry Miller, "Fiscal 1962 Laser Funding Completed," *Aviation Week & Space Technology*, 9 July 1962, p. 42; James W. Canan, *The Super Warriors: The Fantastic World of Pentagon Superweapons* (New York: Weybright & Talley, 1975), pp. 267–268.

22. Interview with David R. Jones, Colonel, USAF, Retired, Albuquerque, New Mexico, 19 September 1988; Ltr, Dr. Arthur H. Guenther, AFSWC/Chief, Material Dynamics Lab, to Lt Col Neuer, SWRPZ; "Optical Maser Research," 12 February 1962.

23. Ibid.

24. Interview with Jones, 19 September 1988; Ltr, Lieutenant Colonel Ira McMann, USAF, Assistant for Program Administration, DCS/Plans & Programs, HQ Office of Aerospace Research, to AFSC (SCTL), "Reassigning ARPA Order 313-62 to AFSWC," 13 March 1962; Ltr, J. P. Ruina, Director, ARPA, to Commander, Office of Aerospace Research, "ARPA Order 313-62," 26 February 1962; Rpt, WLREL-68-008, "Advanced Development Plan: Gas Dynamic Laser Weapon," September 1967, pp. 1-1–2-6.

25. Rpt, AFSC, "High-Energy Gas Laser Applications Master Plan," February 1972, p. B-2; John Bosma, "Space and Strategic Defensive Orientation: Project Defender," *Defense Science & Electronics*, September 1983, pp. 58–65.

26. Rich conducted a follow-on study to the initial ARPA Order 313-62. That involved a $73,000 ARPA-funded contract with RCA starting in March 1963 to increase the number of various materials exposed to laser radiation. The focus was on the interaction of laser energy with solids. Interview with Jones, 19 September 1988; Abstract of Work Statement [Contract AF29(601)-5845], 1st Lt John C. Rich, Research Directorate, AFSWC, "Theoretical Radiation Opacity Study," 27 March 1963; Biography (USAF), Dr. Petras V. Avizonis, July 1980.

27. Interview with Avizonis, 16 September 1988; Notes on Article for KAFB *Nucleus*, n.a., "Center Assembles Top Level Team for Laser Research," 21 May 1962; Memo for Record, Jones, "Lasers," 19 February 1962; Minutes of Cocoa Sub-Group on Laser Research, Capt Marvin C. Atkins, 16 March 1962; Letter, 1st Lt Donald C. Wunsch, AFSWC/SWRPL, to SWR, "Trip Report," 29 March 1962.

28. Austrian physicist Hans Thirring, writing in Britain's *New Scientist*, challenged the idea that lasers would be the magical arrow to shoot down incoming warheads. He calculated destroying such targets would require delivering a beam of 807 kilowatts, power levels not realistic or practical to his thinking. (See "Death to Death Rays," *Time*, 28 June 1963, p. 36.) There were also later critics who questioned the efficacy of laser weapons in the tactical arena. *The New York Times* reported on 6 March 1969 (p. 35) that a hand-held pulsed laser research instrument developed by Maser Optics, Inc. of Boston and delivered to the Army's Frankfort Arsenal in Philadelphia proved ineffective. The Army stated the device's "ability to ignite objects and to detonate explosives" had been evaluated and "determined to be totally insufficient for initiation of a development program for a destructive instrument." Interview with Donald L. Lamberson, Major General, USAF, Office of the Assistant Secretary of the Air Force for Acquisition, Pentagon, Washington, D.C., 12 January 1989; Interview with Avizonis, 16 September 1988; Rpt, RTD-TDR-63-3070, Marvin C. Atkins *et al.* "AFWL Progress on PROJECT SEASIDE as of 20 July 1963," November 1963, pp. iii, 1, 41–44; Petras V. Avizonis and T. Farrington, "Internal Self-Damage of Ruby and Nd-Glass Lasers," *Applied Physics Letters*, Vol. 7, No. 8, pp. 205–206.

29. The forerunner of the solid-state laser was the solid-state maser invented by Harvard Professor Nicolaas Bloembergen in 1956. He succeeded applying the maser principle to solid paramagnetic salts (cryogenically cooled crystals).

    Gerry demonstrated much more power could be gained from an optical cavity by extracting the beam of laser energy perpendicular to the gas flow as opposed to extracting photons along the flow line. In that way, one could get rid of the waste heat more easily. Gas laser devices could remove waste heat energy at a rate ten times higher than solid-state devices that could only remove heat by conduction through the solid material to the walls of the device. Also, scaling lasers to higher powers could be achieved without using a several-hundred-yard-long device. This was important to make the laser device more compact in weight and volume, and to make it more cost effective. Gerry eventually left AVCO to become director of ARPA's high-energy laser technology program. Gerry, Arthur Kantrowitz, director of AERL, and AERL researchers Donald A. Leonard and Jack Wilson were considered "coinventors" of the gas dynamic laser and jointly hold the patent. AFWL Paper, "A Thumbnail Sketch of the US HEL Weapons Program," n.d., p. 1; Rpt, WLREL-69-025, "Advanced Development Program: High Energy Laser Program, Program 644A—Task II," 24 February 1969, pp. A1–A3; Rpt, WLRE-67-170, "Advanced Development Plan: Gas Dynamic Laser Weapon," September 1967, pp. A-1–A-2; Rpt, Scholtz, pp. 6–23; Rpt, Office of the Director of Defense Research and Engineering, "In-House Program in Optical Masers," 2 January 1968; C. Kumar N. Patel, "Selective Excitation Through Vibrational Energy Transfer and Optical Laser Action in N_2-CO_2," *Physical Review Letters*, Vol. 13, 1964, pp. 617–619; Edward T. Gerry, "Gasdynamic Lasers," *IEEE Spectrum*, November 1970, pp. 51–58; John D. Anderson, Jr., *Gas Dynamic Lasers: An Introduction* (New York: Academic Press, 1976), p. 2.

30. Axelrod was in charge of the gas dynamic group at AFWL, which had a total of less than ten civilians and military assigned. Rpt, WLRE-67-170, pp. ii, 1-1, 2-1–2-24. Compare the expecta-

tions for high-energy lasers set out in this document with those outlined in AFSC West Coast Study Facility, SCL-6-66-14, "Final Report: Technology Applications Study, High Energy Laser Program (HELP)," April 1967, "Volume I: Highlights," pp. 2–5, and "Volume II: Study Reports," pp. II-1–II-10.

31. Interview with Avizonis, 16 September 1988; Interview with Howard Leaf, Lieutenant General, USAF, Retired, McLean, Virginia, 14 August 1990.

32. Interview with Keith Gilbert, Colonel, USAF, Retired, Albuquerque, New Mexico, 31 August 1988; Interview with Avizonis, 16 September 1988; Interview with Murray Hill, Air Force Weapons Laboratory, Albuquerque, New Mexico, 23 September 1988; Interview with Harry I. Axelrod, Hughes Aircraft Company, El Segundo, California, 27 January 1993; Rpt, WLRE-67-133, "Advanced Development Plan: Solid State Laser Weapon," September 1967, pp. 1-1–1-4.

33. Interview with Avizonis, 16 September 1988.

34. Interview with Avizonis, 16 September 1988; Headquarters, United States Air Force, Development Directive 104-1, 3 October 1968. See also Rpt, WLREL-68-008, "Advanced Development Plan: Gas Dynamic Laser Weapon," 15 February 1968 and Rpt, WLREA-68-137, "Advanced Development Plan, Solid State Laser Weapon," February 1968, for revisions to the 1967 ADPs because of Development Directive 104. Ltr, Jones, AFWL/CC to AFSC/CC, "Recommendation for Award of Unit Decoration, 1 January 1967 to 31 December 1967," 2 August 1968; Report of 16 April 1968 Meeting on "White Paper," n.a., 15 May 1968; Rpt, Colonel Raymond A. Gilbert, AFWL/CC, "AFWL Significant Events," 10 August 1966; William J. Beane, "The High-Energy Laser: Strategic Policy Implications," *Strategic Review*, Winter 1977, p. 101.

35. AFAL's and AFWL's laser outlooks were different. AFAL emphasized electronics programs and tended to think in terms of avionics applications for lasers, i.e., laser designators, illuminators, countermeasures, and gyros. On the other hand, AFWL was more interested in lasers as potential weapon systems, a category the ALL fit into. In testimony to a Congressional Subcommittee on Science, Technology, and Space on 12 January 1980 in Albuquerque, Secretary of the Air Force Hans Mark recalled AFWL had become the center of early laser research because in his mind "the Weapons Laboratory was probably the best Air Force technology development center and the prospects of applying lasers would in the end be consistent with the other areas of strategic weapons technology that are being developed here at the Laboratory." Interview with Raymond A. Gilbert, Brigadier General, USAF, Retired, Albuquerque, New Mexico, 23 February 1993; Interview with Jones, 19 September 1988; Interview with Hill, 23 September 1988; Ltr, Brigadier General Gilbert, AFSC/DOL, to AFAL (AFG/Col James L. Dick), "High Energy Lasers," 13 June 1968; Ltr, Colonel Dick, Director, AF Avionics Lab, to AFWL (WLG/Colonel David R. Jones), "High Energy Laser Program," 1 July 1968; Ltr, Avizonis, AFWL Senior Research Physicist, to Leaf, "Gas Laser Advanced Development Program," 5 March 1968; Trip Report, Avizonis, "Visit to DOL, HQ USAF, Asst Sec Harry Davies, and ARPA, 20–21 May 1968."

36. Not all at AFAL, especially at the worker level, favored moving the laser work to AFWL. Later, in the early 1970s, Lamberson tried to patch up some of the bad feelings that had developed between the Avionics Lab and AFWL. Realizing the enormous complexity of the airborne pointer and tracker, he offered this portion of the ALL program to the Avionics Lab. However, they "wanted the whole thing [entire ALL program]," Lamberson recalled, "which was unfortunate because AFWL had to invest tremendous resources in pointing and tracking by hiring 150–200 people to work that issue." Interview with Lamberson, January 1989; Interview with Gilbert, 23 February 1993; Ltr, Colonel Dick, Director, AF Avionics Lab, to AFWL (WLG, Colonel David R. Jones), "High Energy Laser Program," 1 July 1968.

37. Interview with Lamberson, 12 January 1989.

38. Interview with Hill, 23 September 1988; Award/Contract (F29601-68-C-109), AFSWC with United Aircraft Corp., Pratt & Whitney Aircraft Division, 10 June 1968; Ltr, Major General Otto J. Glasser, Assistant DCS/R&D, HQ USAF, to Director, ARPA, "Support for United Aircraft Corporation Laser Program," 20 April 1968; Rpt, AFWL-TR-69-47, Edward A. Pinsley *et al.*,

United Aircraft Corp., "XLD-1 Development Program: Final Technical Report," 1 June 1969, pp. 1–10; Rpt, AFWL, Harry I. Axelrod, "A 'White Paper' on the Air Force Laser," April 1972, p. 1. For the best background on United Technologies/Pratt & Whitney work on lasers, see Rpt, AFWL-TR-70-175, Missimer *et al.*, Pratt & Whitney Corp., "Program for XLD-1 Development: Final Technical Report," Vol. I, "Experimental and Analytical Investigations," February 1971, pp. 5–13; Vol. II, "Experimental and Analytical Investigations: Appendices," February 1971, pp. 1–8 for gain measurements and pp. 211–220 for unstable resonator work; and Vol. III, "Task VII-Advanced System Study," March 1971, for progress on work eventually culminating in the Split Stage Demonstrator and the GDL integration into the ALL. For the impressive power levels and early dates these levels were achieved, see Vol. I above, pp. 2–3, and Rpt, PWA FB 70-513, R. Coar *et al.*, Pratt & Whitney Corp., "XLD Follow-On Program, Volume I: Technical Proposal," 27 April 1970, p. 7.

39. Interview with Jones, 19 September 1988; AFWL Paper, "A Thumbnail Sketch of the US HEL Weapons Program," n.a., n.d., pp. 1–2; Rpt, ARTO-74-2, Colonel Donald L. Lamberson, "Program Management Plan: Advanced Radiation Technology Office," 1974, p. 19; Interim Report of the USAF Scientific Advisory Board Ad Hoc Committee on Laser Technology, May 1974, p. 3; Rpt, AFSC Directorate of Laboratories, "Five Year Plan for the Development of Laser Applications and Technology," August 1967, p. 4; Ltr, Avizonis, AFWL Senior Research Physicist, to Leaf, "Gas Laser Advanced Development Program," 5 March 1968.

CHAPTER 2. FIRST LASER GROUND DEMONSTRATIONS

1. Engineering and fabrication of the GDL (the future tri-service laser) took place under the supervision of AVCO's Systems Division at Wilmington, Massachusetts. Interview with Jones, 19 September 1988; Rpt, WLREL-68-008, "Advanced Development Plan," 15 February 1968, pp. 1-1, 2-2, 2-7–2-14; Rpt, AFWL, Harry I. Axelrod, "A 'White Paper' on the Air Force Laser," April 1972, p. 1; Rpt, WLREL-69-032, "Advanced Development Plan," 24 February 1969, pp. D-1–D-13; Ltr, Capt John W. Duemmel, Director of Information, "Groundbreaking for AFWL Laser Optics Building," 26 June 1967; U.S. Air Force News Release, Office of Information, KAFB, NM, "Ground Broken for New Laser Building at Kirtland," 30 June 1967.

2. Groundbreaking for the $441,000, 13,000-square-foot Laser Optics Laboratory took place on 30 June 1967. This horseshoe-shaped facility allowed AFWL's 15-man Laser Section to use ruby and glass pulsed lasers and a carbon dioxide continuous beam laser to study the interaction of laser radiation with a variety of materials to determine survivability and vulnerability levels of those materials. The facility also had a 200-foot underground tunnel to simulate various atmospheric conditions to investigate beam propagation problems, one of the most serious drawbacks of any laser system. Colonel David Jones, the new Weapons Lab commander, presided at the ceremony. Interviews with Kyrazis, 20 June and 8 August 1985; Interview with Avizonis, 16 September 1988; Interviews with Lamberson, 12 and 13 January 1989; Rpt, Axelrod, p. 1; Memo for the Chairman, Defense Science Board, Dr. John S. Foster, Jr., "Defense Science Task Force: EIGHTH CARD," 3 July 1968.

3. Foster selected AFWL to lead the TSL program because he knew Air Force research in lasers was well ahead of that of the Army and Navy. Scholtz was chosen mainly because he outranked Axelrod. A fighter pilot most of his career, Scholtz was new to the research and development environment of AFWL. Consequently, with his technical expertise and experience with lasers, Axelrod did most of the work and, in fact, acted as chairman. Interview with Axelrod, 27 January 1993; Rpt, Axelrod, pp. 1–2; "Laser Advances May Evolve New Weapons," *Aviation Week & Space Technology*, 9 March 1970, pp. 209, 211.

4. Air Force Development Directive 104-1, dated 3 October 1968, authorized work to begin on GDL research at AFWL. Interviews with Lamberson, 12 and 13 January 1989; Interview with Avizonis, 16 September 1988; Interview with Hill, 23 September 1988; Interview with Gilbert, 31 August

1988; Program Memorandum for the Accelerated High Energy Laser Program P.E. 63602F, September 1970.

5. The XLD-1 produced a high-temperature mixture of nitrogen, carbon dioxide, and water vapor by means of combustion of carbon monoxide and hydrogen in oxygen with the addition of gaseous nitrogen. There were expected failures as well as successes in the development of the XLD-1. For example, in May 1968 a nitrogen valve failed to open, causing the system to overheat, resulting in damage to the combustion chamber and nozzles. Plus, beam quality of the XLD-1 was poor—like a flashlight—in the early days. Interviews with Lamberson, 12 and 13 January 1989; Interview with Avizonis, 16 September 1988; Interview with Axelrod, 27 January 1993; Rpt, Axelrod, p. 3; Rpt, AFWL-TR-69-47, E. A. Pinsley *et al.*, Pratt & Whitney Aircraft Corp., "XLD-1 Development Program," June 1969, pp. 1–10; Rpt, AFWL-TR-70-175, William C. Missimer *et al.*, Pratt & Whitney Corp., "Program for XLD-1 Development, Final Technical Report, Vol I, Experimental and Analytical Investigation," February 1971, pp. 2–3, 5.

6. Air Force Development Directive DD-104-2, dated 6 January 1969, provided an initial $2.7 million to fund the EIGHTH CARD program. At that time, AFWL had only 33 people working on various projects of its laser program. Interviews with Lamberson, 12 and 13 January 1989; Interview with Thomas J. Dyble, Lieutenant Colonel, USAF, Air Force Space Technology Center, Albuquerque, New Mexico; 1 August 1988; Interview with Hill, 23 September 1988; Rpt, Axelrod, p. 3; Memo, Richard J. Casey, GAO Denver Regional Office, to Col Lamberson, AFWL/AR, "Summary of Air Force High Energy Program," 3 October 1973.

7. Interview with Larry Sher, Air Force Weapons Laboratory, Albuquerque, New Mexico, 11 October 1988; Interview with Hill, 23 September 1988; Rpt, AFWL-TR-73-94, "Optics for High-Power Laser Systems, Final Report for Period February 1969 to June 1973," November 1973, pp. 1–2, 4–7; Rpt, AFCMD, Walter A. Brown, "R&D Contracts Status Report as of 30 Jun 70," 1970, p. 6; Rpt, Brown, "R&D Contracts Status Report as of 25 Sep 70," 1970, p. 5; Rpt, Brown, "R&D Contracts Status Report as of 24 Jan 71," 1971, p. 3; Rpt, Brown, "R&D Contracts Status Report as of 30 Jun 71," 1971, p. 2.

8. The AFWL Commander, Colonel David R. Jones, endorsed Colonel Scholtz's plan for development of an airborne laser. Jones's Management Plan covering EIGHTH CARD program work at AFWL, dated 13 February 1969, advocated designing and building an airborne prototype to be used against aircraft and air-to-air and ground-to-air missiles. Interview with Axelrod, 27 January 1993; Interview with Avizonis, 16 September 1988; Letter Contract, F29601-69-C-0058, Air Force Special Weapons Center with Hughes Aircraft Company for "Optics for High Power Systems," effective 3 February 1969; Statement of Work PR L9-644A-106, 20 January 1969; Sole Source Justification for PR L9-644A-106, 24 January 1969; Rpt, AFWL, Colonel David R. Jones, "The Air Force Weapons Laboratory Management Plan: EIGHTH CARD Program," 13 February 1969, p. 5; Rpt, Colonel John C. Scholtz, Jr., "The Air Force High Energy Laser Program: Part II" (Air War College Professional Study No. 4442), April 1971, pp. 4–6, 11–13, 15–16.

9. Interviews with Lamberson, 12 and 13 January 1989 and 7 December 1995.

10. Interview with Dyble, 1 August 1988; Interview with John C. Rich, Colonel, USAF, Retired, Danbury, Connecticut, 10 January 1989; Interview with Hal A. Shelton, Colonel, USAF, Retired, Albuquerque, New Mexico, 21 December 1989; Rpt, Thomas J. Dyble, "Peace Through Light: The Airborne Laser Laboratory" (Air Command & Staff College Rpt No. 83-0615), February 1983, pp. 4–5; Rpt, AFWL, Jake Spidle and Demos T. Kyrazis, "Getting Started: The Air Force High Energy Laser Program" (unpublished paper), pp. 24–25.

11. Interview with Hill, 23 September 1988; Rpt, Axelrod, pp. 7–8.

12. Interview with Avizonis, 16 September 1988; Interview with Lamberson, 12 January 1989; Interview with Dyble, 1 August 1988; Rpt, AFWL, Col Robert W. Rowden, "AFWL Significant Events," 31 August 1971; Rpt, Axelrod, pp. 30–33.

13. Interview with Avizonis, 16 September 1988; Interview with Gilbert, 31 August 1988; Interview with Lamberson, 12 January 1989; Interviews with Kyrazis, 20 June and 8 August 1985; Rpts, Rowden, "AFWL Significant Events," 28 February, 30 April, and 30 June 1971; Rpt, Axelrod, pp. 63–67.

14. Memo for the Record, Axelrod, "Colonel Rowden's Visit to AVCO/ Wilmington," 1 September 1970; Ltr, J. R. Dempsey, Vice President, Group Executive, AVCO Corporation Government Products Group, to Col Walter A. Brown, USAF DCS/Procurement, "Contract Number F29601-69-C-0083," 22 September 1970.

15. Interview with Lamberson, 12 January 1989; Rpt, Axelrod, pp. 10–11; Ltr, Rowden, AFWL/CC, to Dr. A. R. Kantrowitz, AVCO-Everett Research Laboratory, "Tri-Service Laser Program," 8 November 1971.

16. Interview with Dyble, 1 August 1988; Interviews with Kyrazis, 20 June and 8 August 1985; Interview with Rich, 10 January 1989; Rpt, Rowden, "AFWL Significant Events," 31 December 1971; Rpt, Axelrod, pp. 6, 28; Memo for the Record, Axelrod, "TSL TD Meeting No. 18—Acceptance of the Air Force Laser," 8 December 1971.

17. Memo for the Record, Axelrod, 8 December 1971; Memo for the Record, "Background Paper—Air Force Laser Milestone," n.d.; Rpts, Rowden, "AFWL Significant Events," 31 December 1971 and 31 January 1972.

18. Interview with Hill, 23 September 1988; Interview with Axelrod, 27 January 1993; Rpt, Axelrod, pp. 18–20; Memo, Casey to Lamberson, 3 October 1973. AVCO tried to remedy management failures on 16 October 1970, with the establishment of a Tri-Service Laser Design Review Team to review the system's mechanical design. See Rpt, ESDM-F300-144, AVCO Corp, Wilmington, Massachusetts, "Technical Review Team Final Report," 16 November 1970.

19. Interview with Thomas Welch, Hughes Aircraft Company, El Segundo, California, 15 May 1990; Interview with Shelton, 21 December 1988; Rpt, AFWL, Donald L. Lamberson *et al.*, "Advanced Radiation Technology," 15 June 1971; Rpt, AFWL-TR-73-94, pp. 1–7.

20. Interview with Shelton, 21 December 1988.

21. The October 1971 date represented an accumulation of data from numerous FTT tracking tests. On the morning of 28 May 1971, the tracker imager on the FTT for the first time tracked on the engines of the T-39. Testing of the tracker continued through the summer. By October, all of the test data confirmed the FTT could consistently track with accuracies of 20 microradians or better. This gave Lamberson the confidence to report the success of the first TSL milestone to Systems Command in October 1971. Ben Johnson and Larry Sher, "The Air Force Field Test Telescope: Tracking and Laser Pointing Test Results," in *Proceedings of the Fifth Conference on Laser Technology*, Vol. III, April 1972, pp. 269–304; Rpt, AFWL-TR-74-225, David R. Dean *et al.*, "Field Test Telescope/Air Force Laser Integration Test Results," September 1974, p. 5; Rpt, AFWL-TR-74-286, James Negro and R. P. Connor, "Field Test Telescope Evaluation Report," June 1975, pp. 5–6, 10–14, chronicles later FTT successes. Rpt, AFWL-TR-73-94, November 1973, pp. 2–4; Rpt, Rowden, "AFWL Significant Events," 31 May 1971.

22. Interview with Lamberson, 17 June 1976.

23. The reason glass mirrors were originally used in the FTT was because Hughes did not have the high-power, water-cooled mirrors ready. Difficulties were encountered in developing these new state-of-the-art mirrors, mainly because they were larger than any previous mirrors Hughes had built. Interview with Axelrod, 27 January 1993; Johnson and Sher, pp. 269–304; Rpt, ARTO-73-1, Lamberson, "Program Management Plan: Advanced Radiation Technology Office," June 1973, pp. 2–3; Briefing Chart, AFWL/ARTO, "TSL Milestones," 7 September 1972.

24. Although Rich rightfully deserved much of the credit, his success depended on all of the AFWL trial-and-error work trying to resolve the TSL technical problems prior to his arrival. Interview with Sher, 11 October 1988; Interview with Dyble, 1 August 1988; Interview with Lamberson, 12 January 1989; Rpt, Axelrod, pp. 27–28.

25. Interview with Keith Gilbert, 31 August 1988; Interview with Edward A. Duff, Colonel, Air Force Space Technology Center, Albuquerque, New Mexico, 13 September 1988; Interview with Rich, 10 January 1989; Interview with Dyble, 1 August 1988; Rpt, Axelrod, pp. 30–33, 41, 63–70.

26. At the time, only the University of Rochester's Institute of Optics and the University of Arizona's Optical Science Center granted degrees in optics. At both universities, these were independent departments separate from the Physics Department.

    By May 1972, the TSL had completed 690 test runs. Wear and tear on the device required it be torn down for large-scale maintenance. This involved overhaul of the nozzle bank, burner, and cavity, as well as checking for minor leaks in the system. Interview with Rich, 10 January 1989; Background Paper, n.a., "Air Force Laser Milestone," n.d. [April 1972]; Rpt, ARTO-73-1, pp. 2, 53–55.

27. Interview with Avizonis, 16 September 1988; Interview with Rich, 10 January 1989; Memo, Casey to Lamberson, 3 October 1973; Rpts, Rowden, "AFWL Significant Events," 29 February, 31 March, 30 April, and 31 May 1972.

28. The General Accounting Office's independent assessment of the TSL program issued in October 1973 recognized the benefit of Lamberson's decision by stating, "AFWL personnel were able to make a success out of AVCO's failure." See rpt, Air Force Audit Agency, "Management of High Energy Laser Program," 26 November 1975, p. 1. Interview with Lamberson, 12 January 1989; Interview with Sher, 11 October 1988; Interview with Dyble, 1 August 1988; Memo, Casey to Lamberson, 3 October 1973.

29. Interview with Axelrod, 27 January 1993; Rpt, AFWL-TR-73-094, pp. iii, 111–133; Rpt, AFWL-TR-74-225, pp. 5–9; Ltr, Major Edward N. Laughlin, Chief, Optics Technology Branch, AFWL, to LRS, "Response to AFSC TWX," 12 September 1972.

30. Interview with Sher, 11 October 1988; Interview with Lamberson, 12 January 1989; Interview with Dyble, 1 August 1988; Rpt, Lamberson *et al.*, "Advanced Radiation Technology," 15 June 1971.

31. Interview with Duff, 13 September 1988; Rpt, Lamberson *et al.*, 15 June 1971.

32. Interview with Dyble, 1 August 1988; Rpt, AFWL-TR-74-225, pp. 5, 24–30; Rpt, AFWL-TR-73-94, pp. 2–4; Rpts, Rowden, "AFWL Significant Events," 31 October, 30 November, and 31 December 1972.

33. Interview with Lamberson, 12 January 1989; Interview with Dyble, 1 August 1988; Rpt, Rowden, "AFWL Significant Events," 31 December 1972.

34. Memo for Col Lamberson, Maj George W. McKemie, Asst. Executive Secretary, USAF Scientific Advisory Board, HQ USAF, 1 January 1973, with attached Outline Report, 20 December 1972.

35. Interview with Gilbert, 31 August 1988; Rpt, AFWL-TR-76-271, Gilbert and Sher, "The Airborne Laser Laboratory Cycle II: Low Power Experiments," May 1977, p. 25; Rpt, Rowden, "AFWL Significant Events," 31 December 1972.

36. Interview with Lamberson, 30 June 1976; Interview with Dyble, 1 August 1988; Interview with Denny Boesen, Lieutenant Colonel, USAF Retired, Albuquerque, New Mexico, 30 September 1988; Interview with Gilbert, 31 August 1988; Interview with Richard V. Feaster, Colonel, USAF, Retired, Arlington, Virginia, 12 January 1989.

37. Rpt, AFWL-TR-74-250, Colonel Richard V. Feaster *et al.*, "*Project DELTA*," October 1974, pp. 3–4, 9–10; Ltr, Col William B. Allison, Chief, Office of Security Review/HQ Air Force, to AFSC/PAS, "Fact Sheet and Film—DELTA," 29 August 1980.

38. Rpt, AFWL-TR-74-250, pp. 10, 13, 21; Rpt, AFWL-TR-74-285, Captain Edward A. Duff *et al.*, "USAF Precision Pointing and Tracking Systems for High Energy Laser Applications—Test Results," December 1975, p. 23. Note that this work was also presented as a paper at the First DOD High Energy Laser Conference, Naval Training Center, San Diego, California, on 1 October 1974.

39. Interview with Dyble, 1 August 1988; Rpt, AFWL-TR-74-250, p. 32; John G. Duffey and Darrell E. Spreen, "Project DELTA," in *Journal of Defense Research, Series A: Strategic Warfare—High Energy Lasers* (Vol. 4A, No. 1), May 1975, pp. 41–47.

40. Rpt, AFWL-TR-74-250, p. 33; Duffey and Spreen, pp. 41–42, 46–47; "DOD Increasing Laser Weapons Development Budget: Airborne Tests Planned," *Aerospace Daily*, 21 March 1975, p. 113; Philip J. Klass, "Laser Destroys Missile in Test," *Aviation Week & Space Technology*, 7 August 1978, pp. 14–16; Ltr, Col John C. Rich, AFWL/AR, to Dr. William L. Lehmann, AFWL/CC, "Declassification of Project DELTA Film," 27 January 1980.

41. Interview with Gilbert, 31 August 1988; Interview with Dyble, 1 August 1988; Interview with Kyrazis, 20 June 1985; Rpt, AFWL-TR-74-250, pp. 33–34.

42. Ltr, Col Allison, Chief, Office for Security Review, Office of Public Affairs, HQ USAF, to AFSC/PAS, "Fact Sheet and Film—DELTA," 29 August 1980, with attached Fact Sheet: Project DELTA.

43. In December 1976, AFWL decommissioned the TSL after it had distinguished itself with "over 2,500 runs for a total of 12,958 seconds of hot gas flow and 4,470 seconds of photon extraction." The highest maximum power output achieved was 288 kilowatts. In 1979, AFWL disassembled and disposed of the TSL as scrap.

    Navy scientists pointed out their shootdown was different from the earlier Air Force and Army tests. The Navy's success was unique because it was the first time a laser was used against small, high-speed targets. Some disputed the "high-speed" aspect of this test as the missile poked along at about 500 miles per hour. Also, the thrust of the experiment was not on laser lethality—the main goal was to determine the accuracy of the pointing and tracking system. Interview with Dyble, 1 August 1988; Interview with Gilbert, 31 August 1988; Interview with Duff, 13 September 1988; Rpt, AFWL/LR, "AFWL Weekly Activity Report," 12 November 1976, p. 1; Ltr, Col John C. Rich, AFWL/AR, to AFWL/CC, "Declassification of Project DELTA Film," 27 June 1980; Ltr, Dyble to Dr. Ward Alan Minge, "Project DELTA," 31 October 1982; Rpt, AFWL-TR-74-250, p. 34; "U.S. Nears Laser Weapon Decisions," *Aviation Week & Space Technology*, 4 August 1980, pp. 48–54; "Industry Observer," *Aviation Week & Space Technology*, 23 August 1976, p. 9; Sandy Graham, "Flying Laser Gun is Kirtland Goal," *Albuquerque Tribune*, 14 October 1978, pp. 1, A-8; Ltr, Rich to Lehmann, "Declassification of Project DELTA Film," 27 January 1980.

44. Interview with Duff, 13 September 1988; Interview with Gilbert, 31 August 1988.

CHAPTER 3. LAUNCHING THE AIRBORNE LASER LABORATORY

1. Interview with Lamberson, 12 January 1989; Interview with Avizonis, 16 September 1988; Interview with Kyrazis, 20 June 1985; Interview with Shelton, 9 December 1988; Interview with Axelrod, 27 January 1993; Rpt, WLRE-67-133, "Advanced Development Plan," September 1967, pp. 1–2.

2. Interview with Lamberson, 12 January 1989.

3. Ibid.; Interviews with Shelton, 9 and 21 December 1989; Interview with Avizonis, 16 September 1988; Interview with Axelrod, 27 January 1993. For examples of early airborne applications see Rpt, AFWL-TR-69-125, Vols. I–IV, D. Green *et al.*, "Airborne Applications Study," October 1969; Rpt, AFWL-TR-70-155, Vols. I–III, Thomas J. Ernst, "Conceptual Design Study of an Airborne Test Bed," May 1971; and Rpt, AFWL-TR-70-156, Vols. I–II, A. I. Masters *et al.*, "Conceptual Design Study for an Airborne Test Bed," April 1971.

4. Avizonis was enhancing AFWL's laser reputation and expertise by presenting papers on the subject. For example, his "Interaction of High Power Optical (Laser) Radiation with Solid Target Surfaces" at the Tri-Service High Power Laser Technology meeting held at Redstone Arsenal on 7 April 1964 was well received. Interview with Avizonis, 16 September 1988; Program, "Tri-

Service Meeting: High Power Laser Technology," Directorate of Research and Development, U.S. Army Missile Command, Redstone Arsenal, Alabama, 7–8 April 1964.

5. Interview with Gilbert, 31 August 1988; Interview with Dyble, 1 August 1988; Interview with Avizonis, 16 September 1988; Requirements Action Directive, Maj Gen Kenneth C. Dempster, DCS/R&D HQ USAF, "Laser Weapon Program," 28 April 1967; Rpt, WLRE-67-133, pp. 1-1-1-4, 2-1–2-5; Rpt, WLREA-68-137, "Advanced Development Plan," February 1968, pp. 1-1–2-4, 3-3–3-4, A-1; Rpt, WLREL-69-032, "Advanced Development Plan," 24 February 1969, pp. 1-1–2-5, A-1. A separate Advanced Development Plan was published for solid-state lasers at AFWL until 1971, when solid-state laser work was transferred to the Air Force Avionics Lab. Ltr, Col David R. Jones, AFWL/CC, to AFSC/CC, "Recommendation for Award of Unit Decoration, 1 January 1967 to 31 December 1967," 14 March 1968; Ltr, Lt Gen W. A. Davis, AFSC/CV, to Maj Gen Marvin Demler, AFSC/RTD, "SCIENCE CUBE Report," 9 March 1967; Rpt, AFSC, West Coast Study Facility, "SCIENCE CUBE," January 1967.

6. Interview with Axelrod, 27 January 1993; AFSC Program Directive 644A-1-67-2, 1 June 1967, transmitting Requirements Action Directive 7-120-(1), 28 April 1967; Rpt, WLRE-67-133, pp. iii, 1-1–1-4, I-5.

7. Lamberson reported that by fiscal year 1970 solid-state lasers were not considered feasible for development into high-power weapons. He believed GDL work would emerge as AFWL's major laser program. Interview with Avizonis, 16 September 1988; Interview with Dyble, 1 August 1988; Rpt, ARTO-73-1, Donald L. Lamberson, "Program Management Plan: Advanced Radiation Technology Office," June 1973, p. 17. For more particulars see the individual Advanced Development Plans for 1967, 1968, 1969, and 1970 for Project 644A, Task I. Also see Headquarters Air Force, Development Directives for the High Energy Laser Program for those same years.

8. Effective 1 July 1971, 644A became 317J and known as the Advanced Radiation Technology Program. Avizonis and Axelrod wrote the GDL 1967 ADP. Interview with Avizonis, 16 September 1988; Interview with Axelrod, 27 January 1993; Rpt, AFSC West Coast Study Facility, Los Angeles, California, "Final Report: Technology Applications Study, High Energy Laser Program (HELP)," Vol. II, April 1967, p. A-1. The gas dynamic laser ADP was authorized by AFSC in AFSC's message to AFWL SCTSW 32032, 7 September 1967, directing AFWL "to prepare documentation for FY69 Advanced Development Program . . . on High Energy Lasers." The first two ADPs, completed in September 1967 and covering solid-state lasers (644A-Task I) and gas dynamic lasers (644A-Task II), were prepared in compliance with CSAF message AFRDDE 87451, dated 1 September 1967. Message, AFSC to AFAL, SCTSW 321781, dated 7 September 1967, called for a meeting at Air Force Systems Command Headquarters on 30 September 1967, "for the purpose of coordinating" the ADP package.

9. Rpt, WLREL-69-025, "Advanced Development Program: High Energy Laser Program, Program 644A-Task II," 24 February 1969, p. D-9; Rpt, WLL-70-010, "Advanced Development Program: High Energy Laser Program, Program 644A-Task II," March 1970; Statement of Work/Award of Contract for contract F29601-69-C-0099 with Hughes Aircraft Company for "Airborne Laser Applications." This contract was completed on 16 October 1969, at a cost of $281,215.00. Statement of Work/Award of Contract for contract F29601-69-C-0087 with the Boeing Company for "Ground Based Laser Applications." This contract was completed on 30 September 1969, at a cost of $221,476.00. Updated/final dollar amounts for these contracts can be found in Rpt, Walter A. Brown, AFCMD, "R&D Contract Status Report as of 30 June 1970," 1970, p. 36. Follow-on efforts by the Air Force Special Weapons Center resulted in the award of two additional contracts: F29601-69-C-0153 with the Hughes Aircraft Company for "Test and Evaluation Study for Airborne Laser Systems" and F29601-69-C-0146 with the Boeing Company for "Test and Evaluation Study for Ground Based Laser Systems." For more information, see Rpt, AFSWC-TR-70-10, W. A. Yates *et al.*, "Test and Evaluation Study for Airborne Laser Systems," 1970; and Rpt, AFSWC-TR-70-6, David E. Chadwick *et al.*, "A Test and Evaluation Study for Ground-Based

Laser Systems," 1970. Each of these studies is a good example of the optimism and enthusiasm the Air Force and private industry had for potential uses of lasers as weapons.

10. Rpt, AFWL-TR-69-125, Vol. I, "Executive Summary," pp. 1–2.

11. Ibid., pp. 2, 10–11; Interview with Raymond C. Saunders, Lockheed Missiles & Space Company, Inc., Albuquerque, New Mexico, 14 September 1988.

12. Interview with Saunders, 14 September 1988; Rpt, AFWL-TR-69-125, Vol. I, pp. 50–52.

13. Rpt, AFWL-TR-69-125, Vol. I, pp. 16–20 and Vol. III, "Recommended Study and Technology Activity," pp. 2–13.

14. Interview with Avizonis, 16 September 1988; Interview with Axelrod, 27 January 1993; Interview with Dyble, 1 August 1988; Interview with Saunders, 14 September 1988; Rpt, AFWL-TR-69-125, Vol. II, "System Concept Description," pp. 2–3.

15. Interview with Lamberson, 13 January 1989; Interviews with Shelton, 9 and 21 December 1988; Interview with Saunders, 14 September 1988. For details on the Boeing contract, see Rpt, AFWL-TR-69-123, Vols. I–V, Raymond C. Saunders, "Technology Application Study of Ground Based Laser Applications," October 1969. Volume I is an excellent summary of the work performed under the contract, whereas Volumes II through V are more theoretical/technical in nature.

16. Interview with Feaster, 12 January 1989; Interview with Saunders, 14 September 1988.

17. Interview with Lamberson, 13 January 1980; Interview with Saunders, 14 September 1988; Interviews with Shelton, 21 and 28 December 1988.

18. Interview with Saunders, 14 September 1988; Interview with Lamberson, 13 January 1989.

19. Interviews with Lamberson, 12 and 13 January 1989.

20. Interview with Lamberson, 12 January 1989; Interview with Shelton, 28 December 1988; Rpt, WLL-70-010, pp. 2–5.

21. For these early vulnerability tests at the AFWL Optics Laboratory, experimenters used continuous-wave carbon dioxide and neodymium glass lasers. Construction on the Optics Lab began on 30 June 1967 and reached completion on 22 April 1968 at a cost of $477,000. With 12,982 square feet of floor space, the new facility was designed to conduct testing of a multikilowatt gas dynamic laser and beam propagation experiments in a 200- × 7-foot underground tunnel. Interview with Avizonis, 16 September 1988; Rpt, AFWL, "Science Laboratory, Laser Optics," 9 September 1968; Rpt, AFWL, Colonel David R. Jones, "AFWL Significant Events," 10 July 1967; Rpts, AFWL, Colonel Robert W. Rowden, "AFWL Significant Events," 31 July 1970 and 30 September 1970; John Ira Petty, "At Kirtland's Laboratory: Laser Weaponry is Nearing Reality," *Albuquerque Journal*, 12 June 1970, pp. A-1, A-8; Rpt, ARTO-73-1, p. 6; Notes, prep. by Keith Gilbert, "Effects & Vulnerability Experiments," 4 March 1994.

22. Interview with Lamberson, 12 January 1989.

23. Interview with Lamberson, 12 January 1989; Rpt, AFWL-TR-70-175, William C. Missimer *et al.*, "Program for XLD-1 Development, Final Technical Report," Vol. I: "Experimental and Analytical Investigations," Vol. II: "Experimental and Analytical Investigations—Appendices," and Vol. III: "Task VII-Advanced System Study," February/March 1971; Rpt, PWA-FP-70-513, R. Coar *et al.*, Pratt & Whitney Corp., "XLD Follow-on Program, Vol. I: Technical Proposal," 27 April 1970, pp. 7–12.

24. Interview with Lamberson, 12 January 1989; Interview with Shelton, 28 December 1988. Precise funding levels can be found in Development Directive 104 and its amendments, and the revised annual Advanced Development Program submissions.

25. Interview with Orpha R. Cunningham, Colonel, USAF, Retired, Los Angeles, California, 15 May 1990; Interview with Lamberson, 12 January 1989; Interview with Avizonis, 16 September 1988.

26. Interview with Lamberson, 12 January 1989; Interview with Avizonis, 16 September 1988; Interview with Shelton, 9 December 1988; Interview with Parsons, 15 December 1988.

27. Statement of Work for Purchase Request 626010-LLL-0067 for a "Conceptual Design Study for an Airborne Test Bed," awarded to Pratt & Whitney Corp. as contract F29601-70-C-0049 [contract

effective 13 April 1970 for $286,538.00] and to AVCO-Everett as contract F29601-70-C-0048 [contract effective 10 April 1970 for $349,758.00]. Research and Technology Work Unit Summaries, "Conceptual Design of an Airborne Test Bed," 28 September 1970 [contract F29601-70-C-0048], and 5 October 1970 [contract F29601-70-C-0049]; Rpt, AFWL. Rowden, "AFWL Significant Events," 31 July 1970.

28. Rpt, AFWL-TR-70-155, Vol. I, "Summary," pp. 1–7; Rpt, AFWL-TR-70-156, Vol. I, "Summary," pp. 1–3, 19–30.

29. Ibid.

30. Interviews with Shelton, 9 and 28 December 1988; Rpt, DD-DR&E(AR) 637, "Advanced Development Plan, Program 644A Task II," March 1971, pp. 1–4, 3-28–3-32.

31. Interview with Shelton, 21 December 1988; Interview with Richard W. Davis, Lieutenant Colonel, USAF, Air Force Weapons Laboratory, Albuquerque, New Mexico, 20 September 1988; Interview with John William (Bill) Dettmer, Colonel, USAF, Retired, Albuquerque, New Mexico, 3 October 1988; Interview with Saunders, 14 September 1988; Interview with Otten, 27 December 1988.

32. Interview with Otten, 27 December 1988; Interview with Shelton, 21 December 1988.

33. Interview with Dettmer, 3 October 1988; Interview with Hill, 23 September 1988; Interviews with Kyrazis, 20 June and 17 October 1985; Interview with Parsons, 15 December 1988; Interviews with Shelton, 9, 21, and 28 December 1988.

34. Interview with Lamberson, 12 January 1989; Interviews with Shelton, 9 and 28 December 1988; Interview with Otten, 27 December 1988; Interview with Axelrod, 27 January 1993; Letter contract between AFWL and Hughes Aircraft Company for "Optics for High Power Systems," contract F29601-69-C-0058, effective 3 February 1969; Statement of Work for Purchase Request L9-644A-106, "Optics for High Power Laser Systems," 20 January 1969; Sole Source Justification for Purchase Request L9644A-106, 24 January 1969. This contract authorized Hughes to develop both ground-based and airborne optics. The Statement of Work required Hughes "to conduct a system and design study of airborne GDL optical systems." Also see the monthly reports on technical progress submitted under this contract for the months of March through November (inclusive) 1971.

35. Interviews with Shelton, 9 and 21 December 1988; Rpt, AFWL-TR-70-175, Vols. I–III, February 1971. Statement of Work and Award/Contract form for Purchase Request L9-1256-124, "Further Development of an Existing High Power Laser System," awarded as letter contract F29601-68-C-0109 to Pratt & Whitney Corp. effective 10 June 1968 and funded for $966,159.00. This contract was superseded by contract F29601-69-C-0069. Rpt, GP 71-177, Missimer *et al.*, Pratt & Whitney, "SSD Design Review Meeting, Pratt & Whitney Aircraft, Florida Research and Development Center, June 14 through June 17, 1971," 1971, pp. 2–36.

36. Rpt, PWA FP 70-513, "XLD Follow-On Program," 27 April 1970, pp. 7–10; Rpt, AFWL-TR-71-129, Vol. I, M. T. Schilling *et al.*, "GDL Technology Confirmation Program," January 1972, p. 1; AFWL-TR-70-175, Vol. I, p. 3.

37. "Laser Progress Boosts USAF Funding," *Aviation Week & Space Technology*, 12 January 1970, p. 16; "Progress On Laser Weapons," *U.S. News & World Report*, 1 October 1973, pp. 41–42. Three years later Dr. Lehmann stated "good progress" is being made in development of lightweight, reliable, and "reasonably efficient" lasers.

38. Interview with Lamberson, 12 January 1989; Interview with Shelton, 29 December 1988; Rpt, AFWL-TR-70-175, Vol. I, p. 3.

39. Interview with Lamberson, 12 January 1989; Interview with Cunningham, 15 May 1990.

40. Ibid.; Interview with Parsons, 15 December 1988; Rpt, AFSC, "High-Energy Gas Laser Applications Master Plan," February 1972, p. 1; Rpt, AFWL, Lamberson *et al.*, "Advanced Radiation Technology: Attachment #2 Program 317J—Funding Schedule," 15 June 1971.

41. Program Memorandum for the Accelerated High Energy Laser Program, PE 63602F, September 1970; Rpt, Lamberson, 15 June 1971.

42. Interview with Lamberson, 12 January 1989; Interviews with Shelton, 9, 21, and 28 December 1988; Interview with Parsons, 15 December 1988; Program Budget Decision 338, AFSC, "High Energy Laser Program: P.E. 63605F," 7 December 1970; Development Directive 104 for High Energy Laser Program, Brig Gen C. H. Bolender, Deputy Director of Development & Acquisition, DCS/R&D, HQ USAF, 3 February 1971; "Program Memorandum for the Accelerated High Energy Laser Program, PE 63602F," September 1970; Memo for the Record, n.a., "Background Paper—Air Force Laser Milestone," n.d.; Rpt, ARTO-73-1, p. 8.

43. Interview with Lamberson, 12 January 1989; Program Memorandum for the Accelerated High Energy Laser Program, PE 63602F, September 1970.

44. Interview with Lamberson, 12 January 1989; Interview with Gilbert, 31 August 1988; Ltr, Brig Gen C. H. Bolender, Deputy Director for Development and Acquisition, DCS/R&D, HQ AF, to SAF/RDL (Dr. Lehmann), "High Energy Lasers," 8 February 1971; Edgar E. Ulsamer, "Laser: A Weapon Whose Time is Near," *Air Force and Space Digest*, December 1970, pp. 28–34; "Next U.S. Superweapon—The Pentagon's 'Light Ray,'" *U.S. News & World Report*, 18 October 1971, pp. 85–87.

45. In 1968, Soviet scientists N. N. Sobdev and V. V. Sokovikov, from the renowned Lebedev Physics Institute, published a scientific paper stating that "different military applications [for carbon dioxide lasers] are possible. In particular, targets can be damaged with the aid of a laser beam if powers on the order of several kilowatts in the continuous mode are reached." See "Laser Weaponry Seen Advancing," *Aviation Week & Space Technology*, 12 January 1970, pp. 16–17. Interview with Lamberson, 12 January 1989; Rpt, AFSC, "High-Energy Gas Laser Applications Master Plan, Vol. I," July 1970, p. 6; Philip J. Klass, "Research Nears Application Level," *Aviation Week & Space Technology*, 14 August 1972, p. 13; Rpt, WLREL-69-025, pp. F1–F2; Rpt, ARTO-73-1, pp. 5–7.

CHAPTER 4. GAINING MOMENTUM

1. Rpt, Major Thomas J. Dyble, "Peace Through Light: The Airborne Laser Laboratory" (Air Command and Staff College Student Report 83-0615, 1983), pp. 1, 20–21; Michael H. Gorn, *Harnessing the Genie: Science and Technology Forecasting for the Air Force 1944–1986* (Washington, D.C.: Office of Air Force History, 1988), pp. 11–13; Thomas M. Coffey, *HAP: The Story of U.S. Air Force and the Man Who Built It* (New York: The Viking Press, 1981), p. 232.

2. Gorn, pp. 13–32; Ltr, H. H. Arnold to Dr. Theodore von Karman, "AAF Long Range Development Program," 7 November 1944; Rpt, SAB, "Toward New Horizons: A Report to General of the Army H. H. Arnold by the AAF Scientific Advisory Group," 15 December 1945; Thomas A. Sturm, *The USAF Science Advisory Board: Its First Twenty Years, 1944–1964* (Washington, D.C.: USAF Historical Division Liaison Office, 1967), pp. 2–14; James A. McDonnell, Jr., "Centennial Tribute to Dr. Theodore von Karman," *Air Force Magazine*, August 1981, pp. 84–86.

3. Gorn, pp. 8, 39, 45–48, 62, 98; Sturm, pp. 22–26; "The Tools: Weapons," *Air Force Magazine: The Golden Anniversary*, August 1957, p. 344.

4. Gorn, pp. 47–48, 59.

5. Ltr, John S. Foster, Jr., Director of Defense Research & Engineering, to Chairman, Defense Science Board (Dr. Millburn), "Defense Science Board Task Force," 3 July 1969; Ltr, Col Harold E. Collins, CDS/R&D, to AFSC (SCTSW—Maj Tony Chiota), "Defense Science Board Task Force," 10 July 1968; Memo, Lt Col Glenn G. Sherwood, DCS/Development Plans, to SCLA (Col Gilbert), "Defense Science Board Task Force on High Energy Laser Program," 16 August 1968.

6. Memo, Sherwood to SCLA (Col Gilbert), "Defense Science Board Task Force on High Energy Laser Program," 16 August 1968.

7. Ltr, G. P. Sutton to SAB Ad Hoc Committee Members, 30 December 1968, with attached draft on High-Energy Lasers, 2 January 1969.

8. Ibid.

9. Ibid.; Ltr, Col David R. Jones, AFWL/CC, to AFSC (BG R. A. Gilbert), "USAF Scientific Advisory Board Ad Hoc Committee Report on Laser Programs, May 1969," 1 August 1969.

10. United States Air Force, Scientific Advisory Board, *Report of the USAF Scientific Advisory Board Spring General Meeting, Kirtland AFB, New Mexico, 15–16 April 1971* (Washington, D.C.: USAF, 1971), pp. 2–5.

11. Ibid., pp. i, 3–5.

12. Ibid., pp. 104–108.

13. Ibid., pp. 1–2, 5–6, 12, 34, 74.

14. Ibid., pp. 77–79; Edgar Adcock, Jr., ed., *American Men and Women of Science* (New York: R. R. Bowker Co., 1977), p. 813.

15. Memo for the Record, Col Donald L. Lamberson, "April Meeting of the SAB," April 1971.

16. Interviews with Lamberson, 12 and 13 January 1989; Interview with Kyrazis, 19 September 1985.

17. Interview with Axelrod, 27 January 1993; Science Advisory Board, *Spring General Meeting*, pp. 1–2, 87–88.

18. Interview with Axelrod, 27 January 1993; Science Advisory Board, *Spring General Meeting*, pp. 3–29, 34–46, 61–71.

19. Interview with Axelrod, 27 January 1993; Science Advisory Board, *Spring General Meeting*, pp. 80–82, 84–85; Jacob Neufeld, *Ballistic Missiles in the United States Air Force, 1945–1960* (Washington, D.C.: Office of Air Force History, 1990), p. 237.

20. Science Advisory Board, *Spring General Meeting*, pp. 1–2, 87–88.

21. Writing about the history of technology in the United States, noted historian Daniel J. Boorstin referred the reader to Arthur C. Clarke's *Profiles of the Future* for an assessment of turning visionary ideas into technological realities. (Clarke also authored *2001: A Space Odyssey*.) Boorstin quoted "Clarke's Law": "When a distinguished but elderly scientist states that something is possible, he is almost certainly right. When he states something is impossible, he is very probably wrong." See Boorstin, *The Republic of Technology: Reflections on our Future Commentary* (New York: Harper & Row, 1978), p. 31. Paul Dickson, "It'll Never Fly, Orville: Two Centuries of Embarrassing Predictions," *Saturday Review*, December 1979, p. 36; Zbigniew Brzezinski, Robert Jastrow, and Max M. Kampelman, "Defense in Space is not 'Star Wars,'" *The New York Times Magazine*, 27 January 1985, p. 48; NASA Information Summaries, "The Early Years: Mercury to Apollo-Soyuz," May 1987, pp. 3, 5–6.

22. Science Advisory Board, *Spring General Meeting*, p. 81.

23. Ibid., p. 102.

24. Interview with Dyble, 1 August 1988; Interview with Feaster, 12 January 1989; Interview with Gilbert, 31 August 1988; Rpt, ARTO 73-1, Donald L. Lamberson, "Program Management Plan: Advanced Radiation Technology Office," June 1973, pp. 1–2; Rpt, ARTO 74-2, Lamberson, "Program Management Plan: Advanced Radiation Technology Office," June 1974, pp. 93–97, 103.

25. Interviews with Lamberson, 12 and 13 January 1989. For a later SAB look at the ALL Program, see United States Air Force, Scientific Advisory Board, *Interim Report of the USAF Scientific Advisory Board Ad Hoc Committee on Laser Technology, May 1974* (Washington, D.C.: USAF, 1974). Appendix A in this report has a good overview of other potential laser applications.

26. Rpt, AFWL, Colonel Robert W. Rowden, "AFWL Significant Events," 31 March 1972.

27. SAC had been open-minded in exploring the possibility of using lasers on aircraft. Major General Douglas T. Nelson, project manager for the B-1 bomber, stated in October 1971, "We are very interested in the laser. We are watching it closely. It is very promising as a possible weapon system for the B-1."

    AFWL continued to follow-up and brief SAC and TAC at every opportunity. General William W. Momyer, TAC's commander, was very responsive and enthusiastic to AFWL's 25 May 1973 briefing covering programs on lasers. Lamberson in a 29 May 1973 message sent to AFSC reported Momyer stated he "directed his staff to more formally pursue the role of the high-energy laser for tactical fighter and other TAC applications. He [Momyer] stated his satisfaction in the technology [laser] progress to date and indicated a desire to phase TAC responsiveness with the accomplishment of future [laser] milestones. My staff will aggressively follow up."

Lamberson's briefing to the SAC vice commander, Lt Gen Martin, on 21 May 1973 covered the conceptual design for installing a laser on a B-52 resulted in no response or commitment to support. Lamberson interpreted this as "little enthusiasm for B-52 defense." Instead, Martin wanted to withhold judgment until he visited AFWL to "look at the [laser] hardware." As it turned out, SAC never made a commitment to lasers. "We frankly couldn't get the time of day in Strategic Air Command headquarters," Lamberson recalled. "I'd go there every year, and they were very polite, and we had good discussions. They would get me into CINCSAC and he would nod benignly and was generally interested, but there was no requirement [for lasers]."

Rpt, Rowden, 31 March 1972; Memo for the Record, ARTO, "ARTO Briefings to SAC and TAC," 21 and 25 May 1973; Message, AFWL to AFSC/CC, "Advanced Radiation Technology Program Status Report," 29 May 1973. "Next U.S. Superweapon—The Pentagon's 'Light Ray,'" *U.S. News & World Report*, 18 October 1971, p. 85; Orr Kelly, "Bomber May Carry Death Ray," *The Evening Star* (Washington, D.C.), 19 May 1971, p. A-8.

28. AFSC Management Plan, "Proposed Laser Engineering and Applications for Prototype Systems Program Office (LEAPS)," March 1971; Ltr, Rowden to AFSC MET (Det 24), "Request for Organizational Changes and Additional Manpower," 26 March 1971.

29. Scholtz's new appointment meant Lamberson's former boss was now working for Lamberson. However, that presented no problems. Lamberson emphasized, "The relations between us were always very, very good." Interviews with Lamberson, 12 and 13 January 1989; Interview with Kyrazis, 20 June 1985; Interview with Axelrod, 27 January 1993; Rpt, AFWL, Lamberson, "Airborne Laser Laboratory Configuration and Status Report as of May 1972," May 1972, pp. 44–47; Rpt, Rowden, "AFWL Significant Events," 31 July 1971; Rpt, Lamberson, 1973, p. 1.

30. Interviews with Lamberson, 12 and 13 January 1989.

31. Ibid.; Rpt, ARTO-73-1, pp. 21–80.

32. ADPs for 644A—Task II for 1967, 1968, 1969, 1970, and 1971.

33. Interviews with Lamberson, 12 and 13 January 1989; Interview with Gilbert, 31 August 1988; Rpt, ARTO-73-1, pp. 1–4, 8–15; Rpt, Dyble, pp. 7–11.

34. Rpt, Lamberson, pp. 2–5.

35. Interview with Lamberson, 12 January 1989; Interview with Shelton, 28 December 1988; Interview with Sher, 11 October 1988; Interview with Otten, 27 December 1988; Rpt, ARTO-73-1, pp. 2–3.

36. Development Directive 104 for High Energy Laser program, Brig Gen C. H. Bolender, Deputy Director of Development & Acquisition, DCS/R&D, HQ USAF, 3 February 1971; Program Memorandum for the High Energy Laser Program (P.E. 63602F), September 1970.

37. Rpt, ARTO-73-1, pp. 10–13.

38. Interview with Gilbert, 31 August 1988; Interview with Otten, 27 December 1988.

39. Interview with Parsons, 15 December 1988.

40. Interview with Avizonis, 18 September 1988.

41. Memo for the Record, n.a., "ARTO Contract Management," 28 October 1975.

42. Interviews with Lamberson, 12 and 13 January 1989; Ltr, Lt Gen J. W. O'Neil, AFSC/VC, to AFWL, "Program Action Directive on the Advanced Radiation Technology Program," 30 June 1972, with attached "Program Action Directive: Establishment of the Advanced Radiation Technology Office," June 1972.

43. Interview with Gilbert, 31 August 1988; Interview with Kyrazis, 20 June 1985; Interview with Lamberson, 12 January 1989; Rpt, Rowden, 31 July 1971; Rpt, Air Force Audit Agency, "Management of Air Force High Energy Laser Program," 16 November 1975, p. 1; Edgar Ulsamer, "Status Report on Laser Weapons," *Air Force Magazine*, January 1972, p. 63; Program

Management Directive, AFSC, "Advanced Radiation Technology Program," 30 June 1972; Rpt, ARTO-73-1, June 1973, pp. 10–17.

44. Interviews with Lamberson, 12 and 13 January 1989; Interview with Shelton, 9 December 1988; Rpt, Air Force Audit Agency, "Management of Air Force High-Energy Laser Program," 26 November 1975, p. 1; Ltr, Colonel Russell K. Parsons, AFWL Chief, Laser Development Division, to AFWL/AR, "AFWL Laser Program," 22 April 1974, with attached questions and answers.

45. Interview with Lamberson, 13 January 1989; Interview with Gilbert, 31 August 1988; Interview with Sher, 11 October 1988; Interview with Jerome T. Janicke, Colonel, USAF, Retired, Albuquerque, New Mexico, 4 October 1988; Rpt, ARTO-73-1, p. 2.

46. Interview with Axelrod, 27 January 1993; Ltr, 1st Lt Paul M. Mebane to Mr. Peyton Robinson, General Dynamics, "Technical Direction on Contracts," 29 August 1972; Memo, Lt Col C. E. Brunson, AFWL/LR, "Airborne Laser Laboratory," 8 May 1972; Rpt, Lamberson, p. 46.

47. Rpt, ARTO-73-1, pp. 15–72.

48. Interview with Gilbert, 31 August 1988.

CHAPTER 5. MOVING AHEAD WITH HARDWARE

1. Interview with Parsons, 15 December 1988; Interview with Lamberson, 12 January 1989.

2. AFSWC's 4900th Test Group flew and maintained these aircraft. Interview with Cunningham, 15 May 1990; Interview with Parsons, 15 December 1988.

3. Ibid.

4. Interviews with Shelton, 9 and 28 December 1988.

5. Interview with Shelton, 21 December 1988.

6. Ibid.; Interview with Shelton, 9 December 1988.

7. Interview with Dettmer, 3 October 1988; Interview with Hill, 23 September 1988; Interviews with Kyrazis, 20 June and 17 October 1985; Interview with Parsons, 15 December 1988; Interview with Shelton, 28 December 1988; Rpt, AFWL, Colonel Donald L. Lamberson, "Airborne Laser Laboratory Configuration and Technology Status Report," March 1973, p. 2; Statement of Work for Installation of a Laser Device into the Airborne Laser Laboratory, AFWL/LRL, 11 October 1973.

8. Interview with Davis, 20 September 1988; Interview with Dettmer, 3 October 1988; Interview with Parsons, 15 December 1988; Interviews with Shelton, 9 and 28 December 1988.

9. Ltr, Col Martin H. Brewer, AFSWC/CC, to AFSC (DOOA), "Request for RDT&E Aircraft," 4 February 1971; Rpt, AFSWC, Oscar S. Ayers *et al.*, "Advanced Radiation Technology," 26 March 1976.

10. AFSWC's responsibility for the KC-135 ended in February 1976, when support for the aircraft transferred to the 4950th Test Wing, headquartered at Wright-Patterson AFB. This action was taken in anticipation of the disestablishment of AFSWC, which occurred in April 1976. After that, Detachment 2, 4950th Test Wing, located at Kirtland, had primary operational responsibility for the ALL. Interview with Cunningham, 15 May 1990; Interview with Parsons, 15 December 1988; Interview with Dettmer, 3 October 1988; Ltr, Lt Col Carl L. Rucker, Chief of Plans and Requirements at AFSWC, to Col Lamberson, "NKC-135A Aircraft," 10 November 1971; Rpt, ARTO-73-1, Lamberson, "Program Management Plan: Advanced Radiation Technology Office," June 1973, p. 73.

11. Interview with Kyrazis, 20 June 1985; Interview with Lamberson, 13 January 1989; Interview with Shelton, 9 December 1988; Interview with Duff, 13 September 1988; Interview with Gilbert, 31 August 1988; Interview with Otten, 27 December 1988; Interview with Janicke, 4 October 1988; Rpt, AFWL-TR-86-01, Vol. I, Raymond V. Wick, "Airborne Laser Laboratory Cycle III: System and Test Descriptions," May 1988, pp. iv–ix; Rpt, AFWL, Lamberson, "Airborne Laser Laboratory Configuration and Technology Status Report," May 1972, pp. 2–4.

12. Interview with Duff, 13 September 1988; Interview with Kyrazis, 20 June 1985; Interviews with Shelton, 21 and 28 December 1988; Rpt, AFWL-TR-86-01, Vol. I, pp. iv–v; Rpt, Lamberson, pp. 2–19; Rpt, AFWL, Colonel Robert W. Rowden, "AFWL Significant Events," 31 July 1970.

13. Interview with Shelton, 21 December 1988; Interview with Lamberson, 12 January 1989; Rpt, AFWL, Harry I. Axelrod, "A 'White Paper' on the Air Force Laser," April 1972, pp. 26–28; Rpt, AFSC, "High Energy Laser Applications Master Plan," February 1972, p. 42; Development Directive 104, Brig Gen C. H. Bolender, HQ USAF, "High Energy Laser Program," 3 February 1971.

14. Interviews with Lamberson, 12 and 13 January 1989; Interview with Shelton, 9 December 1988; Statement of Work for contract F29601-69-C-0058 with Hughes Aircraft Company for "Development and Production of a Field Test Telescope," 23 January 1969; Rpt, AFWL-TR-73-94, John R. Goos, "Optics for High-Power Laser Systems," November 1973, pp. 1–10, 152–153; Rpt, AFWL, Lamberson, "Airborne Laser Laboratory Configuration and Technology Status Report," March 1973, p. 8.

15. Interview with Duff, 13 September 1988; Interview with Boesen, 30 September 1988; Interview with Otten, 27 December 1988; Rpt, ARTO-73-1, pp. 39, 57; Rpt, AFWL-TR-71-30, S. J. Novak, "Airborne Pointing and Tracking System Conceptual Design Final Report," June 1973, pp. ii, 1–27; Rpt, AFWL-TR-73-94, pp. 1–110; Rpt, Lamberson, pp. 8–18; Rpt, AFWL, Lamberson, "Airborne Laser Laboratory Configuration and Technology Status Report," May 1972, pp. 14–23; Briefing, Hughes Aircraft Company to the Airborne Test Bed Configuration Review Meeting at AFWL, 25 May 1971.

16. The entire APT system, to include the turret, control, electronics, instrumentation consoles, and hydraulics, would weigh 4240 pounds according to first estimates. As it turned out, this was about 1500 pounds too low. Interview with Duff, 13 September 1988; Interview with Boesen, 30 September 1988; Interview with Gilbert, 31 August 1988; Rpt, AFWL-TR-71-30, pp. 1–49; Rpt, Lamberson, pp. 18–23.

17. Statement of Work for contract F29601-71-C-0058 with Hughes Aircraft Company for "Airborne Pointing and Tracking System—Phase I, Design" and Contract/Award forms for contract F29601-71-C-0058 with Hughes Aircraft Company for "Airborne Pointing and Tracking System—Phase I, Design," 26 February 1971.

18. The first four monthly reports were: Rpt, P71-146, Hughes Aircraft, "Monthly Progress Report, Period Covered 1 March 1971 through 28 March 1971"; Rpt, SDN F-60192, Hughes Aircraft, "Monthly Progress Report, Period Covered 1 April 1971 through 30 April 1971"; Rpt, P71-230, Hughes Aircraft, "Monthly Progress Report, Period Covered 1 May 1971 through 31 May 1971"; Rpt, P71-308, Hughes Aircraft, "Monthly Progress Report, Period Covered 1 June 1971 through 31 July 1971." Specifically see pp. 1, 3 of this last document for results of the preliminary design review.

19. Interview with Shelton, 21 December 1988.

20. Rpt, P71-308, pp. 1, 5–8; Rpt, P71-362, Hughes Aircraft, "Monthly Progress Report, Period Covered 1 August 1971 through 31 August 1971," pp. 1–18; Rpt, P71-410, Hughes Aircraft, "Monthly Progress Report, Period Covered 1 September 1971 through 30 September 1971," pp. 1–7; Rpt, P71-452, Hughes Aircraft, "Monthly Progress Report, Period Covered 1 October 1971 through 31 October 1971," pp. 1–2, 13–16.

21. Interview with Duff, 13 September 1988; Interview with Dyble, 1 August 1988; Interview with Rich, 10 January 1989; Message, AFWL (Lt Col C. E. Brunson) to Secretary of Defense (Advanced Research Projects Agency), "Weekly Progress Report," 11 November 1971; Rpt, AFWL-TR-73-94, pp. 1–7; Award/Contract Form and Statement of Work for contract F29601-72-C-0029 with Hughes Aircraft Company for "Final Design, Fabrication and Test of an Airborne Pointing and Tracking System," 4 November 1971.

22. Lieutenant Colonel Denny Boesen, who had worked on the APT for nearly 8 years, lamented over the final disposition of the solid mahogany prototype. Boesen recalled the wooden model "sat out

in the scrap yard here [Kirtland AFB] in the rain and sunshine for years" until it was sold as salvage. Interview with Sher, 11 October 1988; Interview with Boesen, 30 September 1988; Interview with Duff, 13 September 1988; Interview with Otten, 27 December 1988; Rpt, AFWL-TR-72-124, Thomas R. Welch *et al.*, "Airborne Pointing and Tracking System," May 1975, pp. 19–28; Rpt, Lamberson, p. 5.

23. Interview with Lamberson, 12 January 1989; Interview with Shelton, 28 December 1988; Rpt, AFWL, Lamberson, "ALL Configuration and Technology Status Report," March 1973, p. 8; Rpt, AFWL-TR-72-124, pp. 1–2.

24. Interview with Lamberson, 12 January 1989; Interview with Shelton, 28 December 1988; Rpt, Lamberson, p. 8.

25. Memo for the Record, Ronald H. Stephens, Kirtland Air Force Base Contracting Officer, "Source Selection (PR FY617-71-10102) Contract F29601-71-C-0064," 12 March 1971.

26. Ibid.

27. Ibid.

28. Rpt, FZP-1187, General Dynamics, Convair Aerospace Division, "Part I—Technical Proposal for a Study and Detailed Design of an Airborne Test Bed," 4 December 1970, pp. 2-13–2-14, 4-3–4-28.

29. Ibid., pp. 2-14–2-16, 5-3–5-11.

30. Statement of Work and later Amendments to Statement of Work for contract F29601-71-C-0064 with General Dynamics (Fort Worth, Texas), for "Study and Design of an Airborne Test Bed," 16 October 1970.

31. Ibid.

32. Ibid.

33. Ibid.

34. Ibid.; Interview with Lamberson, 12 January 1989; Interview with Shelton, 28 December 1988; Rpt, AFWL, Lamberson, "Airborne Laser Laboratory Configuration and Status Report," May 1972, p. 9.

35. Statement of Work/Amendments, contract F29601-71-C-0064, "Study and Design of an Airborne Test Bed," 16 October 1970.

36. The numbers refer to inches aft of a reference point established near the aircraft nose. Rpt, 599FW0023-I, T. Peyton Robinson, General Dynamics, "Preliminary Design Analysis Report of an Airborne Test Bed (Phase I)," 1 September 1971, pp. 4-153–4-161; Rpt, General Dynamics, "R&D Contract Status Rpt (F29601-71-C-0064)," 28 April 1972, pp. 2-1–2-4. For progress on APT, see Hughes Aircraft, "Monthly Progress Reports" for April–September 1971.

37. Interview with Shelton, 28 December 1988; Interview with Duff, 13 September 1988; Rpt, 599FW0023-II, Robinson, General Dynamics, "Final Design Analysis Report of an Airborne Test Bed (Phase II)," 30 December 1971, pp. 117–134; Rpt, 599FW0023-I, pp. 4-95, 4-116–4-126.

38. Interview with Lamberson, 12 January 1989; Interview with Shelton, 28 December 1988; Interview with Ed Laughlin, Lieutenant Colonel, USAF, Retired, Las Cruces, New Mexico, 4 April 1990; Interview with Edgar A. O'Hair, Jr., Lieutenant Colonel, USAF, Retired, Lubbock, Texas, 26 March 1990.

39. Ibid.

40. Rpt, AFWL-TR-74-52, J.M. Fitts *et al.*, "Prototype Airborne Pointing and Tracking System," September 1974, Vol. I: "Lightweight Engineering Techniques and Concepts," pp. 33–41, and Vol. II, "Schedule and Concept Verification Plan," pp. 1–17; Rpt, FZM-5762, General Dynamics, "Preliminary ATB Mock-Up Panel Description and Operation," 20 July 1971, pp. 1–82.

41. Rpt, 599FW0023-I, pp. 4-127–4-131; Rpt, FZM-5762, pp. 1–82.

42. Interviews with Shelton, 9 and 28 December 1988; Rpt, 599FW0023-I, pp. 1-1–1-2, 2-1–2-6.

43. Rpt, 599FW0023-II, pp. 1–28.

44. Rpt, 599FW0023-II, pp. 72–78; Rpt, AFWL, Lamberson, "Airborne Laser Laboratory Configuration and Technology Status Report," March 1973, p. 16.

45. Ibid., pp. 7–8; Interview with Shelton, 28 December 1988.

46. Interview with Kyrazis, 3 November 1991; Rpt, 599FW0023-II, p. 108.

47. Handover from the OAD to APT was accomplished in Cycle I to prove it could be done. Interviews with Kyrazis, 8 August and 12 and 19 September 1985; Interview with Shelton, 28 December 1988; Rpt, 599FW0023-II, pp. 10–11, 135–137.

48. Interviews with Kyrazis, 12 and 19 September 1985; Interview with Shelton, 28 December 1988; Rpt, 599FW0023-II, pp. 193–198.

49. Interviews with Kyrazis, 20 June and 17 October 1985; Interview with Shelton, 21 December 1988; Interview with Steve Coulombe, Lieutenant Colonel, USAF, Phillips Laboratory, Albuquerque, New Mexico, 1 February 1994; Rpt, 599FW0023-II, pp. 198–203.

CHAPTER 6. AERODYNAMICS AND SAFETY

1. Interview with Kyrazis, 20 June 1985; Interviews with Shelton, 21 and 28 December 1988; Interview with Otten, 27 December 1988.

2. Second Lieutenant Leonard J. Otten had been assigned to the HEL group at AFWL headed by Captain Axelrod during the pre-TSL days. Earlier he had served as the AVCO MK-5B project officer. Interview with Kyrazis, 20 June 1985; Interviews with Lamberson, 12 and 13 January 1989. See also F. L. Smith, "Passive Control Effects on Flow Separation Around a Protuberance at High Subsonic Speed" (unpublished master's thesis, University of New Mexico, 1985).

3. Interview with Lamberson, 18 April 1984; Interview with Shelton, 21 December 1988.

4. The title of Kyrazis's Ph.D. dissertation was "A Numerical Modeling of Fluid Dynamics Stability of Hagen-Poiseuille Flow." Interview with Kyrazis, 20 June 1985; Biography (USAF), "Demos T. Kyrazis."

5. Interview with Shelton, 21 December 1988.

6. Ibid.; John W. Dettmer and Demos T. Kyrazis, "Overview of the Airborne Laser Laboratory Program," in *Proceedings of the Third DOD High Energy Laser Conference,* July 1979, p. 380; Rpt, AFWL-TR-86-01, Vol. I, Raymond V. Wick, "Airborne Laser Laboratory Cycle III: System and Test Descriptions," May 1988, p. 1.

7. Biography (USAF), "Colonel Leonard J. Otten III, as of August 1988."

8. Interview with Lamberson, 18 April 1984.

9. Interview with Shelton, 21 December 1988; Interview with Kyrazis, 20 June 1985.

10. Interview with Kyrazis, 20 June 1985.

11. Ibid.

12. Ibid.

13. Ibid.

14. Ibid.; Interview with Shelton, 21 December 1988.

15. Ibid.; Interview with Shelton, 28 December 1988.

16. The diagnostic plane (#371) used in support of the ALL originally was a "nuclear observer" aircraft assigned to the Air Force Special Weapons Center at Kirtland AFB. Interview with Shelton, 21 December 1988; Interview with Otten, 27 December 1988.

17. Interview with Kyrazis, 20 June 1985; Interview with Parsons, 15 December 1988; Rpt, FZE-875, J. E. Hanes, General Dynamics, "HAVE CHARITY Experimental Aerodynamic Stability and Control Characteristics," April 1969, pp. 1, 144–145; Rpt, FZA-442, E. L. Mauzy, General Dynamics, "Experimental Lift, Drag and Flow Field Effects of Large Protuberances Attached to a C-135B Model," March 1969, pp. 46–50.

18. Interview with Kyrazis, 20 June 1985; Interview with James A. Davis, Lieutenant Colonel, USAF, Retired, Canoga Park, California, 14 May 1990; Interview with Parsons, 15 December 1988.

19. Ibid.; Memo for the Record, Donald A. Buell to Director, Aeronautics and Flight Mechanics, "Visit to Kirtland Air Force Base Concerning KC-135 External Telescope Mount," 21 July 1971.

20. Francis Wenham, one of the founders of the Aeronautical Society of Great Britain (now the Royal Aeronautical Society) built the world's first wind tunnel in 1871 to test airfoils. Great Britain, the United States, Austria, and Russia by the mid-1890s had all built their own wind tunnels to conduct scientific studies on the theory of flight. See Richard P. Hallion, *Test Pilots: The Frontiersmen of Flight* (Revised Edition), 1988. The Engineering Division at McCook Field (now Wright-Patterson AFB), Dayton, Ohio, built a wind tunnel in 1918 measuring 22 feet long, 14 inches in diameter at the choke of the tunnel, and flaring out to 80 inches in diameter at the exit. The purpose of this device was to calibrate airspeed instruments and to study the effects of airflow on airfoil sections. Airspeeds of 453 miles per hour could be generated in the tunnel to assist in interpreting aerodynamic laws. The tunnel was mounted on a massive wooden base (21 feet × 6 feet) constructed from 2100 pieces of quality propeller walnut. This early wind tunnel is on display at the Air Force Museum in Dayton. Interview with Kyrazis, 20 June 1985; Statement of Work/Amendments, Contract F29601-71-C-0064, "Study and Design of an Airborne Test Bed," 16 October 1970.

21. Briefing, General Dynamics to AFWL personnel, "Contract F29601-71-C-0064," 5 May 1971.

22. Ibid.; Interview with Otten, 27 December 1988.

23. Rpt, FZA-454, D. Bergman, General Dynamics, "Airborne Test Bed Free Jet Test Results (Summary and Analysis)," September 1971, pp. 1–22, 85–87.

24. For test plans, summaries, and analyses, see the following General Dynamics/Convair Aerospace Division reports: Rpt, FZT-196, Bergman, "ATB Free Jet Test Results. Model and Test Information Report, ATB Program, 0.035-Scale NKC-135A Force Model, AEDC 16-Foot Propulsion Wind Tunnel," 1971, and "Addendum I," 6 October 1971; Rpt, FZT-198, "Wind Tunnel Data Report, ATB Program, AEDC PWT Test TF-266," 1971, and "Addendum I," 23 November 1971; Rpt, FZA-455, "Airborne Test Bed: Wind Tunnel Test Results," 1971; Rpt, FZE-1150, "Airborne Test Bed: Stability and Control Wind Tunnel Test Results," 1971; Rpt, FZS-190, "Loads and Structural Analyses, .30-Scale Airborne Test Bed, AEDC 16-Foot Transonic Tunnel," 1971, and "Addendum I," 13 December 1971, and "Addendum II," 25 August 1972; Rpt, FZT-203, Vols. I and II, "Model and Test Information Report, ATB Program, 0.3-Scale APT Turret and Fairing Model, AEDC 16-Foot Propulsion Wind Tunnel," 1971; Briefing, AFWL, John Otten, "Free Jet Tests," n.d.

25. Interviews with Kyrazis, 20 June 1985, 7 and 19 January 1989; Memo for the Record, 21 July 1971; Dettmer and Kyrazis, p. 380; Rpt, AFWL-TR-86-01, p. 1; Rpt, AFWL, Donald L. Lamberson, "Airborne Laser Laboratory Configuration and Technology Status Report," May 1972, p. 17.

26. Interview with Kyrazis, 20 June 1985; Dettmer and Kyrazis, p. 380; Rpt, AFWL-TR-86-01, p. 1.

27. Interview with Otten, 27 December 1988; Ltr, Gen James Ferguson to AFSC Divisions, Centers, SAMSO, and Laboratories, "Use of AEDC Test Facilities," 25 February 1970.

28. Interview with Otten, 27 December 1988; Interview with O'Hair, 26 March 1990; Memo for the Record, 21 July 1971; Memo for the Record, Buell to Director, Aeronautics and Flight Systems, "Report of Meetings at Air Force Weapons Laboratory, Kirtland Air Force Base, Albuquerque, New Mexico, on August 10 and 11, 1971, Concerning the Proposed C-135 Turret," 16 August 1971.

29. Interview with Otten, 27 December 1988; Interviews with Shelton, 21 and 28 December 1988; Interview with Kyrazis, 20 June 1985.

30. Interviews with Shelton, 21 and 28 December 1988.

31. Ltr, Ferguson to AFSC Divisions, Centers, SAMSO, and Laboratories, "Use of AEDC Test Facilities," 25 February 1970; Ltr, Ferguson to AFSC Divisions, Centers, SAMSO, and Laboratories, "Use of AEDC Test Facilities," 10 April 1969; Ltr, Maj Gen John B. Hudson to AFSC Divisions, Centers, SAMSO, and Laboratories, "Use of AEDC Test Facilities," 24 August 1970; AFSCR 806, "NASA Research and Development Test and Evaluation Support," 28 July 1969.

32. Interview with Otten, 27 December 1988; Interview with Kyrazis, 20 June 1985.

33. Ltr, Dr. Hans Mark to O'Hair, ["NASA–Air Force Cooperative Test Program Relating to ATB"], 22 October 1971.
34. Interview with Otten, 27 December 1988; Interviews with Shelton, 21 and 28 December 1988; Rpt, Lamberson, p. 17.
35. Ltr, Ronald H. Puent to many recipients, "Joint NASA–USAF Preliminary Test Planning Meeting," 26 October 1971.
36. Memo for the Record, Buell to Director, NASA-Ames, "Meeting on November 2–3, 1971, Concerning the C-135 Turret Tests to be Conducted in the 14-Foot Wind Tunnel," 11 November 1971.
37. Interviews with Otten, 27 December 1988 and 1 November 1990; Rpt, AFWL-TR-73-17, Vol. I, Leonard J. Otten and James A. Davis, "0.3-Scale Open Port ALL Turret Wind Tunnel Test Results: Overall Results," April 1973, pp. 3, 12–14, 190–192.
38. Interview with Otten, 27 December 1988; Interview with Kyrazis, 20 June 1985.
39. Interview with Otten, 27 December 1988; Rpt, AFWL, Colonel Robert W. Rowden, "AFWL Significant Events," 31 January 1972; Memo for the Record, 11 November 1971; AFWL-TR-73-17, Vol. I, pp. ii, 1–12, 159–164.
40. There were three sources causing the beam to degrade. One was shock waves. The second was aerodynamically induced turbulence in the boundary layer of the aircraft. The other was inviscid flow, outside the shock wave and turbulent flow region, which produced a weak lens and caused the beam to defocus. Rpt, Lamberson, pp. 14–17; Rpt, AFWL, Lamberson, "ALL Configuration and Technology Status Report," March 1973, pp. 80–84; Rpt, Kyrazis, "Tunnel Testing for the Airborne Laser Laboratory," 28 August 1972; Ltr, Mark to Professor Abe Hertzberg, no subj, 19 September 1972, with attached report, "Status Report on Turret Testing to the Airborne Laser Laboratory."
41. Interview with Otten, 2 June 1994; Ltr, Mark to Hertzberg, 19 September 1972; Dettmer and Kyrazis, pp. 379–380.
42. Ibid.
43. Ibid.
44. Interview with Lamberson, 18 April 1984; Interview with Kyrazis, 20 June 1985; Interview with Otten, 27 December 1988; Interviews with Shelton, 21 and 28 December 1988; Rpt, Lamberson, pp. 80–81; Rpt, AFWL, Lamberson, "Airborne Laser Laboratory Configuration and Status Report," May 1972, p. 17.
45. Interview with Kyrazis, 17 October 1985; Memo for the Record, AFWL, "History of the Development of Laser Windows Supporting the ALL Program," [1974], p. 1.
46. Interview with Kyrazis, 7 June 1990; Memo for the Record, [1974], pp. 1–4; Rpt, AFWL-TR-74-311, D.L. Sullivan *et al.*, "Airborne Pointing and Tracking System Low-Power Window," October 1975, pp. 1–5; Rpt, Lamberson, p. 24; Rpt, AFWL, Lamberson, "Airborne Laser Laboratory Configuration and Technology Status Report," March 1973, pp. 46–51.
47. Interview with Kyrazis, 7 June 1990; Rpt, AFWL-TR-74-311, p. 15; Rpt, Dr. A. M. Lovebee, Director, "AF Materials Lab Significant Events," 30 June 1972.
48. Change Order P00003 to contract F29601-71-C-0058 with Hughes Aircraft Company for "Airborne Pointer and Tracker," 25 June 1971; Rpt, AFWL-TR-74-311, pp. 2–4.
49. Memo for the Record, [1974], pp. 2–4; Rpt, AFWL-TR-74-311, pp. 1, 15; Rpt (draft), AFWL, "Low-Power Window," 31 October 1974; Rpt, AFWL, Rowden, "AFWL Significant Events," 30 September 1972; Rpt, Dr. A. M. Lovelace, Director, "AF Materials Lab Significant Events," 30 June 1972.
50. Interviews with Kyrazis, 20 June and 17 October 1985; Statement of Work for contract F29601-71-C-0147 with Raytheon for "Conceptual Engineering Study of Material Windows for High-Power Infrared Lasers," 15 September 1971; Memo for the Record, [1974], pp. 2–3.
51. By May 1973, Raytheon could produce zinc selenide in fairly large sheets (15 inches long × 24 inches wide × 0.35 inch thick). Producing the material was expensive, costing $50–100 per square

inch of zinc selenide. But this was not the finished product. The zinc selenide had to be treated and undergo a costly polishing process before it was ready for the ALL. Statement of Work for contract F29601-72-C-0119 with Raytheon for "Exploratory Development of Material Windows for Lasers," 15 November 1972; Statement of Work for contract F29601-74-C-0069 with Raytheon for "High-Power Window Conceptual Design," 6 May 1974; Rpt, AFWL, "Laser Window Development Program Status Review," 20 November 1972; Rpt, Lamberson, p. 50.

52. Interview with Kyrazis, 7 June 1990; Statement of Work for contract F29601-72-C-0122 with the University of Dayton Research Institute for "Design and Fabrication of the Laser Window Test Apparatus," 4 January 1972; Rpt, Lamberson, pp. 53–54.

53. Sole Source Justification for "Low Power Window Fabrication," 21 December 1972.

54. Rpt, AFWL-TR-74-311, pp. 2, 196–197; Ltr, W. D. Pittman, "The Aerospace Corporation," 13 August 1973; Progress Report #1, "Low Power Window Fabrication Project, 29 June–15 July 1973," in official R&D file folder for contract F29601-73-C-0099.

55. Interview with Boesen, 30 September 1988; Memo for the Record, [1974], pp. 3–5; Rpt, AFWL-TR-74-311, pp. 35–40; Rpt, Lamberson, p. 62.

56. Memo for the Record, [1974], pp. 4–6; Progress Report #3, "Low Power Window Fabrication Project, 31 July–31 August 1973," in official R&D file folder for contract F29601-73-C-0099; Progress Report #4, "Low Power Window Fabrication Project, 1 September–30 September 1973," in official R&D file folder for contract F29601-73-C-0099; Progress Report #5, "Low Power Window Fabrication Project, 1 October–31 October 1973," in official R&D file folder for contract F29601-73-C-0099; Progress Report #9, "Low Power Window Fabrication Project, 1 February–28 February 1974," in official R&D file folder for contract F29601-73-C-0099; Progress Report #14, "Low Power Window Fabrication Project, 1 July–31 July 1974," in official R&D file folder for contract F29601-73-C-0099; Trip Report, AFWL, Capt Alan D. Blackburn, "Low-Power Window," 21 March 1974; Trip Report, AFWL, Maj Otis A. Prater, "Low-Power Window," 14 February 1974; Letter Rpt, Air Force Cambridge Research Laboratories, Solid State Science Laboratory, "Evaluation of Raytheon CVD ZnSe," 20 March 1974, pp. 1, 8–10.

57. Progress Report #12, "Low Power Window Fabrication Project, 1 May–31 May 1974," in official R&D file folder for contract F29601-73-C-0099; Memo for the Record, [1974], pp. 3–6; Rpt, Lamberson, p. 65.

58. Rpt, AFWL-TR-74-311, p. 14; Memorandum Proposal, University of Dayton Research Institute, "Additional Effort Under Contract F29601-73-C-0124 to Instrument, Test, and Evaluate the Aluminum Sector Low-Power Window System," 26 November 1973; Trip Report, Blackburn, 7–8 January 1974.

59. Memo for the Record, [1974], pp. 4–7; Memo for the Record, AFWL/LRE, Blackburn and Capt John S. Loomis, AFWL/LRE, "ALL Subsystems Review Meeting," 28 February 1974.

CHAPTER 7. PREPARING FOR CYCLE I

1. Rpt, AFSWC, Oscar S. Ayers, "4900th Test Group (Flight Test): Test Plan," June 1973, pp. 1–12.

2. Design of the aircraft modification came under contract F29601-71-C-0064; modification was accomplished under F29601-72-C-0082. Rpt, T. Peyton Robinson, General Dynamics, "R&D Contract Status Report, Airborne Laser Laboratory (Contract No. F29601-72-C-0082) 9 March–15 April 1972," pp. 2.0–2.5; Memo for the Record, E. C. Johnson, "Minutes of ATB Coordinating Committee," 21 March 1972.

3. Ltr, Roger Shinnick to David Nienow, "Airborne Test Bed," 7 March 1972; Memo for the Record, Ronald Puent, "Costs of Airborne Test Bed Modifications," 22 October 1974; Purchase Request, "ALL Modification Cost Growth," 27 March 1974; Rpt, AFSWC, Ayers, "Advanced Radiation Technology," 26 September 1972.

4. Interview with Shelton, 21 December 1988; Appendix A to Contract F29601-72-C-0082, General Dynamics, "Work Specification IRAN KC-135 Type Aircraft, AFSWC," 20 November 1971, pp.

1–3, 8–13; Statement of Work for contract F29601-72-C-0082, General Dynamics, "Airborne Test Bed Modification and IRAN," 10 January 1972; Rpt, FZP-1347-I, Revision A, General Dynamics, "Technical Proposal for the Airborne Test Bed Modification and IRAN, Part I—Technical Proposal," 17 February 1972, pp. 2-26–2-27, 2-61, 2-66, 3-14, 4-1, 5-1–5-3, 5-8.

5. Interview with Davis, 20 September 1988; Statement of Work for contract F29601-72-C-0082, 10 January 1972; Rpt, FZP-1347-I, pp. 2-6–2-27; Rpt, Robinson, "R&D Contract Status Report, Airborne Laser Laboratory (Contract No. F29601-72-C-0082), 16 April–14 May 1972," pp. 2.0–2.5; Memo for the Record, L. C. Lockert, "Minutes of ATB Coordinating Meeting for 11 April 1972," 12 April 1972.

6. Statement of Work for contract F29601-72-C-0082, 10 January 1972; Memo for the Record, 12 April 1972.

7. Interview with Shelton, 28 December 1988.

8. Rpt, Robinson, "R&D Contract Status Report, Airborne Laser Laboratory (Contract No. F29601-72-C-0082), 9 March–15 April 1972," p. 3.0; Rpt, Robinson, "R&D Contract Status Report, Airborne Laser Laboratory (Contract No. F29601-72-C-0082), 16 April–14 May 1972," pp. 3.0–4.0; Rpt, Robinson, "R&D Contract Status Report, Airborne Laser Laboratory (Contract No. F29601-72-C-0082), 15 May–18 June 1972," pp. 3.0–4.0; Rpt, Robinson, "R&D Contract Status Report, Airborne Laser Laboratory (Contract No. F29601-72-C-0082), 17 July–20 August 1972," pp. 3.0–4.0; Rpt, Robinson, "R&D Contract Status Report, Airborne Laser Laboratory (Contract No. F29601-72-C-0082), 21 August–17 September 1972," pp. 3.0–4.0; Rpt, Robinson, "R&D Contract Status Report, Airborne Laser Laboratory (Contract No. F29601-72-C-0082), 18 September–15 October 1972," pp. 3.0–4.0; Rpt, Robinson, "R&D Contract Status Report, Airborne Laser Laboratory (Contract No. F29601-72-C-0082), 16 October–19 November 1972," pp. 3.0–4.0.

9. Rpt, Robinson, "R&D Contract Status Report, Airborne Laser Laboratory (Contract No. F29601-72-C-0082), 19 June–16 July 1972," pp. 1.0–5.2; Rpt, Robinson, "R&D Contract Status Report, Airborne Laser Laboratory (Contract No. F29601-72-C-0082), 21 August–17 September 1972," pp. 1.0–5.2; Rpt, Robinson, "R&D Contract Status Report, Airborne Laser Laboratory (Contract No. F29601-72-C-0082), 18 September–15 October 1972," pp. 1.0–5.1; Rpt, Robinson, "R&D Contract Status Report, Airborne Laser Laboratory (Contract No. F29601-72-C-0082), 16 October–19 November 1972," pp. 1.0–5.1; Rpt, Robinson, "R&D Contract Status Report, Airborne Laser Laboratory (Contract No. F29601-72-C-0082), 20 November–17 December 1972," pp. 1.0–5.1; Rpt, Robinson, "R&D Contract Status Report, Airborne Laser Laboratory (Contract No. F29601-72-C-0082), 18 December 1972–28 January 1973," pp. 1.0–5.1.

10. Rpt, Robinson, "R&D Contract Status Report, Airborne Laser Laboratory (Contract No. F29601-72-C-0082), 16 April–14 May 1972," pp. 2.5–4.0.

11. Rpt, Robinson, "R&D Contract Status Report, Airborne Laser Laboratory (Contract No. F29601-72-C-0082), 19 June–16 July 1972," pp. 3.0–4.0.

12. Rpt, Robinson, "R&D Contract Status Report, Airborne Laser Laboratory (Contract No. F29601-72-C-0082), 21 August–17 September 1972," pp. 2.0–5.2.

13. Rpt, FZP-1347-I, pp. 2-37–2-38; Rpt, AFWL, Donald L. Lamberson, "Airborne Laser Laboratory Configuration and Technology Status Report," May 1972, p. 14.

14. Rpt, Robinson, "R&D Contract Status Report, Airborne Laser Laboratory (Contract No. F29601-72-C-0082), 15 May–18 June 1972," pp. 2.0–2.5; Rpt, Robinson, "R&D Contract Status Report, Airborne Laser Laboratory (Contract No. F29601-72-C-0082), 19 June–16 July 1972," pp. 1.0–2.5, 5.1; Rpt, Robinson, "R&D Contract Status Report, Airborne Laser Laboratory (Contract No. F29601-72-C-0082), 17 July–20 August 1972," pp. 2.0–2.5.

15. Interview with Shelton, 28 December 1988; Rpt, AFWL, Lamberson, "Airborne Laser Laboratory Configuration and Technology Status Report," March 1973, p. 8.

16. Ibid.

17. Ibid.; Rpt, Robinson, Dynamics, "R&D Contract Status Report, Airborne Laser Laboratory (Contract No. F29601-72-C-0082), 20 November–17 December 1972," pp. 1.0–2.5, 5.1; Memo for the Record, Lockert, "Minutes of ALL Coordination Meeting for 19 December 1972," 19 December 1972; Rpt, AFWL, Colonel Robert W. Rowden, "AFWL Significant Events," 31 December 1972.

18. GD's Dr. Bill Steekin was responsible for the design of the Cycle I fairings. AFWL's Jim Van Kuren designed the Cycle III fairings. Rpt, Robinson, "R&D Contract Status Report, Airborne Laser Laboratory (Contract No. F29601-72-C-0082), 21 August–17 September 1972," pp. 2.5, 5.1; Rpt, Robinson, "R&D Contract Status Report, Airborne Laser Laboratory (Contract No. F29601-72-C-0082), 18 September 1972–15 October 1972," pp. 1.0–2.5, 5.1.

19. Rpt, Robinson, "R&D Contract Status Report, Airborne Laser Laboratory (Contract No. F29601-72-C-0082), 16 October–19 November 1972," pp. 1.0, 2.5; Rpt, Robinson, "R&D Contract Status Report, Airborne Laser Laboratory (Contract No. F29601-72-C-0082), 18 December 1972–28 January 1973," pp. 1.0, 2.5, 5.1.

20. Rpt, Robinson, "R&D Contract Status Report, Airborne Laser Laboratory (Contract No. F29601-72-C-0082), 15 May–18 June 1972," p. 2.5; Memo for the Record, Lockert, "Minutes of ATB Coordination Meeting for 25 May 1972," 25 May 1972; Memo for the Record, Lockert, 12 April 1972.

21. Rpt, FZP-1347-I, pp. 2-37, 2-39–2-41.

22. Ibid.; Rpt, Robinson, "R&D Contract Status Report, Airborne Laser Laboratory (Contract No. F29601-72-C-0082), 19 June–16 July 1972," p. 2.5; Rpt, Robinson, "R&D Contract Status Report, Airborne Laser Laboratory (Contract No. F29601-72-C-0082), 17 July–20 August 1972," p. 2.5; Rpt, Robinson, "R&D Contract Status Report, Airborne Laser Laboratory (Contract No. F29601-72-C-0082), 21 August–17 September 1972," p. 2.5.

23. Interview with Shelton, 28 December 1988; Rpt, FZP-1347-I, p. 2-40.

24. Rpt, FZP-1347-I, pp. 2-48–2-49; "0082 History," [isolated viewgraph slide from unidentified briefing]; Memo for the Record, Lockert, "Minutes of ALL Coordination Meeting for 15 August 1972," 16 August 1972.

25. Interview with Shelton, 28 December 1988.

26. Rpt, FZP-1347-I, p. 2-56; Rpt, Robinson, "R&D Contract Status Report, Airborne Laser Laboratory (Contract No. F29601-72-C-0082), 9 March–15 April 1972," pp. 2.5, 4.0; Rpt, Robinson, "R&D Contract Status Report, Airborne Laser Laboratory (Contract No. F29601-72-C-0082), 15 May–18 June 1972," pp. 2.5, 4.0, 5.2; Rpt, Robinson, "R&D Contract Status Report, Airborne Laser Laboratory (Contract No. F29601-72-C-0082), 19 June–16 July 1972," pp. 2.5, 5.1–5.2; Rpt, Robinson, "R&D Contract Status Report, Airborne Laser Laboratory (Contract No. F29601-72-C-0082), 17 July–20 August 1972," pp. 2.5, 5.1; Rpt, Robinson, "R&D Contract Status Report, Airborne Laser Laboratory (Contract No. F29601-72-C-0082), 21 August–17 September 1972," pp. 2.5, 5.1.

27. Memo for the Record, Lockert, "Minutes of ALL Coordination Meeting for 8 August 1972," 9 August 1972; Memo for the Record, Lockert, "Minutes of ALL Coordination Meeting for 5 October 1972," 6 October 1972.

28. Rpt, FZP-1347-I, pp. 2-45–2-48; Rpt, Lt Col Tracy A. Scanlan, "4900th Test Group Test Plan: ALL ECS Capacity Flight Test," 19 September 1974.

29. Interview with Shelton, 28 December 1988; Memo for the Record, Lockert, "Minutes of ALL Coordination Meeting for 5 December 1972," 7 December 1972; Rpt, Robinson, "R&D Contract Status Report, Airborne Laser Laboratory (Contract No. F29601-72-C-0082), 9 March–15 April 1972," p. 2.5; Rpt, Robinson, "R&D Contract Status Report, Airborne Laser Laboratory (Contract No. F29601-72-C-0082), 16 April–14 May 1972," p. 2.5; Rpt, Robinson, "R&D Contract Status Report, Airborne Laser Laboratory (Contract F29601-72-C-0082), 21 August–17 September 1972," p. 2.5; Rpt, Robinson, "R&D Contract Status Report, Airborne Laser Laboratory (Contract

No. F29601-72-C-0082), 18 September–15 October 1972," pp. 2.5, 5.1; Rpt, Robinson, "R&D Contract Status Report, 16 October–19 November 1972," pp. 1.0, 2.5.

30. Rpts, AFSWC, Ayers, "Advanced Radiation Technology," 15 February 1973 and 17 July 1974.

31. Interview with Kyrazis, 3 November 1991; Rpt, Lamberson, pp. 4–5; Rpt, AFWL, Rowden, "AFWL Significant Events," 31 October 1972.

32. Interview with Otten, 27 December 1988; Rpt, Robinson, "R&D Contract Status Report, Airborne Laser Laboratory (Contract F29601-72-C-0082), 16 October–19 November 1972," pp. 1.0, 2.2, 5.1; R&D Management Rpt, AFWL, "Advanced Radiation Technology," 18 December 1972.

33. Interviews with Otten, 27 December 1988 and 2 June 1994; R&D Management Rpt, 18 December 1972.

34. Interview with Kyrazis, 3 November 1991.

35. Ibid.

36. Aeronautical Systems Division evaluated the engineering feasibility of installing the fence by analyzing instrumentation data and film obtained from earlier test flights as well as aerodynamic calculations derived from the NASA-Ames wind-tunnel experiments. ASD concluded adding the fence was a good engineering solution. Interview with Kyrazis, 3 November 1991; Rpt, Robinson, "R&D Contract Status Report, Airborne Laser Laboratory (Contract No. F29601-72-C-0082), 16 October–19 November 1972," pp. 1.6, 2.5; Rpt, Robinson, "R&D Contract Status Report, Airborne Laser Laboratory (Contract No. F29601-72-C-0082), 18 December 1972–28 January 1973," p. 1.0; Rpts, Ayers, AFSWC, "Advanced Radiation Technology," 18 December 1972 and 15 February 1973; Rpt, Lamberson, p. 4.

37. Later, when the ALL returned from GD to Kirtland, two other flight tests took place confirming the radar tail extension was stable. These tests took place prior to the ALL's departure for Edwards AFB to recertify the flight performance of the aircraft. Interviews with Shelton, 9, 21, and 28 December 1988; Rpt, Lamberson, pp. 4–5.

38. Interview with Otten, 27 December 1988; Rpt, FTC-TR-73-22, Bill R. Boxwell, Air Force Flight Test Center, Edwards AFB, California, "Limited Performance and Flying Qualities Tests and Structural Demonstration of the Airborne Laser Laboratory NKC-135A Aircraft," 1973, pp. ii–iii, 1–6; Rpt, Lamberson, pp. 4–7.

39. Interview with Kyrazis, 3 November 1991; Rpt, FTC-TR-73-22, pp. 7–20, 51.

40. Interview with Otten, 27 December 1988; Rpt, FTC-TR-73-22, pp. 1–9, 14–16.

41. Rpt, FTC-TR-73-22, pp. 17–20, 51.

42. Ibid., pp. iii, 17–20.

43. Ibid., pp. 1–14; Interview with Otten, 27 December 1988; Rpt, Lamberson, pp. 4, 82.

44. Rpt, AFFDL/FYS-73-10, James J. Olsen, Actg Chief, Aerospace Dynamics Branch, Wright-Patterson AFB, Ohio, "Test Report on the Acoustic Survey of the Airborne Laser Laboratory," 26 June 1973, pp. 1–5.

45. Rpt, FTC-TR-73-22, pp. iii, 17–20; Rpt, Lamberson, p. 20.

46. Rpt, Lamberson, pp. 4–5.

CHAPTER 8. TRACKING AERIAL TARGETS

1. Interview with Shelton, 28 December 1988; Interview with Laughlin, 4 April 1990; Telecon with Ernie Endes, Honeywell's Sperry Defense System Division, Albuquerque, New Mexico, 30 March 1990.

2. Interview with Laughlin, 4 April 1990; Telecon with Endes, 30 March 1990.

3. Telecon with Endes, 30 March 1990; Rpt, AFWL, Donald L. Lamberson, "Airborne Laser Laboratory Configuration and Technology Status Report," March 1973, pp. 11–16.

4. Ibid.; Interview with Raymond V. Wick, Air Force Weapons Laboratory, Albuquerque, New Mexico, 13 April 1990.

5. Interview with Laughlin, 4 April 1990; Rpt, Lamberson, pp. 8–10; Rpt, AFWL-TR-86-01, Vol. I, Raymond V. Wick, "Airborne Laser Laboratory Cycle III: Systems and Test Descriptions," May 1988, p. v.

6. Interview with Laughlin, 4 April 1990; Telecon with Endes, 30 March 1990.

7. Ibid.

8. Ibid.; Interview with John Gromek, Colonel, USAF, Retired, Albuquerque, New Mexico, 21 March 1990; Rpt, Lamberson, p. 8.

9. Interview with Laughlin, 4 April 1990; Telecon with Endes, 30 March 1990; Rpt (AF Form 111), AFSWC, Oscar S. Ayers, "Advanced Radiation Technology," 30 October 1973.

10. Interview with Laughlin, 4 April 1990; Rpt, Lamberson, pp. 10–16.

11. Interview with Gromek, 21 March 1990; Interview with Laughlin, 4 April 1990; Interview with Wick, 13 April 1990; Telecon with Endes, 30 March 1990; Rpt, AFWL-TR-74-285, Edward A. Duff, James E. Negro, and James F. Russell, "USAF Precision Pointing and Tracking Systems for High Energy Laser Applications—Test Results," December 1975, pp. 19–20.

12. Rpt, AFWL-TR-74-285, pp. 18–23.

13. Ibid., pp. 18–23; Interview with Kyrazis, 20 June 1985; Interview with Laughlin, 4 April 1990; Rpt, AFWL-TR-86-01, pp. iv–v; Rpt (AF Form 111), Ayers, "Advanced Radiation Technology," 31 May 1973; Rpt, Lamberson, p. 8.

14. Interview with Gromek, 21 March 1990; Interview with Shelton, 28 December 1988.

15. Interview with Gromek, 21 March 1990; Interviews with Kyrazis, 8 August, 12 September, and 17 October 1985.

16. Interview with Gromek, 21 March 1990.

17. Interview with Boesen, 20 September 1988; Rpt, 4900th Test Group, Robert Greenberg, "4900th Test Group (Flight Test) Supplemental Test Plan: ALL Target Test Plan," 29 August 1973, pp. 1–17; Rpt, AFWL-TR-74-285, pp. 21–23; Rpt, AFSWC, Ayers, "4900th Test Group (Flight Test): Test Plan," May 1973, pp. 20–21, 42–43.

18. Interview with Gilbert, 31 August 1988; Interview with Ben F. Johnson, Lieutenant Colonel, USAF, Retired, Albuquerque, New Mexico, 11 May 1990.

19. Ibid.

20. Ibid.; Interviews with Shelton, 9 and 28 December 1988.

21. Interviews with Kyrazis, 20 June and 12 September 1985.

22. Rpt, AFWL-TR-74-285, pp. 7–10; Rpt, AFWL-TR-76-271, Keith Gilbert and Larry Sher, "The Airborne Laser Laboratory Cycle II: Low Power Experiments," May 1977, p. 52.

23. Interview with Sher, 11 October 1988; Interview with Kyrazis, 19 September 1985; Interview with Boesen, 30 September 1988; Interview with Johnson, 19 May 1990; Rpt, AFWL-TR-83-5, Sher *et al.*, "Lessons Learned from the Airborne Laser Laboratory," June 1983, pp. 54–56; Rpt, AFWL-TR-74-285, p. 48.

24. Interview with Duff, 13 September 1988; Interview with Boesen, 20 September 1988; Rpt, AFWL-TR-83-5, pp. 43–46, 54–56, 60; Rpt, AFWL-TR-76-271, p. 32.

25. Interview with Boesen, 20 September 1988; Interview with Duff, 13 September 1988; Interview with Kyrazis, 8 August 1985; Rpt, AFWL-TR-74-285, pp. 18–22; Rpt, AFWL-TR-83-5, pp. 7, 60–62; Sher and Ben F. Johnson, "Pointing and Tracking for High Energy Laser Systems," in *Journal of Defense Research, Series A: Strategic Warfare—High Energy Lasers*, May 1975, pp. 259–282; Rpt, AFWL, Richard V. Feaster, "Airborne Laser Laboratory Cycle III Test Plan: Revision I," 5 June 1978, p. 4; Col John C. Rich, "Air Force Overview," in *Proceedings of the Third DOD High-Energy Laser Conference*, July 1979, p. 50; Rpt, AFWL, L.J. Otten and D.L. Boesen, "The Airborne Laser Laboratory: A Program Review," n.d.

26. Rpt, AFWL-TR-74-285, pp. 22–23.

27. Interview with Lamberson, 12 January 1989; Rpt, AFWL-TR-86-01, p. v.

CHAPTER 9. LOW-POWER LASER GOES AIRBORNE

1. Memo for the Record, AFWL, Capt Timothy J. Graves, "Flight Information from Cycle IIA and Cycle IIB," n.d.
2. Interview with Kyrazis, 19 September 1985; Interview with Boesen, 30 September 1988; Rpt, AFWL, Ben F. Johnson, "Airborne Laser Laboratory Cycle II Experiment Plan," December 1974, pp. 1–3, 22; Rpt, AFWL-TR-76-271, Keith K. Gilbert and Larry Sher, "The Airborne Laser Laboratory Cycle II: Low Power Experiments," May 1977, pp. 25–27; Rpt #1, W. W. Short *et al.*, Applied Technology Associates, Inc., Albuquerque, New Mexico, "Preliminary Analysis of Cycle IIB Data from the Airborne Laser Laboratory," April 1976, p. 26.
3. Interview with Lamberson, 12 January 1989; Interview with Boesen, 29 October 1993; Ltr, Col Otis A. Prater, Commander, 4900th Test Group, to Col Russell K. Parsons (AFWL/LR), "ALL Cycle II Test Plan," 9 November 1973; Rpt, ARTO-73-1, Donald L. Lamberson, "Program Management Plan: Advanced Radiation Technology Office," June 1973, p. 40.
4. Rpt, AFWL-TR-76-271, pp. 29–30; Donald L. Lamberson, "Overview of the Air Force High Energy Laser Program," in *Second DOD High Energy Laser Conference Proceedings*, November 1976, pp. 31–42; Rpt, Johnson, pp. 120a–121.
5. Rpt, Johnson, pp. 1–11, 14–15, 41–57.
6. Interview with Gilbert, 31 August 1988; Interview with Boesen, 30 September 1988; Interview with Dr. Dale Holmes, Rocketdyne Division, Rockwell International, Canoga Park, California, 16 May 1990; Rpt, AFWL-TR-76-271, pp. 41–43; Rpt, AFSWC/FTET, Oscar S. Ayers, "4900th Test Group Plan: Cycle II Off-Range Missions Laser Propagation Tests," 9 April 1975, p. 1.
7. Interviews with Boesen, 30 September 1988 and 4 June 1990; Rpt, AFWL-TR-76-271, p. 44.
8. Interview with Boesen, 6 March 1991.
9. Interviews with Boesen, 30 September 1988 and 4 June 1990; Rpt, AFWL-TR-76-271, pp. 50–51.
10. Interview with Kyrazis, 8 August 1985; Interview with James A. Davis, 14 May 1990; Interview with Gromek, 17 March 1990; Interview with Dick Applegate, Hughes Aircraft Company, El Segundo, California, 15 May 1990.
11. Interview with Kyrazis, 8 August 1985; Interview with Davis, 14 May 1990; Modifications to the ALL were made under contract F29601-73-C-0109. Rpt, AFSWC, David E. Holcombe, "4900th Test Group Plan: ALL IPAS Cycle IIA," 19 September 1974; Rpt, AFWL-TR-86-01, Vol. I, Raymond V. Wick, "Airborne Laser Laboratory: System and Test Descriptions," May 1988, p. v; Telefax, 4900th Test Group, KAFB, to AFSC, Andrews AFB, "Request for Temporary Release for Flight Renewal," 21 January 1975; Interview with Gromek, 17 March 1990; Interview with Applegate, 15 May 1990.
12. Interview with Holmes, 16 May 1990; Rpt, Johnson, pp. 14–15.
13. Rpt, AFWL-TR-76-271, p. 56.
14. A dichroic beam combiner wrapped the visible helium–neon beam around the invisible CO_2 beam. Rpt, AFWL-TR-76-271, pp. 44–45; Rpt, Holcombe, 19 September 1974.
15. Rpt, ARTO-73-1, pp. 58–59; Rpt, AFWL-TR-76-271, pp. 44–45; Rpt, Holcombe, 19 September 1974; Rpt, Ayers, pp. 3, 6.
16. Interview with Boesen, 6 March 1991; Rpt, Johnson, pp. 51–56.
17. Interview with Johnson, 11 May 1990; Rpt, Johnson, pp. 27–30a.
18. Measurements taken were not only to characterize Cycle II conditions, but were also used as a data base for directing Cycle III requirements in defining future beam control concepts, such as advanced tracking algorithms.
    Before AFWL settled on the name Advanced Radiation Test Facility, this structure was first called the Boresight and Calibration Range and the Armament Research Test Facility. ARTF served as a dedicated facility to permit assembly and disassembly, maintenance, modifications, and testing of the various subsystems of the ALL. The ARTF complex consisted of the laser test cell (a separate building), an optics laboratory and clean room, an ALL loading pad, maintenance

shops, a fuel farm, and a test range with four target sites. Burns & Roe Architects & Engineers designed the ARTF. Wood Construction Company, the phase I contractor, started work on 5 March 1973, and finished on 5 November 1974. Contractor for phase II was Rutherford Construction Company who began work on 12 November 1973, and finished in September 1975. Cost of ARTF amounted to $5.7 million. Interview with Kyrazis, 8 August 1985; Rpt, AFWL-TR-76-271, p. 34; Rpt, Johnson, p. 17; Rpt, AFWL, Lamberson, "Airborne Laser Laboratory Configuration and Status Report," May 1972, p. 44; Ltr, Col Gustav J. Freyer, AFWL/CC, to AFSC/DEP, "Report of House Appropriations Committee Surveys and Investigations," 30 January 1976, with six attachments.

19. From 10 to 20 November 1973, the Wright-Patterson team conducted ground vibration tests using a shaker on the ALL. The purpose of the tests was to determine whether results (e.g., structural resonance response, mechanical impedance) for the modified ALL differed from the ground vibration effects for an unmodified KC-135A aircraft. Addition of the turret and fairings contributed to higher vibrational effects on the ALL, but these were not significant enough to threaten crew or aircraft safety. Interview with Davis, 14 May 1990; Interview with Boesen, 6 March 1991; Rpt, WPAFB, Duane M. Davis, "Test Report on the Ground Vibration Survey of the Cycle I Airborne Laser Laboratory," 11 March 1974, pp. 1–6.

20. Interview with Boesen, 30 September 1988; Rpt, Johnson, pp. 22–26.

21. Interview with Kyrazis, 8 August 1985; Rpt, Johnson, pp. 63, 125–126; Ltr, Lt Col William E. Burke, Chief, AFSWC Requirements Division, to AFWL/PRP, "AFWL Airborne Laser Lab, JON 317JAHOO," 11 July 1974; Msg, AFSC, Andrews AFB, to 4900th Test Group, KAFB, "AFWL Modifications to NC-135A 60-0371," 10 May 1974; Rpt, AFSWC, Ayers, "4900th Test Group Test Plan: ALL Sandia Optical Range," 27 August 1974, pp. 10–11.

22. Use of 371 did not come cheaply. Under the scheme of industrial funding mandated by the government, AFWL had to pay the 4900th several thousand dollars each time 371 flew in support of the ALL experiments. When flying the ALL and diagnostic plane at White Sands Missile Range, it was not unusual for AFWL to pay the Army, who ran the range, several thousand dollars an hour to cover such expenses as radar coverage, cleared air space, and so forth. As an example, AFSWC charged AFWL $322,522 for FY75 to cover "Flying Hour Cost." When AFSWC was disestablished in 1976, responsibility for 371 transferred from AFSWC to the 4950th Test Wing at Aeronautical Systems Division at Wright-Patterson AFB. Interview with Gilbert, 31 August 1988; Ltr with Atch, Lamberson to AFWC/XO, "Request for an Additional Target Aircraft," 5 June 1974; Ltr, Kyrazis to AFSWC/FTOOM, "Request for Additional Target Aircraft," 6 June 1974; Rpt, Col Leonard R. Sugarman, AFSWC, "AFSWC Management Plan," 11 October 1974.

23. Ltr, Maj Jerald N. Jensen to "See Distribution," "371 Modification Meeting," 19 June 1974; Ltr, Lamberson to AFSWC/XO, 5 June 1974; Ltr, Lt Col James G. Espey, Deputy Assistant for ALL, to "See Distribution," "Cycle II Experiments," 10 June 1974.

24. Ltr, Burke to AFWL/LR, "Use of NC-135, S/N 371," 9 July 1974; Ltr, Col Sugarman (AFSWC/XOP), to TG/CC, "NC-135 Aircraft #60-371," 23 July 1974; Rpt, Burke, AFSWC, "Advanced Radiation Technology," 17 July 1974.

25. Interview with Kyrazis, 8 August 1985; Interview with Gromek, 21 March 1990; Rpt, Johnson, pp. 125–126; Memo for the Record, Espey, "Reflections On ALL Flights 72 and 73," 18 September 1975.

26. Interview with Johnson, 11 May 1990.

27. Rpt, Johnson, pp. 123–124.

28. Ibid., p. 10; Interview with Boesen, 30 September 1988; Rpt, AFWL-TR-76-271, pp. 32–33, 115, 134, 314–319; Ltr, Espey to LRP (Capt Graves), "Cycle II Schedule," 22 November 1974.

29. Interview with Otten, 2 June 1994; Interview with Kyrazis, 4 November 1994.

30. Rpt, Johnson, p. 31; Rpt, AFSWC, "4900th Test Group Test Plan: ALL—Cycle II Qualification Test Plan for Operational APT and Material Window Configuration," n.d.

31. Rpt, Johnson, pp. 32–34a; Rpt, AFWL-TR-76-271, p. 55.

32. Rpt, AFWL-TR-76-271, pp. 34, 107, 113.

33. Prior to this time, a series of airworthiness flight tests (Cycle IIA-1) took place to assess only the IPAS subsystem. The main objective was to measure aircraft structural bending and how that affected the IPAS. Once this checked out, the laser and APT were installed on the plane, followed by Cycle IIA flight testing. Interview with Davis, 14 May 1990; Memo for the Record, Graves, n.d.; Rpt, Johnson, pp. 20–21a; Rpt, Air Force Audit Agency, "Management of Air Force High Energy Laser Program," 26 November 1975, p. 2; Memo for the Record, Graves, "ALL Flight Schedule," 9 December 1974.

34. Transcript of Speech, Kyrazis, "ALL Departure from AFWL," 4 May 1988.

35. Based on the ALL's performance during Cycle I, higher headquarters was confident that Cycles II and III would be successful. Anticipating that the ALL would eventually shoot down an air-to-air missile, Systems Command realized it had to begin a formal study to come up with a realistic operational mission for the laser. Factors to be considered were required operational capabilities, technical risks, costs, and time required to build and test hardware. To be in a good bargaining position with the operating commands once Cycle III finished, AFSC directed AFWL to conduct a Laser Weapon Mission Analysis (LWMA) from 23 July to 14 November 1974. The LWMA study group presented its findings to AFSC in January 1975 and in a modified report to Headquarters Air Force in July. A defensive air-to-air laser weapon capable of intercepting air-to-air and ground-to-air missiles was selected as the best mission for a laser. Laser-armed planes would essentially perform self-defense and/or escort missions. Development of an integrated system demonstration device for specific aircraft was approved by Systems Command on March 26, 1976. This gave AFWL the authority to start research in this area which became known as the Short Range Applied Technology (SRAT) program. See "Program Management Plan: Short Range Applied Technology Program," AFWL/PO, 1 March 1976. Interview with Gilbert, 31 August 1988.

36. Interview with Boesen, 30 September 1988; Interview with Gilbert, 31 August 1988.

37. Interview with Boesen, 6 March 1991.

38. Many contributed to 371 diagnostics. However, Dr. Al Saxman was the prime linchpin responsible for designing and fielding of this multidimensional diagnostic subsystem. In addition, he served as the 371 Tech Director for most of the Cycle IIB missions that occurred later in Cycle II. Interview with Kyrazis, 8 August 1985; Interview with Gilbert, 31 August 1988; Interview with Boesen, 30 September 1988; Rpt, AFWL-TR-76-271, p. 134; Memo, Daniel F. Creedon, FAA/Air Route Traffic Control Center, Albuquerque, New Mexico, "Special Airborne Laser Lab Flights from Kirtland AFB," 4 October 1975.

39. Interview with Kyrazis, 8 August 1985; Rpt, AFWL-TR-76-271, pp. 309–314; Lamberson, "Overview of the Air Force High Energy Laser Program," p. 33; Rpt, Johnson, p. 131.

40. Interview with Kyrazis, 8 August 1985.

41. Twenty-two flights involved testing the APT; 11 were flown open port and 11 were flown using the low-power window. Open-port testing reached completion in June 1975. Interview with Davis, 14 May 1990; Rpt, AFWL-TR-76-271, pp. 34, 134; Memo for the Record, Graves, n.d.; Rpt, AFSWC, Maj Ronald L. Hager, "4900th Test Group Test Plan, ALL Cycle II: APT Open Port With No Input Window Test," 19 December 1975, p. 1 and Attachment 2.

42. Interview with Boesen, 30 September 1988; Interview with Gilbert, 31 August 1988; Rpt, AFWL-TR-76-271, pp. 33, 148–154, 174, 195–215; Rpt, Johnson, pp. 35–50.

43. The diameter of the beam at the target is limited by the laws of physics. The main central lobe of the beam can be no smaller than 2.44 times the wavelength of the light times the range to the target divided by the transmitting aperture diameter (the diameter of the primary mirror). This meant that if the range remained constant, larger optics would be required to focus longer-wavelength lasers to a small spot on target. Another way to look at it was if the size of the diameter of the optics remained constant, say 0.6 meter, then the diameter of the beam on target would be the same if a 1-micron laser wavelength was focused at 100 kilometers, or a 10-micron laser was

focused to a target at 10 kilometers. Interview with Gilbert, 31 August 1988; Interviews with Kyrazis, 12 and 19 September 1985; Rpt, AFWL-TR-76-271, p. 174.

44. Interview with Davis, 14 May 1990; Interview with Gilbert, 31 August 1988; Interview with Boesen, 29 October 1993; Rpt, AFWL-TR-76-271, pp. 33, 136; Rpt, Johnson, pp. 120a–121; Rpt, Ayers, "4900th Test Group Test Plan," p. 1; Notes, Kyrazis, "Cycle II Atmospheric Constituent Measurement Contract," n.d.; Ltr, Hans Mark to Dr. Robert Greenberg, Office of the Director of Defense Research and Engineering, "ALL Cycle II," 21 July 1976.

45. Rpt, AFWL-TR-76-271, p. 34; Rpt, Johnson, pp. 66–113; Rpt, AFSWC, Ayers, "4900th Test Group Test Plan, Cycle II Window Fairing and Aero Fence Qualification Flight Tests," 1 May 1975, pp. 8–10, 12–14; Ltr, Kyrazis to AFSWC, "APT Cycle II External Mods," 22 March 1974.

46. Interview with Gilbert, 31 August 1988; Rpt, AFWL-TR-76-271, pp. ii, 32, 461.

47. Interview with Kyrazis, 12 September 1985.

48. Interviews with Boesen, 30 September 1988 and 29 October 1990.

49. Ibid.; Rpt, AFWL-TR-76-271, p. 238.

50. Interview with Boesen, 4 June 1990; Rpt, AFWL-TR-76-271, pp. 56, 238–256.

51. Interviews with Boesen, 30 September 1988 and 29 October 1990.

52. In May 1977, the 4950th Test Wing at Wright-Patterson AFB modified aircraft 371 so a number of airborne propagation experiments could be performed. One goal was to measure the absorption of hydrogen fluoride/deuterium fluoride laser beams in the upper atmosphere. A second goal was to measure the diffuse spreading of the beam as it propagated from an aircraft. Test results helped to improve the data base for laser propagation in general and more specifically were useful in predicting beam behavior in the turbulent flow region next to the ALL aircraft for the Cycle III testing. These beam propagation experiments on 371 made up what was known as Cycle II.5. Interviews with Boesen, 30 September and 29 October 1990; Rpt, AFWL-TR-76-271, p. 31; Ltr, Espey to "See Distribution," "Start of Cycle IIB," 8 August 1975; Memo for the Record, Espey, "Reflections on ALL Flights 72 and 73," 18 September 1975; Memo for the Record, Espey, "Costs for the Support," 29 September 1975; Rpt, Leonard J. Otten III, "Flight Test Plan, Airborne Laser Laboratory: Cycle II.5," n.d. [Feb 77].

53. Interviews with Boesen, 30 September 1988 and 4 June 1990; Interview with Gilbert, 31 August 1988; Rpt, Johnson, p. 11.

54. Rpt, AFWL-TR-76-271, pp. 71–72, 115–117.

55. During the final flights of Cycle IIB, the flight crew experienced some terrifying moments shortly after takeoff. During climb out from Kirtland AFB, the ALL's engines would mysteriously "roll back" from their climb power settings to idle power. First the two inboard engines would lose power in unison and remain in idle for several seconds. As they returned to climb power, the two outboard engines would repeat the same frightening sequence. All of this was done without any throttle changes by the pilots. Between flights, AFWL technicians tried everything in the book to locate and solve the problem. Although they replaced several suspected components, they were unable to eliminate the uncommanded roll backs. Once the Cycle IIB flights were completed, the ALL went to GD for Cycle III modifications. There, GD workers tried to solve the vexing engine roll back problem by removing the engine fuel supply system from the aircraft and sending the components to Air Force laboratories for detailed inspection. The source of the problem was unique, as it had never appeared before in the Air Force tanker fleet. A fuel line hose had delaminated. After a mission, the engine shutdown sequence would create a partial vacuum in the hose, which in turn sucked air into it. The hose was at a high point in the fuel supply system and the air remained trapped there until the next flight. Fuel was not routed through the hose until the pilots changed fuel settings during climb out. At that time, the air pocket moved through the fuel lines, out the wings, into the inboard engines, and then into the outboard engines, causing the roll backs. The leak in the hose was undetectable when it finally was pressurized with fuel because the delaminated section of the hose acted as a one-way valve; air could leak in but fuel could not

leak out. Rpt, AFWL-TR-76-271, p. 72; Briefing, AFWL, Lamberson, "Milestone Review of Cycle II," 8 June 1976; Rpt, Johnson, pp. 19–21a; Rpt, AFWL-TR-86-01, p. vi.

56. Rpt, AFWL-TR-76-271, p. 72; Briefing, Lamberson, "Milestone Review of Cycle II." 8 June 1976; Rpt, Johnson, pp. 6–9.

57. Rpt, AFWL-TR-76-271, pp. 32, 235–237; Rpt, Johnson, pp. 57–60.

58. Rpt, AFWL-TR-76-271, pp. 67–69.

59. Ibid., pp. 32, 67–69; Briefing, Lamberson, 8 June 1976.

60. Ibid.

61. Rpt, Short, p. 3; Rpt, AFWL-TR-76-271, p. 33; Briefing, Lamberson, 8 June 1976.

62. Interview with Gilbert, 31 August 1988.

63. Ibid.

64. Rpt, AFWL-TR-76-271, pp. ii, 34, 257–258; Briefing, Lamberson, 8 June 1976; Rpt, AFWL-TR-80-06, p. vi.

65. The Air Force High Energy Laser Milestone Assessment Board reviewed and verified the successful completion of Cycle II in June 1976. Transcript of Speech, Kyrazis, "ALL Departure from AFWL," 4 May 1988.

66. Ltr, Mark to Greenberg, Office of the Director of Defense Research and Engineering, "ALL Cycle II," 21 July 1976.

CHAPTER 10. FIRST STEPS TO READY HIGH-POWER LASER

1. Ltr, Lt Col Samuel M. Guild, Jr., Chief, Engineering Div, FTEM (4900th Test Group), to FTET (Mr. Ayers), "Airborne Laser Laboratory Class II Modification Documentation, Part I, Modification Proposal for Cycle III," 6 June 1974; Rpt, No. 599ABQ3002, W. L. Jones, Senior Design Engineer, General Dynamics, "R&D Equipment Installation and Checkout Plan," 24 June 1977, pp. 1-1–1-2.

2. Originally, AFWL planned to hire Hughes to build the automated dynamic alignment system because of that company's demonstrated competence with optics gained from the TSL program. Plus, Hughes had conducted some early design studies for the ADAS. In the end, award of the ADAS contract went to Perkin-Elmer based on their superior technical proposal and lower projected cost. Later in the Cycle III buildup, higher performance beam-steering mirrors were built by Hughes. Still later, field support of the ADAS was combined with that for the APT, and Hughes was awarded the support contract. Ltr, Col Russell K. Parsons, AFWL, Chief, Laser Development Division, to AFWL/AR/CC, "AFWL Laser Program," 22 April 1974; Rpt, No. 599ABQ3002, pp. 1-1–1-2.

3. John W. Dettmer and Demos T. Kyrazis, "Overview of the Airborne Laser Program," in *Proceedings of the Third DOD High Energy Laser Conference*, July 1979, pp. 379–396.

4. Rpt, AFWL-TR-86-01, Vol. I, Raymond V. Wick *et al.*, "Airborne Laser Laboratory: System and Test Description," May 1988, pp. 4–5.

5. Mainly because of funding problems with HELRATS combined with delays in completing the ALL program on schedule, Air Force Systems Command eventually canceled Cycle IV. Rpt, AFWL, "Airborne Laser Laboratory: Cycle III Test Plan (Executive)," 5 June 1978, pp. 1–7; Ltr, Feaster, to "See Distribution," "ALL Cycle III Test Plan," 22 November 1977; John C. Rich, "Air Force Overview," in *Proceedings of the Third DOD High-Energy Laser Conference*, July 1979, pp. 47, 52–53; Rpt, AFWL-TR-80-102, Vol. I, Paul H. Merritt, "Airborne Laser Laboratory Test Cell Report," January 1981, pp. 14–15.

6. Rpt, AFWL, "Airborne Laser Laboratory Cycle III Test Plan (Executive)," 31 October 1977, pp. 3–5, 93–94; Rich, pp. 47, 52–53.

7. Rpt, AFWL-TR-86-01, p. 3.

8. Ibid., pp. 3–4; Dettmer and Kyrazis, pp. 380–381; Rpt, AFWL, "R&D Equipment Installation and Checkout Plan," 24 June 1977, pp. 5–7.

9. Interview with Kyrazis, 4 November 1994.

10. Ibid.; Interview with Dettmer, 3 October 1988; Resume, Colonel John W. Dettmer, 1984.

11. When Dettmer first arrived at Kirtland he actually headed up a 4950th "operating location" (OL-AA) that later became Detachment 2. This unit was designed to maintain the ALL and the ALL diagnostic aircraft, in addition to the "Big Crow" aircraft (Army program), and two readiness-to-test aircraft in support of the Department of Energy. The key people responsible for getting Detachment 2 under way were Captain Hal Rhoades, Staff Sergeant Jim "Augie" Augustine (ALL flight mechanic), Sergeant Bob Hoppenrath, Staff Sergeant Ken Vanderwall (ALL crew chief), and Jim Switzer. Interview with Kyrazis, 4 November 1994; Interview with Dettmer, 3 October 1988; Resume, Colonel John W. Dettmer, 1983.

12. Interview with Kyrazis, 4 November 1994.

13. Ibid.; Interview with Dettmer, 3 October 1988; Rpt, "Airborne Laser Laboratory: Cycle III Test Plan (Executive)," 5 June 1978, pp. 1–10.

14. Interview with Kyrazis, 4 November 1994; Telecon with Philip P. Panzarella, Electronic Systems Center, Hanscom AFB, Massachusetts, 24 July 1995; Biography (USAF), Philip P. Panzarella, 25 April 1995.

15. Interview with Chief Master Sergeant James T. Augustine, Phillips Laboratory, Albuquerque, New Mexico, 20 July 1994; Telecon with Panzarella, 24 July 1995.

16. Interview with Kyrazis, 4 November 1994; Telecon with Panzarella, 24 July 1995.

17. Interview with Augustine, 20 July 1994; Telecon with Panzarella, 24 July 1995.

18. Rpt, AFWL-TR-75-257, Part 1 of 2, James E. Bell, "Airborne Laser Laboratory, Cycle III, Final Installation Design Report," May 1977, pp. 1–2.

19. GD received $3 million for its engineering design work under AFWL contract F29061-74-C-0114. Rpt, AFWL-TR-75-257, pp. 1–2.

20. Interview with Kyrazis, 12 September 1985; Rpt, AFWL, "Interim Progress Review: ALL Cycle III Airplane Modification & Installation Contract," 9 December 1976, p. 26; Rpt, AFWL-TR-86-01, p. 3.

21. The ALL was operated and maintained by the 4950th Test Wing, which was part of ASD. Interview with Lamberson, 12 January 1989.

22. Interview with Kyrazis, 17 October 1985; J. Schaefer *et al.*, "Airborne Laser Laboratory GDL Status Update," in *Proceedings of the Third DOD High Energy Laser Conference*, July 1979, pp. 231–233; Rpt, AFWL-TR-75-257, pp. 13, 147, 309.

23. Interview with Kyrazis, 12 September 1985; Rpt, AFWL-TR-75-257, p. 12; Rpt, "Airborne Laser Laboratory Cycle III Test Plan (Executive)," 31 October 1977, pp. 10–11; Rpt, AFWL-TR-86-01, pp. 5–9.

24. Interview with Lamberson, 12 January 1989; Rpt, AFWL-TR-75-257, pp. 376–385; Interview with Kyrazis, 12 September 1985; Rpt, AFWL-TR-86-01, pp. 7–8.

25. Interview with Kyrazis, 17 October 1985; Interview with Dettmer, 3 October 1988; Rpt, AFWL-TR-75-257, pp. 211–214, 293, 309–313, 397, 419–422; Rpt, AFWL, pp. 10–11; Rpt, AFWL-TR-86-01, pp. 5–6.

26. Interview with Kyrazis, 4 November 1994.

27. Ibid.

28. Interviews with Kyrazis, 12 September and 17 October 1985; Rpt, AFWL-TR-75-257, pp. 22, 153–158, 359, 482–488; Rpt, AFWL-TR-86-01, p. 5.

29. Rpt, AFWL-TR-75-257, pp. 482–488.

30. Ibid., pp. 359–361; Interview with Kyrazis, 12 September 1985; Dettmer and Kyrazis, p. 380.

31. Interview with Otten, 27 December 1988.

32. Interview with Otten, 27 December 1988; Ltr, Guild to FTOD, "Rerouting of Aircraft Control Cables in NKC-135A 2123," 18 March 1974; Rpt, AFWL-TR-75-257, pp. 320–323.

33. Interview with Kyrazis, 12 September 1985; Rpt, AFWL-TR-75-257, pp. 320–327.

34. Rpt, AFWL-TR-75-257, pp. 353, 385–386, 395.

35. Ibid., pp. 328–333; Interview with Kyrazis, 17 October 1985; Dettmer and Kyrazis, p. 380; "Fort Worth Modified Aircraft is Key Part of Laser Development," *General Dynamics World*, March 1981.

36. Interview with Dettmer, 3 October 1988; Interview with Augustine, 20 July 1994; Rpt, AFWL-TR-75-257, pp. 124–134.

37. Interview with Kyrazis, 12 September 1985.

38. Interview with Augustine, 20 July 1994; Interview with Dettmer, 3 October 1988; Telecon with Panzarella, 24 July 1995.

39. Interview with Kyrazis, 17 October 1985; Rpt, AFWL-TR-75-257, pp. 479–481.

40. Msg, 4950th TW WPAFB OH/CC to HQ AFSC, "Airborne Laser Laboratory/4950th Test Wing," 28 July 1977.

41. Interview with Otten, 27 December 1988; Rpt, AFWL-TR-86-01, p. 7.

42. Dettmer and Kyrazis, p. 381; Rpt, AFWL-TR-86-01, p. 7; Addendum No. 2 to 28 June 1977 Report of Airborne Laser Laboratory Cycle III Flight Readiness Review, by Executive Independent Review Team, BG Richard K. Saxer, Chairman, 17 November 1978, pp. 1, 3–5.

43. AFWL selected Pratt & Whitney because it had developed the XLS laser (an improved version of the XLD-1) which was the free world's only 500-kilowatt GDL. The SSD would be modeled after the XLS. Interview with Otten, 30 August 1991; Rpt, AFWL-TR-76-277, Joseph F. Zmuda and John A. Carpenter, "Fabrication of an Advanced Gas Dynamic Laser," September 1977, pp. 29–31; Award/Contract Statement (F29601-71-C-0097), AFWL with Pratt & Whitney, 3 June 1971; Sole Source Justification Statement, Col Charles J. Avery, AFWL/CV, 29 March 1971; Statement of Work (F29601-71-C-0097), "GDL Technology Confirmation Program," 30 April 1971; Rpt, AFWL, Donald L. Lamberson, "Airborne Laser Laboratory Configuration and Technology Status Report," March 1973, pp. 2, 26–27; Memo, Richard J. Casey, General Accounting Office, Denver Regional Office, to Colonel Lamberson, AFWL/ARTO, "Summary of Air Force High Energy Laser Program," 3 October 1973.

44. Interview with Roger Wahl, Rocketdyne Division, Rockwell International, Canoga Park, California, 14 May 1990; Ltr, A. R. Weldon, Assistant Secretary, Pratt & Whitney, to AFSWC (Mr. S. Jorgensen), "GDL Technology Confirmation Program," 12 November 1971; Rpt, Pratt & Whitney, "Proposal For The Split Stage Demonstrator Unstable Resonator," 15 November 1971, pp. II-1–II-2.

45. The combustor was regeneratively cooled by circulating cold liquid CO around the combustor. As the liquid CO absorbed heat from the combustor at ambient temperature, it changed to a gas and was then injected into the combustor to burn to start the gas flow moving down the distribution manifold. Interview with Bob Hindy, Rocketdyne Division, Rockwell International, Canoga Park, California, 14 May 1990; Rpt, AFWL-TR-80-102, pp. 24–26; Rpt, William C. Missimer *et al.*, Pratt & Whitney, "SSD Design Review Meeting," 14–17 June 1971, p. 65; Rpt, AFWL-TR-86-01, pp. 16–17.

46. The SSD was configured for the gases to flow upward as opposed to downward for the GDL that eventually went in the ALL. Therefore, the SSD nozzle array formed the ceiling rather than the floor of the distribution manifold as in the ALL setup. The lasing principle, however, was the same for both devices. Interview with Kyrazis, 17 October 1985; AFWL-TR-76-277, p. 30.

47. This movement of the flow was analogous to releasing air from a balloon. Air inside a blown-up balloon possesses a great deal of pressure but no velocity. Once the balloon is punctured, air squirts out at great velocity. Interview with Kyrazis, 17 October 1985; Interview with Wahl, 14 May 1990; Interview with Otten, 30 August 1991; Rpt, AFWL-TR-76-277, pp. 125–147.

48. Interview with Otten, 30 August 1991; Interview with Wahl, 14 May 1990; Rpt, AFWL-TR-86-01, pp. 16–17.

49. Rpt, AFWL-TR-71-129, Vol. III, M. T. Schilling *et al.*, "GDL Confirmation Program," September 1972, pp. iii, 160, 167.

50. One of the problems identified with the SSD was the generation of shock waves within the cavity caused by aerodynamic imperfections in the shape and smoothness of the nozzle array and cavity sidewalls. This introduced turbulence in the cavity which degraded beam quality and stressed the structural integrity of the cavity walls. Also, as the temperatures and pressures changed in the nozzle array with the flow, the nozzles tended to expand and contract which altered the width of the opening between nozzles for the flow to pass through. This affected the high tolerances demanded of the system and required continual adjustments of the spacing of the nozzles to achieve peak operating efficiency. Interview with Wahl, 14 May 1990; Rpt, FP 72-119, Pratt & Whitney, "Program For Repair and Testing of the SSD," 11 August 1972, pp. 1–47.

51. Interview with Otten, 30 August 1991; Rpt, AFWL, Colonel Robert W. Rowden, "AFWL Significant Events," 31 October 1972.

52. Rpts, Rowden, "AFWL Significant Events," 31 October and 30 November 1972; Rpt, ARTO-73-1, Lamberson, "Program Management Plan: Advanced Radiation Technology Office," June 1973, p. 19; Rpt, Lamberson, "Airborne Laser Laboratory Configuration," pp. 2, 26.

53. Interview with Lamberson, 13 January 1989; Rpt, Rowden, 30 November 1972; Rpt, ARTO-73-1, June 1973, p. 25.

CHAPTER 11. FUELING THE LASER

1. Rpt, AFWL, Barron Oder, "Index of Contracts in WL/HO Archives: Arranged Alphabetically by Contractor," June 1990; Rpt, AFWL-TR-76-277, Part I of II, Joseph F. Zmuda and John A. Carpenter, "Fabrication Of An Advanced Gas Dynamic Laser," September 1977, pp. 9–30; "Boeing NKC-135 Modified as Laser Laboratory," *Aviation Week & Space Technology*, 30 April 1973, p. 22.

2. Interview with Otten, 18 December 1991; Rpt, AFWL-TR-80-102, Vol. I, Paul H. Merritt *et al.*, "Airborne Laser Laboratory Test Cell Report: Summary Results," January 1981, pp. 14–15; Rpt, AFWL, L.J. Otten and D.L. Boesen, AFWL, "The Airborne Laser Laboratory: A Program Review," n.d.

3. Rpt, AFWL-TR-80-102, pp. 71–79.

4. Interview with Kyrazis, 19 September 1985.

5. Construction of the ARTF cost approximately $4 million from emergency military construction program funds. AFWL personnel began moving into the building in April 1975. Dick Frosch led an AFWL team who worked with the contractor McDonnell Douglas on the design and construction of the fuel farm behind the ARTF hangar. Rpt, AFWL-TR-80-102, p. 71.

6. Interview with Richard MacCutcheon, Master Sergeant, USAF, Retired, Albuquerque, New Mexico, 23 December 1991.

7. Interview with Chuck Edwards, Chief Master Sergeant, USAF, Retired, Albuquerque, New Mexico, 31 August 1995.

8. Interview with Kyrazis, 19 September 1985; Interview with Dyble, 4 November 1993; John W. Dettmer and Demos T. Kyrazis, "Overview of the Airborne Laser Laboratory," in *Proceedings of the Third DOD High Energy Laser Conference,* July 1979, p. 381.

9. Interview with MacCutcheon, 21 December 1991; Interview with Kyrazis, 19 September 1985; Interview with Dyble, 4 November 1993.

10. Interview with Dyble, 4 November 1993.

11. Interview with MacCutcheon, 21 December 1991; Interview with Kyrazis, 19 September 1985; Rpt, AFWL-TR-80-102, p. 35.

12. The Control and Instrumentation Subsystem consisted of the fluid flow controls, shutoff and check valves, pressure relief and regulator valves, and computation and data processing devices to ensure the safe operation of the FSS. Rpt, AFWL-TR-76-277, pp. 39–49, 69, 286–295, 325–337 and Part II of II, pp. 341–342, 411–416; Dettmer and Kyrazis, pp. 380–382; Rpt, AFWL, Donald L. Lamberson, "ARTO Progress Report for October 1975," 5 November 1975.

13. The combustor was regeneratively cooled by circulating liquid carbon monoxide around the combustor. As the cold (77 degrees Kelvin or several hundred degrees below zero Fahrenheit) liquid CO absorbed heat from the combustor at room temperature, the liquid CO changed to the gas carbon monoxide. This gas was then injected into the combustor to burn and start the gas flow moving down the distribution manifold.

    This same principle was applied to cool the laser device and diffuser. Liquid nitrogen circulated through the nozzle channels to regeneratively cool the sidewalls of the device and also circulated around the diffuser to cool its sidewalls. As the liquid nitrogen picked up heat, it turned into gaseous nitrogen, which was subsequently fed into the downstream chamber of the combustor. Adding nitrogen at that point to the mixture of nitrous oxide and carbon monoxide already burning created the right proportion of gases to produce ideal lasing conditions in the cavity or unstable oscillator. The nitrogen did not lase, but it exhibited the same frequency as the CO_2. When an excited nitrogen molecule collided with a CO_2 molecule, energy transferred from the nitrogen to the CO_2, which helped to keep the CO_2 energized in its population inversion state to sustain the lasing action. To illustrate, think of the CO_2 and nitrogen as two billiard balls. As the moving nitrogen strikes the stationary CO_2, the impact causes the nitrogen to transfer its energy to the CO_2, the latter thus moving to a higher energy level.

    Regenerative cooling techniques on the ALL represented an advancement over the SSD, which was an uncooled system. Interview with Kyrazis, 12 September 1985; Interview with Hindy, 14 May 1990; Rpt, AFWL-TR-76-277, pp. 288, 307–314 and Part II of II, pp. 462–479; Rpt, Lamberson, 5 November 1975.

14. The cooled ALL diffuser had a start rate of about 10 percent better than the uncooled SSD diffuser. Interview with Don Teasdale, Captain, USAF, Retired, Albuquerque, New Mexico, 13 January 1992; Rpt, AFWL-TR-76-277, p. 287 and Part II of II, pp. 341–342, 411–412, 503–507.

15. Interview with Joseph F. Zmuda, Lieutenant Colonel, USAF, Retired, Albuquerque, New Mexico, 25 February 1992.

16. Ibid.

17. Ibid.

18. Ibid.; J. Schaefer et al., "Airborne Laser Laboratory GDL Status Update," in *Proceedings of the Third DOD High Energy Laser Conference*, July 1979, p. 232; Rpt, AFWL-TR-78-168, William C. Hurley et al., "ALL/GDL Cycle III Support Services," February 1980, pp. 221–222—see pp. 213–235 for a listing of individual test runs in the test cell.

19. As a safety precaution, nitrogen and helium gases and liquid nitrogen were used during the cold flow testing to simulate the real reactant of toxic liquid carbon monoxide that would be used in the system during the follow-on hot flow testing. Cold flow tests simulated flow characteristics, temperatures, and the densities of toxic liquid carbon monoxide.

    As a further check on the structural integrity of the FSS storage tanks, Pratt conducted lengthy pressurization tests to ensure they did not rupture or leak. Each tank was pressurized and left to sit overnight while a Pratt technician monitored the tank's performance. This overnight testing went on for weeks, which demonstrated the tanks could hold the pressures required to run the FSS. Interview with MacCutcheon, 23 December 1991; Rpt, AFWL-TR-79-215, Paul H. Merritt, Arthur L. Pavel, and Richard A. Frosch, "Airborne Laser Laboratory Cycle III/Phase 3 Test Report," December 1980, pp. 13–20.

20. Flodyne Controls, Inc., Murray Hill, New Jersey, manufactured all of the flight valves. Because of the very high precision demanded in the timing of the valves, Flodyne was the only contractor to bid on the valves. Amendment No. 6 to Statement of Work for Contract F29601-73-C-0117, AFWL/LRL, 4 August 1976; Interview with MacCutcheon, 23 December 1991; Interview with Teasdale, 13 January 1992.

21. Interview with MacCutcheon, 23 December 1991.

22. Pratt's meticulous recordkeeping was part of their quality control. Nothing went on the FSS until approved by the Pratt test cell engineer who ensured the change was first made on the wooden

mock-up. MacCutcheon recalled that if you wanted a light bulb changed, they wrote that down in their manual confirming they had directed a technician to accomplish that task. "After the technician did it, he'd come back and sign it off!" Interview with MacCutcheon, 23 December 1991; Rpt, AFWL-TR-76-277, p. 66.

23. The FSS supply tanks were capable of storing water at a pressure of 800 psi. The system could regulate the flow to deliver water at a rate up to 36.3 pounds per second to the mirrors in the device, optical bench, ADAS, and APT. Captain Don Teasdale, who was AFWL's lead project officer during the FSS testing, explained water flowed through the system for only 7.6 seconds, which meant all of the water had to get to the mirrors quickly under high pressure to cool them. Because of the water hammer effect, several mirrors were broken during testing in the test cell. Teasdale stated that one of the valves from the water tanks was opening too fast "sending a slug of water down the line and into the mirror and breaking it." To correct this, Teasdale's group "slowed the opening of that valve up and allowed it to go down the line a little slower. We also had to slow down the closing of it, because a water hammer effect is probably more critical on the closing side than the opening side. When you close the valve, if you close it real fast and then there's an air pocket behind the flow, because the fluid has momentum, that slug of fluid is going down the line." Schaefer, pp. 233–235; Amendment No. 12 to Statement of Work, "Fabrication of an Advanced GDL," AFWL/LRL, 22 August 1974.

24. Schaefer, pp. 233–234; Rpt, AFWL-LR-TN-79-002, Peter D. McQuade, "An Investigation of Flow-Induced Vibrations of the Water-Cooled Laser Mirrors of the U.S. Air Force Airborne Laser Laboratory," 20 April 1979, pp. 1–7, 18; Rpt, AFWL-ARL-TN-83-10, Peter D. McQuade *et al.*, "Water Flow Induced Jitter Investigation Airborne Laser Laboratory Cycle III and IV," June 1979, pp. 27–43.

25. Before installing the FSS on the airplane, the FSS was transported to the Manzano Laboratory where it was completely refurbished. All valves and lines were checked, some were replaced, and the entire FSS was cleaned. Interview with Kyrazis, 19 September 1985; Interview with Teasdale, 13 January 1992; Rpt, AFWL, Ted Clements, "Technology Activity Information Flow," 13 October 1978.

26. The TCM used in the test cell did not require a boom. The boom supported the lines that carried the fuels into the aircraft. Rpt, AFWL-TR-79-215, pp. 7–21.

27. Rpt, AFWL-TR-79-215, p. 21.

28. Ibid., pp. 21–24.

29. Ibid.; Interview with Teasdale, 13 January 1992.

30. The Cycle III fairing, High Energy Laser Radar Acquisition and Tracking System (HELRATS), and dummy APT were flown on the ALL for the FSS flight tests. HELRATS, designed to track multiple targets, was eventually discarded because of problems in getting it to work properly and because of insufficient funding. Rpt, AFWL-TR-79-215, pp. 11–12, 24.

31. AFWL-TR-79-215, pp. 10–12, 24–27.

32. Ibid., pp. 25–28, 103–108; Interview with Kyrazis, 19 September and 17 October 1985; Interview with MacCutcheon, 21 December 1991; Interview with Joe Hoerter, Lieutenant Colonel, USAF, Retired, El Segundo, California, 16 May 1990.

33. Interview with Kyrazis, 19 September 1985; Rpt, AFWL-TR-79-215, pp. 11–12, 28–36, 103–108; Dettmer and Kyrazis, p. 381.

34. Interview with Kyrazis, 4 November 1994; Interview with Otten, 2 June 1994.

35. Ibid.; Interview with Dettmer, 3 October 1988.

36. Interview with Kyrazis, 4 November 1994; Interview with Otten, 2 June 1994.

37. Interview with Kyrazis, 4 November 1994.

38. Ibid.

39. Ibid.; Interview with Otten, 2 June 1994.

40. Interview with Kyrazis, 4 November 1994; Biography (USAF), Major General Donald L. Lamberson, September 1988.

41. Award Citation of Distinguished Service Medal for Colonel Donald L. Lamberson, AFWL/ARP, 28 February 1978.

42. Nomination for the Air Force Association Citation of Honor for Colonel Donald L. Lamberson, General George S. Brown, Commander, Systems Command, 1971.

43. Award (narrative) for Distinguished Service Medal for Brigadier General Donald L. Lamberson, 1978.

44. Interviews with Lamberson, 12 and 13 January 1989 and 7 December 1995.

45. Letter of Instruction, HQ AFSC to AFWL, "Personnel Changes," 10 March 1978; Biography (USAF), Colonel John C. Rich, Office of Information, KAFB, January 1978.

46. Later, in 1982 Systems Command offered Rich a general's position at Aeronautical Systems Division at Wright-Patterson AFB. By that time, Rich's personal goal of becoming a general had "faded" as he began exploring the job opportunities in private industry. Interview with Rich, 10 January 1989.

47. Interview with Rich, 10 January 1989; Ltr, Lt General Richard C. Henry, Commander, Space Division, to Colonel John C. Rich, "Laser Contributions," 24 November 1982.

48. Interview with Kyrazis, 12 September 1985; Report No. P74-519, Hughes Aircraft Company, Culver City, California, "Subsystem Design Analysis Report Technology Breadboard Report," December 1974, pp. 1-1–8-1.

49. Rpt, AFWL-TR-75-241, Vol. III, Burton O'Neil *et al.*, "Subsystem Design Analysis Report Technology Breadboard Report," February 1976, pp. 1-1–3-6, 6-1–6-5.

50. Interview with Kyrazis, 12 September 1985.

51. Interview with Kyrazis, 12 September 1985; Interview with Paul H. Merritt, Phillips Laboratory, Albuquerque, New Mexico, 2 March 1992; Rpt, AFWL-TR-77-109, Frank J. Briscoe, "Airborne Dynamic Alignment System Fabrication Program," January 1980, pp. 9–14, 123–124.

52. Rpt, AFWL-LR-TN-78-001, Lanny Larson *et al.*, "ADAS Test Cell Characterization Report," [May 1978], pp. 1–8, 15; Rpt, AFWL-TR-86-01, Vol. I, Raymond V. Wick, "Airborne Laser Laboratory Cycle III: Systems and Test Descriptions," May 1988, pp. 10–11.

53. Interview with Merritt, 2 March 1992; Paul H. Merritt *et al.*, "Laboratory Testing of the Airborne Laser Laboratory Pointing and Tracking System," in *Proceedings of the Third DOD High Energy Laser Conference*, July 1979, p. 741; Rpt, AFWL-LR-TN-78-001, L. Larson *et al.*, "ADAS Test Cell Characterization Report," May 1978, pp. 22–24.

54. Interview with McCutcheon, 23 December 1991; Interviews with Kyrazis, 20 June and 12 September 1985; Robert L. Van Allen *et al.*, "High Power Testing of the Airborne Dynamic Alignment System (ADAS)," in *Proceedings of the Third DOD High Energy Laser Conference*, July 1979, pp. 741–750; Rpt, AFWL-TR-86-01, pp. 11–12.

55. Schaefer, pp. 231–232; AFWL-TR-80-102, pp. 104–105, 124.

56. Interview with Merritt, 2 March 1992; Van Allen, pp. 741–750; Rpt, AFWL-TR-77-109, pp. 42–47; Rpt, AFWL-TR-80-102, pp. 31–33.

57. Interview with Merritt, 2 March 1992; Rpt, AFWL-TR-77-109, pp. 123–124.

58. Interview with Merritt, 2 March 1992; Rpt, AFWL-LR-TN-78-001, p. 15.

59. Interview with Parsons, 2 April 1992; Interview with Wick, 2 April 1992; Interview with Lamberson, 12 January 1989.

60. Perkin-Elmer bid on the follow-on ADAS contract but was not selected. AFWL required each bidder to meet strict technical specifications for the mirrors, but Perkin-Elmer was unsure they could attain those technical specifications. They wanted more leeway and took the position that the technical specifications should be "goals" to shoot for. That was unacceptable to AFWL and put Perkin-Elmer out of the running for additional ADAS work. Rpt, FR-97-77-853R, Hughes Aircraft Company, "AABISM Final Report: APT/ADAS Beam Interface Steering Mechanism Subsystem," August 1979, pp. 1-1–1-3, 2-1.

61. Rpt, FR-97-77-853R, pp. 1-1–3-5.

62. Rpt, AFWL-TR-80-102, p. 33.

63. Interview with Boesen, 5 January 1994; Notes on Cycle III, Boesen, 5 January 1994; Rpt, AFWL-TR-86-01, pp. vii, 25; Rpt, AFWL-TR-80-102, p. 129.

64. Ibid.

65. Interview with Boesen, 5 January 1994; Notes on Cycle III, Boesen, 5 January 1994; Van Allen, p. 739.

66. Interview with Boesen, 5 January 1994; Notes on Cycle III, Boesen, 5 January 1994; Dettmer and Kyrazis, pp. 379–380; Rpt, AFWL-TR-80-102, pp. 29, 126; Rpt, AFWL-TR-86-01, pp. vii, 10; Joseph Albright, "U.S. Readies Powerful Laser Weapon," *Atlanta Journal*, 10 March 1980, p. 1.

67. Interview with Boesen, 5 January 1994; Notes on Cycle III, Boesen, 5 January 1994.

68. At one point in the planning process, AFWL considered using a water-cooled primary mirror on the Cycle III ALL. The mirror was developed by Garrett AiResearch Corporation and, during initial checkout at the contractor's plant, it leaked coolant. The leaks were small and around the edges of the mirror. Garrett AiResearch and AFWL decided to use a chemical substance added to the water to try to plug the leaks, similar to using Stop Leak in an automobile radiator. Unfortunately, it did not work on the ALL primary; the chemical plugged all of the cooling channels, effectively destroying the cooled mirror. As it turned out, AFWL studies showed that the uncooled primary mirror was capable of handling the HEL beam for Cycle III. Dettmer and Kyrazis, p. 380; AFWL-TR-86-01, pp. vii, 10–11, 24.

69. Van Allen, pp. 733–735.

70. Ibid., pp. 735–739; Rpt, AFWL-TR-80-102, p. 119.

71. Schaefer, p. 233.

72. Rpt, AFWL-TR-80-102, pp. 14, 64–68 and Vol. II, "System Description—Meteorology, Diagnostics, Instrumentation," pp. 68, 306–313.

73. Rpt, AFWL-TR-86-01, p. 13; Rpt, AFWL-TR-80-102, Vol. I, pp. 1–2, 64–65; Vol. II, pp. 64–66.

74. Rpt, AFWL-TR-80-102, Vol I, pp. 1–2.

75. The wall was constructed 15 degrees off perpendicular from the beam line of sight from the test cell. This was done to reduce glint reflections off the wall back to the APT in the test cell. Stationary target boards were affixed to the wall for static shots. Rpt, AFWL-TR-80-102, pp. 75–92.

76. The target board consisted of an array of 256 IR detectors aligned in 16 horizontal and vertical rows. With the help of multiplexers, individual detector output was summed and converted to a digital signal. This total data picture was telemetered to two receiver sites—one at the instrumentation van at the downrange site and also to the diagnostic aircraft parked next to the test cell—where it was recorded and processed to determine the beam quality. Rpt, AFWL-TR-80-102, pp. 92–96 and Vol. II, pp. 83–84; Work Unit Monitor Report, AFWL, Jamie Londono, "Cycle III Downrange Beam Diagnostics," 24 August 1984, pp. 373–376.

77. Rpt, AFWL-TR-80-102, Vol. I, pp. 75–80.

78. Rpt, AFWL-TR-80-102, Vol. II, pp. 305–328.

79. Ibid., pp. 117–123.

80. Ibid., pp. 469–470.

81. Twenty minutes after the missile was irradiated, an identical high-power laser shot was made against a Plexiglas target placed in front of the AIM-9B. Burn patterns on the Plexiglas confirmed the compactness and lethality of the beam. Rpt, AFWL-TR-80-102, pp. 117–123, 469–473.

82. The focusing mirror was mounted on a stand 6.67 meters to the rear of the drone. It was used to recoup the loss of on-target intensity from thermal blooming as the beam traveled from the test cell to the downrange target. Rpt, AFWL-TR-80-102, pp. 123–130, 473–475.

83. Rpt, AFWL-TR-80-102, Vol. I, pp. 1–3, 141–145.

84. Another problem was that the laser produced a beam that included four different wavelengths, not just the 10.6-micron beam that was planned. This caused problems with the optical system, especially with the tracker that was tuned to reject the strong 10.6-micron reflections from the target. The other wavelengths were not rejected by the tracker and, consequently, overloaded the

tracker when the laser beam reflected from a target. The problem was caused by the high gain of the optical cavity at the unwanted wavelengths. AFWL solved the problem by putting a complex coating on the cavity mirrors to reduce the cavity gains at all but the 10.6-micron wavelength. Rpt, AFWL-TR-80-102, pp. 16–19, 129–150.

CHAPTER 12. HIGH-POWER LASER GOES AIRBORNE TO ENCOUNTER AIR-TO-AIR MISSILES

1. Rpt, AFWL-TR-81-151, Leonard J. Otten III, "Airborne Laser Laboratory Cycle III/Phase 4," December 1981, p. I-1; Ltr, Kari J. Fielder, CMD/PA, to AFWL/AR (Maj Gamble), "Review of Attached Document," 12 May 81, with attached document on ALL.
2. Rpt, AFWL, L.J. Otten and D.L. Boesen, "The Airborne Laser Laboratory: A Program Review," n.d.
3. Ibid.; Rpt, AFWL-TR-81-151, pp. i–iii, I-1; Talking Paper on Airborne Laser Technology, AFSC/DLWM, Capt Thomas J. Dyble, "Airborne Laser Technology (ALL)," 28 July 1980.
4. Rpt, AFWL-TR-81-151, p. i.
5. Talking Paper, Dyble, 28 July 1980; Rpt, Otten and Boesen, n.d.
6. Agenda, Senate Committee on Commerce, Science & Transportation—Science, Technology & Space Subcommittee Hearings on Laser Applications & Technology, 12 January 1980; "Future Uses of Lasers Are Outlined to Panel," *Albuquerque Journal*, 13 January 1980, p. A-5.
7. Major General George F. Keegan resigned his job as chief of the Air Force Intelligence Agency in 1977, warning the United States was not investing enough in beam weapons development at a time when the Soviets were accelerating their program. Keegan, who some believed to be an alarmist, stated, "The development of beam weapons is even more momentous than the development of the atomic bomb. Its implications for the security of the free world, in this decade, are so awesome that they are beyond the political comprehension of this government and most leaders of the free world." See "U.S. and Soviet Race to Perfect 'Death Ray,'" *Virginia Beach Monitor*, 12 August 1980. Statement of Senator Harrison "Jack" Schmitt, Hearings on Laser Technology and Applications, 12 January 1980; "Airborne Laser Lab to Shoot Down Missiles, AF Secretary Says," *Aerospace Daily*, 17 January 1980, p. 84.
8. Statement of The Honorable Hans Mark, Secretary of the Air Force, Science, Technology & Space Subcommittee Hearings on Laser Applications & Technology, 12 January 1980.
9. Robert Frank Futrell, *Ideas, Concepts, Doctrine: A History of Basic Thinking in the United States Air Force, 1907–1964* (Montgomery, Alabama: Air University, 1974), p. 21; John F. Shiner, *Foulois and the U.S. Army Air Corps, 1931–1935* (Washington, D.C.: Office of Air Force History, 1983), pp. 17–21; Charles D. Bright, ed., *Historical Dictionary of the U.S. Air Force* (New York: Greenwood Press, 1992), pp. 389–391.
10. Statement of The Honorable Hans Mark, 12 January 1980; Lecture, "Technology and the Strategic Balance," delivered by Dr. Hans Mark to the Students and Faculty of The Naval War College, 6 May 1981, pp. 27–28; Interview with Hans Mark, University of Texas, Austin, 29 September 1992.
11. Interview with Mark, 29 September 1992; Statement of Hans Mark, Testimony at Hearing 12 January 1980 Before Senator Schmitt's Subcommittee, Albuquerque, New Mexico; "Airborne Laser Lab to Shoot Down Missiles, AF Secretary Says," *Aerospace Daily*, 17 January 1980, p. 84; "Air Force Plans Laser Weapon Use," *Fort Worth Star-Telegram*, 9 March 1980, p. 13A; "Future Uses of Lasers Are Outlined to Panel," *Albuquerque Journal*, 13 January 1980, p. A-5.
12. John Noble Wilford, "Laser Weapons to be Tested at White Sands Range," *New York Times*, 3 March 1980, p. A-16; "Air Force Plans Laser Weapon Use," p. 13A; Richard Burt, "U.S. Says Russians Develop Satellite-Killing Laser," *New York Times*, 22 May 1980, pp. A1, A9; "New Laser Weapon In Russia?," *Albuquerque Tribune*, 22 May 1980, pp. A-1, A-12.

13. The Army's program manager for the Ballistic Missile Defense program, Major General Grayson D. Tate, Jr., echoed Slay's caution. Tate told a congressional committee he estimated the development of high-energy lasers for defense was "at least a decade away." Edgar Ulsamer, "A More Liberal, Avant Garde R&D Program," *Air Force Magazine*, June 1980, pp. 42–43; Robert C. Toth, "Laser Will Attempt to Shoot Down Missile," *Los Angeles Times*, 16 March 1980, p. 4; "SAC Chief Says Laser Bomber Defense Faces Many Hurdles," *Defense Daily*, 7 August 1980, p. 187; "BMD Chief: Defense Lasers 10 Years Away," *Defense Daily*, 7 August 1980, p. 187.

14. "U.S. Effort Redirected to High Energy Lasers," *Aviation Week & Space Technology*, 28 July 1980, pp. 50–57; "Report Urges Major Increase in Space Laser Weapon Program," *Defense Daily*, 8 August 1980, p. 194; "Weapons Laboratory Aids Beam Effort," *Aviation Week & Space Technology*, 4 August 1980, pp. 57–59; Robert C. Toth, "War in Space," *Science*, September/October, pp. 76–77.

15. The Air Force exhibit at Farnborough was the first at an air show in a decade. William H. Gregory, "Crosscurrents at Farnborough," *Aviation Week & Space Technology*, 8 September 1980, p. 11; "Air Force Research Programs Displayed," *Aviation Week & Space Technology*, 13 October 1980, pp. 58–59; Briefing Chart, "ALL Shown at Farnborough," n.d.; Ltr, Col Charles D. Cooper, Chief, Media Relations Division, Headquarters Air Force to AFSC/PAM, "Public Affairs Guidelines, Airborne Laser Laboratory Program," 29 April 1980; Ltr, Col John C. Rich, AFWL/Deputy for Advanced Radiation Technology, "Publish Release of Air Force HEL Activities," 17 September 1980.

16. Rpt, AFWL-TR-81-151, pp. II-3–II-5.

17. Ibid.

18. Ibid.

19. During the flight tests at this time, a combination of missions were carried out. For example, during the third week of April, the ALL flew to Wright-Patterson to check for high pressure leaks in the FSS. Also, diagnostic aircraft 371 accompanied the ALL on this flight to test and maintain the telemetry link between the two aircraft. Rpt, AFWL-TR-81-151, pp. i–ii, II-17–II-18, III-2–III-3; Ltr, Major Lynn L. Gamble, AFWL/ARTO, to AFWL/CC, "AR Weekly Activity Report (14–18 April 1980)," 21 April 1980.

20. Rpt, AFWL-TR-81-151, pp. i–ii, III-4–III-5; Attachment (Chart: ALL Cycle III-Phase IV Flight Tests), to letter, HQ AFSC/DLWM to HQ AFSC/CST, "Significant Events Report, 26 Aug–2 Sep," 2 September 80; Rpt, Otten and Boesen, n.d.; Talking Paper, Dyble, 28 July 1980.

21. Rpt, AFWL-TR-81-151, pp. III-4–III-5; Msg, AFSC to AFWL, "Outstanding Issues on Airborne Laser Laboratory Cycle III, Phase IV Flight Test Program," 28 July 80; Memo for Record, Dyble, "Significant ALL Test Dates," 22 December 1981.

22. Rpt, AFWL-TR-81-151, p. I-4.

23. Ltr, HQ AFSC/DLWM to HQ AFSC/CST, "Significant Events Report, 26 Aug–2 Sep," 2 September 1980, with attached chart.

24. Ltr, Col Thomas R. Ferguson, AFSC/Asst Dir of Science & Technology, to HQ AFSC/CST, "Significant Events Report, 26 Aug–2 Sep," 2 September 1980; Rpt, AFWL-TR-81-151, pp. III-4–III-5; Ltr, Gamble, AFWL/ARTO, to AFWL/CC, "AR Weekly Activity Report (25 Aug–5 Sep 80)," 9 September 1980; Viewgraph, AFWL/ARTO, Phase 4 Flight Summary, 1981.

25. Rpt, AFWL-TR-81-151, p. ii.

26. Ibid., pp. i–iii, I-2–I-4, III-73; Interview with Otten, 1 November 1990.

27. Ibid.

28. Rpt, AFWL-TR-81-151, December 1981, pp. I-2–I-3.

29. Ltr, Col Demos T. Kyrazis, AFWL/Laser Development Division, to AFWL/ARL (Each Individual), "Phase 5 Schedule," 20 October 1980.

30. Ibid.

31. Rpt, Otten and Boesen, n.d.

32. Memo for Record for AFSC/DLWM, Dyble, "ALL and TDU-HEL-X Status Reports," 30 December 1980; Talking Paper on Airborne Laser Laboratory, AFSC, Dyble, 20 February 1981; Ltr, Gamble, AFWL/ARTO, to AFWL/PR, "AR Weekly Activity Report (29 Sep–3 Oct 80)," 8 October 1980; Ltr, Gamble to AFWL/PR, "AR Weekly Activity Report (17–21 Nov 80)," 26 November 1980; Ltr, Gamble to AFWL/PR, "AR Weekly Activity Report (15–19 Dec 80)," 23 December 1980.

33. Log of Phase 5 Test Shots, AFWL, Paul Merritt, December 1983; Chart, AFWL/AR, "ALL Phase 5 Tests: Ground Operations," 3 January–3 June 1981.

34. Various diagnostic instruments in the clean room assessed system performance, e.g., jitter, near-field intensity. Most of the beam energy was dumped into a calorimeter near the turret for safety reasons and to avoid beam thermal blooming along the 1-kilometer path to the downrange target behind the ARTF. Diagnostics at the downrange site measured beam quality by measuring the burn patterns in thick acrylic plastic to calculate time-averaged performance in the far-field. Workers dubbed the clean room the "white elephant" for several reasons: it was white, big, and had a duct system along one side that resembled the trunk of an elephant. Looking at this large white structure, and using a little imagination, the clean room did take on the shape of a white elephant. Interview with Janicke, 4 October 1988; Interview with Harrison H. Schmitt, former U.S. Senator, Albuquerque, New Mexico, 15 January 1993; Memo for Record for HQ AFSC, Dyble, "ALL Flight Schedule," 8 January 1981; Rpt, Otten and Boesen, n.d.

35. Interview with Janicke, 4 October 1988; Ltr, Gamble to AFWL/PR, "AR Activity Report (12–16 Jan 81)," 21 January 1981; Log of Phase 5 Tests Shots, Merritt, December 1983; Memo for Record, AFWL/AR, "Information gathered via telephone from Capt Barnes," 30 July 1985; Transcript, News Conference with Senator Schmitt and Dr. Mark in Albuquerque, New Mexico, 15 January 1981; "Air Force Fires HEL at Full Power From Aircraft on Ground," *Aerospace Daily*, 19 January 1981, p. 83; "Mark, Schmitt Discuss Laser, ABM Ties," *Aerospace Daily*, 26 January 1981, pp. 114–115; USAF News Release (AFSC #81-12), Air Force Systems Command, "Laser Tests," 3 February 1981; Memo for Record, Dyble, 22 December 1981.

36. Interview with Boesen, 30 September 1988; Interview with Janicke, 4 October 1988; Matt Mygatt, "'Star Wars' May Become Earth Wars With New Air Force Laser Weapon," *Philadelphia Inquirer*, 17 January 1981, p. 3.

37. Interview with Schmitt, 15 January 1993; Interview with Boesen, 10 September 1988; Interview with Janicke, 4 October 1988; Transcript, Senator Harrison Schmitt's Remarks to the Albuquerque Press Club, 15 January 1981; "Washington Roundup: Space Defense," *Aviation Week & Space Technology*, 29 December 1980, p. 13; Jim Bradshaw, "Laser Tests at Kirtland A 'Milestone,'" *Albuquerque Journal*, 16 January 1981, p. 1; "A Milestone," *KAFB Focus*, 23 January 1981, p. 9.

38. Transcript of News Conference with Senator Schmitt and Dr. Mark, 15 January 1981; Special Wire Service News, SAF/PA, "Laser Weapon," 16 January 1991; *Newsreview*, Air Force Systems Command, "Airborne Laser Lab Undergoes Initial Test-Firing Successfully," 27 February 1981, p. 1; Lem Famiglietti, "Laser Weaponry Still 10 Years Away," *Air Force Times*, 6 April 1981, p. 23.

39. Rpt, Otten and Boesen, n.d.; Rpt (draft), AFWL, "Airborne Laser Laboratory Cycle III/ Phase 5: Final Report," 27 July 1984, pp. 99–107.

40. Ibid.

41. Originally, AFWL planned to use a high-energy tow target (a modified TDU) dubbed HEL-X because of its large size. However, instabilities of this target during recovery operations posed major safety problems to the tow aircraft when attempting to reel in the HEL-X when the aerial experiments were completed. A number of evaluation tests revealed that when the TDU-30 was reeled back to the F-4, the target would go into violent oscillation about 10 feet from the aircraft. Consequently, the HEL-X was eliminated as a platform for laser aerial demonstrations and replaced by a less expensive—approximately $1500—and more stable LCTT. Several versions

of the LCTT were used depending on the specific airborne mission. Interview with Janicke, 4 October 1988; Interview with Otten, 25 February 1993; Rpt (draft), pp. 74–75, 99; Rpt (draft), AFSC/DLWM, Dyble, "ALL Test Report," 19 May 1981; Ltr, Gamble to AFWL/PR, "AR Activity Report (19–23 Jan 81)," 27 January 1981; Philip J. Klass, "Electroplating Use Expands to Weapons," *Aviation Week & Space Technology*, 26 October 1981, p. 91.

42. Interview with Janicke, 4 October 1988; Interview with Otten, 25 February 1993; Rpt (draft), AFWL, pp. 74–75; Rpt (draft), AFSC/DLWM, Dyble, "ALL Test Report," 19 May 1981; Ltr, Gamble to AFWL/PR, "AR Activity Report (19–23 Jan 81)," 27 January 1981.

43. Interview with Otten, 25 February 1993; Rpt (draft), "Airborne Laser Laboratory Cycle II," pp. 14–20, 76–80, 97–107.

44. For the BQM-34 simulated tests, conical scan tracking with the high-energy laser beam was not used. Interview with Otten, 25 February 1993; Rpt (draft), "Airborne Laser Laboratory Cycle III," pp. 14, 19–20, 94–96, 107.

45. Interview with Otten, 25 February 1993; Rpt (draft), "Airborne Laser Laboratory Cycle III," p. 107; Rpt, AFWL-TR-86-01, Vol. I, Raymond V. Wick, "Airborne Laser Laboratory: System and Test Descriptions," May 1988, p. 84.

46. Interview with Otten, 25 February 1993; Talking Paper, Dyble, 20 February 1981; Rpt, AFWL, Merritt, "Flight Test Schedule," n.d.; Viewgraph, AFWL/ARTO, "ALL Phase 5 Tests: Flight Operations," 1981; Ltr, Gamble, ARTO, to AFWL/PR, "Weekly Activity Report (23–27 Feb 81)," 2 March 1981.

47. Memorandum for AFSC/DLWM, Dyble, "Successful ALL Deployment to Edwards AFB," 10 April 1981; Flight Test Schedule, Merritt, n.d.; Viewgraph, AFWL/ARTO, 1981; Ltr, Gamble to AFWL/PR, "PR Weekly Activity Rpt (2–6 March 1981)," 10 March 1981.

48. Interview with Janicke, 4 October 1988; Memorandum for AFSC/DLWM, Dyble, 10 April 1981; Memorandum for HQ USAF/RDQ, Dyble, "Airborne Laser Laboratory Test Program Status," 1 May 1981.

49. Memorandum for AFSC/DLWM, Dyble, 10 April 1981; Memo, AFSC, "Command Event Summary," 13 April 1981.

50. Ibid.; Memorandum for HQ USAF/RDQ, Dyble, 1 May 1981; Ltr, Gamble to AFWL/PR, "AR Weekly Activity Report (6–10 April 1981)," 15 April 1981.

51. Interview with Otten, 25 February 1993; Memorandum for AFSC/DLWM, Dyble, "On-Site Review of ALL Status," 1 May 1981.

52. Interview with Dettmer, 15 September 1995.

53. Interview with Otten, 25 February 1993; Rpt, Merritt, n.d.; Viewgraph, AFWL/ARTO, 1981; Rpt, AFWL-TR-86-01, pp. 81–85; Ltr, Gamble to AFWL/PR, "AR Activity Report (20 April–2 May 1981)," 6 May 1981.

54. Memorandum for AFSC/DLWM, Dyble, "ALL Status Report," 20 May 1981; Memorandum for ODUSDRE/R&AT, AFSC/DLWM, Dyble, "Long Duration Power Extraction from Airborne Laser Laboratory," 21 May 1981; Memorandum for HQ AFSC/DLWM, Dyble, "Bugs in ALL Mirror Cooling System," 16 April 1981; Ltr, Gamble to AFWL/PR, "AR Weekly Activity Report (11–15 May 1981)," 20 May 1981.

55. Ibid.

56. Rpt, Merritt, n.d.

57. Interview with Otten, 25 February 1993; Rpt, AFWL-TR-86-01, pp. 4–5.

58. Interview with Otten, 25 February 1993.

59. Interview with Janicke, 4 October 1988; Interview with Boesen, 30 September 1988; Rpt, AFWL-TR-86-01, p. 269; Ltr, Demos T. Kyrazis, AFWL/ARTO, to HQ AFSC/DL, "ALL Self-Defense Deployment Summary," 8 June 1981.

60. Ltr, Kyrazis to HQ AFSC/DL, "ALL Self-Defense Deployment Summary," 8 June 1981; AFWL-TR-86-01, p. 407; Ltr, Maj Gen Jasper A. Welch, Jr., Asst DCS/Research, Development and

Acquisition, to AF/CC *et al.*, "Item of Interest—Airborne Laser Laboratory (Summary Report 1)," 12 June 1981.

61. Ltr, Kyrazis to HQ AFSC/DL, "ALL Self-Defense Scenario Summary," 8 June 1981; Viewgraph, AFWL/ARTO, 1981; Rpt, Merritt, n.d.; Staff Summary Sheet, Capt Rand, SAF/PAMS, to AF/CC and SAF/OS, "Airborne Laser Laboratory (ALL) News Queries," 23 June 1981.

62. Glints from other parts of the missile impeded the acquisition and track functions of the conical scan. If glint returns faded in and out, then conical scan broke lock on the target. Interview with Otten, 25 February 1993; Interview with Dettmer, 3 October 1988; Ltr, Kyrazis to HQ AFSC/DL, 8 June 1981; Viewgraph, AFWL/ARTO, 1981; Rpt, Merritt, n.d.; Staff Summary Sheet, Rand, 23 June 1981; Rpt, Otten and Boesen, n.d.; "Laser Weapon Flunks a Test," *Laser Focus*, Vol. 1, No. 7, July 1981, p. 4.

63. Ltr, Kyrazis to HQ AFSC/DL, 8 June 1981; Ltr, Welch to AF/CC *et al.*, 12 June 1981; Viewgraph, AFWL/ARTO, 1981; Talking Paper on Airborne Laser Laboratory, AFSC, Dyble, 31 December 1981.

64. Interview with Otten, 25 February 1993; Ltr, Kyrazis to HQ AFSC/DL, "ALL Self-Defense Deployment Summary," 8 June 1981; Memorandum for SAF/OS, Lt Gen Hans H. Driessnack, Assistant Vice Chief of Staff, "Significance of Recent Airborne Laser Laboratory Flight Tests," 25 June 1991; Msg, AFWL to AFSC, "Blueline Status Report of Airborne Laser Laboratory Program," 171724Z Jun 81; Msg, HQ AFSC to AFCMD/KAFB, "Q&As on Airborne Laser Laboratory," 24 June 1981.

65. Interview with Kyrazis, 12 February 1986; Interview with Otten, 25 February 1993.

66. Interviews with Kyrazis, 19 September 1985, 12 February 1986; Interview with Boesen, 30 September 1988; Ltr, Kyrazis to HQ AFSC/DL, "Summary," 8 June 1981.

CHAPTER 13. SETBACKS AND DELAYS

1. Memorandum for HQ AFSC/DL, Captain Thomas J. Dyble, "Public Release Plan for ALL Cycle III," 5 May 1981.

2. Memo for Record, AFSC, Colonel Thomas R. Ferguson, "Proposed Press Release on Airborne Laser Laboratory," 21 May 1981, with attached press release and fact sheet.

3. "Air Force Laser Flunks 1st Test," *Albuquerque Tribune*, 2 June 1981, p. 1; "U.S. Laser Weapon Fails Test on Air-to-Air Missile," *Albuquerque Journal*, 3 June 1981, p. 1; George C. Wilson, "In Test Aloft, Air Force Fails to Stop Sidewinder," *Washington Post*, 3 June 1981, p. A6; "Laser Fails to Destroy Missile," *Aviation Week & Space Technology*, 8 June 1981, p. 63; "Hit or Miss, Laser Fails Test," *New York Times*, 7 June 1981, p. 4E.

4. Interview with Dyble, 13 April 1993.

5. Ibid.

6. Ibid.; Memo for Record, Captain Dyble, "Telecon with George Wilson," *The Washington Post*, 3 June 1981; Staff Summary Sheet, HQ USAF, Major General Robert D. Russ, "Significance of Recent Airborne Laser Laboratory Flight Tests," 18 June 1981.

7. Airey also served as Chairman of the High Energy Laser Review Group (HELRG), which periodically reviewed the progress of government-sponsored laser research and development programs. He had been a proponent of laser research and development efforts under both the Carter and Reagan administrations. Interview with Dyble, 5 November 1993; News Release—Question and Answer, SAF/PAMS, Capt Val Elbow, "Airborne Laser Laboratory," 2 June 1981; Revised News Release, Jack Pulmers/Action Officer, "Airborne Laser Laboratory," 4 June 1981; "Laser Testing," *Aerospace Daily*, 8 June 1981, p. 201.

8. Msg, HQ AFSC to PA/AFCMD, "Q & A's on Airborne Laser Laboratory," 24 June 1981; Stewart Lyle, "Laser fails again, but Air Force says test is 'better,'" *Albuquerque Journal*, 22 June 1981, p. 1; "Laser Engaged Target But Then Developed Mechanical Problems, AF Says," *Aerospace Daily*, 9 June 1981, p. 212; Msg, HQ AFCMD to HQ AFSC, "Laser Focus Magazine Queries

About Laser Tests," 18 June 1981; Staff Summary Sheet, Capt Elbow, SAF/PAMS, "Airborne Laser Laboratory News Queries," 23 June 1981; "Laser Shot Missile '75%' in New Test, Defense Says," *Baltimore Sun*, 23 June 1981, p. 5.

9. Memo for SAF/OS, Lieutenant General Hans H. Driessnack, "Significance of Recent Airborne Laser Laboratory Flight Tests," 25 June 1981.

10. Interview with Robert T. Marsh, General, USAF, Retired, Arlington, Virginia, 13 August 1990; Msg, Brig Gen Peter W. Odgers, HQ AFSC/TE to AFWL/AR, "Airborne Laser Laboratory (ALL) Technical Assistance Review," 24 June 1981; "ALL Review," *Aerospace Daily*, 20 July 1981, p. 97.

11. By the start of FY82, a total of $632.2 million had been committed to the Airborne Laser Laboratory. Interview with General Marsh, August 13, 1990; Interview with Rich, 10 January 1989; Talking Paper, AFSC, Capt Bernard P. Carey, "Airborne Laser Laboratory: Technical Assistance Review," 20 July 1981; Briefing, Colonel John C. Rich to AFSC Commander (Gen Marsh), "Airborne Laser Laboratory Status and Options," 5 January 1982.

12. Rpt, HQ AFSC/TE, Odgers, "Final Report of the Airborne Laser Laboratory Technical Assistance Review Group (6–17 July 1981)," 21 July 1981; Memorandum for HQ AFSC/DLWM, Capt Dyble, "Summary of ALL Technical Assistance Review," 20 July 1981.

13. Rpt, Odgers, 21 July 1981.

14. Memorandum for HQ AFSC/DLWM, Dyble, 20 July 1981; Rpt, Odgers, 21 July 1981.

15. Ibid.

16. Ibid.

17. Ibid.

18. Ibid.

19. Ibid.

20. Ibid.; Interview with Kyrazis, 12 February 1986.

21. Ibid.

22. Memorandum for HQ AFSC/DL, Dyble, 20 July 1981, with attached TAR Team Exec Summary.

23. Ibid.

24. ARTO Weekly Activity Reports, Major Lynn L. Gamble, 28 July, 12, 19, and 26 August 1981; Memos for DL, Capt Dyble, "ALL Ground Test," 21 and 28 July 1981; Talking Paper on Airborne Laser Laboratory, AFSC, Capt Dyble, 21 October 1981.

25. Interview with Kyrazis, 12 February 1986; ARTO Weekly Activity Reports, Gamble, 2, 16, and 23 September 1981.

26. ARTO Weekly Activity Report, Gamble, 20 October 1981.

27. By the spring of 1982, the dampers showed more positive results in reducing vibrations. During this time, AFWL had removed the APT and ADAS from the plane to rework the ADAS optical realignment. Many of the ADAS deficiencies were attributed to Perkin-Elmer who initially built the ADAS, but which was subsequently taken over by Hughes. In a report to higher headquarters in November 1981, AFWL identified a number of Perkin-Elmer flaws in the ADAS to include loose fittings which resulted in water leaks, incorrect O-rings found at four locations, inconsistencies between hardware and drawings, electrical connectors and wiring on water pressure abort switches incorrectly installed, and failure to conduct all required quality control inspections. Memo for HQ AFSC/DL, Capt Dyble, "Follow-up on 16 Oct 81 ALL Ground Test," 23 October 1981; Msg, AFWL to HQ AFSC, "ALL Ground Test, 7 Nov 81," 9 November 1981; Memo for AFSC/DL, Capt Dyble, "ALL Ground Test," 8 November 1981; Memo for DL, Capt Dyble, "ALL Tests," 10 November 1981.

28. Msg, AFWL to HQ AFSC, "ALL Flight Test," 18 November 1981; Memo for HQ AFSC/DL, Capt Dyble, "ALL Flight Test, 19 November 1981," 19 November 1981; Msg, AFWL to HQ AFSC, "ALL Flight Test, 19 November 1981," 19 November 1981; Memo for ODUSDRE/R&AT (Dr. Airey), Capt Dyble, "Airborne Laser Laboratory Flight, 19 Nov 81," 19 November 1981.

29. Msg, AFWL to AFSC, "ALL Flight Tests, 30 November Through 2 December 81," 3 December 1981; Memo for DL, Capt Dyble, "Command Briefing Item on ALL Flights, 30 Nov–2 Dec 81," 3 December 1981; Msg, AFWL to AFSC, "ALL Flight Tests," 27 November 1981.

30. Memorandum for HQ AFSC/DL, Capt Dyble, "ALL Daily Schedule and Daily Problems," 9 December 1981; Memorandum for HQ AFSC/DL, Capt Dyble, "Hot Flash—ADAS is Alive!," 10 December 1981; Point Paper on Airborne Laser Laboratory, AFSC, Capt Dyble, 11 December 1981; Msg, AFWL to AFSC/DL, "ALL Ground Test," 14 December 1981; Msg, AFWL to AFSC/DL, "ALL Ground Test, 15 Dec 81," 17 December 1981; Point Paper on Airborne Laser Laboratory, Capt Dyble, 20 December 1981.

31. Msg, AFWL to HQ AFSC/DL, "ALL Ground Test," 22 December 1981.

32. Ibid.; Interview with Boesen, 30 September 1988; Briefing, Rich to Marsh, 5 January 1982.

33. Interview with Boesen, 30 September 1988; Interview with Rich, 10 January 1989; Memorandum, General Brien D. Ward, AFSC/DL, "Hot Flash—Airborne Laser Laboratory Mirror Failure," 22 December 1982; Rpt, AFWL-TR-86-01, Vol. II, Raymond V. Wick, "Airborne Laser Laboratory: System Performance and Test Results," May 1988, p. 291.

34. Interview with Boesen, 30 September 1988; Rpt, AFSC/TE, Brig Gen Odgers, "Final Report of the Airborne Laser Laboratory Technical Assessment Review Group II," 20 February 1982; Rpt, AFWL-TR-86-01, pp. 292–303.

35. Talking Paper on Airborne Laser Laboratory, AFSC, Capt Dyble, 31 December 1981.

36. Briefing, Rich to Marsh, 5 January 1982; ARTO Weekly Activity Report, 4–8 January 1982, Col Rich, 13 January 1982.

37. Ibid.; Point Paper on Airborne Laser Laboratory, AFSC/DLWM, Capt Dyble, 31 December 1981.

38. Ibid.; Staff Summary Sheet, AFSC/DL, Colonel Hugh T. Bainter, "Actions on Airborne Laser Laboratory Effort," 19 January 1982; Talking Paper on Airborne Laser Laboratory, Dyble, 31 December 1981.

39. Staff Summary Sheet, Bainter, 19 January 1982; Msg, AFWL to HQ AFSC, "Airborne Laser Laboratory Effort," 11 January 1982; Msg, CMS Taylor/LGMW to AFWL, "Aircraft Movement Directive," 12 January 1982; Msg, HQ AFSC (Brig Gen Ward) to AFWL, "Airborne Laser Laboratory Effort," 15 January 1982; ARTO Weekly Activity Report, 11–15 January 1982, Gamble, 19 January 1982.

40. Msg, HQ AFSC (Brig Gen Ward) to AFWL, 15 January 1982; Memorandum for HQ AFSC/DL, Capt Dyble, "ALL Status Report," 1 February 1982.

41. Msg, HQ AFSC to AFWL, "Airborne Laser Laboratory Technical Assistance Review," 26 January 1982; Rpt, Odgers, 20 February 1982; ARTO Weekly Activity Report, Maj Vincent E. Dellamea, "Airborne Laser Laboratory," 7 April 1982.

42. Memorandum for AFSC/DL, Capt Dyble, "ALL Technical Assistance Review II, 18–21 February 1982," 2 March 1982; Rpt, Odgers, 20 February 1982; Memorandum for AFSC/CV (Lt Gen Bond), AFSC/DL, Maj Gen Ward, "Airborne Laser Laboratory Mirror Status & Program Schedule," 25 February 1982.

43. Interview with Dettmer, 3 October 1988; Rpt, Odgers, 20 February 1982.

44. Memorandum for AFSC/DL, Dyble, 2 March 1982; Rpt, Odgers, 20 February 1982.

45. Interview with Janicke, 4 October 1988; Rpt, Odgers, 20 February 1982; Ltr, General Robert T. Marsh, AFSC/CC, to Dr. Allen Puckett, Chairman, Hughes Aircraft Company, "ALL," 17 March 1982.

46. Staff Summary Sheet, AFSC/DL, Col Hugh T. Bainter, "Hughes Aircraft Support of Airborne Laser Laboratory," 23 February 1982; Ltr, Puckett to General Marsh, "ALL," 12 April 1982.

47. Memorandum for AFSC/CV (Lt Gen Bond), Ward, 25 February 1982.

48. Memorandum for AFSC/DL, Capt Dyble, "ALL Command Briefing Item," 11 May 1982; Briefing, AFSC/TE, Brig Gen Odgers, "Airborne Laser Laboratory Technical Assistance Review II, Kirtland Air Force Base, New Mexico," 12–13 May 1982; Memorandum for HQ AFSC/DL, Capt Dyble, "ALL TAR Briefing," 10 June 1982.

49. Briefing, Odgers, 12–13 May 1982.

50. Ibid.; Memorandum for AFSC/DL, Dyble, 11 May 1982; ARTO Weekly Activity Report, Dellamea, 2 June 1982.

51. A complex, but extremely important component of the advanced tracker, the FLIR sensor package consisted of a Cassegrainian telescope, an 11-element × 6-row cryogenically cooled scanned HgCdTe detector array, a preamplifier, a video processor which included scan conversion circuits, radiometer electronics, and automatic focus systems. The output of the FLIR sensor package was directed to the Digital Correlation Real-Time Processor for processing. ARTO Weekly Activity Report, 26–30 July 1982, Dellamea, 3 August 1982; Msg, AFWL to HQ AFSC, "ALL Status Report For 26 July Through 30 July 1982," 29 July 1982; Notes, Lt Col Steve Coulombe, PL/TM, "FLIR," 25 February 1994; Rpt (draft), AFWL/ARTO, "Airborne Laser Laboratory Cycle III/Phase 5 Final Report," 27 July 1984, p. 41.

52. Ibid.

53. Program, Retirement Ceremony of Colonel Demos T. Kyrazis, 28 May 1982.

54. Interview with Kyrazis, 22 June 1995; Background paper on ALL, Major Tom Dyble, 21 November 1982.

55. Interview with Kyrazis, 12 February 1986.

56. Ltr, Maj Gen Brien D. Ward, Director of Laboratories, AFSC, "Retirement," 19 May 1982.

57. Ltr, General Marsh to Colonel Kyrazis, "Retirement," 24 May 1982.

58. ARTO Weekly Activity Report, 26–30 July 1982, Dellamea, 3 August 1982.

59. Ltr, Captain Anthony J. Sobol, AFWL/ARTO, to AFWL/PR, "Weekly Activity Report, 13–17 September 1982," 21 September 1982; Rpt (draft), AFWL/ARTO, 27 July 1984, pp. 74–92.

60. Ltr, Sobol to AFWL/PR, 21 September 1982.

61. Ibid.; Ltr, Major Dellamea, AFWL/AR, to HQ AFSC/DL, "Monthly Activity Report (September 1982)," 1 October 1982; Ltr, Major Dellamea, AFWL/AR, to AFWL/PR, "Weekly Activity Report, 6–10 December 1982," 15 December 1982; Ltr, Major Dellamea, AFWL/AR, to AFWL/PR, "Weekly Activity Report, 13–17 December 1982," 21 December 1982.

62. These new two-axis steerable mirrors combined to form the airborne beam control system/airborne dynamic alignment system beam interface mechanism (ABBISM 1 and 2). Rpt (draft), AFWL/ARTO, pp. 40–41.

63. Ltr, Major Dellamea, AFWL/ARTO, to AFWL/PR, "Weekly Activity Report, 30 Aug–3 Sep 1982," 7 September 1982; Ltr, Sobol, to AFWL/PR, 21 September 1982; Ltr, Dellamea to HQ AFSC/DLW, 1 October 1982.

64. Rpt, AFWL-TR-86-01, pp. 304–305.

65. Ibid., pp. 306–312.

66. Ibid., pp. 305–306, 313–314; Ltr, Major Dellamea, AFWL/AR, to AFWL/PR, "Weekly Activity Report, 18–22 Oct 82," 25 October 1982.

67. Technical Notebook on ALL activities, Captain Coulombe, AFWL/AR, 2 February 1982–2 November 1984; Ltr, Major Dellamea, AFWL/AR, to AFWL/PR, "Weekly Activity Report, 4–8 October 1982," 12 October 1982; Ltr, Dellamea to AFWL/PR, 25 October 1982; Ltr, Captain Sobol, AFWL/AR, to AFWL/PR, "AFWL Weekly Activity Report, Week Ending 13 Nov 82," 15 November 1982; Ltr, Major Dellamea, to AFWL/PR, "Weekly Activity Report, 22–26 November 1982," 1 December 1982.

68. Interview with Boesen, 30 September 1988.

69. Rpt, AFWL-TR-86-01, Vol. I, Raymond V. Wick, "Airborne Laser Laboratory: System and Test Descriptions," May 1988, pp. 73–75; Ltr, Dellamea to AFWL/PR, 15 December 1982; Ltr, Dellamea to AFWL/PR, 21 December 1982.

70. The main portion of the beam was absorbed in a noninstrumented beam dump located in the auxiliary environmental control system or "white elephant" surrounding the APT on the ALL. Rpt, AFWL-TR-86-01, pp. 75–80.

71. Rpt (draft), AFWL/ARTO, 27 July 1984, pp. 7–8.

72. Rpt, AFWL-TR-86-01, pp. 75–80.

73. ARTO Weekly Activity Report, 3–7 January 1983, Captain Sobol, 12 January 1983; ARTO Weekly Activity Report, 10–14 January 1983, Sobol, 19 January 1983; Weekly Activity Report, 17–21 January 1983, Sobol, 26 January 1983; ARTO Weekly Activity Report, 7–11 February 1983, Sobol, 15 February 1983; ARTO Weekly Activity Report, 14–18 March 1983, Sobol, 22 March 1983; ARTO Weekly Activity Report, 21–25 March 1983, Sobol, 28 March 1983; Rpt, AFWL-TR-86-01, Vol. II, pp. 266–269, 413–415; Technical Notebook on ALL Activities, Coulombe, 2 February 1982–2 November 1984.

74. ARTO Weekly Activity Report, 21–25 March 1983, Sobol, 28 March 1983; Rpt, AFSC/TE, Odgers, "Airborne Laser Laboratory Technical Assistance Review Team III Report, 9–11 May 1983," [11 May 1983].

75. Rpt, AFWL-TR-86-01, Vol. I, pp. 81–86 and Vol. II, pp. 269–271.

76. Rpt, Odgers, [11 May 1983], with attached message from General Odgers to General Marsh.

77. Rpt, AFWL-TR-86-01, Vol. I, pp. 81–86 and Vol. III, Raymond V. Wick, "Airborne Laser Laboratory: Tracking and Target Interaction," May 1988, p. 73.

78. Rpt, Odgers, [11 May 1983].

79. Ibid.; Rpt, "Airborne Laser Laboratory: Mission No. 15032, 20 April 1983," n.d.; Rpt, AFWL, Major Robert L. Van Allen, "Airborne Laser Laboratory Phase 5 Technical Report: Part 2 of 2, Air-to-Air Engagement Results, Briefing to Industry," 11 April 1984; Ltr, Major General Brian D. Ward, AFSC/DL, to AF/RDQ, "Airborne Laser Flight Tests," 13 May 1983; Rpt, AFWL-TR-86-01, Vol. I, pp. 84–87 and Vol. III, p. 76.

80. Ltr, Ward to AF/RDQ, 13 May 1983.

81. Interview with Marsh, 13 August 1990; Biography (USAF), Major General Peter W. Odgers, October 1982; Message, AFFTC to AFSC, "Airborne Laser Laboratory Technical Assistance Review III," 14 May 1983.

82. There were a total of four Technical Assistance Reviews (TARs) during the ALL program: TAR I, July 81; TAR II, February 82; TAR IIA, May 82; and TAR III, May 83. Rpt, Odgers, [11 May 1983].

83. Interview with Otten, 26 May 1974; Rpt, Odgers, [11 May 1983].

84. Rpt, Odgers, [11 May 1983].

85. Ibid.; Message, AFFTC to AFSC, 14 May 1983; Ltr, Ward to AF/RDQ, 13 May 1983.

CHAPTER 14. SUCCESS: TWO SHOOTDOWNS

1. Interview with Janicke, 4 October 1988.

2. Ibid.; Interview with Otten, 26 May 1994; Ltr, Maj Gen Jasper A. Welch, Jr., Asst DCS/Research, Development and Acquisition, HQ USAF, to AF/CC *et al.*, "Item of Interest," 26 May 1983.

3. Interview with Otten, 26 May 1994.

4. Ibid.

5. Ibid.; Ltr, Michael E. Flynn, Scientific/Technology Advisor, HQ USAF, DCS/Research, Development and Acquisition, to AF/RDQ, "Status of Airborne Laser Laboratory (ALL) Flight Test," 24 May 1983.

6. Interview with Otten, 26 May 1994; Interview with Janicke, 4 October 1988; Rpt, AFWL-TR-86-01, Vol. I, Raymond V. Wick, "Airborne Laser Laboratory: System and Test Descriptions," May 1988, pp. 8–9.

7. Interview with Otten, 2 June 1994.

8. Ibid.; Interview with Janicke, 4 October 1988.

9. Interview with Otten, 26 May 1994.

10. Ibid.; Rpt, AFWL-TR-86-01, p. 87; "Airborne Laser Lab: 'Five For Five' in Sidewinder Tests," *Aerospace Daily*, 28 July 1983, p. 149.

11. Again, the firing boxes were established as a precaution to make sure that the AIM-9B did not intercept the ALL and that the A-7 had sufficient time to get out of the way of the laser beam. All of this had to be accomplished while abiding by strict safety limitations imposed by the boundaries of the test range. Interview with Otten, 26 May 1994; Interview with Janicke, 4 October 1988; Rpt, AFWL-TR-86-01, p. 4.

12. Interview with Otten, 26 May 1994; Rpt, AFWL-TR-86-01, Vol. II, Raymond V. Wick, "Airborne Laser Laboratory: System Performance and Test Results," May 1988, pp. 269–272.

13. Interview with Otten, 26 May 1994; Interview with Janicke, 4 October 1988; Rpt, AFWL-TR-86-01, pp. 264, 269–274; Rpt, AFWL-TR-86-01, Vol. III, Raymond V. Wick, "Airborne Laser Laboratory: Tracking and Target Interaction," May 1988, pp. 19–37.

14. Interview with Otten, 26 May 1994; Rpt, AFWL-TR-86-01, Vol. II, pp. 269–270.

15. Interview with Otten, 26 May 1994; Messages, AFWL to AFSC, "ALL Mission Report," 17, 19, 24, and 25 May 1983.

16. Interview with Otten, 26 May 1994; Ltr, Flynn to AF/RDQ, 24 May 1983; "Airborne Laser Lab: 'Five For Five' in Sidewinder Tests," *Aerospace Daily*, 28 July 1983, p. 149; Memo for AF/RDQ, Flynn, "Status of ALL Flight Test," 24 May 1983; Berl Brechner, "Chase," *Air & Space Smithsonian*, April/May 1986, p. 62.

17. Interview with Otten, 26 May 1994; Interview with Janicke, 4 October 1988; Ltr, Welch to AF/CC, 26 May 1983; Video, "Airborne Laser Laboratory Laser Flight," a.k.a. "ALL Bird," 1 March 1985.

18. Interview with Otten, 26 May 1994; Interview with Janicke, 4 October 1988; Ltr, Welch to AF/CC, 26 May 1983; Rpt, AFWL-TR-86-01, p. 275; "Airborne Laser Lab: 'Five For Five' in Sidewinder Tests," 28 July 1988, p. 149.

19. Interview with Otten, 26 May 1994; Rpt, AFWL-TR-86-01, Vol. III, pp. 68–69.

20. Interview with Otten, 26 May 1994; Rpt, AFWL-TR-86-01, pp. 70, 80.

21. Aligning all of the aircraft into the proper firing boxes continued to be a tedious and time-consuming process. For the 26 May tests, the ALL and diagnostic aircraft remained airborne for 2 hours and 15 minutes. Two T-38s (target aircraft furnished by the 6510th Test Wing) each flew for approximately 1 hour. Two A-7 shooter aircraft also supported the 26 May mission. Interview with Otten, 26 May 1994; Rpt, AFWL-TR-86-01, Vol. II, p. 417.

22. Interview with Otten, 26 May 1994; Interview with Janicke, 4 October 1988.

23. Interview with Otten, 26 May 1994.

24. Ibid.; Rpt, AFWL-TR-86-01, Vol. III, p. 72.

25. Interview with Otten, 26 May 1994; Interview with Janicke, 4 October 1988; Rpt, AFWL-TR-86-01, p. 80; "Sidewinder Destroyed in Airborne Laser Test," *Aviation Week & Space Technology*, 8 August 1983, p. 66.

26. Interview with Otten, 26 May 1994.

27. Ibid.; AFWL-TR-86-01, Vol. II, p. 417.

28. Interview with Otten, 26 May 1994; Rpt, AFWL-TR-86-01, Vol. III, pp. 67–68, 80; "Airborne laser defeats missiles launched at it," Air Force Systems Command *Newreview*, 26 August 1983, p. 1; Video, "Airborne Laser Laboratory Laser Flight," 1 March 1985.

29. Rpt, AFWL-TR-86-01, Vol. II, p. 259.

30. Ibid.

31. Interview with Otten, 26 May 1994; AFWL-TR-086-01, Vol. I, pp. 259–278; Charles Lamar, "Results of the Airborne Laser Laboratory Tests Against the AIM-9B," in *Journal of Defense Research*, May 1986, pp. 253, 256, 259, 272; "Airborne Laser Lab downs missiles," *Laser Focus*, September 1983, p. 82.

32. Interview with Janicke, 4 October 1988.

33. Ibid.

34. Ibid.; "Airborne Laser Lab Finally Kills Missiles," *Lasers & Applications*, September 1983, p. 24.

35. Interview with Otten, 26 May 1994.

36. Ibid.; Message, AFWL to AFSC, "ALL High Power Flight Test," 3 June 1983.

37. The bellows failure was caused by an acoustic resonance generated when the venting gases passed over the bellow, the same phenomenon one gets by blowing over the top of an open bottle. NASA had also experienced this type of failure in some of its bellows. Interview with Otten, 26 May 1994; Message, AFWL to AFSC, 3 June 1983; Log of ALL Missions, 30 December 1980–9 December 1983, AFWL, Paul H. Merritt; Rpt, AFWL-TR-86-01, Vol. II, pp. 161–166.

38. Interview with Janicke, 4 October 1988; Rpt, AFWL-TR-86-01, pp. 170–178; ARTO Weekly Activity Report, 13–17 June, Captain Mark S. Rabinowitz, 21 June 1983; ARTO Weekly Activity Report, 8–13 August 1983, Rabinowitz, 15 August 1983.

39. USAF News Release (AFSC #83-79), "Airborne Laser Laboratory Experiment," 25 July 1983.

40. Ibid.; "Airborne laser downs missiles," *Focus* (KAFB), 29 July 1983, p. 1.

41. Interview with Janicke, 4 October 1988; Interview with Boesen, 30 September 1988; Walter Andrews, "Airborne laser beam used successfully by Air Force," *Washington Times*, 26 July 1983, p. 2; Loring Wirbel, "N.M. laser weapons face exotic challenge," *Albuquerque Tribune*, 28 November 1983, pp. A-1, A-4; "Airborne Laser Lab Finally Kills Missiles," September 1983, p. 24; "Laser Passes Key Test—Missiles KOd," *San Francisco Chronicle*, 26 July 1983, pp. 1, 6; "Laser Test: 'Star Wars' air defense system cripples 5 missiles," *Albuquerque Tribune*, 26 July 1983, p. A-6.

42. "Plane's Laser Hits Missile," *Albuquerque Journal*, 26 July 1983, p. C-8; "Laser Test: 'Star Wars' air defense system cripples 5 missiles," 26 July 1983, p. A-6.

43. "Airborne Laser Disables Missiles in Air Force Test," *Washington Post*, 26 July 1983, p. A5; Andrews, p. 2; Robert C. Toth, "Air Force 'Defeats' Speedy Missiles in Test: Airborne Result Called 'Major Milestone' in Weapon Development," *Los Angeles Times*, 26 July 1983, pp. 1, 9; "Laser Passes Key Test—Missiles KOd," *San Francisco Chronicle*, 26 July 1983, p. 1; "Laser Disables Missile," *Aviation Week & Space Technology*, 1 August 1983, p. 23; "Sidewinder Destroyed in Airborne Laser Test," *Aviation Week & Space Technology*, 8 August 1983, p. 66.

44. The Navy had invested in developing and testing several large laser systems over the years to include the Navy Advanced Chemical Laser (late 1970s) and the Mid-Infrared Advanced Chemical Laser and Sea Lite Beam Director (early 1980s). "Directed energy weapons: where are they headed?," *Physics Today*, August 1983, pp. 17–20; "Surveying military programs," *Photonics Spectra*, November 1983, p. 73.

45. Interview with Spreen, 31 August 1994; Interview with Boesen, 12 August 1994; R. L. Rudkin *et al.*, "Ship-Based Laser Weapons for Battle Group Defense," in *Journal of Defense Research: Special Issue 86-1*, May 1986, pp. 121–133.

46. Ibid.

47. USAF News Release (AFSC #83-130), "Air Force/Navy ALL Tests," 29 November 1983; A. Goroch *et al.*, "The Near-Ship and Ambient Marine Environment," in *Journal of Defense Research: Special Issue 86-1*, May 1986, pp. 121–123.

48. Interview with Coulombe, 6 February 1995; Interview with Boesen, 12 August 1994.

49. Ibid.; Charles D. Bright, ed., *Historical Dictionary of the U.S. Air Force* (New York: Greenwood Press, 1992), p. 589.

50. Rpt, AFWL-TR-86-01, Vol. I, pp. 84–85; Msg, AFWL/ARL to HQ AFSC, "ALL Flight Test Report," 1 September 1983.

51. Interview with Coulombe, 6 February 1995; Interview with Boesen, 12 August 1994.

52. Interview with Coulombe, 5 February 1995.

53. Interview with Boesen, 12 August 1994; Log of ALL Missions, 30 December 1980–9 December 1983, Merritt.

54. Interview with Boesen, 12 August 1994; Msg, AFWL/AR to HQ AFSC/DL, " ALL Flight Test Report," 1 September 1993.

55. Ibid.; Rpt, AFWL-TR-86-01, p. 84.

56. Interview with Boesen, 12 August 1994; Ltr, Lt Gen Robert D. Russ, DCS Research, Development and Acquisition, HQ USAF, to AF/CC *et al.*, "Item of Interest," 11 October 1983; Log of ALL Missions, 30 December 1980–9 December 1983, Merritt.

57. Interview with Boesen, 12 August 1994; Interview with Coulombe, 6 February 1995.

58. Ibid.; Bright, p. 589.

59. Interview with Boesen, 12 August 1994; Interview with Coulombe, 6 February 1995.

60. Interview with Boesen, 12 August 1994; Log of ALL Missions, 30 December 1980–9 December 1983, Merritt.

61. Interview with Coulombe, 6 February 1995; Log of ALL Missions, 30 December 1980–9 December 1983, Merritt.

62. Interview with Coulombe, 6 February 1995; Rpt, AFWL, L.J. Otten and D.L. Boesen, "The Airborne Laser: A Program Review," n.d.; Rpt, AFWL-TR-86-01, p. 89; Rpt, AFWL-TR-86-01, Vol. IV, Raymond V. Wick, "Airborne Laser Laboratory," May 1988, p. 94.

63. Ibid.

64. Interview with Boesen, 12 August 1994.

65. Ibid.; Interview with Coulombe, 6 February 1995; Ltr, Russ to AF/CC, 11 October 1983; USAF News Release (AFSC #83-130), 29 November 1983; Rpt, AFWL-TR-86-01, p. 191.

66. Interview with Boesen, 12 August 1994; Ltr, Russ to AF/CC, 11 October 1983.

67. Rpt, AFWL-TR-86-01, p. 94; Rpt, Otten and Boesen, n.d.

68. Interview with Boesen, 12 August 1994; Interview with Coulombe, 6 February 1995.

69. Ibid.; Rpt, AFWL-TR-86-01, p. 94.

70. Interview with Janicke, 4 October 1988.

71. Interview with Boesen, 12 August 1994.

72. Ibid.; Interview with Coulombe, 6 February 1995; USAF News Release (AFSC #83-130), 29 November 1983.

73. Ibid.; Video, "Airborne Laser Laboratory Laser Flight," 1 March 1985; Ltr, Russ to AF/CC, 11 October 1983.

74. Ibid.; Interview with Janicke, 4 October 1988; Rpt, AFWL-TR-86-01, p. 94.

75. USAF News Release (AFSC #83-130), 29 November 1983.

76. Ibid.; Video, "Airborne Laser Laboratory Laser Flight," 1 March 1985.

77. Interview with Marsh, 13 August 1990.

78. Ibid.

79. Ibid.; Interview with Dettmer, 11 April 1995.

80. Msg, AFSC/CC (Gen Marsh) to AFWL/CC (Col Johnson), "Airborne Laser Flight Experiments," 7 October 1983.

81. "Laser Lab shoots down drone over ocean," *Focus* (KAFB), 9 December 1983, p. 5.

82. Interview with Marsh, 13 August 1990; Interview with Boesen, 12 August 1994; Interview with Janicke, 4 October 1988; Interview with Dettmer, 10 April 1995.

83. Interview with Dettmer, 10 April 1995; Interview with Coulombe, 6 February 1995; Talking Paper on ALL Testing/Future of ALL, AFSC/DLWM, Major Haynes, 14 September 1983; "Conference Urges Laser Program Termination," *Aviation Week & Space Technology*, 15 August 1983, p. 21; "Airborne Laser Lab Finally Kills Missiles," September 1983, p. 24.

84. Interview with Coulombe, 6 February 1995; Ltr, Lt Col Otten to Det 2, 4950th TW/CC, "Deferment of Phase Inspection for NKC-135A," 30 August 1983; Briefing Chart, AFWL, "Airborne Laser Laboratory Phasedown," n.d. [September 1983].

85. During this phasedown period, the ALL had flown to the West Coast to look at strategic missiles launched from Vandenberg AFB, California. The purpose of these tests was to collect tracking data on ICBMs for use in evaluating high-energy laser engagements of strategic missiles—a key to some of the Strategic Defense Initiative Organization's ideas. Interview with Coulombe, 6 February 1995; Msg, HQ AFSC to SD, "Interim Form 56 for PE 63605F," 31 October 1983; Rpt,

AFWL-TR-84-112, Stephen A. Coulombe, "Aero-Optics Modification Airworthiness Test Report NKC-135A, USAF S/N 55-3123," February 1985, pp. 1–17, 59.

86. Briefing, AFWL, Lt Col Alexander J. Halber and Lt Col Otten, "Airborne Laser Laboratory Thrust Phase-Down Plan," September 1983.

87. Ibid.

88. Ibid.; Interview with Coulombe, 6 February 1995; Ltr, Col Tony M. Johnson, AFWL/CC, to AFAA Area Audit Office, "Internal Controls," 30 March 1984.

89. Msg, HQ AFSC to ALAFSC/CS, "Dynamic Laser Technology Test Bed," 26 February 1985.

90. Interview with Boesen, 12 August 1994; Interview with Coulombe, 6 February 1995; Rpt, Otten and Boesen, n.d.

91. Ltr, Maj John W. Koch, Commander, Det 2, 4950th Test Wing, to 4950TESTW/MA, "Storage Plan for NKC-135A 55-3123, Airborne Laser Laboratory," 30 August 1984.

92. Ltr, Col J.P. Amor, AFWL/CC, to Col Joseph R. Johnson, AFSTC/CC, "Airborne Laser Laboratory," 19 April 1988; Placard, Air Force Museum, "Boeing NKC-135A 'Stratotanker' (Airborne Laser Lab/ALL)," 2 May 1991.

93. USAF News Release (AFSC #88-20), "Laser Aircraft Makes Last Flight," 4 May 1988; Lawrence Spohn, "Historical flying laser will be retired," *Albuquerque Tribune*, 3 May 1988, pp. 1, A2.

94. Transcript of Demos Kyrazis speech delivered at ALL flyaway, 4 May 1988.

95. Ibid.

96. Ibid.

97. Steve Coulombe, "Airborne Laser Lab makes last flight," *Leading Edge*, July 1988, p. 7.

EPILOGUE

1. Newscast (KOAT-TV, Albuquerque), "ALL Flyaway," 5 May 1988.

2. Ltr, Lt Gen James A. Abrahamson, Director, Strategic Defense Initiative Organization, to Airborne Laser Laboratory Personnel, "Letter of Commendation," 29 April 1988.

3. "Airborne Laser (ABL) For Theater Missile Defense," in *Phillips Laboratory Success Stories, 1993–1994*, PL/HO, May 1995, pp. 52–53; "Airborne Laser Program Moves Forward," in *Phillips Laboratory Success Stories, 1995*, PL/HO, May 1996, pp. 52–53.

4. Ibid.

5. "Laser aircraft contract goes to Boeing," *Kirtland Focus*, 15 November 1996, pp. 1, 4; Lawrence Spohn, "Phillips hires Boeing to build laser-firing jet," *Albuquerque Journal*, 13 November 1996, p. A2; Jeff Cole, "Boeing, TRW, Lockheed Win a Pact to Build Laser Weapons," *Wall Street Journal*, 13 November 1996, p. 1/B4; Video of ABL Contract Award Ceremony held in the Pentagon (in possession of Colonel Richard D. Tebay, System Program Director for ABL, Kirtland AFB, New Mexico), 12 November 1996.

6. Ibid.; Richard Parker, "Air Force Taps 3 Firms for Laser Project," *Albuquerque Journal*, 13 November 1996, p. C3.

7. Speech by Colonel Tebay delivered at 13th Annual Airborne Laser Laboratory Reunion, Tanoan Country Club, Albuquerque, New Mexico, 23 November 1996.

8. David H. DeVorkian, *Science With A Vengeance: How the Military Created the US Space Sciences After World War II* (Berlin: Springer-Verlag, 1992), p. 10.

9. Briefing Chart, AFWL/AR, "Airborne Laser Laboratory Accomplishments," November 1983.

Glossary

AAS *(Autoalignment Sensor)* An autocollimator assembly located near the base of the APT to align the beam properly as it passed through the APT.

ABCS *(Airborne Beam Control System)* Optics on Airborne Laser Laboratory (ALL) for directing the beam from the laser device to the airborne pointing and tracking (APT) system.

ABL *(Airborne Laser)* Second-generation airborne laser under development by the Air Force's Space & Missile Systems at Los Angeles AFB and Phillips Laboratory, Kirtland AFB, New Mexico.

ABM *(Antiballistic Missile)* Weapon system designed to counteract/intercept ballistic missiles.

ADAS *(Airborne Dynamic Alignment System)* AFWL contracted Perkin-Elmer (Norwalk, Connecticut) to develop the ADAS designed to steer the beam from the laser device inside the airplane to the APT on top of the fuselage.

ADCOM *(Aerospace Defense Command)* A specified command component of North American Air Defense Command (NORAD) with mission of operational control of aerospace defense forces.

ADP *(Advanced Development Plan)* Road map to show how research goals will be accomplished for a particular program.

AERL *(AVCO-Evertt Research Laboratories)* Private corporation that proposed the development of the gas dynamic laser to DOD as a revolutionary new weapon; contractor who built the TSL.

AFAL *(Air Force Avionics Laboratory)* Located at Wright-Patterson AFB, Dayton, Ohio.

AFB *(Air Force Base)*

AFFTC *(Air Force Flight Test Center)* Evaluated and certified safety and flightworthiness of ALL aircraft; located at Edwards AFB, California.

AFSC *(Air Force Systems Command)* Headquarters to the Air Force Weapons Laboratory, located at Andrews AFB, Maryland.

AFSWC *(Air Force Special Weapons Center)* 1 April 1952–1 April 1976, Kirtland AFB, New Mexico (precursor of Air Force Weapons Laboratory).

AFWL *(Air Force Weapons Laboratory)* Located at Kirtland AFB, New Mexico, from May 1963 to December 1990; was lead laboratory for developing and operating a high-power CO_2 gas dynamic laser and precision pointing and tracking system aboard the ALL.

ALL *(Airborne Laser Laboratory)* Modified NKC-135 aircraft equipped with a CO_2 laser and precision pointer and tracker to demonstrate ability of laser to engage and destroy aerial targets.

ALPE *(Airborne Laser Propagation Experiment)* ALPE had no direct connection with the ALL, except that the ALL was a convenient and available resource with the right diagnostic equipment to conduct propagation experiments.

AMT *(Aligned Mirror Train)* Optical components used to steer the laser beam from the floor of the aircraft up through the APT so the beam could exit into the atmosphere.

APT *(Airborne Pointing and Tracking)* system. Hardware assembly consisting of a tracker and telescope to focus and point the laser beam toward an aerial target.

ARPA *(Advanced Research Projects Agency)* Created on 7 February 1958, ARPA was the first government organization to sponsor laser research and development work. In 1972, ARPA's name was changed to the Defense Advanced Research Projects Agency (DARPA).

ARTF *(Advanced Radiation Test Facility)* AFWL's hangar facility designed exclusively for the ALL testing included test-cell and optics laboratory.

ARTO *(Advanced Radiation Technology Office)* Formed in 1972 at AFWL to lead laser research and development effort.

ASD *(Aeronautical Systems Division)* Located at Wright-Patterson AFB, Ohio. ASD owned fleet of test planes used to conduct scientific experiments in support of various Air Force research and development programs.

ATB *(Airborne Testbed)* NKC-135 aircraft, with laser and optics installed, designed to generate, focus, and direct a laser beam to intercept an aerial target. ATB was early name of laser aircraft; later changed to ALL.

ATLAS *(Advanced Technology Laser Applications Study)* Study that evaluated potential application of ground-based and airborne lasers.

AWACS *(Airborne Warning and Control System)* E-3A aircraft capable of surveillance of all air vehicles, manned or unmanned, at all altitudes above all kinds of terrain.

BAS *(Beam Angle Sensor)* Located in the ADAS alignment assembly above the optical bench. The BAS measured angular beam misalignment so corrections could be made to keep beam on desired path.

BDAS *(Breadboard Dynamic Alignment System)* Hughes assembled the BDAS at its Culver City laboratory, where it underwent integration and low-power testing starting in October 1974. Although the BDAS was not airworthy and was not the same configuration as the real ADAS, the optical train in both devices was similar.

BQM-34 Navy's self-propelled aerial target vehicle ALL engaged over the Pacific Ocean as part of the ship defense scenario.

C/DAS *(Control Data Acquisition System)* Monitored all aspects (e.g., pressure, temperature, flow) of the FSS/GDL.

Cold Flow Inert gases not ignited are passed through system to obtain pressure readings.

CVD *(Chemical Vapor Deposition)* Chemical process used to produce zinc selenide in a solid form to be used in the development of the low-power window.

CW *(Continuous Wave)* Refers to the power output of a laser device as continuous as opposed to pulsed.

DARPA *(Defense Advanced Research Projects Agency)* In 1993, DARPA reverted back to its original name, ARPA, as a result of the deemphasis on "Defense" driven by the near elimination of the Soviet threat.

DCRP *(Digital Correlation Real-Time Processor)* A processor electronics box.

DDR&E *(Deputy Director for Research and Engineering)* At the Department of Defense in the Pentagon.

DELTA *(Drone Experimental Laser Test and Assessment)* Project DELTA officially started on 13 August 1973. The objective of the experiment was to demonstrate that the integrated TSL and FTT ground system could acquire, track, and destroy an aerial target in a realistic, dynamic environment.

DF *(Deuterium Fluoride)* One type of chemical laser.

DOD *(Department of Defense)*

DSB *(Defense Science Board)* An independent body of civilian scientific experts outside the government who advised the Department of Defense on a variety of technical matters, including lasers. The DSB operated one level above the SAB and reported to DDR&E in the Office of the Secretary of Defense.

ECS *(Environmental Control System)* Vapor-cycle refrigeration system used to maintain a temperature of 70 degrees in the aircraft.

ECU *(Environmental Control Unit)* An evaporative, vapor-cycle air conditioning system designed to maintain the proper temperatures inside the ALL for the laser system to operate at peak performance.

EDL *(Electric Discharge Laser)* Instead of combusting fuels to generate a laser beam, electricity was injected into the device to excite CO_2 molecules to produce a low-power CO_2 beam.

EIRT *(Executive Independent Review Team)* ASD, who owned the ALL, instituted a series of EIRTs to periodically conduct safety inspections on the plane and its equipment.

FESS *(Fire/Explosion/Sensing/Suppression System)* A fire detection and suppression system installed on the ALL that was capable of sensing and extinguishing fires quickly.

FLIR *(Forward-Looking Infrared)* An 8- to 12-micron wavelength infrared imaging camera.

FSS *(Fluid Supply System)* Designed and developed by Pratt & Whitney (West Palm Beach). The FSS contained the laser fuels (gases and liquids), the essential ingredients to create the right conditions for lasing to occur.

FTT *(Field Test Telescope)* First-generation telescope (developed by Hughes) used as part of Tri-Service Laser program to focus and direct a ground-based laser beam to intercept an aerial target.

GD *(General Dynamics Corporation, Fort Worth)* GD was selected to perform extensive structural modifications to the ALL.

GDL *(Gas Dynamic Laser)* The gas dynamic laser derived its name from the high-speed flow and manipulation of gases resulting in population inversion of photons that produced laser light.

GILPAR *(Guidelines Identification Program for Antimissile Research)* A government program that laser research was grouped under.

GOFSS *(Ground-Only FSS)* AFWL decided to build a Ground-Only FSS to match the performance characteristics of the flightworthy FSS.

HEL *(High-Energy Laser)* A laser with a power output of 20 kilowatts or higher.

HELRATS *(High-Energy Laser Radar Acquisition and Tracking System)* Advanced system that AFWL planned to install on the ALL aircraft to carry out Cycle IV experiments. Because of technical problems with HELRATS combined with funding constraints and delays in completing the ALL program on schedule, AFSC eventually canceled phase IV.

HELRG *(High-Energy Laser Review Group)* Independent group of scientific experts who evaluated progress of ALL program.

Hot Fire Gases ignited and passed through system and photons extracted to produce a beam.

Hot Flow Gases ignited and passed through system, but no photons extracted.

ICBM *(Intercontinental Ballistic Missile)* A missile launched from the ground that travels through space and reenters the atmosphere to deliver its warhead on a ground target.

ILS *(Input Laser System)* Located in the forward midsection of the ALL. The foundation of the ILS was a large optical bench anchored to the floor of the aircraft.

IPAS *(Interplatform Alignment System)* A major subsystem of the Input Laser System.

IR *(Infrared)* Light emitted at 1.3 to 3.8 microns in the electromagnetic spectrum.

IRAN *(Inspect and Repair as Necessary)* Routine depot-level examination of all parts of an airplane to ensure the aircraft had not suffered any degradation that could affect its flightworthiness.

KAFB *(Kirtland Air Force Base)* Located in Albuquerque, New Mexico, and home of the Air Force Weapons Laboratory that was responsible for development of the Airborne Laser Laboratory.

KCAS *(Knots Calibrated Air Speed)*

LCTT *(Low-Cost Tow Target)* An F-4 aircraft reeled out the LCTT on a 5000-foot cable so the tow target was positioned behind and below the F-4. During testing, ALL tracked and directed beam to tow targets.

LITE *(Large Integrated Telescope)* Used to focus and collect data on beam characteristics.

LEAPS *(Laser Engineering and Applications for Prototype Systems)* AFWL laser division established in July 1972 to assess mission applications of lasers.

LWMA *(Laser Weapon Missile Analysis)* AFWL study group tasked to evaluate feasibility of various laser applications for future Air Force missions.

MCWS *(Mirror Cooling Water System)* Water-cooled mirrors capable of reflecting and directing laser beam through the ALL beam control system.

Micron A unit of length equal to one-millionth of a meter.

Microradian A radian is an angle of just less than 60 degrees (57.295 degrees). A microradian is a millionth of a radian, or 0.000057 degree.

MOPA *(Master Oscillator Power Amplifier)* Component of ground-based laser system that generated the output beam of the desired quality, power, and wavelength.

NASA-Ames *(National Aeronautics and Space Administration-Ames)* Research center at Moffett Field, California (near San Francisco), where wind-tunnel experiments were performed.

Nearly diffraction limited The beam spread in angles at the minimum rate theoretically possible.

OAD *(Optical Acquisition Device)* Designed to hand off target acquisition and tracking data electronically to the APT.

OCAMA *(Oklahoma City Air Material Area)* Every few years the Air Force required each of its KC-135s to undergo routine IRAN as a safety precaution at OCAMA.

OPTICS *(Oscura Peak Tracker Investigation and Comparison Series)* Proof-of-concept experiments on the advanced tracker took place as part of the OPTICS program in 1975. North Oscura Peak is located at White Sands Missile Range in New Mexico.

OUSDR&E *(Office of the Under Secretary of Defense for Research and Engineering)* Office responsible for planning, funding, and research for advancing laser technology.

PDM *(Programmed Depot Maintenance)* The ALL and diagnostic aircraft had this scheduled and comprehensive maintenance performed at Tinker AFB, Oklahoma.

RAD *(Requirements Action Directive)* Air Force document authorizing work to proceed on a specific research and development program.

RASTA *(Radiation Augmented Special Test Apparatus)* One of AVCO's early gas lasers.

RCAT *(Radio Controlled Aerial Target)* A drone that flew at approximately 200 miles per hour over a 4-mile racetrack course at SOR as part of the DELTA program to test ground-based laser.

SAB *(Scientific Advisory Board)* Established in May 1946 under the U.S. Army Air Forces to replace the Scientific Advisory Group (SAG) that had evaluated scientific research and development efforts during WWII. The SAB was the forerunner of the USAF Scientific Advisory Board, created 8 months after the Air Force became a separate service on 18 September 1947.

SAC *(Strategic Air Command)* Air Force command responsible for development of long-range strike force of bombers and intercontinental ballistics capable of delivering nuclear or conventional weapons.

SAG *(Scientific Advisory Group)* Formed in December 1944 in the Pentagon.

SDI *(Strategic Defense Initiative)* President Reagan's revolutionary strategic plan announced in March 1983 to develop space-based systems to destroy incoming missiles in their suborbital or terminal phase of flight—popularly known as Star Wars.

SDIO *(Strategic Defense Initiative Organization)* DOD organization tasked to implement SDI.

SOR *(Sandia Optical Range)* Laser test site located on Kirtland AFB. Name later changed to Starfire Optical Range in the 1980s.

SSD *(Split Stage Demonstrator)* A parallel two-stage laser device was envisioned for installation in the ALL. The SSD was one-half the planned aircraft laser, hence the name "split" stage demonstrator, used to conduct initial device testing.

SSFAN *(Steady-State Flow Analysis)* A tool used to evaluate the fluid mechanics characteristics of the MCWS.

TAC *(Tactical Air Command)* Air Force command responsible for deploying fighter aircraft to achieve and maintain air superiority.

TAR *(Technical Assistance Review)* Team chaired by Brigadier General Peter Odgers to thoroughly review the ALL test program prior to proceeding with airborne tests to engage missiles and drones.

TCM *(Transfer Control Module)* Regulated and monitored the flow of fluids delivered from the ground tank farm to the FSS tanks inside the plane.

Test Cell A laboratory with sophisticated instrumentation and control equipment for checking out and testing each ALL component. Once that was done, components were integrated and evaluated as one complete ALL system.

TOW *(Tube-launched, optically tracked, wire-guided)* The Army's antitank missile.

TSL *(Tri-Service Laser)* First ground-based laser to shoot down aerial targets.

UDRI *(University of Dayton Research Institute)* UDRI designed and built a laser window test apparatus to conduct testing on small zinc selenide samples. UDRI confirmed that zinc selenide was indeed the best window material to use on the ALL.

UHF *(Ultra High Frequency)* Any frequency between 300 and 3000 megacycles per second.

XLD *(Experimental Laser Device)* Pratt & Whitney's early gas dynamic laser that produced an output beam of 77 kilowatts in April 1968.

Bibliography

BOOKS

Adcock, Edgar, Jr., ed. *American Men and Women of Science*. New York: R. R. Bowker Co., 1977.

Anderberg, Bengt, and Wolbarsht, Myron L. *Laser Weapons: The Dawn of a New Military Age*. New York: Plenum Press, 1992.

Anderson, John D., Jr. *Gasdynamic Lasers: An Introduction*. New York: Academic Press, 1976.

Baucom, Donald R. *The Origins of SDI, 1944–1983*. Lawrence: University Press of Kansas, 1992.

Berry, F. Clinton, Jr. *Inventing the Future*. New York: Brassey's Inc., 1993.

Bertolotti, M. *Masers and Lasers: An Historical Approach*. Bristol, Great Britain: Adam Higler, Ltd., Techno House, 1983.

Boorstin, Daniel J. *The Republic of Technology: Reflections on Our Future Commentary*. New York: Harper & Row, 1978.

Bright, Charles D., ed. *Historical Dictionary of the U.S. Air Force*. New York: Greenwood Press, 1992.

Bromberg, Joan Lisa. *The Laser in America, 1950–1970*. Cambridge, Massachusetts: The MIT Press, 1991.

Bruce-Briggs, B. *The Shield of Faith: A Chronicle of Strategic Defense from Zeppelins to Star Wars*. New York: Simon & Schuster, Inc., 1988.

Buenneke, Richard H., Jr., ed. *Guide to the Strategic Defense Initiative*. Arlington, Virginia: Pasha Publications, Inc., 1986.

Canan, James W. *The Super Warriors: The Fantastic World of Pentagon Superweapons*. New York: Weybright & Talley, Inc., 1975.

Canan, James. *War in Space*. New York: Harper & Row, 1982.

Carroll, John M. *The Story of the Laser*. New York: E. P. Dutton, 1964.

Coffey, Thomas M. *HAP: The Story of U.S. Air Force and the Man Who Built It*. New York: The Viking Press, 1981.

Cunningham, Ann Marie, and Fitzpatric, Mariana. *Future Fire: Weapons for the Apocalypse*. New York: Warner Books, Inc., 1983.

Drell, Sidney D., Farley, Philip J., and Holloway, David. *The Reagan Strategic Defense Initiative: A Technical, Political, and Arms Control Assessment*. Cambridge, Massachusetts: Ballinger Publishing Co., 1985.

Futrell, Robert Frank. *Ideas, Concepts, Doctrine: A History of Basic Thinking in the United States Air Force, 1907–1964*. Montgomery, Alabama: Air University, 1974.

Goldman, Leon. *Applications of the Laser*. Melbourne, Florida: Krieger, 1982.

Gorn, Michael H. *Harnessing the Genie: Science and Technology Forecasting for the Air Force 1944–1986*. Washington, D.C.: Office of Air Force History, 1988.

————. *The Universal Man: Theodore von Karman's Life in Aeronautics*. Washington, D.C.: Smithsonian Institution Press, 1992.

Gray, Colin S. *American Military Space Policy: Information Systems, Weapon Systems and Arms Control*. Cambridge, Massachusetts: Abt Books, 1982.

Hecht, Jeff. *Beam Weapons: The Next Arms Race*. New York: Plenum Press, 1984.

————. *The Laser Guidebook*. New York: McGraw–Hill, Inc., 1986.

————, and Teresi, Dick. *Laser: Supertool of the 1980s*. New York: Ticknor & Fields, 1982.

Hitz, C. Breck. *Understanding Laser Technology: An Intuitive Introduction to Basic and Advanced Laser Concepts*. Tulsa, Oklahoma: Pennwell Publishing Co., 1985.

Hogg, Christopher A., and Sucsy, Lawrence G. *Masers and Lasers*. Cambridge, Massachusetts: Maser/Laser Associates, 1962.

Kamal, A. K. *Laser Abstracts*. New York: Plenum Press, 1964.

Karas, Thomas. *The New High Ground: Strategies and Weapons of Space-Age War*. New York: Simon & Schuster, Inc., 1983.

Kevles, Daniel J. *The Physicists: The History of a Scientific Community in Modern America*. New York: Vintage Books, 1977.

Lakoff, Sanford, and York, Herbert F. *A Shield in Space?* Berkeley: University of California Press, 1989.

Larsen, Egon. *Lasers Work Like This*. London: J.M. Dent & Sons, 1972.

Lengyel, Bela A. *Lasers*. New York: John Wiley & Sons, Inc., 1971.

McAleese, Frank G. *The Laser Experimenter's Handbook*. Blue Ridge Summit, Pennsylvania: TAB Books, 1979.

McNeil, William H. *The Pursuit of Power*. Chicago: University of Chicago Press, 1982.

Marshall, Samuel L. *Laser Technology and Applications*. New York: McGraw–Hill, Inc., 1968.

Maurer, Allan. *Lasers: Light Wave of the Future*. New York: Arco Publishing, Inc., 1982.

Muncheryan, Hrand M. *Principles & Practice of Laser Technology*. Blue Ridge Summit, Pennsylvania: TAB Books, 1983.

Myrabo, Leik, and Dean, Ing. *The Future of Flight*. New York: Baen Enterprises, 1985.

Neufeld, Jacob. *Ballistic Missiles in the United States Air Force, 1945–1960*. Washington, D.C.: Office of Air Force History, 1990.

Payne, Keith B., ed. *Laser Weapons in Space: Policy and Doctrine*. Boulder, Colorado: Westview Press, 1983.

Peebles, Curtis. *Battle for Space*. New York: Beaufort Books, Inc., 1983.

Purcell, John. *From Hand Ax to Laser*. New York: Vanguard Press, 1982.

Schawlow, Arthur L., ed. *Lasers and Light*. San Francisco: W. H. Freeman & Co., 1969.

Scientific Staff of the Fusion Energy Foundation. *Beam Defense: An Alternative to Nuclear Destruction*. Fallbrook, California: Aero Publishers, Inc., 1983.

Bibliography

Shiner, John F. *Foulois and the U.S. Army Air Corps, 1931–1935.* Washington, D.C.: Office of Air Force History, 1983.

Siegman, Anthony E. *An Introduction to Lasers and Masers.* New York: McGraw–Hill Book Co., 1971.

———. *Lasers.* Mill Valley, California: University Science Books, 1986.

Stares, Paul B. *The Militarization of Space: U.S. Policy, 1945–1984.* Ithaca, New York: Cornell University Press, 1984.

Stenholm, Stig. *Foundations of Laser Spectroscopy.* New York: John Wiley & Sons, Inc., 1984.

Sturm, Thomas A. *The USAF Science Advisory Board: Its First Twenty Years, 1944–1964.* Washington, D.C.: USAF Historical Division Liaison Office, 1967.

Tirman, John, ed. *The Fallacy of Star Wars.* New York: Vintage Books, 1984.

Walker, Jearl, ed. *Light and Its Uses: Making and Using Lasers, Holograms, Interferometers, and Instruments of Dispersion.* San Francisco: W. H. Freeman & Co., 1980.

Wells, H. G. *The War of the Worlds.* London: Octopus Books Limited, 1898.

Zukav, Gary. *The Dancing Wu Li Masters: An Overview of the New Physics.* New York: Bantam New Age Books, 1984.

ARTICLES

Anonymous. "Airborne laser beam ready for test." *Albuquerque Tribune*, 30 May 1981, p. A-4.

———. "Airborne laser defeats missiles launched at it." *Newsreview (AFSC)*, 26 August 1983, p. 1.

———. "Airborne Laser Disables Missiles in Air Force Test." *Washington Post*, 26 July 1983, p. A5.

———. "Airborne laser downs missiles." *KAFB Focus*, 29 July 1983, p. 1.

———. "Airborne Laser Lab downs missiles." *Laser Focus/Electro-Optics*, September 1983, p. 82.

———. "Airborne Laser Lab." *Aerospace Daily*, 16 January 1984, p. 74.

———. "Airborne Laser Lab Finally Kills Missiles." *Lasers & Applications*, September 1983, p. 24.

———. "Airborne Laser Lab: 'Five For Five' in Sidewinder Tests." *Aerospace Daily*, 28 July 1983, p. 149.

———. "Airborne Laser Lab to Shoot Down Missiles, AF Secretary Says." *Aerospace Daily*, 17 January 1980, p. 84.

———. "Airborne Laser Lab Undergoes Initial Test-Firing Successfully." *Newsreview*, 27 February 1981, p. 1.

———. "Air Force begins regular test-firing of a prototype antiaircraft laser." *Laser Focus*, January 1972, pp. 12–13.

———. "Air Force Fires HEL at Full Power From Aircraft on Ground." *Aerospace Daily*, 19 January 1981, p. 83.

———. "Air Force Laser Flunks 1st Test." *Albuquerque Tribune*, 2 June 1981, p. 1.

———. "Air Force Plans Laser Weapon Use." *Fort Worth Star-Telegram*, 9 March 1980, p. 13A.

———. "Air Force Readies Airborne Lab For High Energy Laser Tests." *Aerospace Daily*, 20 April 1973, p. 289.

———. "Air Force Research Programs Displayed." *Aviation Week & Space Technology*, 13 October 1980, pp. 58–59.

———. "ALL Review." *Aerospace Daily*, 20 July 1981, p. 97.

———. "Already Dozens of Civilian Uses for the Laser." *U.S. News & World Report*, 18 October 1971, pp. 86–87.

———. "A Milestone." *KAFB Focus*, 23 January 1981, p. 9.

———. "Antisatellite Laser Weapons Planned." *Aviation Week & Space Technology*, 16 June 1980, p. 244.

———. "Archimedes' Weapon." *Time*, 26 November 1973, p. 60.

———. "Army Belittles Efficacy on Laser 'Ray' Weapon." *New York Times*, 6 March 1969, p. 35.

———. "Avco Describes Gas-Dynamic System That Attains 60-Kilowatt Pulses." *Laser Focus*, July 1970, pp. 16, 18.

———. "Base scientist lauded." *KAFB Focus*, 22 September 1972, p. 7.

———. "BMD Chief: Defense Lasers 10 Years Away." *Defense Daily*, 7 August 1980, p. 187.

———. "Boeing NKC-135 Modified as Laser Laboratory." *Aviation Week & Space Technology*, 30 April 1973, p. 22.

———. "Conference Urges Laser Program Termination." *Aviation Week & Space Technology*, 15 August 1983, p. 21.

———. "Congress Seeks Earliest On-Orbit Laser Weapon Deployment Plan." *Defense Daily*, 25 June 1980, p. 284.

———. "'Death Ray' To Get First Aerial Test in N.M." *Albuquerque Journal*, 23 March 1980, p. E-10.

———. "Death to Death Rays." *Time*, 28 June 1963, p. 36.

———. "Directed energy weapons: where are they headed?" *Physics Today*, August 1983, pp. 17–20.

———. "DOD Funds $15 Million for Laser R&D." *Aviation Week & Space Technology*, 8 April 1963, p. 75.

———. "DOD Increasing Laser Weapons Development Budget: Airborne Tests Planned." *Aerospace Daily*, 21 March 1975, p. 113.

———. "Elaborate test equipment supports weapons studies at Kirtland AFB." *Laser Focus*, August 1972, pp. 12–15.

———. "Electronics: Death to Death Rays." *Time*, 28 June 1963, p. 36.

———. "Enlarged laser portions of arms budget are sailing through Congress unscathed." *Laser Focus*, September 1972, p. 12.

———. "First airborne laser test fails." *Albuquerque Tribune*, 2 June 1981, pp. A-1, A-6.

———. "First Working Optical Maser." *Aviation Week & Space Technology*, 11 July 1960, p. 38.

———. "Flying Laser Lab Knocks Out 5 Missiles in Test." *Air Force Times*, 8 August 1983, p. 1.

———. "Fort Worth Modified Aircraft is Key Part of Laser Development." *General Dynamics World*, March 1981. p. 3.

———. "Future Uses of Lasers Are Outlined to Panel." *Albuquerque Journal*, 13 January 1980, p. A-5.

———. "High Energy Laser Supplementals Asked By Air Force, Army, Navy." *Aerospace Daily*, 7 May 1973, pp. 37–38.

———. "Hill Told Of Breakthrough In Short Wavelength Laser Technology." *Aerospace Daily*, 21 March 1983, pp. 117–118.

———. "Hilltop Facility in New Mexico Testing The Air Force's Highest-Energy Lasers." *Laser Focus*, January 1971, pp. 14–16.

———. "Hit or Miss, Laser Fails Test." *New York Times*, 7 June 1981, p. 4E.

———. "Industry Observer." *Aviation Week & Space Technology*, 23 October 1961, p. 19.

———. "Industry Observer." *Aviation Week & Space Technology*, 23 August 1976, p. 9.

———. "Kirtland to survey snags in airborne laser weapon." *Albuquerque Tribune*, 24 June 1981, p. A-3.

———. "Laser Advances May Evolve New Weapons." *Aviation Week & Space Technology*, 9 March 1970, pp. 209–211.

———. "Laser As Weapon Eyed for New B-1." *Washington Post*, 20 May 1971, p. A-13.

———. "Laser Beam Succeeds In Destroying Missiles." *New York Times*, 27 July 1983, p. A16.

———. "Laser Disables Missile." *Aviation Week & Space Technology*, 1 August 1983, p. 23.

———. "Laser Engaged Target But Then Developed Mechanical Problems, AF Says." *Aerospace Daily*, 9 June 1981, p. 212.

———. "Laser Fails to Destroy Missile." *Aviation Week & Space Technology*, 8 June 1981, p. 63.

———. "Laser Knocks Out Missiles In Test." *Electronic Engineering Times*, 15 August 1983, p. 1.

———. "Laser Lab shoots down drone over ocean." *KAFB Focus*, 9 December 1983, p. 5.

———. "Laser Passes Key Test--Missiles KOd." *San Francisco Chronicle*, 26 July 1983, pp. 1, 6.

———. "Laser Progress Boosts USAF Funding." *Aviation Week & Space Technology*, 12 January 1970, p. 16.

———. "Laser Shot Missile '75%' in New Test, Defense Says." *Baltimore Sun*, 23 June 1981, p. 5.

———. "Laser Space Weapons in R&D." *Electronics*, 10 November 1961, p. 81.

———. "Laser Testing." *Aerospace Daily*, 8 June 1981, p. 201.

———. "Laser Test: 'Star Wars' air defense system cripples 5 missiles." *Albuquerque Journal*, 26 July 1983, p. A-6.

———. "Laser Weapon Flunks A Test." *Laser Focus*, 7 July 1981, p. 4.

———. "Laser Weaponry Seen Advancing." *Aviation Week & Space Technology*, 12 January 1970, pp. 16–17.

———. "Laser weaponry way down the road, Keyworth says." *Albuquerque Tribune*, 29 June 1981, p. C-6.

———. "Laser Weapons—Fancy or Fact?" *Astronautics & Aeronautics*, February 1972, pp. 12–13.

———. "Laser Weapons: Funding All Options." *Astronautics & Aeronautics*, May 1976, pp. 13–15.

———. "Laser weapons still far off." *Laser Focus*, July 1972, p. 4.

———. "Light Ray—Fantastic Weapon of the Future?" *U.S. News & World Report*, 2 April 1962, pp. 47–50.

———. "'Major milestone' in laser weapons tests." *Science News*, 6 August 1983, p. 96.

———. "Mark, Schmitt Discuss Laser, ABM Ties." *Aerospace Daily*, 26 January 1981, pp. 114–115.

———. "New laser gives 6 kW—It's done with mirrors and preheated gas." *Microwaves*, July 1970, p. 20.

———. "New Laser Weapon In Russia?" *Albuquerque Tribune*, 22 May 1980, pp. A-1, A-12.

———. "Next U.S. Superweapon—The Pentagon's 'Light Ray.'" *U.S. News & World Report*, 18 October 1971, pp. 85–87.

———. "Now, the Death Ray?" *Time*, 4 September 1972, p. 46.

———. "Optical Maser R&D." *Aviation Week & Space Technology*, 12 February 1962, p. 32.

———. "Plane's Laser Hits Missiles." *Albuquerque Journal*, 26 July 1983, p. C-8.

———. "Power And Precision From Lasers, Perhaps More Than You Can Use." *Automotive Engineering*, September 1970, pp. 38–41.

———. "Progress On Laser Weapons." *U.S. News & World Report*, 1 October 1973, pp. 41–42.

———. "Report Urges Major Increase in Space Laser Weapon Program." *Defense Daily*, 8 August 1980, p. 194.

———. "Retailers Want More Efficient Bar Codes." *Albuquerque Tribune*, 30 March 1992, p. 3 (Business Outlook Section).

———. "Russia Said To Outspend U.S. on 'Death Ray.'" *Albuquerque Journal*, 21 August 1980, p. D-4.

———. "SAC Chief Says Laser Bomber Defense Faces Many Hurdles." *Defense Daily*, 7 August 1980, p. 187.

———. "Sidewinder Destroyed in Airborne Laser Test." *Aviation Week & Space Technology*, 8 August 1983, p. 66.

———. "Significant ALL Test." *Aerospace Daily*, 6 December 1979, pp. 173–174.

———. "Soviet laser seen by '83." *Albuquerque Tribune*, 3 March 1982, p. B1.

———. "Soviets gain in lasers." *Albuquerque Tribune*, 20 August 1980, pp. A-1, A-16.

———. "Star Wars comes to Farnborough." *Observer*, 7 September 1980, p. 3.

———. "Surveying military programs." *Photonics Spectra*, November 1983, p. 73.

———. "Tardy Approach to High Energy Laser Weapon Development?" *Air Force Magazine*, April 1981, pp. 18–19.

———. "Technology News: Airborne Laser Lab downs missiles." *Laser Focus*, September 1983, p. 82.

———. "The First Laser Weapon." *Newsweek*, 30 September 1974, p. 51.

———. "The Laser Whammy." *Time*, 12 January 1976, p. 39.

———. "The Tools: Weapons." *Air Force Magazine: The Golden Anniversary*, August 1957, pp. 343–359.

———. "USAF/Boeing airborne laser laboratory." *Aviation Week & Space Technology*, 15 June 1981, p. 34.

———. "USAF, Navy Laser Destroys Drone." *Aviation Week & Space Technology*, 5 December 1983, p. 26.

———. "USAF will start testing laser guns for jet fighters." *San Antonio Star*, 24 February 1980, p. 13.

———. "U.S. And Soviets Race to Perfect 'Death Ray.'" *Virginia Beach Monitor*, 12 August 1980, p. 1.

———. "U.S. Effort Redirected to High Energy Lasers." *Aviation Week & Space Technology*, 28 July 1980, pp. 50–57.

———. "U.S.: Lasers Score Against Sidewinder." *Defense & Foreign Affairs Daily*, 29 July 1983, p. 2.

———. "U.S. Laser Weapon Fails Test on Air-to-Air Missile." *Albuquerque Journal*, 3 June 1981, p. 1.

———. "U.S. Nears Laser Weapon Decisions." *Aviation Week & Space Technology*, 4 August 1980, pp. 48–54.

———. "Vaporizing Rays May Be Installed On New Bomber." *Albuquerque Journal*, 20 May 1971, p. A-10.

———. "Washington Roundup: Space Defense." *Aviation Week & Space Technology*, 29 December 1980, p. 13.

———. "Weapons Laboratory Aids Beam Effort." *Aviation Week & Space Technology*, 4 August 1980, pp. 56–59.

Adams, Melissa. "Kirtland lab perfecting laser direction from source." *Albuquerque Tribune*, 22 May 1981, p. A-3.

———. "Kirtland may benefit from Senate 'death ray' report." *Albuquerque Tribune*, 19 November 1980, pp. A-1, A-14.

Albright, Joseph. "Lethal Lasers: Military Ready to Test Powerful Light Beams." *Atlanta Journal*, 10 March 1980, pp. 1A, 14A.

Albright, Nelson. "Laser Weapons—How Close Are We?" *Popular Science*, March 1972, pp. 64–66, 142.

Andrews, Walters. "Airborne laser beam used successfully by Air Force." *Washington Times*, 26 July 1983, p. 2.

Avizonis, Petras V., and Farrington, T. "Internal Self-Damage of Ruby and Nd-Glass Lasers." *Applied Physics Letters*, Vol. 7, No. 8, 15 October 1965, pp. 205–206.

Barkan, Robert. "Buck Rogers and The Deadly Laser." *New Republic*, 8 April 1972, p. 9-F.

Batezel, Anthony Lynn. "Best Kept Secrets." *Airman*, April 1982, pp. 43–48.

Beane, William J. "The High-Energy Laser: Strategic Policy Implications." *Strategic Review*, Winter 1977, pp. 100–107.

Bernstein, Peter J. "U.S. steps up research in laser weaponry." *Atlanta Journal and Constitution*, 28 February 1982, p. 38-A.

Bierman, John. "'Death Ray' Work in Russia Stirs U.S. Weapons Activity." *Washington Star*, 3 August 1980, pp. 1, 8.

Bloom, Arnold L. "Optical Pumping." In *Lasers and Light*, ed. Arthur L. Schawlow, pp. 46–54. San Francisco: W. H. Freeman & Co., 1969.

Blundell, William E. "Air Force Laboratory Simulates the Effects of Nuclear Explosions." *Wall Street Journal*, 22 January 1963, p. 1.

Boffey, Philip M. "Laser Weapons: Renewed Focus Raises Fears and Doubts." *New York Times*, 9 March 1982, pp. C-1, C-5.

Boraiko, Allen A. "'A Splendid Light': Lasers." *National Geographic*, March 1984, pp. 334–363.

Bosma, John. "Space and Strategic Defensive Orientation: Project Defender." *Defense Science & Electronics*, September 1983, pp. 58–65.

Bradshaw, Jim. "Laser Tests at Kirtland A 'Milestone.'" *Albuquerque Journal*, 16 January 1981, p. 1.

Brechner, Berl. "Chase!" *Air & Space Smithsonian*, April/May 1986, pp. 58–60.

Breen, Walter M. "The Laser: Its Function and Its Future." *Air University Review*, May/June 1975, pp. 59–67.

Brezinski, Zbigniew, Jastrow, Robert, and Kampelman, Max M. "Defense in Space is not 'Star Wars.'" *New York Times Magazine*, 27 January 1985, p. 48.

Bromberg, Joan Lisa. "Amazing Light." *Invention and Technology*, Spring 1992, pp. 18–26.

———. "The Birth of the Laser." *Physics Today*, October 1988, pp. 26–30.

Browne, Malcolm W. "Weapon That Fights Missiles Could Alter World Defense Focus." *New York Times*, 4 December 1978, pp. A1, D11.

Buchanan, Patrick J. "The Soviets have made a scary breakthrough in weaponry." *Philadelphia Inquirer*, 9 August 1980, p. 8.

Burt, Richard. "New Laser Weaponry is Expected to Change Warfare in the 1980s." *New York Times*, 10 February 1980, pp. 1, 54.

———. "U.S., Kremlin Explore Use of Laser Weapons." *San Diego Union*, 16 March 1980, p. A-10.

———. "U.S. Says Russians Develop Satellite-Killing Laser." *New York Times*, 22 May 1980, Section A, pp. 1, 9.

Cady, Steven E. "Beam Weapons In Space." *Air University Review*, May/June 1982, pp. 33–39.

Chatham, George N. "ZAP—Energy Weapons And Aircraft." *Astronautics & Aeronautics*, April 1970, pp. 16, 21.

Chester, Arthur N. "Chemical lasers: a status report." *Laser Focus*, November 1971, pp. 25–29.

Claus, A. C. "Archimedes' Burning Mirrors." *Applied Optics*, October 1973, p. A14.

Covault, Craig. "Space Command Seeks Asat Laser." *Aviation Week & Space Technology*, 21 March 1983, pp. 18–19.

Davis, Monte. "Is There A Laser Gap?" *Discover*, March 1981, pp. 62–66.

Day, Bonner. "Progress on Energy Beam Weapons." *Air Force Magazine*, July 1979, pp. 62–65.

Dickson, Paul. "It'll Never Fly, Orville: Two Centuries of Embarrassing Predictions." *Saturday Review*, December 1979, p. 36.

Doe, Charles. "Laser Weapon Promising, Yet Elusive." *Air Force Times*, 28 March 1983, p. 23.

Douglas, John H. "High-Energy Laser Weapons." *Science News*, 3 July 1976, pp. 11–13.

Duffey, John G., and Spreen, Darrell E. "Project DELTA." *Journal of Defense Research, Series A: Strategic Warfare—High Energy Lasers*, May 1975, pp. 41–47.

Duke, Bob. "Laser Weapons to Counter Soviets." *Albuquerque Tribune*, 27 March 1980, p. A-1.

Duran, David. "New laser facility slated." *Albuquerque Tribune*, 6 September 1973, pp. A-1, A-10.

Edson, Lee. "The Advent of the Laser Age." *New York Times*, 26 March 1978, Section 6, pp. 35–46.

Elson, Benjamin M. "USAF Weapons Lab Mission Expanded." *Aviation Week & Space Technology*, 29 January 1979, pp. 212–216.

Famiglietti, Leonard. "Airborne Laser Laboratory Destroys Navy Drones." *Air Force Times*, 19 December 1983, p. 32.

———. "Conferees Want Directed-Energy Office." *Air Force Times*, 5 September 1983, p. 23.

———. "Laser Director Wants More Industry Input." *Air Force Times*, 16 August 1982, pp. 23–24.

———. "Laser Weaponry Still 10 Years Away." *Air Force Times*, 6 April 1981, p. 23.

Feder, Baraby J. "Lasers, Finally." *New York Times*, 2 August 1981, p. C-9.

Gay, Lance. "Kirtland's discontinued Laser Lab shoots down target cruise missile." *Albuquerque Tribune*, 30 November 1983, p. 1.

Gerry, Edward T. "Gasdynamic Lasers." *IEEE Spectrum*, November 1970, pp. 51–58.

Gorman, P.F. "What the High Technology Edge Means." *Defense 83*, June 1983, pp. 22–25.

Graham, Sandy. "Flying laser gun is Kirtland goal." *Albuquerque Tribune*, 14 October 1978, pp. A-1, A-8.

Gregory, William H. "Crosscurrents at Farnborough." *Aviation Week & Space Technology*, 8 September 1980, p. 11.

———. "The Great Laser Battle—Continued." *Aviation Week & Space Technology*, 14 June 1982, p. 13.

Hecht, Jeff. "High-Energy Lasers for Weapons Applications." *Military Electronics/Countermeasures*, August 1982, pp. 73–75.

———. "Lasers as a commercial technology." *Laser Focus World*, May 1994, pp. 81–83.

Heilmeir, George. "Contributions to Science and Technology by the Advanced Research Projects Agency (ARPA) 1957–1970." *Commander's Digest*, 7 October 1976, p. 2.

Herriott, Donald R. "Applications of Laser Light." In *Lasers and Light*, ed. Arthur L. Schawlow, pp. 313–321. San Francisco: W. H. Freeman & Co., 1969.

Hoffman, Fred S. "Air Force Killer Beam Ready For Testing Against Missile." *Albuquerque Journal*, 30 May 1981, p. A-1.

———. "U.S., Soviets in Race Toward Laser Superkill." *Albuquerque Journal*, 28 October 1976, p. B-4.

Horton, Alan. "Talk grows about U.S.–Russia race for 'death ray.'" *Albuquerque Tribune*, 13 January 1976, p. B-5.

Hotz, Robert. "The real star wars." *Space World*, August/September 1982, pp. 157–189.

Jaszka, Paul R. "The Laser and its Military Applications: A Primer." *Defense Electronics*, July 1989, pp. 80–83.

Javan, A., *et al*. "Population Inversion and Continuous Optical Maser Oscillation in a Gas Discharge Containing a He-Ne Mixture." *Physical Review Letters*, 1 February 1961, pp. 106–110.

Klass, Philip J. "Advanced Weaponry Research Intensifies." *Aviation Week & Space Technology*, 18 August 1975, pp. 34–39.

———. "Chemical Laser Takes Weapons Lead." *Aviation Week & Space Technology*, 21 August 1978, pp. 38–47.

———. "Electroplating Use Expands in Weapons." *Aviation Week & Space Technology*, 26 October 1981, pp. 91–95.

———. "Laser Destroys Missile in Test." *Aviation Week & Space Technology*, 7 August 1978, pp. 14–16.

———. "Major Hurdles for Laser Weapons Cited." *Aviation Week & Space Technology*, 9 July 1973, pp. 38–42.

———. "More Laser Weapon Funds Sought." *Aviation Week & Space Technology*, 14 May 1973, pp. 12–13.

———. "New laser weapon in Russia?" *Albuquerque Journal*, 22 May 1980, pp. A-1, A-12.

———. "New Soviet Weapon Cited in Bid for Military Rise." *New York Times*, 5 March 1982, p. A-4.

———. "Power Boost Key to Feasibility." *Aviation Week & Space Technology*, 21 August 1972, pp. 32–40.

———. "Research Nears Application Level." *Aviation Week & Space Technology*, 14 August 1972, pp. 12–15.

———. "Special Report: Laser Thermal Weapons." *Aviation Week & Space Technology*, 21 August 1972, pp. 32–40.

———. "Special Report: Laser Weapons." *Aviation Week & Space Technology*, 18 August 1975, pp. 34–39.

Lamberson, Donald L. "DoD's Directed Energy Program: Its Relevance To Strategic Defense." *Defense 83*, June 1983, pp. 16–21.

Leith, Emmett N., and Upatnieks, Juris. "Photography by Lasers." In *Lasers and Light*, ed. Arthur L. Schawlow, pp. 339–350. San Francisco: W. H. Freeman & Co., 1969.

Lempicki, Alexander, and Samelson, Harold. "Liquid Lasers." In *Lasers and Light*, ed. Arthur L. Schawlow, pp. 255–263. San Francisco: W. H. Freeman & Co., 1969.

Lengyel, Bela A. "Evolution of Masers and Lasers." *American Journal of Physics*, March 1966, pp. 903–913.

List, William F. "Letters to the Editor: High-Energy Laser." *Aviation Week & Space Technology*, 2 August 1982, p. 106.

Lucero, David. "Lasers Face Test As Arms." *Albuquerque Journal*, 7 September 1973, pp. A-1–A-2.

Lyons, Richard D. "Physicists Hear Of Strong Laser." *New York Times*, 30 April 1970, p. 20.

Lytle, Stewart, "Crash program on lasers urged." *Albuquerque Journal*, 3 June 1980, pp. A-1, A-6.

———. "Engineering complexities slow air laser program." *Albuquerque Tribune*, 4 June 1981, pp. A-1, A-13.

———. "Kirtland laser to be tested." *Albuquerque Tribune*, 3 April 1980, p. A-2.

———. "Laser Fails again, but Air Force says test is 'better.'" *Albuquerque Journal*, 22 June 1981, p. 1.

———. "Secret Senate debate to decide whether U.S. enters laser race." *Albuquerque Tribune*, 20 June 1980, p. A-12.

McCord, Laurie. "Is Kirtland developing a science-fiction ray gun?" *Albuquerque Tribune*, 1 May 1973, pp. A-1, A-6.

McDonnell, James A. Jr., "Centennial Tribute to Dr. Theodore von Karman." *Air Force Magazine*, August 1981, pp. 84–86.

Maher, Charles. "Couple Guilty of Selling High-Technology Optics." *Los Angeles Times*, 13 December 1980, pp. 1, 12 (section C, part II).

Maiman, T.H. "Stimulated Optical Emission in Fluorescent Solids. I. Theoretical Considerations." *Physical Review*, 15 August 1961, pp. 1145–1150.

———. "Stimulated Optical Radiation in Ruby." *Nature*, 6 August 1960, pp. 493–494.

———, *et al.* "Stimulated Optical Emission in Fluorescent Solids. II. Spectroscopy and Stimulated Emission in Ruby." *Physical Review*, 15 August 1961, pp. 1151–1157.

Manchester, Harland. "Light of Hope—Or Terror." *Reader's Digest*, February 1963, pp. 97–100.

Meinel, Carolyn. "A Laser Weapon Fizzles." *Technology Review*, August/September 1983, pp. 79–80.

———. "Laser Weapon Update." *Technology Review*, November/December 1983, pp. 84–85.

Middleton, Drew. "Mass-Produced Precision Guided Weapons are Said to be Revolutionizing Military Doctrine and Tactics." *New York Times*, 23 February 1976, p. 15.

———. "Powerful Lasers Reported Bound For American and Soviet Arsenals." *New York Times*, 16 February 1977, p. A2.

Mielenz, Klaus D. "That Burning Glass." *Applied Optics*, February 1974, pp. A14, A16, 452.

Miles, Marvin. "Laser Weapons Seen Available for Use Within Next Six Years." *Los Angeles Times*, 28 August 1972, Part II, p. 1.

Miller, Barry. "Aerospace, Military Laser Uses Explored." *Aviation Week & Space Technology*, 22 April 1963, pp. 54–69.

———. "Fiscal 1962 Laser Funding Completed." *Aviation Week & Space Technology*, 9 July 1962, p. 42.

———. "Government Funds Boost Laser Research." *Aviation Week & Space Technology*, 12 March 1962, pp. 229–232.

———. "Optical Maser May Aid Space Avionics." *Aviation Week*, 18 July 1960, pp. 96–99.

———. "Optical Systems Have Space Potential." *Aviation Week*, 14 December 1959, pp. 87–91.

———. "U.S. Begins Laser Weapons Programs." *Aviation Week & Space Technology*, 26 March 1962, pp. 41–45.

———. "U.S. Plan to Accelerate Laser Development Spurs Market." *Aviation Week & Space Technology*, 21 August 1967, pp. 92–93.

Miller, Stewart E. "Communication by Laser." In *Lasers and Light*, ed. Arthur L. Schawlow, pp. 323–331. San Francisco: W. H. Freeman & Co., 1969.

Mims, Forrest M., III. "A super energy laser is on the way." *Science Digest*, August 1972, pp. 25–29.

———. "Huge Laser Weapon Is Under Development." *Albuquerque News*, 3 February 1972, p. 2WE.

————. "Laser Death Ray." *Saga*, June 1971, pp. 14–17, 82.

————. "The Evolution of Revolutionary Laser Weapons." *Air Force Magazine*, June 1972, pp. 54–58.

————. "Toward New Horizons In USAF Weapons." *Air Force Magazine*, July 1993, pp. 74–78.

Mygatt, Matt. "'Star Wars' May Become Earth Wars With New Air Force Laser Weapon." *Philadelphia Inquirer*, 17 January 1981, p. 3.

Osmundsen, John A. "Light Amplification Claimed by Scientist." *New York Times*, 8 July 1960, pp. 1, 7.

O'Toole, Thomas. "Mass. Firm Reports Strong Laser Beam." *Washington Post*, 30 April 1970, p. A-22.

————. "Reagan Interested in Speeding Development of Space-Based Laser." *Washington Post*, 26 December 1980, p. 3.

Parsons, Russ. "KAFB to make laser weapon." *Albuquerque Tribune*, 16 April 1982, pp. A-1, A-8.

Patel, C. K. N. "High-Power Carbon Dioxide Lasers." In *Lasers and Light*, ed. Arthur L. Schawlow, pp. 265–275. San Francisco: W. H. Freeman & Co., 1969.

————. "Selective Excitation Through Vibrational Energy Transfer and Optical Laser Action in N_2-CO_2." *Physical Review Letters*, Vol 13, 1964, pp. 617–619.

Pennington, Keith S. "Advances in Holography." In *Lasers and Light*, ed. Arthur L. Schawlow, pp. 351–359. San Francisco: W. H. Freeman & Co., 1969.

Petty, John Ira. "Base Operates Laser Range." *Albuquerque Journal*, 25 August 1971, pp. 1, A-7.

————. "Laser Weaponry Is Nearing Reality." *Albuquerque Journal*, 12 June 1970, pp. 1, A-8.

————. "Laser Weapons May Be a Reality Reports Indicate." *Albuquerque Journal*, 15 January 1970, pp. A-1, A-8.

Pincus, Walter. "Soviets Said Choosing Lasers as Next Lap in Arms Race." *Washington Post*, 23 August 1979, p. A2.

Raeburn, Paul. "'Star Wars' Defense Possible, Experts Believe." *Albuquerque Journal*, 2 May 1983, p. B-5.

Rawles, James W. "Directed Energy Weapons: Battlefield Beams." *Defense Electronics*, August 1989, pp. 47–54.

————. "Lasers: The Battlefield Tools of Tomorrow Are Here." *Defense Electronics*, July 1989, pp. 73–74, 76–78.

Ripley, Anthony. "Laser is Stirring Imagination of Weapon Scientists." *New York Times*, 16 October 1971, p. 15.

————. "Laser work at Kirtland—possibilities staggering." *Albuquerque Tribune*, 27 October 1971, pp. A-1, A-13.

Robinson, Clarence A., Jr. "Developing Beam Weapons," *Aviation Week & Space Technology*, 7 November 1983, p. 11.

————. "Technology Spurs Weapon Gains." *Aviation Week & Space Technology*, 29 January 1979, pp. 45–56.

Roland, Alex. "Science and War." *Osiris*, 1985, pp. 247–272.

Rudkin, R. L. *et al.* "Ship-Based Laser Weapons for Battle Group Defense." *Journal of Defense Research: Special Issue 86-1*, May 1986, pp. 121–133.

Salem, Nancy. "First high-energy weapon made in Albq with A-bomb secrecy." *Albuquerque Tribune*, 31 March 1980, p. A-2.

————. "Labs in N.M. helping to light way for myriad uses of laser." *Albuquerque Tribune*, 13 December 1979, pp. A-1, A-6.

Schawlow, Arthur L. "Advances in Optical Masers." In *Lasers and Light*, ed. Arthur L. Schawlow, pp. 234–245. San Francisco: W. H. Freeman & Co., 1969.

————. "Laser Light." In *Lasers and Light*, ed. Arthur L. Schawlow, pp. 282–290. San Francisco: W. H. Freeman & Co., 1969.

————. "Lasers: The Practical and the Possible." *Stanford Magazine*, Spring/Summer 1979, pp. 24–29.

————. "Optical Masers." In *Lasers and Light*, ed. Arthur L. Schawlow, pp. 224–233. San Francisco: W. H. Freeman & Co., 1969.

————, and Townes, Charles H. "Infrared and Optical Masers." *Physical Review*, December 1958, pp. 1940–1949.

Seidel, Robert W. "From Glow to Flow: A History of Military Laser Research and Development." *Historical Studies in the Physical and Biological Sciences*, 1987, pp. 111–147.

————. "How The Military Responded To The Laser." *Physics Today*, October 1988, pp. 36–43.

Simms, D. L. "Archimedes and the Burning Mirrors of Syracuse." *Technology and Culture*, Vol. 18, 1977, pp. 1–24.

————. "More On That Burning Glass of Archimedes." *Applied Optics*, May 1974 pp. A15–A16.

Smernoff, Barry J. "Strategic and Arms Control Implications of Laser Weapons." *Air University Review*, January/February 1978, pp. 38–50.

————. "The Strategic Value of Space-Based Laser Weapons." *Air University Review*, March/April 1982, pp. 2–17.

Smith, George S. "The Early Laser Years at Hughes Aircraft Company." *IEEE Journal of Quantum Electronics*, June 1984, pp. 577–584.

Spohn, Lawrence. "Historic flying laser lab will be retired." *Albuquerque Tribune*, 3 May 1968, pp. 1, A2.

Stockton, Bill. "Secret Air Force lab near Albq site of laser weapons research." *Albuquerque Tribune*, 27 July 1972, pp. 1, A-5.

Thompson, Barry L. "'Directed Energy' Weapons and the Strategic Balance." *Orbis*, Fall 1979, pp. 697–709.

Toth, Robert C. "Air Force 'Defeats' Speedy Missiles in Test: Airborne Result Called 'Major Milestone' in Weapon Development." *Los Angeles Times*, 26 July 1983, pp. 1, 9.

————. "Laser Will Attempt to Shoot Down Missile." *Los Angeles Times*, 16 March 1980, pp. 4, 10-F.

————. "Shortcomings of Laser Weapons." *Laser Focus*, July 1983, p. 8.

————. "War in Space." *Science*, September/October 1980, pp. 74–80.

Tsipis, Kosta. "Laser Weapons." *Scientific American*, December 1981, pp. 51–57.

Ulsamer, Edgar. "Affordability + Performance: USAF's R&D Goal." *Air Force Magazine*, March 1976, pp. 31–36.

————. "A More Liberal, Avant Garde R&D Program." *Air Force Magazine*, June 1980, pp. 42–47.

————. "Exotic New Weapons: Reality or Myth?" *Air Force Magazine*, September 1977, pp. 124–129.

————. "Laser: A Weapon Whose Time is Near." *Air Force Magazine*, December 1970, pp. 28–34.

————. "Looking Ahead: Aerospace Planes and Airborne Lasers?" *Air Force Magazine*, December 1982, p. 24.

————. "Status Report on Laser Weapons." *Air Force Magazine*, January 1972, p. 63.

————. "The Big Laser Debate." *Air Force Magazine*, June 1982, pp. 20, 23–24.

————. "The Long Leap Toward Space Laser Weapons." *Air Force Magazine*, August 1981, pp. 58–64.

Wilford, John Noble. "Laser Weapons to be Tested at White Sands Range." *New York Times*, 3 March 1980, p. A-16.

Wilson, George C. "Air Force Leaders Skeptical of Laser Threat." *Washington Post*, 10 March 1982, p. A-7.

————. "Air Force Research Chief Is Calm About Prospect of Soviet Laser Weapon." *Washington Post*, 23 April 1982, p. A3.

————. "Experimental Laser Weapons Weighed by Pentagon Scientist." *Washington Post*, 17 May 1977, p. A-14.

————. "In Test Aloft, Air Force Fails to Stop Sidewinder." *Washington Post*, 3 June 1981, p. A6.

Wirbel, Loring. "N.M. laser weapons face exotic challenge." *Albuquerque Tribune*, 28 November 1983, pp. A-1, A-4.

————. "1984 budget kills Kirtland's airborne laser." *Albuquerque Tribune*, 28 November 1983, pp. A-1, A-4.

Index